Probability Theory and Stochastic Modelling

Volume 96

The Probability Theory and Stochastic Modelling series is a merger and continuation of Springer's two well established series Stochastic Modelling and Applied Probability and Probability and Its Applications. It publishes research monographs that make a significant contribution to probability theory or an applications domain in which advanced probability methods are fundamental. Books in this series are expected to follow rigorous mathematical standards, while also displaying the expository quality necessary to make them useful and accessible to advanced students as well as researchers. The series covers all aspects of modern probability theory including

- Gaussian processes
- Markov processes
- Random Fields, point processes and random sets
- Random matrices
- Statistical mechanics and random media
- Stochastic analysis

as well as applications that include (but are not restricted to):

- Branching processes and other models of population growth
- Communications and processing networks
- Computational methods in probability and stochastic processes, including simulation
- Genetics and other stochastic models in biology and the life sciences
- Information theory, signal processing, and image synthesis
- Mathematical economics and finance
- Statistical methods (e.g. empirical processes, MCMC)
- Statistics for stochastic processes
- Stochastic control
- Stochastic models in operations research and stochastic optimization
- Stochastic models in the physical sciences.

More information about this series at http://www.springer.com/series/13205

Søren Asmussen • Mogens Steffensen

Risk and Insurance

A Graduate Text

Søren Asmussen
Department of Mathematics
University of Aarhus
Aarhus, Denmark

Mogens Steffensen
Department of Mathematical Sciences
University of Copenhagen
Copenhagen, Denmark

ISSN 2199-3130   ISSN 2199-3149   (electronic)
Probability Theory and Stochastic Modelling
ISBN 978-3-030-35178-6   ISBN 978-3-030-35176-2   (eBook)
https://doi.org/10.1007/978-3-030-35176-2

Mathematics Subject Classification (2010): 91B30, 91-01, 91-02, 91B70, 60J27, 62P05

This Springer imprint is published by the registered company Springer Nature Switzerland AG.
The registered company address is: Gewerbestrasse 11, 6330 Cham, Switzerland

Preface

This book covers some of the main aspects of insurance mathematics, parts of risk management and financial risk, and some relevant probabilistic tools. The view is theoretical and practical implementation issues are only sporadically touched upon. Accordingly, the emphasis is on probability rather than statistics.

The text has grown out of courses given by Asmussen for masters students in statistics and/or math-finance in Aarhus under titles like *Insurance Mathematics for Non-Actuaries* or *Risk*, and by Steffensen in Copenhagen for actuarial students at various levels. One main use is therefore as a teaching tool and some suggestions on how to use it as such are given below. A further aim is to provide a reference text for the state-of-the-art of the area, and parts are at research level, in particular in Chaps. VII, X, XII and XIII. The reader needs knowledge of probability at masters level, including martingales and (for the life insurance parts) basic stochastic calculus. However, a general feeling for probability is more important than knowledge of specific topics.

Insurance mathematics serves to make the sharing of risk and smoothing of consumption in a population efficient and fair. Risk sharing and consumption smoothing are fundamentally beneficial to both the individual and society when organized on a solid and professional ground. Insurance mathematics is traditionally divided into the areas of life and non-life, where we include the mathematics of pensions in life insurance. One key difference is the typical term of the contract. Contracts in life insurance typically cover a period of 50 years or more, including savings for retirement and spending during retirement whereas contracts in non-life insurance typically cover just a year or so. This has some important consequences. Life insurance, to a larger extent than non-life insurance, (a) is heavily regulated so that customers can trust their insurance company or pension fund 'over the lifecycle'; (b) has national or regional differences because they are an integral part of a political pension system; (c) draw on methods from economics and finance, both in product design and management, because time value of money and financial sustainability 'over the lifecycle' is crucial. In practice, there do exist short term contracts in life insurance. They draw on non-life methods and may even be spoken of as non-life contracts. And vice versa. In recent decades an international harmonization of solvency and accounting rules has brought the two areas closer together as these rules are, to an increasing extent, formulated to cover both areas without further specificities. Also, both non-life and life insurance are becoming more and more integrated with financial mathematics and risk management. All of these areas draw heavily on probabilistic tools such as sums of random variables, stochastic orderings, extreme value theory and the study of dependence, and these therefore form an indispensable background.

So, on one hand, it does make sense to present some tools and patterns of thinking from each area in separate chapters. But, on the other hand, it also makes sense to present those chapters together with other chapters with tools and patterns of thinking that they share and, above all, in one book.

At the textbook level, Bowers, Gerber, Hickman and Nesbitt [31] cover both non-life and life insurance mathematics, and much of the probability background of these areas is in Rolski, Schmidli, Schmidt and Teugels[153]. For non-life insurance mathematics, some further main general references are (in alphabetical order), Daykin, Pentikainen and Pesonen [54], Denuit, Marechal, Pitrebois and Walhin [57], Kaas, Goovaerts, Dhaene and Denuit [97], Klugman, Panjer and Willmot [100], Mikosch [122] and Sundt [166]. For life insurance mathematics, see in particular Dickson, Hardy and Waters [62], Gerber [80], Koller [102] and Møller and Steffensen [128]. The introductory aspects of financial mathematics presented here are inspired by Björk [26] but the integration into life insurance mathematics covers to some extent that by Møller and Steffensen [128] and beyond. A key reference in risk management is McNeil, Frey and Embrechts [116] and there is also much important mathematical material in Rüschendorf [154]. The term *risk* is of course much broader than the selection of topics we cover, as illustrated in, say, Aven, Baraldi, Flage and Zio [15], Klüppelberg, Straub and Welpe [101] and Woo [175]; some topics that are related to those of this book but that we do not cover are operational risk and model risk. More specialized references are given in the text, most often for a section at the end or for a whole chapter at the end of the first or final section. Our reference list is, however, somewhat selected given that searches on the web today are a main tool for an overview of the current state of an area.

For a single course, the book is obviously too long and condensed and one would need to make a selection based on the general purpose and the topics of interest. Also, specific prerequisites of the students decide the pace at which the material can or shall be covered. A course in non-life insurance could start with Chap. I (for many groups of students Sects. 1 and 5 could be skipped) and go on to parts of Chaps. II–IV, VIII, XI (not necessarily in that order). For life insurance, one would start with Chap. V, which in itself makes a course on basic life insurance mathematics based on stochastic processes (Sect. A.3 of the Appendix could also be needed). Some prerequisites in financial mathematics are needed to integrate Chap. VI. The material in Chaps. VII and XII are for an advanced audience but can also be used as teaching material for masters students. A risk or probability oriented course could start with Sect. 2 of Chap. I and Sect. 1 of Chap. X, to go on to parts of Chaps. III, IX, X.

We gratefully acknowledge the numerous useful comments on details provided by students over the years. Jevgenijs Ivanovs lectured on a draft of the risk part in Aarhus in 2018 and 2019, and Mogens Bladt on a draft of the life insurance part in Copenhagen in 2019. This not only helped to improve the details but also provided some material that is now incorporated. Hansjörg Albrecher read large parts of the book and his comments were very helpful as well.

Aarhus, Denmark   Søren Asmussen
Copenhagen, Denmark   Mogens Steffensen
August 2019

Contents

Notation

Internal Reference System

The chapter number is specified only if it is not the current one. As examples, Proposition 1.3, formula (5.7) or Section 5 of Chapter IV are referred to as Proposition IV.1.3, IV.(5.7) and Section IV.5, respectively, in all chapters other than IV, where we write Proposition 1.3, formula (5.7) (or just (5.7)) and Section 5.

The end of a proof is marked by $\square$, the end of an example or remark by $\Diamond$.

Special Typeface

d	differential as in $\mathrm{d}x$, $\mathrm{d}t$, $F(\mathrm{d}x)$; to be distinguished from a variable or constant d, a function $d(x)$ etc.
e	the base $2.71\ldots$ of the natural logarithm; to be distinguished from e which can be a variable or a function.
i	the imaginary unit $\sqrt{-1}$; to be distinguished from a variable i (typically an index).
$\mathbb{1}$	the indicator function, for example $\mathbb{1}_A$, $\mathbb{1}_{x \in A}$, $\mathbb{1}\{x \in A\}$, $\mathbb{1}\{X(t) > 0 \text{ for some } t \in [0, 1]\}$.
O, o	the Landau symbols. That is, $f(x) = \mathrm{O}\big(g(x)\big)$ means that $f(x)/g(x)$ stays bounded in some limit, say $x \to \infty$ or $x \to 0$, whereas $f(x) = \mathrm{o}\big(g(x)\big)$ means $f(x)/g(x) \to 0$. If more variables than x are involved and it is needed to specify that the limit is over x, we write $\mathrm{o}_x(g(x, y))$ etc.
π	$3.1416\ldots$; to be distinguished from π which is often used for a premium rate or other.
$\mathcal{N}(\mu, \sigma^2)$	the normal distribution with mean μ and variance σ^2.

Probability, expectation, variance, covariance are denoted $\mathbb{P}$, $\mathbb{E}$, $\mathbb{V}$ar, $\mathbb{C}$ov. The standard sets are $\mathbb{R}$ (the real line $(-\infty, \infty)$), the complex numbers $\mathbb{C}$, the natural numbers $\mathbb{N} = \{0, 1, 2, \ldots\}$, the integers $\mathbb{Z} = \{0, \pm 1, \pm 2, \ldots\}$.

Matrices and vectors are most often denoted by bold typeface, C, Σ, x, α etc., though exceptions occur. The transpose of A is denoted A^{T}.

Miscellaneous Mathematical Notation

$\overset{\text{a.s.}}{\to}$	a.s. convergence.
$\overset{\mathbb{P}}{\to}$	convergence in probability.
$\overset{\mathcal{D}}{\longrightarrow}$	convergence in distribution.
$\overset{\mathcal{D}}{=}$	equality in distribution.
$|\cdot|$	in addition to absolute value, also used for the number of elements (cardinality) $|S|$ of a set S, or its Lebesgue measure $|S|$.
$\mathbb{E}[X; A]$	$\mathbb{E}[X \mathbb{1}_A]$.
$\sim$	usually, $a(x) \sim b(x)$ means $a(x)/b(x) \to 1$ in some limit like $x \to 0$ or $x \to \infty$, but occasionally, other possibilities occur. E.g. $X \sim \mathcal{N}(\mu, \sigma^2)$ specifies X to have a $\mathcal{N}(\mu, \sigma^2)$ distribution.
$\approx$	a different type of asymptotics, often just at the heuristical level.
$\propto$	proportional to.
$\widehat{F}[\cdot]$	the m.g.f. of a distribution F. Thus $\widehat{F}[is]$ is the characteristic function at s. Sometimes $\widehat{F}[\cdot]$ is also used for the probability generating function of a discrete r.v. See further Sect. A.11 of the Appendix.
$e^{-\int_s^t \mu}$	shorthand for $\exp\{-\int_s^t \mu(v)\,dv\}$.

Standard Distributions

- Normal: As is common, Φ is used for the c.d.f. of the standard normal distribution $\mathcal{N}(0, 1)$ and $\varphi(x) = e^{-x^2/2}/\sqrt{2\pi}$ for the density.
- Exponential: The parametrization we use is the rate λ, not (as often in the statistical literature) the mean. That is, the density is $\lambda e^{-\lambda x}$, $x > 0$, and the mean is $1/\lambda$. When we speak of the standard exponential distribution, $\lambda = 1$.
- gamma(α, λ): is defined by the density

$$\frac{\lambda^\alpha}{\Gamma(\alpha)} x^{\alpha - 1} e^{-\lambda x}, \quad x > 0.$$

Special cases are the Erlang distribution with p stages, defined for $p = 1, 2, \ldots$ as the gamma(p, λ), and the χ^2 or χ_f^2 distribution with f degrees of freedom, the distribution of $X_1^2 + \cdots + X_f^2$ where $X_1, \ldots, X_f$ are independent standard normal; it is gamma($f/2, 1/2$).

- Weibull: this occurs in two variants, one on the positive line defined by the tail $\overline{F}(x) = \mathbb{P}(X_+ > x) = \mathrm{e}^{-x^\beta}$, $x \geq 0$ (a scaling may occur leading to $\overline{F}(x) = \mathrm{e}^{-ax^\beta}$) and one on the negative line being defined by the c.d.f. $\mathbb{P}(X_- \leq x) = \mathrm{e}^{-(-x)^\beta}$, $x \leq 0$. Thus $X_- \overset{\mathscr{D}}{=} -X_+$.
- Standard Brownian motion; denoted W, one with possibly non-zero drift μ and/or possibly non-unit variance constant σ^2 by W_{μ, σ^2}.

Exceptions to all of this occur occasionally!

Abbreviations

BM	Brownian motion
c.g.f.	Cumulant generating function (the log of the m.g.f.)
c.d.f.	Cumulative distribution function, like $F(x) = \mathbb{P}(X \leq x)$
CLT	Central limit theorem
ES	Expected shortfall
HJB	Hamilton–Jacobi–Bellman
i.i.d.	Independent identically distributed
i.o.	Infinitely often
l.h.s.	Left-hand side
LLN	Law of large numbers
MDA	Maximum domain of attraction
m.g.f.	Moment generating function $\widehat{F}[\alpha] = \mathbb{E}\mathrm{e}^{\alpha X}$
ODE	Ordinary differential equation
PDE	Partial differential equation
r.h.s.	Right-hand side
r.v.	Random variable
s.c.v.	Squared coefficient of variation
SDE	Stochastic differential equation
VaR	Value-at-Risk
w.l.o.g.	Without loss of generality
w.p.	With probability

Others occur locally.

Chapter I: Basics

1 Actuarial Versus Financial Pricing

The last decades have seen the areas of insurance mathematics and mathematical finance coming closer together. One reason is the growing linking of pay-outs of life insurances and pension plans to the current value of financial products, another that certain financial products have been designed especially to be of interest for the insurance industry (see below). Nevertheless, some fundamental differences remain, and the present section aims at explaining some of these, with particular emphasis on the principles for pricing insurance products, resp. financial products.

In insurance, *expected values* play a major role. For example, let a claim $X \geq 0$ be the amount of money the insurance company has to pay out for a fire insurance on a given house next year (of course, $\mathbb{P}(X = 0)$ is close to 1!). The insurance company then ideally charges $H(X) = \mathbb{E}X$ in premium plus some *loading*, that is, an extra amount to cover administration costs, profit, risk etc. (different rules for the form of this loading are discussed in Sect. 3). The philosophy behind this is that charging premiums smaller than expected values in the long run results in an overall loss. This is a consequence of the law of large numbers (LLN). In its simplest form it says that if the company faces n i.i.d. claims $X_1, \ldots, X_n$ all distributed as X, then the aggregate claim amount $A = X_1 + \cdots + X_n$ is approximately $n\mathbb{E}X$ for n large. Therefore, if the premium H is smaller than $\mathbb{E}X$, then with high probability the total premiums nH are not sufficient to cover the total aggregate claims A.

This argument carries far beyond this setting of i.i.d. claims, which is of course oversimplified: even in fire insurance, individual houses are different (the area varies, a house may have different types of heating, thatched roof or tiles, etc), and the company typically has many other lines of business such as car insurance,

© Springer Nature Switzerland AG 2020
S. Asmussen, M. Steffensen, *Risk and Insurance*, Probability Theory and Stochastic Modelling 96, https://doi.org/10.1007/978-3-030-35176-2_1

accident insurance, life insurance, etc. Let the claims be $X_1, X_2, \ldots$ Then the asymptotics

$$\frac{X_1 + \cdots + X_n}{\mathbb{E}X_1 + \cdots + \mathbb{E}X_n} \to 1 \tag{1.1}$$

holds under weak conditions. For example, the following elementary result is sufficiently general to cover a large number of insurance settings

Proposition 1.1 *Let $X_1, X_2, \ldots$ be non-negative r.v.s. Write $S_n = X_1 + \cdots + X_n$, and assume $\mathbb{E}S_n \to \infty$ and*

$$\sum_{i=1}^{n} \mathbb{V}\mathrm{ar}\, X_i = o\big((\mathbb{E}S_n)^2\big), \quad \sum_{1 \le i < j \le n} \mathbb{C}\mathrm{ov}(X_i, X_j) = o\big((\mathbb{E}S_n)^2\big)$$

as $n \to \infty$. Then (1.1) holds in the sense of convergence in probability.

Proof By Chebycheff's inequality,

$$\mathbb{P}\big(|S_n/\mathbb{E}S_n - 1| > \epsilon\big) = \mathbb{P}\big(|S_n - \mathbb{E}S_n| > \epsilon \mathbb{E}S_n\big) \le \frac{\mathbb{V}\mathrm{ar}\, S_n}{\epsilon^2 (\mathbb{E}S_n)^2}$$

$$= \frac{1}{\epsilon^2 (\mathbb{E}S_n)^2} \Big[\sum_{1}^{n} \mathbb{V}\mathrm{ar}\, X_n + 2 \sum_{1 \le i < j \le n} \mathbb{C}\mathrm{ov}(X_i, X_j)\Big] \to 0 + 0.$$

$\square$

For insurance arrangements running over longer time it is traditional to incorporate *discounting* when computing expected values. Consider for example a pension plan for a 50 year-old man consisting in a payment of €z (say $z = 100,000$) to the insured at age 70, provided he is still alive, in return for a premium payment h_0 at age 50. The company can then put the premium in the bank. Assuming a fixed interest rate r (say $r = 5\%$), the investment has grown to $h_0 e^{20r}$ when the man turns 70, and if τ_{50} is his remaining lifetime, the claim is $z \mathbb{1}_{\tau_{50} > 20}$. The principle that the premium should at least be the expected pay-out therefore in this setting means $h_0 \ge e^{-20r} z \, \mathbb{P}(\tau_{50} > 20)$.

Premiums calculated as exact (that is, without a loading) expected values according to the above principles are usually referred to as *pure premiums* or *net premiums*.

We now turn to financial mathematics, more precisely the classical problem of option pricing. The prototype of an option is a European call option, which gives the holder the right (but not the obligation) to buy a certain amount of a given asset at the price K at time T (the *maturity time*). For example, the asset can be crude oil, and the option then works as an insurance against high oil prices. More precisely, if $S(T)$ is the market price of the specified amount of oil at time T and $S(T) < K$, the holder will not exercise his right to buy at price K, but will buy at the market at price $S(T)$. Conversely, if $S(T) > K$, the holder will exercise. He then covers K of the price himself, whereas the rest $S(T) - K$ is covered by the option. Thus, he makes a gain of $S(T) - K$ compared to buying at the market price. Therefore $[S(T) - K]^+$ is the value of exercising the option relative to buying at market price. What price Π should one pay to acquire this option at time $t = 0$?

Let again r be the interest rate. Actuarial thinking would lead to the price $e^{-rT}\mathbb{E}[S(T) - K]^+$ of the option because the LLN predicts that this on average favors neither the buyer nor the seller, assuming that both trade a large number of options over time. Economists take a different starting point, the principle of *no arbitrage*, which states that the market balances itself in such a way that there is no way of making money without risk (no *free lunches*). The 'market' needs specification in each case, but is often a world with only two objects to invest in at time $0 \le t < T$: the bank account yielding a risk-free return at short rate r and the underlying asset priced at $S(t)$. Indeed, this leads to a different price of the option, as we will now demonstrate via an extremely simple example.

Example 1.2 We consider a binomial European call option with payoff $(S(1) - K)^+$, where $S(1)$ is thought of as a stock price after one unit of time. We will assume that there are only two possible values for $S(1)$ and (w.lo.g.) that $S(0) = 1$. That is, $\mathbb{P}(S(1) = u) = p$, $\mathbb{P}(S(1) = d) = q = 1 - p$ (up and down), where $d < u$.

An investor with an initial wealth of w_0 is given only two possibilities, to invest at $t = 0$ in the asset (referred to as the *stock* in the following) or in a bank account at fixed interest rate r. If a_1 is the volume of stocks he invests in and a_2 the amount he puts in the bank, we thus have

$$w_0 = a_1 S(0) + a_2 = a_1 + a_2, \quad w_1 = a_1 S(1) + a_2 e^r,$$

where w_1 is the wealth at $T = 1$. We allow a_1, a_2 to be non-integer and to attain negative values, which corresponds to borrowing and shortselling.[1] The pair (a_1, a_2) is referred to as a *portfolio*.

A portfolio is said to *hedge* the option if its return w_1 is exactly the payout $(S(1) - K)^+$ of the option no matter how $S(1)$ comes out. An investor would therefore be indifferent as to whether he puts his money in the portfolio or in the option, so that the price of the option Π should equal the cost of establishing the hedge portfolio, i.e. $\Pi = w_0$. Writing $V_u = (u - K)^+$ (the payoff if the stock goes up) and $V_d = (d - K)^+$, the requirement $w_1 = (S(1) - K)^+$ then means that $V_u = a_1 u + a_2 e^r$, $V_d = a_1 d + a_2 e^r$. This is two linear equations with two unknowns a_1, a_2, and the solution is $a_1 = (V_u - V_d)/(u - d)$, $a_2 = (uV_d - dV_u)/(u - d)e^r$.

---

[1] The meaning of $a_2 < 0$ is simply that $-a_2$ is borrowed from the bank, whereas $a_1 < 0$ means that the investor has taken the obligation to deliver a volume of $-a_1$ stocks at time $T = 1$ (shortselling).

Thus

$$\Pi = w_0 = \frac{V_u - V_d}{u - d} + \frac{u V_d - d V_u}{(u - d)e^r}. \tag{1.2}$$

Probably the most surprising feature of this formula is that p does not enter. Intuitively, one feels that the option of buying a stock for a price of K at $T = 1$ is more attractive the larger p is. But this is not reflected in (1.2).

The market is said to *allow arbitrage* if it is possible to choose a_1, a_2 such that $\mathbb{P}(e^{-r}w_1 \geq w_0) = 1$ and $\mathbb{P}(e^{-r}w_1 > w_0) > 0$.[2] It is easy to see that $d \leq e^r \leq u$ is a necessary condition for the market to be free of arbitrage.[3] Thus introducing $p^* = (e^r - d)/(u - d)$, $q^* = 1 - p^* = (u - e^r)/(u - d)$, we have $0 < p^* < 1$ in an arbitrage-free market, and it is easy to see that (1.2) can be rewritten

$$\Pi = w_0 = e^{-r}[p^*V_u + q^*V_d] = \mathbb{E}^*[e^{-r}(S(1) - K)^+], \tag{1.3}$$

where $\mathbb{E}^*$ denotes expectation with p replaced by p^*. Thus, the price has the same form as the naive guess $\mathbb{E}[e^{-r}(S(1) - K)^+]$ above, but under a different probability specification. One can further check that $\mathbb{E}^*S(1) = S(0)e^r = e^r$. That is, in $\mathbb{E}^*$-expectation the stock behaves like the bank account. For this reason, p^* is referred to as the *risk-neutral probability*. Equivalently, $\{e^{-rt}S(t)\}_{t=0,1}$ is a martingale under the risk-neutral measure (also called the martingale measure).   ◊

The same procedure applies to more than one period and to many other types of options and other models for $\{S(t)\}$. If $\Phi(S_0^T)$ is the payout at maturity where $S_0^T = \{S(t)\}_{0 \leq t \leq T}$, the principle is:

1. Compute risk-neutral parameters such that $\{e^{-rt}S(t)\}$ is a martingale under the changed distribution $\mathbb{P}^*$ and such that[4] $\mathbb{P}$ and $\mathbb{P}^*$ are equivalent measures on the space of paths for $\{S(t)\}$.
2. Set the price as $\Pi = \mathbb{E}^*[e^{-rT}\Phi(S_0^T)]$.

In some cases, such a risk-neutral P^* exists and is unique; in other cases, there may be many or none, obviously creating some difficulties for this approach to option pricing.

For continuous-time formulations, the canonical model for the asset price process $\{S(t)\}_{0 \leq t \leq T}$ is a geometric Brownian motion, i.e., $\{\log S(t)\}$ is Brownian motion

---

[2]Note that w_0 units of money at time 0 has developed into $w_0 e^r$ units at time 1 if put in the bank. Therefore, the fair comparisons of values is the ordering between $e^{-r}w_1$ and w_0, not between w_1 and w_0.

[3]For example, if $d > e^r$, an arbitrage opportunity is to borrow from the bank and use the money to buy the stock.

[4] We do not explain the reasons for this requirement here!

(BM) with drift μ and variance constant σ^2, say. This model is called the *Black–Scholes model* and we return to its details in Sect. VI.1.

The main reason for the principles for price calculations in insurance and finance being different is the non-existence of a market for insurance contracts. You cannot sell the fire insurance designed especially for your house to an owner of a different house, and also life insurance contracts are issued to an individual.

There are, however, examples where the two principles come together. For a basic example, consider a *unit-linked* (also called *equity-linked*) insurance, that in the pension plan example above could mean that the amount z is not fixed, but equal to a random economical quantity Z, say a European option on a stock market index. The mortality risk is then typically priced by traditional actuarial expected value reasoning and the financial risk via the no arbitrage principle, which means that the net premium becomes

$$\mathrm{e}^{-20r}\mathbb{E}^* Z \cdot \mathbb{P}(\tau_{50} > 20).$$

An example of an option of interest for the insurance industry which is being increasingly traded is a weather-based option. For example, the option could pay a certain amount provided the maximum wind speed during a specific winter 2020–21 exceeds 130 km/h and nothing otherwise. Buying such an option, the company hedges some of its risk on storm damage insurance, so that the option could be seen as an alternative to reinsurance. Another type of option aims at hedging *longevity risk*, the risk faced by an insurance company issuing pension plan arrangements because of people living longer. For example, such an option could pay a certain amount if the average mortality of 70–75 year-old males in a population at expiry is less than half of the same mortality at the time the option is issued.

Notes
Björk [26] is an excellent reference for financial mathematics at the level of this book. Some general references refining the LLN types of questions related to Proposition 1.1 are the books by Chow and Teicher [48] and Petrov [139].

2 Utility

The area of *utility theory* or *expected utility theory* is based on two axioms for an individual's rationales in financial matters:

A There is a function $u(x)$, the *utility function*, such that $u(x)$ represents the individual's subjective value of holding the amount x;

B When choosing between two decisions $\mathcal{D}_1, \mathcal{D}_2$ resulting in holdings Y_1, Y_2 that are random variables, the individual will choose the one with the higher expected utility rather than higher expected value. That is, he chooses $\mathcal{D}_1$ if $\mathbb{E}u(Y_1) > \mathbb{E}u(Y_2)$, $\mathcal{D}_2$ if $\mathbb{E}u(Y_2) > \mathbb{E}u(Y_1)$ and is indifferent if $\mathbb{E}u(Y_1) = \mathbb{E}u(Y_2)$.

Unquestionably, most individuals will have difficulties in quantifying their utility function. Nevertheless, some qualitative properties appear universal. If $x_2 > x_1$, it is more valuable to hold x_2 than x_1, and so one should have $u(x_2) \geq u(x_1)$. That is, u should be non-decreasing. The value of adding $a = \text{€}\,10{,}000$ to your fortune is smaller if the fortune is $x_2 = \text{€}\,1{,}000{,}000$ than if it is $x_1 = \text{€}\,500$. This gives

$$u(x_1 + a) - u(x_1) \geq u(x_2 + a) - u(x_2) \quad \text{for } a > 0,\ x_1 < x_2. \tag{2.1}$$

Equation (2.1) is almost the same as requiring u to be concave. The additional requirements are minor, and in particular, it suffices that u is strictly increasing, which is assumed in the following (in fact, even non-decreasing suffices).

Depending on the context, one may also assume $u(0) = 0$, since adding or subtracting a constant from u does not change the conclusion that some decision has larger utility than another one. In fact, working with the utility function $a + bu(x)$ almost always leads to the same conclusions as working with $u(x)$.

A real-valued r.v. is referred to in the following as a *risk*; we thereby do not require a specific sign of X like $X \geq 0$ as in some parts of the literature. Jensen's inequality gives at once:

Theorem 2.1 *For any risk X and any u satisfying* (2.1),

$$\mathbb{E}u(X) \leq u(\mathbb{E}X). \tag{2.2}$$

Conversely, (2.2) can be used to characterize utility functions:

Proposition 2.2 *Let v be a function satisfying $\mathbb{E}v(X) \leq v(\mathbb{E}X)$ for any risk X. Then v is concave.*

Proof Given $x < y$ and $\theta \in [0, 1]$, we must show

$$v\big(\theta x + (1 - \theta)y\big) \geq \theta v(x) + (1 - \theta)v(y).$$

This follows by taking X to be x w.p. θ and y w.p. $1 - \theta$. $\qquad\square$

Remark 2.3 In economics, one sometimes says that X is better than Y in *second-order stochastic dominance* if every risk taker that is risk averse prefers X to Y, that is, if $\mathbb{E}u(x + X) \geq \mathbb{E}u(x + Y)$ for every increasing concave function u and all x. In gambling terms, gamble X has second-order stochastic dominance over gamble Y if X is more predictable (i.e. involves less risk) and has at least as high a mean. $\quad\Diamond$

Example 2.4 Consider a homeowner with fortune x facing two decisions: paying H for a fire insurance covering his house during the next year or running the (small) risk that the house will burn, causing damages of cost X. Depending on his choice, his fortune after a year will be $x - H$ or $x - X$. By Theorem 2.1 with X replaced by $x - X$, we have that $\mathbb{E}u(x - X) < u(x - \mathbb{E}X)$, and so by Axiom B the homeowner prefers to take the insurance if $H = \mathbb{E}X$. Accordingly, he is prepared to pay a somewhat higher premium.

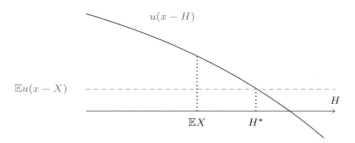

Fig. 1 Utility premium

The customer's *utility premium* H^* is defined as the premium making the homeowner indifferent to insuring or not. That is, H^* is defined by

$$u(x - H^*) = \mathbb{E}u(x - X). \tag{2.3}$$

cf. Fig. 1. The utility premium H^* can alternatively be characterized as the highest premium the insured is prepared to pay. The insurance company typically charges a loading to expected values, so it will only offer insurance with $H > \mathbb{E}X$. The homeowner accepts this premium if it provides at least the same expected utility as not insuring, i.e. if $u(x - H) \geq \mathbb{E}u(x - X)$. Note that the set of such H is an interval of the form $[0, \overline{H}]$ with $\mathbb{E}X$ as interior point, and so a range of premiums larger than $\mathbb{E}X$ is acceptable for the homeowner.   ◇

Remark 2.5 If the decision taker is a gambler and X the gain (prize minus stakes), he will always prefer the amount $\mathbb{E}X$ (usually negative so that we talk of a payment) in cash rather playing the game *provided his utility function is concave*, and this decision does not depend on whether, as typical, $\mathbb{E}X < 0$ or he is in the rare case $\mathbb{E}X > 0$ of a favorable game. Thus the utility function of a typical gambler (a risk lover) cannot be concave: for him, the prospective win of € 10,000 on a lottery outweighs paying its price € 1 even if the odds are far below 1/10,000.   ◇

Motivated by this discussion one calls a risk taker with concave utility function *risk averse*.

Example 2.6 The utility function is a subjective quantity. Virtually all individuals and institutions will be unable to write their utility function u up in closed form, though qualitatively they may admit that they are risk averse and so u should be concave (and increasing). The most popular examples in the literature are:

- exponential utility, $u(x) = b - ce^{-ax}$ where $a > 0$ and $c > 0$ to ensure that the function is increasing and concave. Often the special case $u(x) = (1 - e^{-ax})/a$ with $a > 0$ is encountered because of its convenient normalisation $u(0) = 0$, $u'(x) = 1$.

- power utility, $u(x) = ax^b + c$ where either $0 < b < 1, a > 0$ or $b < 0, a < 0$. Because of problems with the definition of x^b for $x < 0$ and with concavity, the domain has to be restricted to $x \geq 0$. $\diamond$

For a concave utility function u and a given risk X, we can now introduce the *buy* or *bid* price π^b of X and the *sell* or *ask* price π^s. The buy price is the largest amount the decision maker would willingly pay to obtain X or, equivalently, making him indifferent to adding $X - \pi^b$ to x. That is,

$$u(x) = \mathbb{E}u(x + X - \pi^b). \tag{2.4}$$

Similarly, the sell price is the smallest amount the decision maker would accept to sell X if he had it or, equivalently, making him indifferent to replacing the risk X by the fixed amount π^s. For this reason, π^s is also called the *cash equivalent, certainty equivalent* or *value* of X. It is given by

$$\mathbb{E}u(x + X) = \mathbb{E}u(x + \pi^s). \tag{2.5}$$

Of course, the seller and the buyer will typically have different utility functions.

Since the decision taker always prefers to exchange X by $\mathbb{E}X$, we have $\pi^s < \mathbb{E}X$. The difference $\pi(x, X)$ is the amount the decision taker is willing to pay to avoid the randomness in X and is called the *risk premium* at level x. Thus $\pi(x, X) = \mathbb{E}X - \pi^s$ and may be defined directly without reference to the cash equivalent by

$$u(x + \mathbb{E}X - \pi(x, X)) = \mathbb{E}u(x + X). \tag{2.6}$$

Remark 2.7 The above definitions of π^b and π^s are the most classical ones, but variants and different interpretations occur according to the context. For example, shortselling in finance as discussed in Sect. 1 amounts to selling risks X one does not have, and this could motivate defining π^s by

$$u(x) = \mathbb{E}u(x - X + \pi^s). \tag{2.7}$$

This is similar to the decision problem of the insurance company that buys (takes over) the risk X from the policy holder. The received premium then takes the role of the buy price, so in this context

$$u(x) = \mathbb{E}u(x - X + \pi^b). \tag{2.8}$$

From the perspective of an insurance holder with loss X, signing an insurance contract amounts to replacing $x - X$ by the selling price of $-X$, which is the idea adopted in Example 2.4. $\diamond$

The traditional way to quantify unwillingness to take risk is in terms of the *absolute risk aversion*, defined at level x as

$$r(x) = r^a(x) = -\frac{u''(x)}{u'(x)} = -\frac{d}{dx}\log u'(x) \tag{2.9}$$

(note that concavity ensures $r(x) \geq 0$). That the concept should depend on x may be understood from the intuition that potential losses may matter less for larger fortunes than for small. That u'' enters is reasonable since $u''(x)$ describes how big the deviation of u from linearity is in a neighborhood of x, i.e. it somehow measures the degree of concavity at x (it also enters in the formulas for the curvature of u at x). Finally, the normalization by $u'(x)$ ensures that the risk aversion is the same for the utility function $a + bu(x)$ (with $b > 0$), which leads to the same comparisons as u.

Example 2.8 For the exponential utility function $u(x) = b - ce^{-ax}$ in Example 2.6, one gets $r^a(x) = a$. Thus the risk aversion is constant and the exponential utility function is in fact the only one with this property. One speaks of constant absolute risk aversion, often abbreviated CARA.

For the power utility function $u(x) = ax^b + c$ in Example 2.6, one gets $r^a(x) = (1-b)/x$ such that $r^r(x) = 1-b$, where the relative risk aversion is defined at level x as

$$r^r(x) = xr^a(x).$$

In the power utility example, one then speaks of *constant relative risk aversion*, often abbreviated CRRA.

Both exponential utility and power utility belongs to a class of utility function called hyperbolic absolute risk aversion, often abbreviated HARA, and defined by

$$r^a(x) = \frac{1}{ax+b}$$

for constants a and b.   ◊

We give two more rigorous justifications of the definition (2.9) of the risk aversion, of which the first is:

Proposition 2.9 *Let X be a risk with $\mathbb{E}X = 0$. Then*

$$\pi(x, \epsilon X) \sim \epsilon^2 \mathbb{V}\text{ar}(X)r(x)/2 \quad as \; \epsilon \downarrow 0.$$

Proof For brevity, write $\pi_\epsilon = \pi(x, \epsilon X)$, $u' = u'(x)$, $u'' = u''(x)$. By a first-order Taylor expansion

$$u(x + \epsilon\mathbb{E}X - \pi_\epsilon) = u(x - \pi_\epsilon) = u(x) - u'\pi_\epsilon(1 + o(1)),$$

(noting that clearly $\pi_\epsilon \to 0$), and by a second-order Taylor expansion

$$\mathbb{E}u(x + \epsilon X) = u(x) + \epsilon u'\mathbb{E}X + u''\epsilon^2\mathbb{E}X^2/2 + o(\epsilon^2)$$
$$= u(x) + u''\epsilon^2\,\mathbb{V}\mathrm{ar}(X)/2 + o(\epsilon^2).$$

Using the definition (2.6) gives

$$-u'\pi_\epsilon(1 + o(1)) = u''\epsilon^2\,\mathbb{V}\mathrm{ar}(X)/2(1 + o(1))$$

and the conclusion.   □

The final justification for taking $-u''(x)/u'(x)$ as a quantitative measure of risk aversion is that overall higher risk aversion is equivalent to overall higher risk premiums:

Theorem 2.10 *Consider two risk takers with utility functions u_1, u_2, defined on the same open interval I, and risk aversions $r_i(x) = -u_i''(x)/u_i'(x)$, $i = 1, 2$. Then $r_1(x) \geq r_2(x)$ for all $x \in I$ if and only if $\pi_1(x, X) \geq \pi_2(x, X)$ for all risks X and $x \in I$ such that $x + X \in I$.*

Proof The 'if' part follows immediately from Proposition 2.9. Conversely, assume $r_1 \geq r_2$. Let $v_1 < v_2$ and $x_1 = u_2^{-1}(v_1)$, $x_2 = u_2^{-1}(v_2)$. Then $x_1 < x_2$ and (2.9) yields

$$\log u_1'(x_2) - \log u_1'(x_1) = -\int_{x_1}^{x_2} r_1(y)\,\mathrm{d}y$$
$$\leq -\int_{x_1}^{x_2} r_2(y)\,\mathrm{d}y = \log u_2'(x_2) - \log u_2'(x_1).$$

Exponentiating both sides gives

$$\frac{u_1'(x_2)}{u_1'(x_1)} \leq \frac{u_2'(x_2)}{u_2'(x_1)} \quad \text{or} \quad \frac{u_1'(x_2)}{u_2'(x_2)} \leq \frac{u_1'(x_1)}{u_2'(x_1)}.$$

This last inequality means $\varphi'(v_2) \leq \varphi'(v_1)$, where $\varphi(v) = u_1(u_2^{-1}(v))$, so we conclude that φ' is decreasing and φ is therefore concave. Now rewriting (2.6) as

$$\pi(x, X) = x + \mathbb{E}X - u^{-1}(\mathbb{E}u(x + X))$$

and letting $Y = u_2(x + X)$ gives

$$
\begin{aligned}
\pi_1(x, X) - \pi_2(x, X) &= u_2^{-1}\big(\mathbb{E}u_2(x + X)\big) - u_1^{-1}\big(\mathbb{E}u_1(x + X)\big) \\
&= u_2^{-1}(\mathbb{E}Y) - u_1^{-1}\big(\mathbb{E}\varphi(Y)\big) \geq u_2^{-1}(\mathbb{E}Y) - u_1^{-1}\big(\varphi(\mathbb{E}Y)\big) \\
&= u_2^{-1}(\mathbb{E}Y) - u_2^{-1}(\mathbb{E}Y) = 0.
\end{aligned}
$$

$\square$

Returning to the above intuition that potential losses may matter less for larger fortunes:

Corollary 2.11 *The following assertions are equivalent:*

(i) *The risk aversion $r(x)$ is non-increasing in x;*
(ii) *For any risk X, the risk premium $\pi(x, X)$ is non-increasing in x;*
(iii) *$u'(x)u'''(x) \geq u''(x)^2$ for all x.*

Proof The equivalence of (i) and (ii) follows from Theorem 2.10 by comparing $u(x)$ with $u(x + a)$, and the equivalence of (i) and (iii) from

$$
\frac{\mathrm{d}}{\mathrm{d}x} r(x) = \frac{\mathrm{d}}{\mathrm{d}x}\left[-\frac{u''(x)}{u'(x)}\right] = -\frac{u'(x)u'''(x) - u''(x)^2}{u''(x)^2}.
$$

$\square$

Exercises

2.1 (Bernoulli 1738) An art collector possessing two pictures of value v_1, v_2 has to send the paintings by special delivery. A shipment has probability p of being damaged. Show, assuming that damage $=$ total damage, that sending the paintings separately results in a greater expected utility than sending them together.

2.2 Consider a risk with a gamma(α, λ) distribution. Show that the utility premium for an insured with an exponential utility function with risk aversion $a \geq \lambda$ equals ∞, and that the same conclusion holds for the insurer with a similar utility function. Discuss why one talks about an 'uninsurable risk'.

2.3 An insurance company has initial wealth $w \geq 0$ and utility function u given by $u(x) = x$ for $x \geq 0$, $u(x) = 2x$ for $x \leq 0$. A claim X has distribution $\mathbb{P}(X = 1) = 1 - \mathbb{P}(X = 0) = 1/10$.

a) For $H \geq 0$, compute $\mathbb{E}u(w + H - X)$ by considering the cases $w + H \leq 1$ and $w + H > 1$ separately.
b) Show that the utility premium H_{Util} is $H_{\mathrm{Util}} = \mathbb{E}X$ when $w > 9/10$.
c) For $w \leq 9/10$, show first that $H_{\mathrm{Util}} \leq 1 - w$ and compute next H_{Util}.

2.4 Let $u(x) = 1 - \mathrm{e}^{-\sqrt{x}}$ for $x > 0$.

a) Show that $u'(x)$ is decreasing and thereby that $u(x)$ is concave and increasing on the interval $(0, \infty)$.
b) An insured has utility function $u(x)$ for $x > 0$ and initial wealth $w = 2$. A claim $X \in (1, 2)$ is distributed as $2 - U^2$ where U is uniform on $(0, 1)$. Find the utility premium H of the insured [you may check your solution with the numerical value $H \approx 1.79$].
c) What is the risk premium of the claim?

Notes

Like any mathematical model, utility theory is an abstraction and simplification of reality. The mathematical correctness and the salience of its axioms do not guarantee that it always reflects human behavior (going into this is an issue in *behaviorial economics*). Nevertheless, the mathematical clarity has helped scientists design experiments to test its adequacy, and to distinguish systematic departures from its predictions.

Utility functions have a long history in economics. The present setting with random outcomes was to a large extent developed by von Neumann and Morgenstern, see e.g. their 1944 book [173]. Föllmer and Schied [77] give an extensive treatment of modern mathematical utility theory. One may note that the term 'risk premium' is often used in a different way in insurance practice, namely as the net premium (the expected value).

Theorem 2.10 is classical and originates from Pratt [143].

3 Premium Rules

The standard setting for discussing premium calculation in the actuarial literature is in terms of a single risk $X \geq 0$ and does not involve portfolios, stochastic processes, etc. Here X is an r.v. representing the random payment (possibly 0) to be made from the insurance company to the insured. A premium rule is then a $[0, \infty)$-valued function H of the distribution of X, often written $H(X)$, such that $H(X)$ is the premium to be paid, i.e. the amount for which the company is willing to insure the given risk. From an axiomatic point of view, the concept of premium rules is closely related to that of risk measures, to which we return in Sect. X.1.

The standard premium rules discussed in the literature (not necessarily the same as those used in practice!) are the following:

- The net premium principle $H(X) = \mathbb{E}X$ (also called *the equivalence principle*). As follows from a suitable version of the CLT that this principle will lead to a substantial loss if many independent risks are insured. This motivates that a loading should be added, as in the next principles:
- The expected value principle $H(X) = (1 + \eta)\mathbb{E}X$, where η is a specified safety loading. For $\eta = 0$, we are back to the net premium principle. A criticism of the expected value principle is that it does not take into account the variability of X. This leads to:
- The variance principle $H(X) = \mathbb{E}X + \eta \operatorname{Var}(X)$. A modification (motivated by $\mathbb{E}X$ and $\operatorname{Var}(X)$ not having the same dimension) is
- The standard deviation principle $H(X) = \mathbb{E}X + \eta\sqrt{\operatorname{Var}(X)}$.

- The principle of zero utility. This was introduced already in Sect. 2, cf. in particular (2.8). Here $u(x)$ is the utility function of the insurance company. The principle amounts to considering the expected utility after the termination of the contract and arguing that the company would be willing to accept a premium $H(X)$ such that its expected utility in the case of insurance is precisely the same as its expected utility in the case of no insurance. If x is the initial wealth, the wealth after termination of the contract is $x + H(X) - X$ in the first case ($H(X)$ has been received, but X has to be paid out) and remains x in the second. Thus, the company's utility premium is given as the solution of

$$u(x) = \mathbb{E}u\big(x + H(X) - X\big), \tag{3.1}$$

which is just (2.8) in a different notation. Put in another way, (3.1) expresses an indifference to the two arrangements from the point of view of the company; for the customer's perspective, see (2.3).

It was noted in Sect. 2 as a consequence of Jensen's inequality that $H(X) \geq \mathbb{E}X$ for the utility principle. If we insert $X - \mathbb{E}X$ in Proposition 2.9, we get

$$H(\epsilon X) \sim \epsilon \mathbb{E}X + \epsilon^2 \operatorname{Var}(X)r(x)/2 \quad \text{as } \epsilon \downarrow 0,$$

where $r(x)$ is the risk aversion, so that the utility principle is close to the variance principle for small claims ϵX.

The most important special case of the principle of zero utility is:

- The exponential principle, which corresponds to $u(x) = (1 - e^{-ax})/a$ where $a > 0$ is the risk aversion. Here (3.1) is equivalent to $1 = e^{-aH(X)}\mathbb{E}e^{aX}$, and we get

$$H(X) = \frac{1}{a} \log \mathbb{E}e^{aX}.$$

Since m.g.f.s are log-convex, it follows that $H_a(X) = H(X)$ is increasing as a function of a. Further, $\lim_{a \downarrow 0} H_a(X) = \mathbb{E}X$ (the net premium principle) and, provided $b = \operatorname{ess\,sup}X < \infty$, $\lim_{a \to \infty} H_a(X) = b$ (the premium principle $H(X) = b$ is called the *maximal loss principle* but is clearly not very realistic).
- The percentile principle. Here one chooses a (small) number α, say 0.05 or 0.01, and determines $H(X)$ by $\mathbb{P}\big(X \leq H(X)\big) = 1 - \alpha$ (assuming a continuous distribution for simplicity). In risk management, such an $H(X)$ goes under the name of the *Value-at-Risk* (VaR) and we discuss it in more detail in Sect. X.1

Some standard axioms proposed for discussing the merits of premium rules are:

1) Non-negative loading, i.e. $H(X) \geq \mathbb{E}X$.
2) $H(X) \leq b$ when $b = \operatorname{ess\,sup}X$ is finite.
3) $H(X + c) = H(X) + c$ for any constant c.
4) $H(X + Y) = H(X) + H(Y)$ when X, Y are independent.

Note that $H(cX) = cH(X)$ is not on the list! Considering the examples above, the net premium principle and the exponential principle can be seen to be the only ones satisfying all four properties. The expected value principle fails to satisfy, e.g., 3), whereas (at least) 4) is violated for the variance principle, the standard deviation principle, and the zero utility principle (unless it is the exponential or net premium principle). For more details, see e.g. Gerber [79] or Sundt [166].

Exercises

3.1 For a standard exponential risk with rate 1 (the density is e^{-x}, $x > 0$), compute the utility premium (3.1) as a function of x and a.

Notes

The material of this section is standard. A somewhat related axiomatic discussion is in Wang, Young and Panjer [174].

4 Reinsurance

Reinsurance means that the company (the *cedent* or *first insurer*) insures a part of the risk at another insurance company (the *reinsurer*). The purposes of reinsurance are to reduce risk and/or to reduce the risk volume of the company.

We start by formulating the basic concepts within the framework of a single risk $X \geq 0$. A reinsurance arrangement is then defined in terms of a function $r(x)$ with the property $0 \leq r(x) \leq x$. Here $r(x)$ is the amount of the claim x to be paid by the reinsurer and $s(x) = x - r(x)$ the amount to be paid by the cedent. The function $s(x)$ is referred to as the *retention function*. The most common examples are the following two:

- Proportional reinsurance $r(x) = (1 - \theta)x$, $s(x) = \theta x$ for some $\theta \in (0, 1)$. Also called *quota share* reinsurance.
- Stop-loss reinsurance $r(x) = (x - b)^+$ for some $b \in (0, \infty)$, referred to as the *retention limit*. The retention function is $x \wedge b$.

Concerning terminology, note that in the actuarial literature the *stop-loss transform* of $F(x) = \mathbb{P}(X \leq x)$ (or, equivalently, of X) is defined as the function

$$b \;\mapsto\; \mathbb{E}(X - b)^+ \;=\; \int_b^\infty (x - b)\, F(\mathrm{d}x) \;=\; \int_b^\infty \overline{F}(x)\, \mathrm{d}x$$

(the last equality follows by integration by parts, see formula (A.1.1) in the Appendix). It shows up in a number of different contexts, see e.g. Sect. VIII.2.1, where some of its main properties are listed.

The risk X is often the aggregate claims amount $A = \sum_1^N V_i$ in a certain line of business during one year; one then talks of *global reinsurance*. However, reinsurance may also be done *locally*, i.e. at the level of individual claims. Then, if N is the number of claims during the period and $V_1, V_2, \ldots$ their sizes, then the amounts paid by reinsurer, resp. the cedent, are

$$\sum_{i=1}^{N} r(V_i), \quad \text{resp.} \sum_{i=1}^{N} s(V_i). \tag{4.1}$$

If $r(x) = (x - b)^+$, it is customary to talk of *excess-of-loss reinsurance* rather than stop-loss reinsurance when the reinsurance is local. The global equivalents of the payments in (4.1) are, of course, $r(A)$, resp. $s(A)$.

It should be noted that the cedent-to-reinsurer relation is exactly parallel to the insured-to-insurance company relation. Namely, it is common that an insured having a claim of size x does not receive x but an amount $r(x)$ somewhat smaller than x. He thus has to cover $s(x) = x - r(x)$, sometimes referred to as his *own risk*, himself, and one talks of $r(x)$ as the *compensation function*. The most common example is $r(x) = (x - b)^+$ (a stop-loss function). Then $s(x) = x \wedge b$, and one talks of a *deductible* or *first risk* for the insured.

Reinsurance is called *cheap* if (informally) the reinsurer charges the same premium as the cedent. A formal definition is most easily given if the reinsurer and the cedent both apply the expected value principle with loadings η_r, η_s. Then cheap reinsurance means $\eta_r = \eta_s$ and can then be seen as a cost-free way of reducing risk. For example, if the arrangement is proportional reinsurance then without or with reinsurance the cedent faces the loss $X - (1 + \eta)\mathbb{E}X$, resp.

$$s(X) + (1 + \eta)\mathbb{E}r(X) - (1 + \eta)\mathbb{E}X \;=\; \theta\big(X - (1 + \eta)\mathbb{E}X\big).$$

But for $x > 0$,

$$\mathbb{P}\big(\theta(X - (1 + \eta)\mathbb{E}X) > x\big) \;=\; \mathbb{P}\big(X - (1 + \eta)\mathbb{E}X > x/\theta\big)$$
$$< \;\mathbb{P}\big(X - (1 + \eta)\mathbb{E}X > x\big).$$

However, in practice reinsurance incurs an extra cost, meaning in the example that $\eta_r > \eta_s$. The rationale for that can be well understood by considering stop-loss reinsurance where the reinsurer covers the large claims which typically give the main contribution to the risk.

Proportional reinsurance, excess-of-loss reinsurance and stop-loss reinsurance covers the examples to be met in this book and in the majority of the theoretical literature. The contracts used in practice are, however, typically much more complex and often involve different arrangements in different layers. Let $0 = \ell_0 < \ell_1 < \cdots < \ell_K < \infty$. Layer 1: $(0, \ell_1]$ is then the part of the claim or the aggregate claims covered by the cedent. In layer 2: $(\ell_1, \ell_2]$, one may for example use proportional reinsurance with factor θ_2 and loading $\eta_{r,2}$ of the reinsurer, in layer 3: $(\ell_2, \ell_3]$ proportional reinsurance specified by θ_3 and $\eta_{r,3}$, and in the final layer (ℓ_K, ∞) stop-loss reinsurance.

Notes

For further discussion of reinsurance in this book, see in particular Sects. VIII.4 and XI.2.2.

A comprehensive recent treatment of reinsurance, including practical and statistical issues, is given in Albrecher, Beirlant and Teugels [3] and further references may be found there.

5 Poisson Modelling

A very common assumption in insurance mathematics is that the number of claims N in a given period follows a Poisson distribution, or, more generally, that the point process given by the claim arrival instants forms a Poisson process. This section motivates such assumptions, and it also surveys some important probabilistic aspects of the Poisson process. The material should be familiar to most readers of this book, but it serves as a refresher and stresses some points of view of particular importance for the rest of the book.

5.1 The Poisson Distribution as a Binomial Limit

We first recall that the concept of convergence in distribution ($\xrightarrow{\mathcal{D}}$) is elementary in the case of $\mathbb{N}$-valued r.v.s: if $N, N_1, N_2, \ldots$ are r.v.s with values in $\mathbb{N}$, then $N_k \xrightarrow{\mathcal{D}} N$ as $k \to \infty$ simply means $\mathbb{P}(N_k = m) \to \mathbb{P}(N = m)$, $k \to \infty$, for all $m \in \mathbb{N}$. An equivalent characterization is convergence of probability generating functions,

$$N_k \xrightarrow{\mathcal{D}} N \quad \Longleftrightarrow \quad \mathbb{E}z^{N_k} \to \mathbb{E}z^N \quad \text{for all } z \in [0,1]. \tag{5.1}$$

The key result in the Poisson-binomial setting is the following:

Theorem 5.1 *Assume that N_k follows a binomial (n_k, p_k) distribution and that $n_k \to \infty$, $n_k p_k \to \lambda$ as $k \to \infty$ for some $0 < \lambda < \infty$. Then $N_k \xrightarrow{\mathcal{D}} N$ where N is a Poisson(λ) r.v.*

Proof 1. For m fixed and $n_k \to \infty$, we have $n_k! \sim n_k^m(n_k - m)!$, and thus

$$\mathbb{P}(N_k = m) = \binom{n_k}{m} p_k^m(1-p_k)^{n_k-m} = \frac{n_k!}{m!(n_k-m)!} p_k^n(1-p_k)^{n_k-m}$$

$$\sim \frac{n_k^m}{m!} p_k^m(1-p_k)^{n_k-m} \sim \frac{\lambda^m}{m!}(1-p_k)^{n_k-m} \sim \frac{\lambda^m}{m!}e^{-\lambda} = \mathbb{P}(N = m),$$

where the last $\sim$ follows from

$$\log(1-p_k)^{n_k-m} = (n_k - m)\big(-p_k - o(p_k)\big) = -\lambda + o(1). \qquad \square$$

Proof 2. For $z \in [0, 1]$ we have

$$\mathbb{E}z^{N_k} = (1 - p_k + p_k z)^{n_k} = \exp\{n_k \log(1 - p_k + p_k z)\}$$
$$= \exp\{n_k(-p_k + p_k z + o(p_k))\} \to \exp\{-\lambda + \lambda z\} = \mathbb{E}z^N .$$

$\square$

Verbally, Theorem 5.1 is often stated the following way: if n is large and p is small, then an r.v. N following the binomial distribution with parameters n, p has distribution close to the Poisson(λ) distribution where $\lambda = np$. This is often referred to as the *law of small numbers*, as opposed to the law of large numbers saying that if p is moderate or large, then $N \approx np$. Here are some main examples where the law of small numbers motivates assuming an r.v. to be Poisson:

1) n is the number of insurance policies, p the probability that a given policy has a claim in a given period, and N is the total number of claims in the period. Typically, the number of policies is large, whereas (at least for types of insurance such as fire insurance) any given one only generates a claim with a small probability.
2) n is the number of kilometers driven by an insured in a year and p is the probability that an accident occurs in a kilometer. N is then the number of accidents during a year of the insured; in a year; the value $\lambda = np$ is typically 5–10% (in Western countries).
3) A radioactive source contains n atoms, p is the probability that any given atom decays and the decay is registered N times on a Geiger counter in a given period. E.g., $n \approx 2.53 \times 10^{21}$ for 1g of Uranium 238 and $p = 10^{-12}$ for a period of length one year.[5]

An abundance of other examples occur in many other areas of application where one performs a large number of Bernoulli trials with a small success probability for each. Some tacitly made assumptions that are necessary for the Poisson conclusion are independence and, in example 1), that the probability p^* of a given policy producing more than one claim in the period is small. More precisely, one needs $np^* = o(1)$ or equivalently $p^* = o(p)$; similar remarks apply to example 2).

5.2   The Poisson Process

By a (simple) point process $\mathcal{N}$ on a set $\Omega \subseteq \mathbb{R}^d$ we understand a random collection of points in Ω [simple means that there are no multiple points]. We are almost

---

[5]In the physics literature, p is usually implicitly specified through the half life $T_{1/2}$ of an atom, that is, the median of its lifetime, which we will later see to be exponential, say at rate μ. This implies $e^{-\mu T_{1/2}} = 1/2$ and $p = 1 - e^{-\mu}$. For ^{238}U, one has $T_{1/2} = 4.5 \times 10^9$ years.

exclusively concerned with the case $\Omega = [0, \infty)$. The point process can then be specified by the sequence $T_1, T_2, \ldots$ of interarrival times such that the points are $T_1, T_1 + T_2, \ldots$ The associated counting process $\{N(t)\}_{t \geq 0}$ is defined by letting $N(t)$ be the number of points in $[0, t]$. Write

$$\mathcal{N}(s, t] = N(t) - N(s) = \#\{n : s < T_1 + \cdots + T_n \leq t\}$$

for the increment of $\{N(t)\}$ over $(s, t]$ or equivalently the number of points in $(s, t]$.

Definition 5.2 $\mathcal{N}$ is a Poisson process on $[0, \infty)$ with rate λ if $\{N(t)\}$ has independent increments and $N(t) - N(s)$ has a $\text{Poisson}(\lambda(t - s))$ distribution for $s < t$.

Here independence of increments means independence of increments over disjoint intervals.

It is not difficult to extend the reasoning behind example 1) above to conclude that for a large insurance portfolio, the number of claims in disjoint time intervals are independent Poisson r.v.s, and so the times of occurrences of claims form a Poisson process. There are, however, different ways to approach the Poisson process. In particular, the infinitesimal view in part (iii) of the following result will prove useful for many of our purposes.

Theorem 5.3 *Let $\mathcal{N}$ be a point process on $[0, \infty)$ and let $\mathscr{F}(t) = \sigma\big(N(s) : 0 \leq s \leq t\big)$ be the natural filtration. Then the following properties are equivalent:*

(i) *$\mathcal{N}$ is a Poisson(λ) process where $\lambda = \mathbb{E}N(1)$;*
(ii) *$\mathcal{N}$ has stationary independent increments and is simple;*
(iii) *As $h \downarrow 0$, it holds that*

$$\mathbb{P}\big(\mathcal{N}(t, t + h] = 1 \,\big|\, \mathscr{F}(t)\big) = \lambda h + o(h), \tag{5.2}$$

$$\mathbb{P}\big(\mathcal{N}(t, t + h] \geq 2 \,\big|\, \mathscr{F}(t)\big) = o(h). \tag{5.3}$$

Note that traditionally, (5.2), (5.3) are supplemented by

$$\mathbb{P}\big(\mathcal{N}(t, t + h] = 0 \,\big|\, \mathscr{F}(t)\big) = 1 - \lambda h + o(h). \tag{5.4}$$

This is, however, a redundant requirement: since the three conditional probabilities sum to 1, any two of (5.2), (5.3), (5.4) implies the third.

Proof (i) $\Rightarrow$ (ii): all that needs to be shown is that a Poisson process is simple. But the expected number of events in an interval of length h that are multiple is bounded by

$$\mathbb{E}\big[\mathcal{N}(0, h]; \, \mathcal{N}(0, h] \geq 2\big] = \sum_{n=2}^{\infty} n e^{-\lambda h} \frac{(\lambda h)^n}{n!} = O(h^2).$$

Therefore the expectation $m = \mathbb{E}M$ of the number M of events in $(0, r]$ that are multiple is bounded by $\lceil r \rceil 2^k O(2^{-2k})$. Letting $k \to \infty$ gives $m = 0$ and so $M = 0$ a.s. The truth of this for all r gives the desired conclusion. □

(i) $\Rightarrow$ (iii): the definition of a Poisson process implies that the l.h.s. of (5.2) equals $\mathbb{P}(N(h) = 1)$, which in turns equals $\lambda h e^{-\lambda h} = \lambda h + o(h)$. Similarly, (5.4) equals $\mathbb{P}(N(h) = 0) = e^{-\lambda h} = 1 - \lambda h + o(h)$. □

(iii) $\Rightarrow$ (i): Define $f_n(t) = \mathbb{P}(N(t) = n)$. We will first show

$$f_n(t) = e^{-\lambda t} \frac{(\lambda t)^n}{n!} \tag{5.5}$$

by induction and thereby that $N(t)$ has the correct distribution.

Take first $n = 0$. In order for no events to occur in $[0, t + h]$, no events can occur in $(0, t]$ or in $(t, t + h]$. Since these two possibilities are independent, we get

$$f_0(t + h) = f_0(t)\mathbb{P}(\mathcal{N}(t, t + h] = 1 \mid \mathcal{N}(0, t] = 0) = f_0(t)(1 - \lambda h + o(h)),$$

where we used (5.2) in the last step. Subtracting $f_0(t)$ on both sides and dividing by h shows that $f_0(t)$ is differentiable with $f_0'(t) = -\lambda f_0(t)$. Since $f_0(0) = 1$, this gives $f_0(t) = e^{-\lambda t}$, as required [an alternative argument is in the proof of (ii) $\Rightarrow$ (iii) below].

Next let $n > 0$ and assume (5.5) to hold for $n - 1$. If n events are to occur in $[0, t + h]$, there must be either $n, n - 1$ or $n - 2, n - 3, \ldots, 0$ events in $[0, t]$. In the last case, 2 or more events have to occur in $(t, t + h]$, and the probability of this is $o(h)$ by (5.3). In the two other cases, we have instead 0 or 1 events in $(t, t + h]$, and so

$$\begin{aligned} f_n(t + h) &= f_n(t)\mathbb{P}(\mathcal{N}(t, t + h] = 0 \mid \mathcal{N}(0, t] = n) \\ &\quad + f_{n-1}(t)\mathbb{P}(\mathcal{N}(t, t + h] = 1 \mid \mathcal{N}(0, t] = n - 1) + o(h) \\ &= f_n(t) - \lambda h f_n(t) + \lambda h f_{n-1}(t) + o(h), \end{aligned}$$

implying $f_n'(t) = -\lambda f_n(t) + \lambda f_{n-1}(t)$. Using the induction hypotheses, the obvious boundary condition $f_n(0) = 0$ and general facts about solutions of differential equations (Sect. A.2 of the Appendix) shows that the solution of this ODE is indeed given by (5.5).

The property of independent Poisson increments follows by an easy extension of the argument. □

(ii) $\Rightarrow$ (iii): obviously, $f_0(t) = \mathbb{P}(N(t) = 0)$ satisfies the functional equation $f_0(t + s) = f_0(t) f_0(s)$ which together with $f_0(0+) = 1$ and f_0 being monotone implies $f_0(t) = e^{-\lambda t}$ for some $\lambda > 0$. In particular, $\mathbb{P}(T_1 > t) = \mathbb{P}(N(t) = 0) = e^{-\lambda t}$. so T_1 is exponential(λ), and (5.2) holds.

The next step is to show

$$\mathbb{P}(T_1 \le t, T_2 > h) \ge \mathbb{P}(T_1 \le t)e^{-\lambda h}. \tag{5.6}$$

To this end, let $I_{i,n}$ be the interval $\big(t(i-1)/2^n, ti/2^n\big]$. Then

$$\mathbb{P}(T_1 \in I_{i,n}, T_2 > h) \geq \mathbb{P}\big(T_1 \in I_{i,n}, N(ti/2^n) = 1, T_2 > h\big)$$
$$\geq \mathbb{P}\big(T_1 \in I_{i,n}, N(ti/2^n) = 1, \mathcal{N}(ti/2^n, ti/2^n + h) = 0\big)$$
$$= e^{-\lambda h}\mathbb{P}\big(T_1 \in I_{i,n}, N(ti/2^n) = 1\big) \geq e^{-\lambda h}\mathbb{P}(T_1 \in I_{i,n}, T_2 > t/2^n).$$

Summing over $i = 1, \ldots, 2^n$ gives

$$\mathbb{P}(T_1 \leq t, T_2 > h) \geq e^{-\lambda h}\mathbb{P}(T_1 \leq t, T_2 > t/2^n)$$

and letting $n \to \infty$ and using $T_2 > 0$ a.s. then gives (5.6). Hence taking $t = h$ we get

$$\mathbb{P}\big(\mathcal{N}(t, t + h] \geq 2\big) = \mathbb{P}(T_1 \leq h, T_1 + T_2 \leq h) \leq \mathbb{P}(T_1 \leq h, T_2 \leq h)$$
$$= \mathbb{P}(T_1 \leq t) - \mathbb{P}(T_1 \leq t, T_2 > h) \leq \mathbb{P}(T_1 \leq h)[1 - e^{-\lambda h}]$$
$$= [1 - e^{-\lambda h}]^2 = O(h^2),$$

so we have verified two of properties (5.2)–(5.4). $\square$

Corollary 5.4 *The interarrival times $T_1, T_2, \ldots$ of a Poisson process form an i.i.d. sequence with marginal exponential(λ) distribution.*

For the proof, see Exercise 5.2.

5.3 Superposition and Thinning

An appealing feature of Poisson modelling is the fact that many operations on Poisson processes again lead to Poisson processes.

One fundamental such operation is *superposition*. Let $\mathcal{N}_1, \ldots, \mathcal{N}_m$ be point processes. Then the superposition $\mathcal{N} = \mathcal{N}_1 + \cdots + \mathcal{N}_m$ is defined as the point process whose set of event times is the union of the set of events for $\mathcal{N}_1, \ldots, \mathcal{N}_m$. See Fig. 2 for an illustration for the case $m = 2$.

Fig. 2 Poisson superposition

Theorem 5.5 *Let* $\mathscr{N}_1, \ldots, \mathscr{N}_m$ *be independent Poisson processes with rates* $\lambda_1, \ldots, \lambda_m$. *Then* $\mathscr{N} = \mathscr{N}_1 + \cdots + \mathscr{N}_m$ *is a Poisson*(λ) *process, where* $\lambda = \lambda_1 + \cdots + \lambda_m$.

Proof It suffices to consider the case $m = 2$. That $\mathscr{N}$ has independent increment is obvious since $\mathscr{N}_1, \mathscr{N}_2$ have so. Let $J_{t,h} = \mathscr{N}(t, t + h]$ and $J_{i,t,h} = \mathscr{N}_i(t, t + h]$. Then

$$
\begin{aligned}
\mathbb{P}(J_{t,h} = 0) &= \mathbb{P}(J_{1,t,h} = J_{2,t,h} = 0) = \mathbb{P}(J_{1,t,h} = 0)\mathbb{P}(J_{2,t,h} = 0) \\
&= \big(1 - \lambda_1 h + \mathrm{o}(h)\big)\big(1 - \lambda_2 h + \mathrm{o}(h)\big) \\
&= 1 - \lambda_1 h - \lambda_2 h + \mathrm{o}(h) = 1 - \lambda h + \mathrm{o}(h), \\
\mathbb{P}(J_{t,h} = 1) &= \mathbb{P}(J_{1,t,h} = 1, \ J_{2,t,h} = 0) + \mathbb{P}(J_{1,t,h} = 0, \ J_{2,t,h} = 1) \\
&= \mathbb{P}(J_{1,t,h} = 0)\mathbb{P}(J_{2,t,h} = 1) + \mathbb{P}(J_{1,t,h} = 0)\mathbb{P}(J_{2,t,h} = 0) \\
&= \big(\lambda_1 h + \mathrm{o}(h)\big)\big(1 - \lambda_2 h + \mathrm{o}(h)\big) \\
&\quad + \big(1 - \lambda_1 h + \mathrm{o}(h)\big)\big(\lambda_2 h + \mathrm{o}(h)\big) \\
&= \lambda_1 h + \lambda_2 h + \mathrm{o}(h) = \lambda h + \mathrm{o}(h).
\end{aligned}
$$

This verifies (5.2), (5.3). □

Proposition 5.6 *In the setting of Theorem 5.5, the probability that a given point of* $\mathscr{N}$ *came from* $\mathscr{N}_i$ *is* λ_i/λ, $i = 1, \ldots, m$.

Proof Let $m = 2$, $i = 1$, and let the point be located at t. Then the probability in question equals

$$
\begin{aligned}
&\mathbb{P}\big(\mathscr{N}_1(t, t + \mathrm{d}t] = 1 \,\big|\, \mathscr{N}(t, t + \mathrm{d}t] = 1\big) \\
&= \frac{\mathbb{P}\big(\mathscr{N}_1(t, t + \mathrm{d}t] = 1, \, \mathscr{N}_2(t, t + \mathrm{d}t] = 0\big)}{\mathbb{P}\big(\mathscr{N}(t, t + \mathrm{d}t] = 1\big)} \\
&= \frac{\lambda_1 \mathrm{d}t\,(1 - \lambda_2 \mathrm{d}t)}{\lambda \mathrm{d}t} = \frac{\lambda_1 \mathrm{d}t}{\lambda \mathrm{d}t} = \frac{\lambda_i}{\lambda}.
\end{aligned}
$$

□

Another fundamental operation on Poisson processes is *thinning*. Thinning with *retention probability* p is defined by flipping a coin w.p. p at each epoch of $\mathscr{N}$ (the coin flippings are i.i.d. and independent of $\mathscr{N}$), and maintaining the event if heads come up, deleting it otherwise. We show in a moment that the resulting point process is Poisson(λp). Reversing the role of p and $q = 1 - p$ shows that the point process of deleted points is Poisson(λq). In fact, the two processes are independent, as follows by the following more general result:

Theorem 5.7 *Let* $\mathscr{N}$ *be Poisson*(λ) *and let* $p_1 + \cdots + p_k = 1$. *Construct new point processes* $\mathscr{N}_1, \ldots, \mathscr{N}_k$ *by performing i.i.d. multinomial* $(p_1, \ldots, p_k)$ *trials at the*

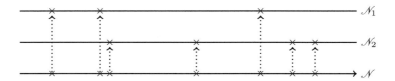

Fig. 3 Poisson thinning

events times of $\mathcal{N}$, such that an event time is allocated to $\mathcal{N}_i$ if alternative i comes up at the event time. Then $\mathcal{N}_1, \ldots, \mathcal{N}_k$ are independent Poisson processes, with rate λp_i for $\mathcal{N}_i$.

The construction is illustrated in Fig. 3.

Theorem 5.7 is an easy consequence of the following lemma:

Lemma 5.8 *Let N be Poisson(λ) and let $(N_1, \ldots, N_k)$ be the outcome of N independent multinomial$(p_1, \ldots, p_k)$ trials. Then $N_1, \ldots, N_k$ are independent and N_i is Poisson(λp_i) distributed.*

Proof Let $(n_1, \ldots, n_k) \in \mathbb{N}^k$ be given and define $n = n_1 + \cdots + n_k$. Then

$$\mathbb{P}(N_1 = n_1, \ldots, N_k = n_k) = \mathbb{P}\big(N_1 = n_1, \ldots, N_k = n_k \mid N = n\big) \cdot \mathbb{P}(N = n)$$

$$= \binom{n}{n_1 \ldots n_k} p_1^{n_1} \cdots p_k^{n_k} \cdot \frac{\lambda^n}{n!} e^{-\lambda}$$

$$= \frac{\lambda^{n_1 + \cdots + n_k}}{n_1! \cdots n_k!} p_1^{n_1} \cdots p_k^{n_k} \exp\{-\lambda(p_1 + \cdots + p_k)\} = \prod_{i=1}^{k} \frac{(\lambda p_i)^{n_i}}{n_i!} e^{-\lambda p_i},$$

where we used $p_1 + \cdots + p_k = 1$. $\qquad\qquad\qquad\qquad\qquad\qquad\qquad\qquad\qquad\square$

Example 5.9 Single men arrive at a bar on average 8 per hour, single women on average 4 per hour and couples on average 4 per hour (we ignore groups of more than two, gay and lesbian couples etc). One out of 4 men and one out of 8 women is a smoker. The bar opens at 5pm. What is the probability p that no smoker has arrived by 7pm?

Solution It is reasonable to model arrivals by Poisson processes (e.g., single men arrive according to a Poisson(8) process) and we also make any appropriate independence assumption. The thinning property then implies that single smoking men arrive according to a Poisson(2) process. Similarly, single smoking women arrive according to a Poisson$(1/2)$ process. The probability that at least one in a couple smokes is $1/4 + 1/8 - 1/4 \times 1/8 = 11/32$, so couples of which at least one smokes arrive according to a Poisson$(11/8)$ process. The superposition property therefore implies that the instances where a smoker arrives form a Poisson(λ) process with $\lambda = 2 + 1/2 + 11/8 = 31/8$. The desired p is the probability

that the first arrival in this process occurs no earlier than within two hours, i.e.
$p = e^{-2\lambda} \approx 4.3 \cdot 10^{-4}$. $\Diamond$

Exercises

5.1 Show that for a Poisson process on $[0, \infty)$, the number of points in disjoints
sets are independent r.v.s.

5.2 Adapt the proof of (5.6) to obtain a similar upper bound and show thereby
Corollary 5.4.

5.3 For a Poisson process $\mathcal{N}$ on $[0, \infty)$ and $y > 0$, define the shifted process $\mathcal{N}_y$
by $N_y(t) = N(t+y) - N(y)$. Adapt the procedure to the case where $\mathcal{N}$ is a Poisson
process on $(-\infty, \infty)$ and $-\infty < y < \infty$.

5.4 Let $T_1, T_{-1}, T_2, T_{-2}, \ldots$ be the epochs of a Poisson(λ) process on $\mathbb{R} =$
$(-\infty, \infty)$, and let $Y_1, Y_{-1}, Y_2, Y_{-2}, \ldots$ be i.i.d. with distribution concentrated on
a finite set $\{y_1, \ldots, y_k\}$. Show that the point process with epochs $T_1 + Y_1, T_{-1} +$
$Y_{-1}, T_2 + Y_2, T_{-2} + Y_{-2}, \ldots$ is again Poisson(λ) on $\mathbb{R}$. [Hint: first use thinning,
next shifting and finally superposition. Using an approximation argument, the result
carries over to a general distribution of the Y_i.]

5.5 Let $X(t)$ be the number of individuals at time t in a population, where
individuals don't die, but produce (single) new children at rate λ. Find similar
infinitesimal relations as (5.2), (5.3), (5.4) for $X(t)$ and compute thereby $\mathbb{P}(N(t) =$
$n)$ for $n = 1, 2, 3$, assuming $X(0) = 1$.

Chapter II: Experience Rating

1 Bayes and Empirical Bayes

Let $X = (X_1, \ldots, X_n)$ be a vector of r.v.s describing the outcome of a statistical experiment. For example, in the insurance context, n can be the number of persons insured for losses due to accidents in the previous year, and X_i the payment made to the ith.

A traditional (frequentists') model is to assume the X_i to be i.i.d. with a common distribution F_θ where θ is an unknown parameter (possibly multidimensional). E.g. in the accident insurance example, one could let b denote the probability that a person has an accident within one year, $b = \mathbb{P}(X_i > 0)$, and one could assume that the cost of the accident has a gamma(α, λ) distribution. Thus the density of X_i is

$$f_{b,\alpha,\lambda}(x) = b\mathbb{1}_{x=0} + (1-b)\frac{\lambda^\alpha x^{\alpha-1}}{\Gamma(\alpha)}e^{-\lambda x}\mathbb{1}_{x>0}$$

w.r.t. the measure defined as Lebesgue measure $\mathrm{d}x$ on $(0, \infty)$ with an added atom of unit size at $x = 0$. Then $\theta = (b, \alpha, \lambda)$, and the conventional statistical procedure would be to compute estimates $\widehat{b}, \widehat{\alpha}, \widehat{\lambda}$ of b, α, λ. These estimates could then be used as basis for computing first the expectation

$$\mathbb{E}_{\widehat{\theta}}X = \mathbb{E}_{\widehat{b},\widehat{\alpha},\widehat{\lambda}}X = (1 - \widehat{b})\widehat{\alpha}/\widehat{\lambda}$$

of X under the estimated parameters, and next one could use $\mathbb{E}_{\widehat{\theta}}X$ as the net premium and add a loading corresponding to one of the premium rules discussed in Sect. I.3. For example, the expected value principle would lead to the premium

$$p = (1 + \eta)(1 - \widehat{b})\widehat{\alpha}/\widehat{\lambda}.$$

© Springer Nature Switzerland AG 2020
S. Asmussen, M. Steffensen, *Risk and Insurance*, Probability Theory
and Stochastic Modelling 96, https://doi.org/10.1007/978-3-030-35176-2_2

Bayesian statistics takes a different view and treats the parameter θ not as completely unknown but assumes that a *prior distribution* (or just a *prior*) on θ is available. That is, one views θ as the outcome of an r.v. Θ given by its density $\pi^{(0)}(\theta)$ w.r.t. some reference measure $\mu(dx)$. In this setting, (Θ, X) is a vector where both the Θ-component and the X component are random; the joint density is $f_\theta(x)\pi^{(0)}(\theta)$, where $f_\theta(x)$ is the density of the observation (or the observation vector) w.r.t. another reference measure $\nu(dx)$, that is, the usual likelihood. It therefore makes sense to define the *posterior distribution* (or just *posterior*) as the distribution of Θ given $X = x$ where x is the observed value of X. By standard facts on conditioning, the posterior has density

$$\pi(\theta) \;=\; \frac{f_\theta(x)\pi^{(0)}(\theta)}{\int f_{\theta'}(x)\pi^{(0)}(\theta')d\mu(\theta')} \tag{1.1}$$

(note that the denominator is the marginal density of X). Equation (1.1) is known as *Bayes's formula*.

For an example of priors, consider a coin coming up heads w.p. θ and let μ be Lebesgue measure on $(0, 1)$. A uniform prior then corresponds to $\equiv 1$ and is *uninformative*, i.e. expresses little or no knowledge of θ. Another standard possibility is a beta(a, b) distribution. For example, if $a = b$ and a is large, the prior then expresses the belief that θ is equally likely to be $1 - t$ and $1 + t, 0 < t < 1/2$, i.e. symmetry, and that we believe θ to be rather close to $1/2$.

If the prior is determined by estimates using other data than X, one talks of *empirical Bayes*. The following example will illustrate the concept and is in fact central for understanding premium calculations using empirical Bayes and related concepts.

Example 1.1 Consider car insurance where x_i is driver D's number of claims in year i. A traditional (and reasonable) statistical model assumes that $X_1, \ldots, X_n$ are i.i.d. Poisson(θ). The Bayesian view is to consider θ as the outcome of an r.v. Θ. A popular choice of prior is a gamma(α, λ) distribution. With this choice, the posterior becomes proportional to

$$f_\theta(x)\pi^{(0)}(\theta) \;=\; \prod_{i=1}^{n} e^{-\theta}\frac{\theta^{x_i}}{x_i!} \cdot \frac{\lambda^\alpha \theta^{\alpha-1}}{\Gamma(\alpha)}e^{-\lambda\theta}\,,$$

which differs from $\theta^{x_1+\cdots+x_n+\alpha-1}e^{-(\lambda+n)\theta}$ only by factors not depending on θ. Therefore the posterior is again a gamma, with α replaced by $x_1 + \cdots + x_n + \alpha$ and λ by $n + \lambda$.

In an empirical Bayes situation, one would have past experience on M other drivers, driver m being in the portfolio for say t_m years and having $y_1^{(m)}, \ldots, y_{t_m}^{(m)}$ claims in these. Each driver would be characterized by his own Poisson intensity θ_m, and the $\theta_1, \ldots, \theta_M$ would be assumed i.i.d. gamma(α, λ). The θ_m are unobserved, but to produce estimates $\widehat{\alpha}, \widehat{\lambda}$ for α, λ, one would compute the natural estimate

$\widehat{\theta}_m = \left(y_1^{(m)} + \cdots + y_{t_m}^{(m)}\right)/t_m$ of θ_m, and treat $\widehat{\theta}_1, \ldots, \widehat{\theta}_M$ as i.i.d. Gamma(α, λ) observations. According to the above, the empirical Bayes posterior of driver D's θ is therefore $\Gamma(x_1 + \cdots + x_n + \widehat{\alpha}, \widehat{\lambda} + n)$.

Assuming that a claim has mean m, the natural choice for the net premium becomes

$$\frac{x_1 + \cdots + x_n + \widehat{\alpha}}{\widehat{\lambda} + n} m \qquad (1.2)$$

(the mean of the gamma(α, λ) distribution is α/λ). Note that as n grows, (1.2) becomes closer and closer to $(x_1 + \cdots + x_n)/n$, which would be the natural estimate of driver D's θ if the knowledge of how a typical driver in the population of drivers behaves was not taken into account. We return to a more precise discussion of this in the next sections. ◇

Example 1.2 The Swiss actuary F. Bichsel was presented in [23] with statistics of the number of accidents by Swiss automobile drivers in 1961. More precisely $N_0 = 103{,}704$ had no accidents, $N_1 = 14{,}075$ had one accident and so on. The maximal number of accidents was 6, corresponding to $N_6 = 2$, and the total number of drivers was $N = N_1 + N_2 + \cdots + N_6 = 119{,}853$ (see Table 1 below for the complete set of data).

Following Bichsel, we will consider two ways to model these data. In both, the number X_i of accidents by a driver has a Poisson distribution. In the first model, we simply assume the drivers to be i.i.d., so that $X_1, \ldots, X_N$ are i.i.d. Poisson(θ). In the second, we adopt the Poisson-gamma model of Example 1.1 so that each driver has his own Poisson parameter θ_n where $\theta_1, \ldots, \theta_N$ are i.i.d. outcomes of sampling from a gamma(α, λ) distribution.

We will address the question of how these two models fit the data. In the first model, the MLE for θ is

$$\widehat{\theta} = \frac{1}{N} \sum_{k=1}^{6} k N_k = 0.1554 \, .$$

In the second, Bichsel used the method of moments to estimate α, λ (of course, one expects MLE to be superior, but the method of moments is computationally simpler, or at least was so in the days of Bichsel). With Y the number of accidents of a typical driver and $\widehat{\sigma}^2 = 0.1880$ the empirical variance, the equations to solve are $\mathbb{E}Y = \widehat{\theta}$, $\mathbb{V}\mathrm{ar}\, Y = \widehat{\sigma}^2$. We get

$$\mathbb{E}Y = \mathbb{E}\big[\mathbb{E}Y \,|\, \Theta)\big] = \mathbb{E}\Theta = \frac{\alpha}{\lambda} \, ,$$

$$\mathbb{V}\mathrm{ar}\, Y = \mathbb{E}\big[\mathbb{V}\mathrm{ar}\, Y \,|\, \Theta)\big] + \mathbb{V}\mathrm{ar}\big[\mathbb{E}(Y \,|\, \Theta)\big] = \mathbb{E}\Theta + \mathbb{V}\mathrm{ar}\, \Theta$$

$$= \frac{\alpha}{\lambda} + \frac{\alpha}{\lambda^2} = \frac{\alpha}{\lambda}\left(1 + \frac{1}{\lambda}\right),$$

Table 1 Swiss claim counter data and fits

	0	1	2	3	4	5	6
Observed	103,704	14,075	1,766	266	45	6	2
Poisson	102,612	15,946	1,239	64	2.49	0.078	0.002
Neg. bin.	103,763	13,918	1,886	256	35	4.76	0.65

which yields $\widehat{\alpha} = 0.9801$, $\widehat{\lambda} = 6.3071$ (note that these parameters are close to the exponential case $\alpha = 1$).

To check the model fit, we also need

$$\mathbb{P}(Y = m) = \int_0^\infty \mathbb{P}(Y = m \mid \Theta = \theta)\mathbb{P}(\Theta \in d\theta)$$

$$= \int_0^\infty e^{-\theta} \frac{\theta^m}{m!} \frac{\lambda^\alpha}{\Gamma(\alpha)} \theta^{\alpha-1} e^{-\lambda\theta} \, d\theta$$

$$= \frac{\lambda^\alpha}{\Gamma(\alpha)m!} \int_0^\infty \theta^{\alpha+m-1} e^{-(\lambda+1)\theta} \, d\theta$$

$$= \frac{\lambda^\alpha}{\Gamma(\alpha)m!} \frac{\Gamma(\alpha+m)}{(\lambda+1)^{\alpha+m}} = \binom{\alpha+m-1}{m} p^\alpha (1-p)^m ,$$

corresponding to a negative binomial distribution with $p = \lambda/(\lambda + 1)$.

The complete set of data and the expected counts under the two models are given in Table 1. We see that the Poisson-Gamma model, corresponding to the negative binomial entry, gives a reasonable fit and is a great improvement upon the Poisson model, not least in the tail.   ◊

1.1 The Bayes Premium

We now turn to the general implementation of Bayesian ideas in insurance. Here one considers an insured with risk parameter $\mathbf{Z}$[1] and an r.v. with distribution $\pi^{(0)}(\cdot)$, with observable past claims $X_1, \ldots, X_n$ and an unobservable claim amount X_{n+1} for year $n + 1$. The aim is to assert which (net) premium the insured is to pay in year $n + 1$.

For a fixed $\boldsymbol{\zeta}$, let $\mu(\boldsymbol{\zeta}) = \mathbb{E}_{\boldsymbol{\zeta}} X_{n+1}$, where $\mathbb{E}_{\boldsymbol{\zeta}}[\cdot] = \mathbb{E}[\cdot \mid \mathbf{Z} = \boldsymbol{\zeta}]$. The (net) *collective premium* H_{Coll} is $\mathbb{E}\mu(\mathbf{Z}) = \mathbb{E}X_{n+1}$. This is the premium we would charge without prior statistics $X_1, \ldots, X_n$ on the insured. The *individual premium* is $H_{\mathrm{Ind}} = \mathbb{E}[X_{n+1} \mid \mathbf{Z}] = \mu(\mathbf{Z})$. This is the ideal net premium in the sense

---

[1] Note that we have changed notation: $\mathbf{Z}$ and its outcomes $\boldsymbol{\zeta}$ are what were denoted Θ, θ earlier. The boldface notation indicates that $\mathbf{Z}$ is allowed to be multidimensional.

of supplying the complete relevant prior information on the customer. The *Bayes premium* H_{Bayes} is defined as $\mathbb{E}\big[\mu(\mathbf{Z}) \mid X_1, \ldots, X_n\big]$. That is, H_{Bayes} is the expected value of X_{n+1} in the posterior distribution.

Note that the individual premium is unobservable because $\mathbf{Z}$ is so; the Bayes premium is 'our best guess of H_{Ind} based upon the observations'. To make this precise, let H^* be another premium rule, that is, a function of $X_1, \ldots, X_n$ and the prior parameters. We then define its loss as

$$\ell_{H^*} = \mathbb{E}[\mu(\mathbf{Z}) - H^*]^2 = \|\mu(\mathbf{Z}) - H^*\|^2$$

where $\|X\| = (\mathbb{E}\, X^2)^{1/2}$ is the L_2-norm (in obvious notation, we write $\ell_{\text{Coll}} = \ell_{H_{\text{Coll}}}$ etc). In mathematical terms, the optimality property of the Bayes premium is then that it minimizes the quadratic loss:

Theorem 1.3 *For any H^*, $\ell_{\text{Bayes}} \leq \ell_{H^*}$. That is,*

$$\mathbb{E}\big(H_{\text{Bayes}} - H_{\text{Ind}}\big)^2 \leq \mathbb{E}\big(H^* - H_{\text{Ind}}\big)^2.$$

Proof This follows from the well-known fact that a conditional expectation is an L_2-projection, see Proposition A.2.   □

Exercises

1.1 Find the posterior distribution if $X_1, \ldots, X_n$ are i.i.d. $\mathcal{N}(\theta, 1)$ and the prior on θ is $\mathcal{N}(\mu, \omega^2)$. Investigate the asymptotic properties of the posterior as 1) $n \to \infty$, 2) $\omega^2 \to \infty$.

1.2 Find the posterior distribution if $X_1, \ldots, X_n$ are i.i.d. Bernoulli(θ) and the prior beta.

1.3 In mathematical finance, it is common to use an inverse gamma as prior for (squared) volatilities. That is, if $X_1, \ldots, X_n$ are i.i.d. $\mathcal{N}(0, \sigma^2)$, it is assumed that $1/\sigma^2$ has a gamma distribution. Show that the posterior is again inverse gamma.

1.4 Perform a goodness-of fit test of the two models in Example 1.2.

1.5 ([43] p. 50) A risk produces either one claim in a year or none, and the payment to the insured in case of a claim is fixed, say € 1,000. Risks fall into three groups (good, medium, bad) such that the probability of a claim is $3\%, 5\%, 10\%$ for the three groups. Further, $\mathbb{P}(\text{bad}) = 5\%$, $\mathbb{P}(\text{medium}) = 30\%$, $\mathbb{P}(\text{good}) = 65\%$.

a) Calculate H_{Ind} and H_{Coll}.

Let $X_1, X_2, \ldots$ be the number of claims in successive years. Calculate H_{Bayes} based upon b) X_1 and c) X_1, X_2.

Notes
Some standard textbooks on Bayesian statistics are Bernardo and Smith [22], Robert [151], and Carlin and Louis [46].

2 Exponential Families and Conjugate Priors

The Poisson-gamma model of Sect. 1 and the exercises are all examples of *conjugate priors*. By this we mean that the posterior belongs to the same parametric family as the prior. This is defined formally as a class $\Pi^{(0)}$ of densities (for example all Gamma densities) such that both the prior $\pi^{(0)}(\theta)$ and the posterior are in $\Pi^{(0)}$.

The main examples of conjugate priors come from exponential families. In its simplest form, an exponential family is a family of densities

$$f_\zeta(x) = \exp\{\boldsymbol{\zeta}^\mathsf{T} \boldsymbol{t}(x) - \kappa(\boldsymbol{\zeta})\} \tag{2.1}$$

w.r.t. some reference measure $\mu(\mathrm{d}x)$ (e.g., Lebesgue measure or counting measure). Here x is the observed quantity, $\boldsymbol{t}(x) \in \mathbb{R}^p$ a function of x and $\boldsymbol{\zeta} \in \mathbb{R}^p$ a parameter (column) vector. In the following, we shall use a generalized form

$$f_\zeta(x) = \exp\left\{ \frac{\boldsymbol{\zeta}^\mathsf{T} \boldsymbol{t}(x) - \kappa(\boldsymbol{\zeta})}{\sigma^2/w} + c(x, \sigma^2/w) \right\}. \tag{2.2}$$

Here ω and σ^2 are to be thought of as constants, not parameters. The interpretation of w is as a weight and of σ^2 as a dispersion parameter (cf. e.g. the treatment of generalized linear models in SAS).

Example 2.1 For the gamma(α, λ) density, we can write

$$\frac{\lambda^\alpha}{\Gamma(\alpha)} x^{\alpha-1} \mathrm{e}^{-\lambda x} = \exp\{-\lambda x + \alpha \log x + \alpha \log \lambda - \log \Gamma(\alpha)\} \cdot x^{-1}.$$

Thus, we are in the framework (2.1) with $\boldsymbol{\zeta} = (-\lambda \ \ \alpha)^\mathsf{T}, \boldsymbol{t}(x) = (x \ \log x)^\mathsf{T}, \kappa(\boldsymbol{\zeta}) = \alpha \log \lambda - \log \Gamma(\alpha)$ and $\mu(\mathrm{d}x) = x^{-1} \mathrm{d}x$. $\qquad \Diamond$

Example 2.2 For the Poisson(λ) density, we can write

$$\mathrm{e}^{-\lambda} \frac{\lambda^x}{x!} = \exp\{x \log \lambda - \lambda\} \cdot \frac{1}{x!}.$$

Again, we are in the framework (2.1) with $\zeta = \log \lambda$, $t(x) = x$, $\kappa(\zeta) = \lambda$ and μ the measure on $\mathbb{N}$ with point masses $1/x!$.

Now assume instead that our observation is Poisson(λw) where w is a 'risk volume' (the mileage class in car insurance, the area in square meters of a building in fire insurance, etc). Denote the observation by X', let $X = X'/w$ and redefine μ to be the measure on $0, 1/w, 2/w, \ldots$ with point mass $1/x!$ at x/w. The density of X w.r.t. μ is then

$$x! \, \mathbb{P}(X' = wx) = \exp\left\{ -\lambda w + wx \log(\lambda w) \right\} = \exp\left\{ \frac{\log \lambda \cdot x - \lambda}{1/w} + xw \log w \right\},$$

which gives the form (2.2) with $\zeta = \log \lambda$, $t(x) = x$, $\kappa(\zeta) = \lambda$, $\sigma^2 = 1$ and $c(x, \sigma^2/w) = xw \log w$. $\diamond$

We have the following main result on conjugate priors in exponential families:

Theorem 2.3 *For a given exponential family of the form* (2.2)*, define*

$$d(\boldsymbol{\theta}, \tau^2) = \log \int \exp\left\{(\boldsymbol{\zeta}^\mathsf{T}\boldsymbol{\theta} - \kappa(\boldsymbol{\zeta}))/\tau^2\right\} d\boldsymbol{\zeta}$$

and let $\Pi^{(0)}$ be the class of all densities of the form

$$\pi^{(0)}_{\boldsymbol{\theta},\tau^2}(\boldsymbol{\zeta}) = \exp\left\{\frac{\boldsymbol{\zeta}^\mathsf{T}\boldsymbol{\theta} - \kappa(\boldsymbol{\zeta})}{\tau^2} - d(\boldsymbol{\theta}, \tau^2)\right\}, \qquad (2.3)$$

where $\boldsymbol{\theta} \in \mathbb{R}^p$ and $\tau^2 > 0$ are parameters such that $d(\boldsymbol{\theta}, \tau^2) < \infty$. If $x_1, \ldots, x_n$ are independent observations from (2.2)*, with the same σ^2 but $w = w_i$ depending on $i = 1, \ldots, n$, then* (2.3) *is a class of conjugate priors. More precisely, if $\boldsymbol{\theta}_0, \tau_0^2$ are the prior parameters, then the posterior parameters $\boldsymbol{\theta}_1, \tau_1^2$ are given by*

$$\frac{1}{\tau_1^2} = \left\{\frac{1}{\tau_0^2} + \frac{1}{\sigma^2}\sum_{i=1}^n w_i\right\}, \quad \boldsymbol{\theta}_1 = \tau_1^2\left\{\frac{\boldsymbol{\theta}_0}{\tau_0^2} + \frac{1}{\sigma^2}\sum_{i=1}^n w_i t(x_i)\right\}. \qquad (2.4)$$

Proof The posterior density is proportional to

$$\pi^{(0)}_{\boldsymbol{\theta}_0,\tau_0^2}(\boldsymbol{\zeta}) \prod_{i=1}^n f_{\boldsymbol{\zeta}}(x_i)$$

$$= \exp\left\{\frac{\boldsymbol{\zeta}^\mathsf{T}\boldsymbol{\theta}_0 - \kappa(\boldsymbol{\zeta})}{\tau_0^2} - d(\boldsymbol{\theta}_0, \tau_0^2)\right\} \exp\left\{\sum_{i=1}^n \frac{\boldsymbol{\zeta}^\mathsf{T} t(x_i) - \kappa(\boldsymbol{\zeta})}{\sigma^2/w_i} + c(x_i, \sigma^2/w_i)\right\}.$$

Omitting terms not depending on $\boldsymbol{\zeta}$, this is the same as

$$\exp\left\{\boldsymbol{\zeta}^\mathsf{T}\left[\frac{\boldsymbol{\theta}_0}{\tau_0^2} + \frac{1}{\sigma^2}\sum_{i=1}^n w_i t(x_i)\right] - \kappa(\boldsymbol{\zeta})\left[\frac{1}{\tau_0^2} + \frac{1}{\sigma^2}\sum_{i=1}^n w_i\right]\right\}.$$

This shows the assertion. $\square$

Example 2.4 For the Poisson(λ) distribution, $\zeta = \log \lambda$, $\kappa(\zeta) = \lambda$. Hence

$$\exp\left\{\frac{\zeta\theta - \kappa(\zeta)}{\tau^2}\right\} = \lambda^{\alpha-1} e^{-r\lambda},$$

where $\alpha = 1 + \theta/\tau^2$, $r = 1/\tau^2$. Up to constants, this is the gamma(α, r) density, showing that the class of conjugate priors is indeed the class of gamma priors, cf. the Poisson-gamma model of Sect. 1. $\diamond$

Proposition 2.5 *In the exponential family setting with $p = 1$, $t(x) = x$, we have $H_{\text{Ind}} = \kappa'(Z)$. If further ζ varies in an interval (ζ_1, ζ_2) such that $\theta_0 \zeta - \kappa(\zeta) \to -\infty$ as $\zeta \downarrow \zeta_1$ or $\zeta \uparrow \zeta_2$, then $H_{\text{Coll}} = \theta_0$ and*

$$H_{\text{Bayes}} = c\overline{X} + (1 - c)H_{\text{Coll}}, \tag{2.5}$$

where

$$\overline{X} = \sum_{i=1}^{n} \frac{w_i}{w_\bullet} X_i, \quad w_\bullet = \sum_{i=1}^{n} w_i, \quad c = \frac{w_\bullet}{w_\bullet + \sigma^2/\tau_0^2}.$$

The constant c is called the *credibility weight*. The formula (2.5) shows that the Bayes premium is a weighted average of $\overline{X}$, a quantity depending on the insured, and H_{Coll}, a quantity depending on the collective. As $n \to \infty$, the credibility weight goes to 1 (provided, of course, that $w_\bullet \to \infty$).

Proof The m.g.f. of X_{n+1} at s given $Z = \zeta$ and $w_{n+1} = w$ is

$$\int_0^\infty e^{sx} f_\zeta(x)\, \mu(dx) = \int_0^\infty e^{sx} \exp\left\{ \frac{\zeta x - \kappa(\zeta)}{\sigma^2/w} + c(x, \sigma^2/w) \right\} \mu(dx)$$

$$= \exp\left\{ \frac{\kappa(\zeta + s\sigma^2/w) - \kappa(\zeta)}{\sigma^2/w} \right\}$$

$$\times \int_0^\infty \exp\left\{ \frac{(\zeta + s\sigma^2/w)x - \kappa(\zeta + s\sigma^2/w)}{\sigma^2/w} + c(x, \sigma^2/w) \right\} \mu(dx)$$

$$= \exp\left\{ \frac{\kappa(\zeta + s\sigma^2/w) - \kappa(\zeta)}{\sigma^2/w} \right\}.$$

Hence

$$\mathbb{E}_\zeta X_{n+1} = \int_0^\infty x f_\zeta(x)\, \mu(dx) = \frac{d}{ds} \int_0^\infty e^{sx} f_\zeta(x)\, \mu(dx) \Big|_{s=0}$$

$$= \frac{\kappa'(\zeta + s\sigma^2/w)}{\sigma^2/w} \frac{d}{ds} s\sigma^2/w \cdot \exp\left\{ \frac{(\kappa(\zeta + s\sigma^2/w) - \kappa(\zeta))}{\sigma^2/w} \right\} \Big|_{s=0},$$

which reduces to $\kappa'(\zeta)$. Therefore $H_{\text{Ind}} = \kappa'(Z)$. We then get

$$H_{\text{Coll}} = \int_{\zeta_1}^{\zeta_2} \mu(\zeta) \pi_{\theta_0, \tau_0^2}^{(0)}(\zeta)\, d\zeta = C \int_{\zeta_1}^{\zeta_2} \kappa'(\zeta) \exp\left\{ \frac{\zeta\theta_0 - \kappa(\zeta)}{\tau_0^2} \right\} d\zeta$$

$$= C \int_{\zeta_1}^{\zeta_2} \big(\theta_0 + (\kappa'(\zeta) - \theta_0)\big) \exp\left\{ \frac{\zeta\theta_0 - \kappa(\zeta)}{\tau_0^2} \right\} d\zeta$$

$$= \theta_0 + C \int_{\zeta_1}^{\zeta_2} \left(\kappa'(\zeta) - \theta_0 \right) \exp\left\{ \frac{\zeta\theta_0 - \kappa(\zeta)}{\tau_0^2} \right\} d\zeta$$

$$= \theta_0 - C\tau_0^2 \left[\exp\left\{ \frac{\zeta\theta_0 - \kappa(\zeta)}{\tau_0^2} \right\} \right]_{\zeta_1}^{\zeta_2} = \theta_0,$$

where $C = \exp\{-d(\theta_0, \tau_0^2)\}$ and we used the boundary condition on $\theta_0\zeta - \kappa(\zeta)$ in the last step.

By Theorem 2.3 and the same calculation as for H_{Coll}, it follows that the Bayes premium is the posterior mean θ_1. Computing this by (2.4) gives

$$\theta_1 = \frac{\theta_0/\tau_0^2 + 1/\sigma^2 \sum_1^n w_i x_i}{1/\tau^2 + 1/\sigma^2 \sum_1^n w_i} = \frac{\theta_0\sigma^2/\tau_0^2 + \sum_1^n w_i x_i}{\sigma^2/\tau_0^2 + \sum_1^n w_i}$$

$$= \frac{1}{w_\bullet} c\theta_0\sigma^2/\tau_0^2 + c\overline{X} = (1-c)\theta_0 + c\overline{X} = (1-c)H_{\mathrm{Coll}} + c\overline{X}.$$

$\square$

Exercises

2.1 Let $X_1, \ldots, X_n$ be i.i.d. $\mathcal{N}(\theta, 1)$. Find an exponential family representation and a conjugate prior.

2.2 Let $X_1, \ldots, X_n$ be i.i.d. $\mathcal{N}(0, \sigma^2)$. Find an exponential family representation and a conjugate prior.

2.3 Let $X_1, \ldots, X_n$ be i.i.d. Bernoulli(θ). Find an exponential family representation and a conjugate prior.

2.4 Let $X_1, \ldots, X_n$ be i.i.d. and Pareto on $[1, \infty)$, that is, with density $\alpha/x^{\alpha+1}$ for $x > 1$. Find an exponential family representation and a conjugate prior.

Notes

Conjugate priors are a classical topic, surveyed for example in Section 5.2 of Bernardo and Smith [22] and Section 3.3 of Robert [151]. A much cited reference is Diaconis and Ylvisaker [60].

3 Credibility Premiums

The individual premium $H_{\mathrm{Ind}} = \mu(Z)$ is what one ideally would charge, but it is not observable since Z is not so. The Bayes premium H_{Bayes} is then the optimal premium based on the observations in the sense that it minimizes the mean square error from H_{Ind}. However, the Bayes premium also has its difficulties: often, the form of the posterior and therefore the expression for H_{Bayes} is intractable.

The idea of *credibility theory* is to refrain from looking for the exact Bayes premium H_{Bayes} defined as the posterior mean, that is, a conditional expectation minimizing the mean square error, but instead restrict attention to linear premium rules. The original reasoning for doing so is that a substantial reduction in computational effort can often be obtained, at a cost of course, namely a larger mean square error. One may, however, note that with today's computers and the modern method of MCMC (Markov Chain Monte Carlo) general Bayesian calculations have become much more accessible. Nevertheless, credibility is a classical topic in insurance mathematics and so we treat it in some detail.

3.1 The Simple Credibility Model

As the simplest example, assume as in Sect. 1.1 that $X_1, X_2, \ldots$ are i.i.d. with distribution F_ζ given $Z = \zeta$, and write

$$\mu(\zeta) = \mathbb{E}[X \mid Z = \zeta], \ \sigma^2(\zeta) = \mathbb{V}\mathrm{ar}[X \mid Z = \zeta], \ \sigma^2 = \mathbb{E}\sigma^2(Z), \ \tau^2 = \mathbb{V}\mathrm{ar} \ \mu(Z).$$

The collective premium is $H_{\mathrm{Coll}} = \mathbb{E}\mu(Z)$ and the Bayes premium is

$$H_{\mathrm{Bayes}} = \mathbb{E}[\mu(Z) \mid X_1, \ldots, X_n].$$

The Bayes premium can be characterized as the orthogonal projection of $\mu(Z)$ onto the space L of all square integrable $\sigma(X_1, \ldots, X_n)$-measurable r.v.s, that is, of all (measurable) functions $\psi(X_1, \ldots, X_n)$ of $X_1, \ldots, X_n$ such that $\mathbb{E}\psi(X_1, \ldots, X_n)^2 < \infty$.

We now define $L_{\mathrm{aff}} \subset L$ as the set of all affine combinations of $X_1, \ldots, X_n$, that is,

$$L_{\mathrm{aff}} = \{a_0 + a_1 X_1 + \cdots + a_n X_n : a_0, a_1, \ldots, a_n \in \mathbb{R}\}.$$

The credibility premium H_{Cred} is defined as the orthogonal projection of $\mu(Z)$ on L_{aff}. In formulas,

$$H_{\mathrm{Bayes}} = P_L \mu(Z), \quad H_{\mathrm{Cred}} = P_{L_{\mathrm{aff}}} \mu(Z).$$

Since $L_{\mathrm{aff}} \subset L$, we have $P_{L_{\mathrm{aff}}} = P_{L_{\mathrm{aff}}} \circ P_L$. If $H_{\mathrm{Bayes}} = P_L \mu(Z)$ is an affine combination of the X_i, i.e. an affine combination of the X_i, it therefore equals $P_{L_{\mathrm{aff}}} \mu(Z) = H_{\mathrm{Cred}}$. This form of the Bayes premium was precisely what was found in the exponential family setting of Sect. 2, cf. the examples and Proposition 2.5. Therefore the Bayes premium and the credibility premium are the same thing there. The following result shows that in general the credibility premium has the same form as was found there. The advantage is the complete generality in that we do not need any assumptions on the F_ζ and the distribution of Z:

Theorem 3.1 *The credibility premium H_{Cred} is given by*

$$H_{\text{Cred}} = c\overline{X} + (1-c)H_{\text{Coll}},$$

where $\overline{X} = (X_1 + \cdots + X_n)/n$, $c = n/(n + \sigma^2/\tau^2)$.

Proof Writing $H_{\text{Cred}} = a_0 + a_1 X_1 + \cdots + a_n X_n$, we must minimize

$$\mathbb{E}[\mu(Z) - a_0 - a_1 X_1 - \cdots - a_n X_n]^2 \tag{3.1}$$

w.r.t. $a_0, a_1, \ldots, a_n$. Setting the derivatives w.r.t. a_0, a_1 equal to 0 yields

$$0 = \mathbb{E}[\mu(Z) - a_0 - a_1 X_1 - \cdots - a_n X_n], \tag{3.2}$$

$$0 = \mathbb{E}\big[[\mu(Z) - a_0 - a_1 X_1 - \cdots - a_n X_n]X_i\big], \quad i = 1, \ldots, n. \tag{3.3}$$

Write $\mu_0 = H_{\text{Coll}} = \mathbb{E}\mu(Z) = \mathbb{E}X_i$. From the structure of our model, it follows by conditioning upon Z that

$$\mathbb{E}[\mu(Z)X_i] = \mathbb{E}[\mu(Z)^2] = \tau^2 + \mu_0^2,$$

$$\mathbb{E}[X_i^2] = \mu_0^2 + \text{Var}\, X_i = \mu_0^2 + \mathbb{E}\sigma^2(Z) + \text{Var}\,\mu(Z) = \sigma^2 + \tau^2 + \mu_0^2,$$

$$\mathbb{E}[X_i X_j] = \mathbb{E}[\mu(Z)^2] = \tau^2 + \mu_0^2, \quad i \neq j.$$

Therefore (3.2), (3.3) become

$$\mu_0 = a_0 + \mu_0 \sum_{j=1}^{n} a_j,$$

$$\tau^2 + \mu_0^2 = a_0\mu_0 + a_i(\sigma^2 + \tau^2 + \mu_0^2) + (\tau^2 + \mu_0^2)\sum_{j \neq i} a_j$$

$$= a_0\mu_0 + a_i\sigma^2 + (\tau^2 + \mu_0^2)\sum_{j=1}^{n} a_j.$$

The last identity shows that a_i does not depend on i so that

$$\mu_0 = a_0 + n\mu_0 a_1,$$

$$\tau^2 + \mu_0^2 = a_0\mu_0 + a_1\sigma^2 + (\tau^2 + \mu_0^2)na_1$$

$$= \mu_0^2 - na_1\mu_0^2 + a_1\sigma^2 + (\tau^2 + \mu_0^2)na_1 = \mu_0^2 + a_1(\sigma^2 + n\tau^2),$$

$$a_1 = \frac{\tau^2}{\sigma^2 + n\tau^2} = \frac{c}{n} = a_2 = \cdots = a_n,$$

$$a_0 = \mu_0(1 - na_1) = \mu_0(1 - c).$$

$\square$

Remark 3.2 The proof of Theorem 3.1 represents the 'brute force' approach and is by no means the fastest or the most elegant one. A simplification is obtained by observing that by symmetry, one has $a_1 = \cdots = a_n$, so therefore the credibility premium must have the form $a_0 + c\overline{X}$. We therefore have to determine c to minimize $\mathbb{E}\big[\mu(Z) - a_0 - c\overline{X}\big]^2$. Setting the derivatives w.r.t. a_0, c equal to 0 and using the moment calculations from the proof of Theorem 3.1 yields

$$
0 = \mathbb{E}\big[\mu(Z) - a_0 - c\overline{X}\big] = \mu_0 - a_0 - c\mu_0 ,
$$

$$
0 = \mathbb{E}\big[\mu(Z) - a_0 - c\overline{X}\big]\overline{X} = \mathbb{E}\big[\mu(Z)X_1\big] - a_0\mu_0 - c\big[\operatorname{Var}\overline{X} + [\mathbb{E}\overline{X}]^2\big]
$$

$$
= \tau^2 + \mu_0^2 - a_0\mu_0 - c\big[(\sigma^2/n + \tau^2) + \mu_0^2\big] .
$$

Solving the first equation for a_0 we get $a_0 = \mu_0(1-c)$, and inserting into the second equation then yields

$$
0 = \tau^2 - c(\sigma^2/n + \tau^2), \quad c = \frac{\tau^2}{\sigma^2/n + \tau^2} = \frac{n}{n + \sigma^2/\tau^2} .
$$

$\diamond$

Remark 3.3 An even quicker argument starts by the same observation as in Remark 3.2, that we have to look for a_0, c minimizing $\mathbb{E}\big[\mu(Z) - a_0 - c\overline{X}\big]^2$. However, the minimizer can only depend on the mean vector and covariance matrix

$$
\boldsymbol{\mu} = \begin{pmatrix} \mu_0 \\ \mu_0 \end{pmatrix}, \quad \text{resp.} \quad \boldsymbol{\Sigma} = \begin{pmatrix} \tau^2 & \tau^2 \\ \tau^2 & \sigma^2/n + \tau^2 \end{pmatrix}
$$

of the vector $\big(\mu(Z)\ \overline{X}\big)$ and therefore remains unchanged if we replace $\big(\mu(Z)\ \overline{X}\big)$ by a bivariate normal random vector $(M^*\ X^*)$ with the same mean vector and covariance matrix. Now if L is the space of square integrable functions of X^*, the standard formulas (A.7.1) for conditioning in the multivariate normal distribution give

$$
P_L M^* = \mathbb{E}[M^* \mid X^*] = \mu_0 + \frac{\tau^2}{\sigma^2/n + \tau^2}(X^* - \mu_0) = (1 - c)\mu_0 + cX^* .
$$

Since this expression is in $L_{\text{aff}} = \{a_0 + a_1 X^*\}$, it must also equal $P_{L_{\text{aff}}}$. Therefore it is the credibility estimator in the normal model and therefore the weights a_0, c must be the same in the given model. $\diamond$

3.2 The General Credibility Model

We now formulate the general question of credibility theory: given a set $X_0, X_1, \ldots, X_n$ of r.v.s with a known second-order structure, what is the best affine predictor of X_0 based upon $X_1, \ldots, X_n$? That is, what are the constants $a_0, , a_1, \ldots, a_n$ which minimize

$$\mathbb{E}\left(X_0 - a_0 - a_1 X_1 - \cdots - a_n X_n\right)^2 ? \tag{3.4}$$

In the credibility setting, we have of course $X_0 = \mu(Z)$.

To answer this question, write $\mu_n = \mathbb{E}X_n, \sigma_{ij}^2 = \mathbb{C}\text{ov}(X_i, X_j)$ (note that we may have $\sigma_{ij}^2 < 0$ when $i \neq j$!) and

$$\boldsymbol{\Sigma}_{10} = \boldsymbol{\Sigma}_{01}^{\mathsf{T}} = \begin{pmatrix} \sigma_{01}^2 \\ \vdots \\ \sigma_{0n}^2 \end{pmatrix}, \quad \boldsymbol{\Sigma}_{11} = \begin{pmatrix} \sigma_{11}^2 & \cdots & \sigma_{1n}^2 \\ \vdots & & \vdots \\ \sigma_{n1}^2 & \cdots & \sigma_{nn}^2 \end{pmatrix}.$$

Further, $\boldsymbol{\mu}_1$ will denote the column vector $(\mu_1 \ldots \mu_n)^{\mathsf{T}}$ and $\boldsymbol{a}_1$ the row vector $(a_1 \ldots a_n)$.

Theorem 3.4 *If $\boldsymbol{\Sigma}_{11}$ is invertible, then the minimizer $\left(a_0^* a_1^* \ldots a_n^*\right) = (a_0^* \ \boldsymbol{a}_1^*)$ of (3.4) is given by*

$$\boldsymbol{a}_1^* = \boldsymbol{\Sigma}_{01} \boldsymbol{\Sigma}_{11}^{-1}, \quad a_0^* = \mu_0 - \boldsymbol{a}_1^* \boldsymbol{\mu}_1. \tag{3.5}$$

The same expression holds in general provided $\boldsymbol{\Sigma}_{11}^{-1}$ is interpreted as a generalized inverse of $\boldsymbol{\Sigma}_{11}$. A different characterization of the minimizer is as the solution of the equations

$$\mu_0 = a_0^* + a_1^* \mu_1 + \cdots + a_n^* \mu_n, \tag{3.6}$$

$$\sigma_{0i}^2 = \sum_{j=1}^{n} a_j^* \sigma_{ij}^2, \quad i = 1, \ldots, n. \tag{3.7}$$

Equations (3.6), (3.7) are commonly referred to as the *normal equations*.

Proof For the second statement, differentiating (3.4) w.r.t. a_0, resp. the a_i, and equating to 0 yields

$$0 = \mathbb{E}\left(X_0 - a_0^* - a_1^* X_1 - \cdots - a_n^* X_n\right),$$
$$0 = \mathbb{E}\left(X_0 - a_0^* - a_1^* X_1 - \cdots - a_n^* X_n\right)X_i, \quad i = 1, \ldots, n,$$

which is the same as

$$\mu_0 = a_0^* + a_1^* \mu_1 + \cdots + a_n^* \mu_n , \tag{3.8}$$

$$\sigma_{0i}^2 + \mu_0 \mu_i = a_0^* \mu_i + \sum_{j=1}^{n} a_j^* (\sigma_{ij}^2 + \mu_i \mu_j), \quad i = 1, \ldots, n . \tag{3.9}$$

Multiplying (3.8) by μ_i and subtracting from (3.9) yields (3.6), (3.7).   □

Again, there is an alternative proof via the multivariate normal distribution: Replace $X_0, X_1, \ldots, X_n$ by a set $X_0^*, X_1^*, \ldots, X_n^*$ of jointly Gaussian r.v.s with the same second-order structure and note that

$$\mathbb{E}\big[X_0^* \,\big|\, X_1^*, \ldots, X_n^*\big] \;=\; \mu_0 + \boldsymbol{\Sigma}_{01} \boldsymbol{\Sigma}_{11}^{-1} (\boldsymbol{X}_1^* - \boldsymbol{\mu}_1) \;=\; a_0^* + \boldsymbol{a}_1^* \boldsymbol{X}_1^* ,$$

where $\boldsymbol{X}_1^* = (X_1^* \; \cdots \; X_n^*)^\mathsf{T}$.

Corollary 3.5 *Let $m > n$ and consider r.v.s $X_0, X_1, \ldots, X_n, X_{n+1}, \ldots, X_m$ such that X_0 is independent of $X_{n+1}, \ldots, X_m$ (or, more generally, uncorrelated with these r.v.s). Then the minimizers $a_0^*, a_1^*, \ldots, a_n^*, a_{n+1}^*, \ldots, a_m^*$ of*

$$\mathbb{E}\big(X_0 - a_0^* X_0 - a_1^* X_1 - \cdots - a_n^* X_n - a_{n+1}^* X_{n+1} - \cdots - a_m^* X_m\big)^2$$

satisfy $a_j^* = 0$, $j = n+1, \ldots, m$, and therefore $a_0^*, a_1^*, \ldots, a_n^*$ are the minimizers of

$$\mathbb{E}\big(X_0 - a_0^* X_0 - a_1^* X_1 - \cdots - a_n^* X_n\big)^2 .$$

Proof Since (3.7) implies

$$\sigma_{0i}^2 \;=\; \sum_{j=n+1}^{m} a_j^* \sigma_{ij}^2 , \quad j = n+1, \ldots, m ,$$

the first statement follows immediately, and the second is then clear.   □

3.3   The Bühlmann Model

The *Bühlmann model* considers a population of M risks, characterized by their risk parameters $Z_1, \ldots, Z_M$. For risk $m = 1, \ldots, M$, a vector $\boldsymbol{X}_m = (X_{1m} \; \cdots \; X_{n_m m})$ (typically claim sizes or claim numbers in consecutive years) is observed, such that given $Z_m = \zeta$, the X_{km} are i.i.d. with distribution F_ζ. The Z_m are assumed i.i.d., and we define $\mu(\zeta), \sigma^2(\zeta)$ etc. as above. Similarly, $H_{\text{Coll}} = \mathbb{E}\mu(Z_m)$. The question is what the credibility premium $H^{(m)}_{\text{Cred}}$ looks like for each m, that is, which affine combination minimizes

$$\mathbb{E}\left(\mu(Z_m) - a_{0m} - \sum_{m=1}^{M}\sum_{k=1}^{n_m} a_{km} X_{km}\right)^2. \tag{3.10}$$

The answer is simple and follows directly from Corollary 3.5: one can just do the calculations for risk m in isolation as if no statistics on risks $m' \neq m$ were available, that is, just use the formulas of the preceding section:

Proposition 3.6 *Define*

$$c_m = n_m/(n_m + \sigma^2/\tau^2), \quad \overline{X}_{\bullet m} = (X_{1m} + \cdots + X_{n_m m})/n_m.$$

Then the mean square error (3.10) is minimized by taking $a_{0m} = (1 - c_m)H_{\text{Coll}}$, $a_{km} = 1/n_m$, $a_{km'} = 0$ for $m' \neq m$. Thus,

$$H^{(m)}_{\text{Cred}} = (1 - c_m)H_{\text{Coll}} + c_m \overline{X}_{\bullet m}. \tag{3.11}$$

Again, there is an alternative proof via the multivariate normal distribution: use the reasoning of Remark 3.3, according to which we can replace the $\mu(Z_m)$, X_{km} with normal r.v.s V^*_m, X^*_{km} such that all means and variances/covariances remain the same. The task is then to show that

$$\mathbb{E}[V^*_m \mid X^*] = (1 - c_m)H_{\text{Coll}} + c_m \overline{V}^*_{\bullet m}, \tag{3.12}$$

where $X^* = (X^*_1 \ldots X^*_M)$. However, the $X_{m'}$ with $m' \neq m$ are independent of X_m, and therefore the $X^*_{m'}$ with $m' \neq m$ are independent of X^*_m. Therefore

$$\mathbb{E}[V^*_m \mid X^*] = \mathbb{E}[V^*_m \mid X^*_m], \tag{3.13}$$

and the analysis of Sect. 3.1 shows that the r.h.s. of (3.13) equals the r.h.s. of (3.12). $\qquad\square$

Remark 3.7 The conclusion that for the calculation of $H^{(m)}_{\text{Cred}}$, one can just forget about the $X_{km'}$ with $m' \neq m$ has a caveat. Namely, the common distribution of the Z_m is typically determined via empirical Bayes where the $X_{km'}$ may play an important role (unless the mass of statistical information is overwhelming compared to that contained in the $X_{km'}$). $\qquad\diamond$

3.4 The Bühlmann–Straub Model

The Bühlmann–Straub model extends the Bühlmann model by allowing different *risk volumes* w_{im}. Examples of risk volumes are allowed maximal mileage in car insurance, area in fire insurance, total amount of persons insured (or total wages) in collective health or accident insurance, and annual premium written by the cedent in reinsurance. In all these examples, it seems reasonable to assume $\mathbb{E}[X_{im} \mid Z_i] = \mu(Z_i)w_{im}$. Further, one can often think of X_{im} as being decomposable into a number of independent parts proportional to the risk volume, which leads to $\mathbb{V}\mathrm{ar}[X_{im} \mid Z_i] = \sigma^2(Z_i)w_{im}$. For technical reasons, it is tradition to replace X_{im} by X_{im}/w_{im}, and the basic assumptions of the Bühlmann–Straub model then become

$$\mathbb{E}[X_{im} \mid Z_i] = \mu(Z_i), \quad \mathbb{V}\mathrm{ar}[X_{im} \mid Z_i] = \sigma^2(Z_i)/w_{im}, \tag{3.14}$$

to which one adds similar independence and conditional independence assumptions as for the Bühlmann model.

To compute the credibility premium in the Bühlmann–Straub model, we may assume $M = 1$ and consider thus $X_0 = \mu(Z)$, $X_1, \ldots, X_n$ with

$$\mathbb{E}[X_i \mid Z = \zeta] = \mu(\zeta), \quad \mathbb{V}\mathrm{ar}[X_i \mid Z = \zeta] = \sigma^2(\zeta)/w_i, \quad i \neq 0.$$

Similar calculations as in the proof of Theorem 3.1 show that the second-order structure is given by

$$\mathbb{E}X_i = \mu_0, \quad \mathbb{V}\mathrm{ar}\, X_i = \sigma^2/w_i + \tau^2, \quad \mathbb{C}\mathrm{ov}(X_i, X_j) = \tau^2$$

for $i = 1, \ldots, n$, where $\mu_0 = \mathbb{E}\mu(Z)$, $\sigma^2 = \mathbb{E}\sigma^2(Z)$, $\tau^2 = \mathbb{V}\mathrm{ar}\, \mu(Z)$, whereas

$$\mathbb{E}X_0 = \mu_0, \quad \mathbb{V}\mathrm{ar}\, X_0 = \tau^2, \quad \mathbb{C}\mathrm{ov}(X_0, X_j) = \tau^2, \quad j \neq 0.$$

Thus, letting $c = a_1^* + \cdots + a_n^*$, $w_\bullet = w_1 + \cdots + w_n$, the normal equations become

$$\mu_0 = a_0^* + c\mu_0, \quad \tau^2 = a_i^*\sigma^2/w_i + c\tau^2,$$

so that $a_i^* = w_i(1 - c)\tau^2/\sigma^2$. Summing over i yields

$$c = w_\bullet(1 - c)\tau^2/\sigma^2, \quad c = \frac{w_\bullet}{w_\bullet + \sigma^2/\tau^2}, \quad a_i^* = \frac{w_i}{w_\bullet + \sigma^2/\tau^2}.$$

Thus

$$a_1^*X_1 + \cdots + a_n^*X_n = c\overline{X} \quad \text{where } \overline{X} = \frac{w_1}{w_\bullet}X_1 + \cdots + \frac{w_n}{w_\bullet}X_n,$$

and the credibility premium for X_0 becomes

$$a_0^* + a_1^*X_1 + \cdots + a_n^*X_n = (1 - c)\mu_0 + c\overline{X}. \tag{3.15}$$

3.5 *Quadratic Loss*

We return to the calculation of the loss functions ℓ_{Coll} etc. defined in Sect. 1.1. Recall for a premium rule H^*, the definition is

$$\ell_{H^*} = \mathbb{E}[\mu(Z) - H^*]^2 = \|\mu(Z) - H^*\|^2 .$$

Theorem 3.8 $\ell_{\mathrm{Bayes}} \leq \ell_{\mathrm{Cred}} \leq \ell_{\mathrm{Coll}}$.

Proof The premiums H_{Bayes}, H_{Cred}, $H_{\mathrm{Coll}} = \mu_0$ are L^2-projections on a decreasing sequence of subspaces, say $L_{\mathrm{Bayes}} \supseteq L_{\mathrm{Cred}} \supseteq L_{\mathrm{Coll}}$, of L^2 (L_{Bayes} is the set of all square integrable functions of $X_1, \ldots, X_n$, $L_{\mathrm{Cred}} = L_{\mathrm{Aff}}$ is the set of all affine functions of $X_1, \ldots, X_n$, and L_{Coll} is the set of all constants). Thus by Pythagoras' theorem (cf. Fig. 1),

$$\ell_{\mathrm{Cred}} = \|\mu(Z) - H_{\mathrm{Cred}}\|^2 = \|\mu(Z) - H_{\mathrm{Bayes}}\|^2 + \|H_{\mathrm{Bayes}} - H_{\mathrm{Cred}}\|^2$$
$$\geq \|\mu(Z) - H_{\mathrm{Bayes}}\|^2 = \ell_{\mathrm{Bayes}} .$$

The proof of $\ell_{\mathrm{Cred}} \leq \ell_{\mathrm{Coll}}$ is similar. $\square$

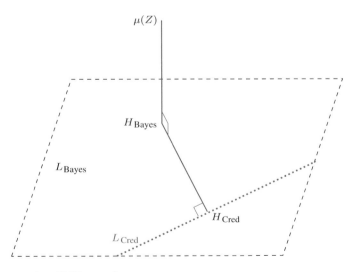

Fig. 1 Bayes and credibility premium

Simple expressions for ℓ_{Cred} and ℓ_{Coll} are available:

Proposition 3.9 *In the setting of the general credibility model in Sect. 3.2, one has*

$$\ell_{\text{Cred}} = \sigma_{00}^2 - \boldsymbol{\Sigma}_{01} \boldsymbol{\Sigma}_{11}^{-1} \boldsymbol{\Sigma}_{10}, \quad \ell_{\text{Coll}} = \sigma_{00}^2.$$

In particular, for the Bühlmann–Straub model $\ell_{\text{Cred}} = (1 - c)\tau^2$, $\ell_{\text{Coll}} = \tau^2$.

Proof To compute ℓ_{Cred}, we may again assume normality and then get

$$\ell_{\text{Cred}} = \mathbb{E}\big(X_0^* - \mathbb{E}[X_0^* \mid X_1^*]\big)^2 = \mathbb{E}\big[\mathbb{E}\big(X_0^* - \mathbb{E}[X_0^* \mid X_1^*]\big)^2 \mid X_1^*\big]$$

$$= \mathbb{E}\operatorname{Var}[X_0^* \mid X_1] = \sigma_{00}^2 - \boldsymbol{\Sigma}_{01} \boldsymbol{\Sigma}_{11}^{-1} \boldsymbol{\Sigma}_{11},$$

$$\ell_{\text{Coll}} = \mathbb{E}(X_0^* - \mu_0)^2 = \operatorname{Var} X_0^* = \sigma_{00}^2.$$

For the Bühlmann–Straub model, $\sigma_{00}^2 = \tau^2$. Further, $\boldsymbol{\Sigma}_{01} \boldsymbol{\Sigma}_{11}^{-1} = (a_1^* \dots a_n^*)$ and hence

$$\sigma_{00}^2 - \boldsymbol{\Sigma}_{01} \boldsymbol{\Sigma}_{11}^{-1} \boldsymbol{\Sigma}_{10} = \tau^2 - (a_1^* \dots a_n^*)(\tau^2 \dots \tau^2)^{\mathsf{T}}$$

$$= \tau^2 - c\tau^2 = (1 - c)\tau^2.$$

$\square$

In general, there is no simple expression for ℓ_{Bayes}. The notable exception is models for which $H_{\text{Bayes}} = H_{\text{Cred}}$, such as exponential families with conjugate priors, cf. Proposition 2.5.

Exercises

3.1 Compute the credibility premium in questions b), c) of Exercise 1.5.

3.2 In many lines of business, a small number of large claims are responsible for a substantial amount of the total claims load. A simple model ignores the claim amounts and considers only the probability Z that a randomly drawn claim is a large claim.

Portfolio information gives the empirical Bayes estimator $\mathbb{E}Z = 2\%$, $\sqrt{\operatorname{Var} Z} = 1\%$. Determine the credibility estimator of Z for a risk having 6 large claims out of a total of 120.

3.3 Let X_1 be Poisson(ζ) given $Z = \zeta$, where the available information on Z is the moments $m_k = \mathbb{E}Z^k$, $k = 1, 2, 3, 4$.

a) Find the credibility estimator of Z based upon X_1 in terms of m_1, m_2.
b) Find the credibility estimator of Z based upon X_1 and $X_2 = X_1(X_1 - 1)$ in terms of m_1, m_2, m_3, m_4. Hint: for a Poisson(λ) r.v. Y,

$$\mathbb{E}Y^{(k)} = \mathbb{E}\big[Y(Y - 1) \dots (Y - k + 1)\big] = \lambda^k.$$

3.4 The lognormal(a, b^2) distribution is defined as the distribution of e^V where $V \sim \mathcal{N}(a, b^2)$.

a) Find the density, mean and variance of the lognormal distribution.

   Let now $X_0, X_1, \ldots, X_n$ be independent lognormal(ζ, b^2) given $z = \zeta$, where Z itself has a $\mathcal{N}(c, d^2)$ distribution, and let $\mu(\zeta) = \mathbb{E}[X_i \mid Z = \zeta]$.

b) Show that the $X_i/\mu(Z)$ are i.i.d. and independent of Z.

c) Find the credibility estimator of X_0 based upon $X_1, \ldots, X_n$ and the Bayes estimator of $\mu(Z)$.

3.5 Show Proposition 3.9 directly (without reference to the multivariate normal distribution).

Notes
Some main references for credibility theory are Bühlmann [42] and Bühlmann and Gisler [43]. There are numerous extensions of the topics treated here, e.g. combination with regression and hierarchical models. As mentioned above, the argument for using linear Bayes estimators instead of those based on the posterior distribution has, however, been somewhat weakened in recent decades since today's computer power makes Markov chain Monte Carlo a feasible alternative.

   Diaconis and Ylvisaker [60] discuss in detail when Bayes and credibility premiums coincide.

4 Bonus-Malus Systems

4.1 Introduction

The main example of bonus-malus systems is car insurance, and we will stick to that setting. The key feature is that the portfolio is divided into K classes, such that the premium H_k in class $k = 1, \ldots, K$ depends on the class, and that the class of a driver is dynamically adjusted according to his claim experience (a driver entering the portfolio is assigned class k_0 for some fixed k_0).

   We will think of the driver's reliability (as estimated by the company) as being a decreasing function of k and thereby H_k as being a increasing function, such that class 1 contains the 'best' drivers and class K the 'worst'. Of course, this may not always reflect reality since a good driver can by randomness have many claims and thereby be allocated a high bonus class.

   For a particular driver, we denote by X_n his bonus class in year n and M_n his number of claims. Many systems then determine the bonus class X_{n+1} in year n on the basis of X_n and M_n alone. That is, $X_{n+1} = \kappa(k, m)$ when $X_n = k$, $M_n = m$. For the whole of the section, one may safely think of X_n as being Poisson(λ), where λ is the risk parameter of the driver under consideration, and of the expected size of a claim as being 1 (this is just a normalization).

Example 4.1 For systems with a small K, a very common system is $-1/+2$ where one moves down one class after a claim-free year and up two classes per claim. In

Table 2 $-1 / +2$ bonus
rules

k	1	2	3	4	5
$\kappa(k, 0)$	1	1	2	3	4
$\kappa(k, 1)$	3	4	5	5	5
$\kappa(k, 2+)$	5	5	5	5	5

Table 3 Norwegian bonus rules

k	1	2	3	4	5	6	7	8	9	10	11	12	13
$\kappa(k, 0)$	1	1	2	3	4	5	6	7	8	9	10	10	10
$\kappa(k, 1)$	3	4	5	6	7	8	9	10	11	12	13	13	13
$\kappa(k, 2)$	5	6	7	8	9	10	11	12	13	13	13	13	13
$\kappa(k, 3)$	7	8	9	10	11	12	13	13	13	13	13	13	13
$\kappa(k, 4)$	9	10	11	12	13	13	13	13	13	13	13	13	13
$\kappa(k, 5)$	11	12	13	13	13	13	13	13	13	13	13	13	13
$\kappa(k, 6+)$	13	13	13	13	13	13	13	13	13	13	13	13	13

Table 4 Dutch bonus rules

k	1	2	3	4	5	6	7	8	9	10	11	12	13	14
H_k	20.0	32.5	35.0	37.5	40.0	45.0	50.0	55.0	60.0	70.0	80.0	90.0	100.0	120.0
$\kappa(k, 0)$	1	1	2	3	4	5	6	7	8	9	10	11	12	13
$\kappa(k, 1)$	6	7	7	8	8	9	10	11	12	13	14	14	14	14
$\kappa(k, 2)$	10	11	11	12	13	14	14	14	14	14	14	14	14	14
$\kappa(k, 3+)$	14	14	14	14	14	14	14	14	14	14	14	14	14	14

formulas, $\kappa(k, m) = \max(1, k - 1)$ if $m = 0$, $\kappa(k, m) = \min(K, k + 2m)$ if $m > 0$.
For $K = 5$, this gives the $\kappa(k, n)$ in Table 2 (here $n = 2+$ means two or more
claims).

Many (but not all!) systems in practical use have a higher number of classes,
typically 12–25. The bonus rules are then most often based on similar ideas as in
the $-1 / +2$ system but with some modifications for k close to 1 or K. For example,
a system which has been used in Norway (see Sundt [166] p. 73) has $K = 13$,
$k_0 = 8$ and the rule that after a claim-free year, one moves down one class (except
from class 12 or 13 where one moves to 10, and from class 1 where one cannot
move down), and that one moves up two classes for each claim when possible. This
gives the rules in Table 3. Another system, introduced by five of the largest Dutch
insurance companies in 1982 (see Kaas et al. [97] p. 127), has $K = 14$, $k_0 = 2$ and
the rules in Table 4. Here H_k are the premiums in percentages of the premium in the
initial class k_0 (such normalized H_k are often denoted *relativities*).   ◇

It is assumed that the numbers M_n of claims by a particular driver are i.i.d.
Poisson(λ) where λ is the outcome of an r.v. Λ. Since

$$X_{n+1} = \kappa(X_n, M_n),\tag{4.1}$$

it follows by the i.i.d. property of the M_n that for a given λ, $\{X_n\}$ is a time-homogeneous Markov chain with state space $\{1, \ldots, K\}$.

In practice, the premium paid is often based upon more detailed information than just the number of claims in the preceding year. Examples of such *rating factors* are age of the driver, weight of the car, age of the car, living area of the driver (rural or urban). In some countries, also risk volume in terms of mileage is used (i.e., full coverage assumes that at most a prespecified number of km is driven).

The initial class k_0 may depend on such rating factors. Also, for a new policy holder who was previously insured by another company, it is customary that the old company will supply some information of the driver's claim experience, from which the initial class is determined.

For the following discussion, one can just think of the bonus system under study as corresponding to some specific combination of such rating factors.

Often, a bonus system has memory longer than one year. For example, it may require two claim-free years to move from class 2 to class 1. Phenomena like this are easily incorporated into the Markov chain set-up, since we can just split up state 2 into two, say $2'$ and $2''$, such that entering the old class 2 corresponds to entering class $2'$. After a claim-free year in class $2'$, class $2''$ is entered, and after a claim-free year in class $2''$, class 1 is entered. If there are claims in any of classes $2'$, $2''$, one is moved to a class $k > 2$. See Example 4.2 for a concrete case.

A study of bonus rules also requires model assumptions for the number of years a driver holds the policy. We return to this point later.

4.2 Loimaranta Efficiency

Consider a fixed bonus system with the $\kappa(k, m)$ and the premiums H_k given. One can then ask how fair the system is, more precisely how well the average premium paid by a driver is adapted to his risk parameter ζ. The ideal is of course proportionality.

To make this question precise, one needs to specify the concept of average premium paid by a driver with risk parameter ζ. The suggestion of Loimaranta [110], upon which most of the later literature is based, is to assume that each driver spends a sufficient number N of years as policy holder that his typical bonus class can be described by the stationary distribution $\pi(\lambda) = \big(\pi_1(\lambda), \ldots, \pi_K(\lambda)\big)$ of the Markov chain (4.1) under the model assumption that the M_n are i.i.d. Poisson(λ). The average premium then becomes

$$H(\lambda) = \sum_{k=1}^{K} \pi_k(\lambda) H_k . \tag{4.2}$$

A further suggestion by Loimaranta [110] is to study $H(\lambda)$ by means of

$$e(\lambda) = \frac{\mathrm{d}\log H(\lambda)}{\mathrm{d}\log\lambda}, \tag{4.3}$$

the *Loimaranta efficiency* or *elasticity*. The intuition behind this definition is that it measures the effect of relative changes in λ on the relative size of $H(\lambda)$. That is, for small h

$$H\bigl(\lambda(1+h)\bigr) \approx H(\lambda)\bigl(1+e(\lambda)h\bigr). \tag{4.4}$$

For computational purposes, note that the chain rule implies

$$e(\lambda) = \frac{\lambda H'(\lambda)}{H(\lambda)} = \frac{\mathrm{d}\log H(\lambda)/\mathrm{d}\lambda}{\mathrm{d}\log\lambda/\mathrm{d}\lambda}. \tag{4.5}$$

Since

$$H\bigl(\lambda(1+h)\bigr) \approx H(\lambda) + \lambda h H'(\lambda),$$

this gives immediately (4.4).

As noted above, $H(\lambda)/\lambda$ should ideally be a constant, say 1, which means $e(\lambda) \approx 1$. This is of course never fulfilled in practice, since at the one extreme, a driver with $\lambda = 0$ can never get lower than the best bonus class 1 where he still has to pay premium $H_1 > 0$, and at the other, a driver with an enormous λ can never get higher than class K and thus there is an upper limit to his premium. In other words,

$$\lim_{\lambda\to 0} \frac{H(\lambda)}{\lambda} = \infty, \quad \lim_{\lambda\to\infty} \frac{H(\lambda)}{\lambda} = 0. \tag{4.6}$$

This means that the very good risks (small λ) will always pay too much and the very bad ones (large λ) too little.

For intermediate values of λ, the picture is less clear cut, but the behavior of the Loimaranta efficiency may help us to understand the unfairnesses imposed by the bonus system. Assume in particular $e(\lambda) < 1$ for all λ (as in, e.g., Example 4.2 below). Then

$$\frac{\mathrm{d}}{\mathrm{d}\lambda}\log\frac{H(\lambda)}{\lambda} = \frac{H'(\lambda)}{H(\lambda)} - \frac{1}{\lambda} = \frac{1}{\lambda}\bigl(e(\lambda) - 1\bigr) < 0.$$

Thus $H(\lambda)/\lambda$ is decreasing in λ. By (4.6), this implies the existence of a unique λ_0 satisfying $H(\lambda_0)/\lambda_0$, such that $H(\lambda) > \lambda$ for $\lambda < \lambda_0$ and $H(\lambda) < \lambda$ for $\lambda > \lambda_0$. This means that risks with $\lambda < \lambda_0$ ('good risks') pay more than they should and risks with $\lambda > \lambda_0$ ('bad risks') less. This confirms and quantifies that the bonus system favors the bad risks.

Example 4.2 ([97] pp. 129 ff.) Consider a system with two premiums levels $a < c$. Such a driver pays c in year $n+1$ if there is at least one claim in years n or $n-1$ and a if there are none. The sequence of bonus classes is then not a Markov chain, but we can construct an auxiliary Markov chain X_n with states 1,2,3 by letting $X_n = 1$ if the driver has no claims in the preceding two years, $X_n = 2$ if he has one claim in year $n-1$ but none in year n, and $X_n = 3$ if he had a claim in year n. The transition matrix is

$$P = \begin{pmatrix} q & 0 & p \\ q & 0 & p \\ 0 & q & p \end{pmatrix}$$

where $p = p(\lambda) = 1 - e^{-\lambda}$ is the probability of one or more claims, $q = 1 - p = e^{-\lambda}$ the probability of none, and the given system corresponds to $H_1 = a$, $H_2 = H_3 = c$.

Writing $\pi_k = \pi_k(\lambda)$, the equilibrium equation $\pi P = \pi$ means

$$\pi_1 = q\pi_1 + q\pi_2,$$
$$\pi_2 = q\pi_3,$$
$$\pi_3 = p\pi_1 + p\pi_2 + p\pi_3.$$

The last equation combined with $\pi_1 + \pi_2 + \pi_3 = 1$ gives $\pi_3 = p$, and so the second yields $\pi_2 = pq$. The first can be rewritten $p\pi_1 = qp_2$ and thus $\pi_1 = q^2$. Therefore

$$H(\lambda) = a\pi_1 + c(\pi_2 + \pi_3) = c(1 - e^{-2\lambda}) + ae^{-2\lambda},$$

and (4.5) implies

$$e(\lambda) = \frac{2\lambda e^{-2\lambda}(c - a)}{c(1 - e^{-2\lambda}) + ae^{-2\lambda}} = \frac{2\lambda(c - a)}{c(e^{2\lambda} - 1) + a}.$$

Since a lower bound for the denominator is $2\lambda c + a \geq 2\lambda c$, we get $e(\lambda) < 1$ for all λ.   ◇

Exercises
4.1 Verify (4.4) and (4.5).

4.3 Bayes Premiums

In the preceding section, we assumed that the set $\kappa(k, m)$ of transition rules as well as the (net) premiums H_k were fixed. We now take only the $\kappa(k, m)$ as fixed and ask how to choose the H_k.

The common suggestion is to pick an insured at random and minimize the mean square distance between his ideal premium Λ and what he actually pays, which is $H_{C(\Lambda)}$, where $C(\lambda)$ is a class picked at random with a distribution which in some sense is typical of the bonus class of a risk sized λ. That is, we have to minimize

$$\mathbb{E}\big[\Lambda - H_{C(\Lambda)}\big]^2 = \sum_{k=1}^{K} \mathbb{E}\big[\Lambda \mathbb{1}\{C(\Lambda) = k\} - H_k \mathbb{1}\{C(\Lambda) = k\}\big]^2$$

w.r.t. $H_1, \ldots, H_K$. This is minimized term by term, and since the minimizer of $\mathbb{E}(X - aY)^2$ is $a^* = \mathbb{E}[XY]/\mathbb{E}Y^2$, it follows that the minimizer is

$$H_k^* = \frac{\mathbb{E}\big[\Lambda;\, C(\Lambda) = k\big]}{\mathbb{P}\big(C(\Lambda) = k\big)} = \mathbb{E}\big[\Lambda \mid C(\Lambda) = k\big]. \tag{4.7}$$

We refer to this as the Bayes premium because it is derived by a similar philosophy as the Bayes premium in Sect. 1.1, though obviously the details are somewhat different. The expression (4.7) shows in particular that

$$\mathbb{E}H^*(\Lambda) = \mathbb{E}\,\mathbb{E}\big[\Lambda \mid C(\Lambda)\big] = \mathbb{E}\Lambda,$$

as it should be.

The distribution of $C(\lambda)$ has so far been left unspecified. Following the lines of Sect. 4.2, an obvious suggestion is to choose $C(\lambda)$ according to the stationary distribution $\pi(\lambda)$,

$$\mathbb{P}\big(C(\lambda) = k\big) = \pi_k(\lambda).$$

However, as noted in Sect. 4.2, in order for this choice to be reasonable it requires the insurance contracts to run sufficiently long that the empirical distribution of time spent in the various bonus classes is close to $\pi(\lambda)$. For example for the Dutch system in Example 4.1, an insured with a very small λ will have $\pi_1(\lambda)$ very close to 1, but can at the earliest reach class $K = 1$ after $k_0 - 1 = 12$ years. Thus, the empirical distribution in the first 12 years is far from $\pi(\lambda)$, and in order for the use of the stationary distribution $\pi(\lambda)$ to be reasonable, one probably needs the average time in the portfolio to be at least 30 years, which is a very long time.

A more refined definition of the typical bonus class of a risk sized λ is to assume the time T as policy holder to be an r.v., say $\mathbb{P}(T > n) = q_n$, $n = 0, 1, 2, \ldots$ The expected fraction of time spent in class k is then

$$p_k(\lambda) = \sum_{n=0}^{\infty} w_n p_{k_0 k}^n(\lambda), \tag{4.8}$$

where $w_n = q_n/\mu$, $\mu = \mathbb{E}T = \sum_0^\infty q_n$. To see this, consider a large number N of insured with the same λ. Their total time spent in the portfolio is of order $N\mu$, their total time in year n of order Nq_n, and their total time in class k in year n of order $Nq_n p_{k_0k}^n(\lambda)$. Thus, the average time in class k in year n is close to $q_n p_{k_0k}^n(\lambda)/\mu$. Summing over n yields (4.8).

The implication of (4.8) is to let $\mathbb{P}(C(\lambda) = k) = p_k(\lambda)$.

Remark 4.3 The argument leading to (4.8) is based on a number of implicit assumptions, for example that T does not depend on λ or the claim experience by the insured (change of company in case of many claims and thereby a low class, etc). $\diamond$

Exercises

4.2 For a two-state Markov chain with transition matrix

$$P = \begin{pmatrix} a & 1-a \\ 1-b & b \end{pmatrix},$$

(a) verify that P can be written as $\mathbf{1}\pi + \lambda h v$ where $\mathbf{1} = (1\ 1)^\mathsf{T}$, $h = (1\ -1)^\mathsf{T}$, $\pi = (\pi_1\ \pi_2)$, $v = (\pi_2\ -\pi_1)$ and $\lambda = a + b - 1$ (π is the stationary distribution). Show thereby that (b) $P^n = \mathbf{1}\pi + \lambda^n h v$.

(c) Now consider a bonus system with two classes $1, 2$, where an insured with risk parameter λ has Poisson(λ) claims in a year, and is moved from 2 to 1 if one or more claims occur, and from 1 to 2 in the case of no claims. Assume further that one always starts in bonus class 1, and that the number of years in the portfolio is a geometric r.v. T, i.e. $\mathbb{P}(T > n) = \rho^n$ for some $\rho < 1$ and $n = 0, 1, \ldots$ Derive formulas for the premiums H_1, H_2 corresponding to (4.8), and give numerical values for $\mathbb{E}T = 5, 10, 25, 1000$. Comment on the interpretation of the limit $\mathbb{E}T \to \infty$.

Notes
Classical references for bonus-malus systems are a series of papers and books by Lemaire and coauthors, e.g. [107–109]. There is also much material in Denuit et al. [57]. The Bayesian view was initiated by Norberg [131], whereas the non-stationary view in (4.8) originates from Borgan, Hoem and Norberg [30] and was further elaborated in Asmussen [7].

Chapter III: Sums and Aggregate Claims

1 Introduction

In this chapter, we study the often encountered problem of assessing the order of magnitude of $\Pi_n(x) = \mathbb{P}(S_n > x)$ for some sequence $X_1, X_2, \ldots$ of r.v.s and $S_n = X_1 + \cdots + X_n$. In many applications, n is an r.v. N rather than deterministic, and the object of interest is then $\Pi_N(x) = \mathbb{P}(S_N > x)$.

In insurance, n may be the portfolio size and $X_1, \ldots, X_n$ the claims. Alternatively, one could consider only the non-zero claims, in which case it would be natural to consider the random sum problem with N a Poisson r.v.; the r.v. S_N with N possibly random goes under the name the *aggregate claims*. In finance, $X_1, \ldots, X_n$ could be the returns of the n assets in the portfolio or the losses on them. One would encounter the random sum problem for example in credit risk, with N the number of defaulted obligors.

In this chapter, we consider only the case where $X_1, X_2, \ldots$ are i.i.d. but return to dependence in Chap. X. In the random sum case, N is taken as independent of the X_i.

If the mean $m = \mathbb{E}X_1$ exists, the roughest estimate of $\Pi_n(x)$ is based on the LLN:

$$\Pi_n(x) \approx \begin{cases} 1 & \text{when } x \ll nm \\ 0 & \text{when } x \gg nm \end{cases}$$

when n is large. If the variance $\sigma^2 = \operatorname{Var} X_1$ is finite, a more precise statement is provided by the CLT. Namely, one writes

$$\Pi_n(x) \approx 1 - \Phi(v) \quad \text{where } v = \frac{x - nm}{\sigma\sqrt{n}}.$$

© Springer Nature Switzerland AG 2020
S. Asmussen, M. Steffensen, *Risk and Insurance*, Probability Theory
and Stochastic Modelling 96, https://doi.org/10.1007/978-3-030-35176-2_3

However, this is only valid when $x - nm$ is of order $\sigma\sqrt{n}$ so that $1 - \Phi(v)$ is a number not too close to 0 or 1. For larger z, $\Pi_n(x)$ is close to 0 and this is in many cases the situation of interest.

There may be various mathematical reasons for such small values of $\Pi_n(x)$: n is moderate (as is often the case of interest) but x very large; x is of larger magnitude than the mean nm; or a combination. The available results on the order of magnitude of $\Pi_n(x)$ depend crucially on which of these situations one is in, as well as on the tail $\overline{F}(x) = \mathbb{P}(X_1 > x)$, more precisely on whether the tail is light or heavy in the sense to be defined next.

The term 'heavy tails' is used in a different meaning in different application areas. In finance, it means heavy compared to the Black–Scholes model, that is, heavier than the normal. According to this definition, the Weibull tail $\overline{F}(x) = e^{-x^\beta}$ is heavy-tailed for $\beta < 2$, say for the exponential distribution that has $\beta = 1$. In contrast, in most other areas, in particular insurance, the benchmark is the exponential distribution rather than the normal so that the Weibull is heavy-tailed only when $\beta < 1$. We follow this tradition here, which roughly leads to calling a distribution F with tail $\overline{F}(x) = 1 - F(x)$ light-tailed if the m.g.f. $\widehat{F}[s] = \int e^{sx} F(dx) = \mathbb{E}e^{X_1}$ is finite for some $s > 0$ and heavy-tailed if it is infinite for all $s > 0$.

Remark 1.1 The distinction between light and heavy tails is closely related to the value of the limiting overshoot probability

$$\ell(a) = \lim_{x\to\infty} \mathbb{P}(X > x + a \mid X > x) = \lim_{x\to\infty} \frac{\overline{F}(x + a)}{\overline{F}(x)}, \qquad (1.1)$$

which exists for all well-behaved F (but certainly not all F!). Namely, $0 \le \ell(a) < 1$ holds for such light-tailed F but $\ell(a) = 1$ for such heavy-tailed ones. $\Diamond$

Remark 1.2 Explicit forms of $\Pi_n(x) = \mathbb{P}(S_n > x)$ do not exist for any of the standard heavy-tailed F. With light tails, there is a handful of explicit examples like the normal, inverse Gaussian or the gamma. In particular, for F gamma(α, λ), the distribution of S_n is gamma$(\alpha n, \lambda)$ and one has

$$\overline{F}(x) = \mathbb{P}(X > x) \sim \frac{\lambda^{\alpha-1} x^{\alpha-1}}{\Gamma(\alpha)} e^{-\lambda x}, \quad \Pi_n(x) \sim \frac{\lambda^{\alpha n-1} x^{\alpha n-1}}{\Gamma(\alpha n)} e^{-\lambda x}$$

as $z \to \infty$, cf. Exercise A.7.1 in the Appendix. With heavy tails, the model example is the Pareto where in the standard form $\overline{F}(x) = \mathbb{P}(X > x) = 1/(1 + x)^\alpha$. Distributions of the same type, i.e. distributions of $aX + b$ with $a > 0$ also go under the name of Pareto, often with special terminology like US Pareto. The c.d.f. or density of a sum is not explicit except for special cases like $n = 2$, $\alpha = 1$. Another important heavy-tailed distribution is the lognormal. $\Diamond$

In addition to finding the asymptotic form of $\Pi_n(x)$, it is of interest to see how $X_1, \ldots, X_n$ behaves in atypical cases; one issue is how much bigger S_n is than x when x is exceeded, i.e. the behavior of the overshoot $S_n - x$. We shall do this by considering $\mathbb{P}^x$, the conditional distribution of $X_1, \ldots, X_n$ given $S_n > x$, and $\xrightarrow{\mathcal{D}^x}$, convergence in $\mathbb{P}^x$-distribution. As a start, we shall consider three simple parametric examples illustrating much of the general picture, taking for simplicity $n = 2$ and assuming F to admit a density $f(x)$. Then $S_2 = X_1 + X_2$ has density

$$f^{*2}(x) = \int_{-\infty}^{\infty} f(y)f(x-y) \stackrel{X \geq 0}{=} \int_0^x f(y)f(x-y)\,dy \qquad (1.2)$$

at x, the conditional density of any of X_1 and X_2 at y given $X_1 + X_2 = x$ is $f(y)f(x-y)/f^{*2}(x)$ for $0 < y < x$ and

$$\Pi_2(x) = \mathbb{P}(X_1 + X_2 > x) = \int_x^{\infty} f^{*2}(y)\,dy. \qquad (1.3)$$

Example 1.3 Let F be standard exponential, i.e. $f(x) = e^{-x}$. Then in (1.2) we have $f(y)f(x-y) = e^{-x}$ and so $f^{*2}(x) = xe^{-x}$ (as also follows from the convolution properties of the gamma distribution). Further, the conditional density given $X_1 + X_2 = x$ becomes $1/x$, so that the pair (X_1, X_2) is distributed as $(xU, x(1-U))$ with U uniform$(0, 1)$. Further, (1.3) gives

$$\Pi_2(x) = \mathbb{P}(X_1 + X_2 > x) = (1+x)e^{-x} \sim xe^{-x}.$$

For the overshoot, this gives in particular that

$$\mathbb{P}^x(S_2 - x > y) = \frac{\Pi_2(x+y)}{\Pi_2(x)} \sim \frac{(x+y)e^{-x-y}}{xe^{-x}} \sim e^{-y} = \mathbb{P}(X^* > y),$$

where X^* is again standard exponential. We also get

$$\mathbb{P}\left(X_1 + X_2 > x, \frac{X_1}{X_1+X_2} \leq u_1, \frac{X_2}{X_1+X_2} \leq u_2\right)$$

$$= \int_x^{\infty} \mathbb{P}\left(X_1 + X_2 \in dy, \frac{X_1}{X_1+X_2} \leq u_1, \frac{X_2}{X_1+X_2} \leq u_2\right)$$

$$= \int_x^{\infty} ye^{-y}\,dy\,\mathbb{P}(U \leq u_1, 1 - U \leq u_2)$$

with U uniform$(0,1)$. Combining, we have altogether that

$$\left(S_2 - x, \frac{X_1}{X_1+X_2}, \frac{X_2}{X_1+X_2}\right) \xrightarrow{\mathcal{D}^x} (X^*, U, 1 - U). \qquad \diamond$$

Example 1.4 Next let F be standard normal. Then by the Mill's ratio estimate (see Sect. A.7.2 of the Appendix)

$$\Pi_2(x) = \mathbb{P}(\mathcal{N}(0,2) > x) = \mathbb{P}(\mathcal{N}(0,1) > x/\sqrt{2}) \sim \frac{1}{x\sqrt{\pi}} e^{-x^2/4}.$$

For the overshoot, this gives

$$\mathbb{P}^x(S_2 - x > y) = \frac{x\sqrt{\pi}}{(x+y)\sqrt{\pi}} e^{-(x+y)^2/4+x^2/4} \sim e^{-y^2/4-xy/2},$$

$$\mathbb{P}^x(x(S_2 - x) > y) \sim e^{-y^2/4x^2-y/2} \sim e^{-y/2}.$$

That is, conditionally $S_2 - x$ is distributed as $2V/x$ with again V exponential. Also, given $S_2 = y$ the pair (X_1, X_2) is distributed as $(y/2+X^*/\sqrt{2}, y/2-X^*/\sqrt{2})$ with X^* standard normal (cf. Sect. A.7.1 of the Appendix). Since the relevant values of y are those of $x + V/x$, we conclude that

$$(x(S_2 - x), X_1 - x/2, X_2 - x) \xrightarrow{\mathcal{D}^x} (2V, X^*/\sqrt{2}, -X^*/\sqrt{2}). \qquad \diamond$$

Example 1.5 Now let F be Pareto with tail $\overline{F}(x) = (1+x)^{-\alpha}$ and density $f(x) = \alpha(1+x)^{-\alpha-1}$. We show first that

$$f^{*2}(x) \sim 2f(x) = \frac{2\alpha}{(1+x)^{\alpha+1}}. \tag{1.4}$$

Indeed,

$$\frac{f^{*2}(x)}{f(x)} = \int_0^x \frac{\alpha}{(1+y)^{\alpha+1}} \frac{\alpha(1+x)^{\alpha+1}}{\alpha(1+x-y)^{\alpha+1}} \, dy$$

$$= 2 \int_0^{x/2} \frac{\alpha}{(1+y)^{\alpha+1}} \frac{(1+x)^{\alpha+1}}{(1+x-y)^{\alpha+1}} \, dy.$$

Here the second fraction is bounded by $2^{\alpha+1}$ and has limit 1 as $x \to \infty$, so dominated convergence gives

$$\frac{f^{*2}(x)}{f(x)} \to 2 \int_0^\infty \frac{\alpha}{(1+y)^{\alpha+1}} \, dy = 2.$$

A further dominated convergence arguments then yields

$$\Pi_2(x) = \int_x^\infty f^{*2}(y) \, dy \sim 2 \int_x^\infty f(y) \, dy = \frac{1}{(1+x)^\alpha}.$$

We can write $f(y)f(x - y) = \alpha^2 \exp\{\alpha g(y \mid x)\}$ where $g(y|x) = -\log(1 + y) - \log(1 + x - y)$ for $0 < y < x$. Elementary calculus shows that g is convex, symmetric around its minimum point $y = x/2$ and with an increase as y approaches one of the boundaries $0, x$ that becomes steeper and steeper as x increases. This indicates that not only can the conditional distribution not be uniform but that one of X_1, X_2 dominates more and more as x becomes large. The precise formulation of this is conveniently expressed in terms of the order statistics $X_{(1)} = \min(X_1, X_2)$, $X_{(2)} = \max(X_1, X_2)$. In fact, we shall show that

$$(X_{(1)}, X_{(2)}/x) \xrightarrow{\mathcal{D}^x} (X_1^*, 1 + X_2^*), \tag{1.5}$$

where X_1^*, X_2^* are independent with the same Pareto distribution F. That is, conditionally $X_{(1)}$ is asymptotically 'normal' and $X_{(2)}$ of order X. This follows since

$$\mathbb{P}(X_{(1)} \le A, (X_{(2)} - x)/x > B) = \frac{1}{\mathbb{P}(X_1 + X_2 > x)} \int_0^A 2f(y)\,dy\,\overline{F}(x + xB)$$

$$\sim \frac{1}{2\overline{F}(x)} \int_0^A 2f(y)\,dy\,\overline{F}(x + xB) = \int_0^A f(y)\,dy\,\frac{(1 + x)^\alpha}{(1 + x + xB)^\alpha}$$

$$\sim \int_0^A f(y)\,dy\,\frac{1}{(1 + B)^\alpha} = \mathbb{P}(X_1^* \le A, X_2^* > B)\,.$$

Altogether, we have

$$\left(\frac{S_2 - x}{x}, X_{(1)}, \frac{X_{(2)} - x}{x}\right) \xrightarrow{\mathcal{D}^x} (X_2^*, X_1^*, X_2^*)\,. \qquad \diamond$$

A qualitative summary of some main differences between light and heavy tails are given in the following three points, assuming F to be sufficiently well-behaved. The quantification involves making orders precise and specifying the exact conditions on F. This has been done for $n = 2$ in three parametric cases in Examples 1.3–1.5 and will be done for heavy tails in Sect. 2 in the framework of subexponential distributions, whereas for light tails we refer to Sect. 4. Consider a fixed general n and the limit $x \to \infty$. Then

1) The order of magnitude of $\mathbb{P}(S_n > x)$ is the same as the order of $\overline{F}(x)$ for heavy tails, where it more precisely is $n\,\overline{F}(x)$. For light tails it is effectively larger.
2) Conditionally, the overshoot $S_n - x$ remains bounded for light tails. More precisely it is asymptotically exponential when F is itself close-to-exponential (the case $0 < \ell(a) < 1$ in Remark 1.1), and goes to 0 for lighter-than-exponential tails (the case $\ell(a) = 0$). For heavy tails, the overshoot grows to ∞.
3) Conditionally, the summands $X_1, \ldots, X_n$ are of the same order for light tails. For heavy tails, one is larger than x in the limit and the rest normal.

In items 1)–3), we looked into the limit $x \to \infty$ with n fixed. The tradition in the light-tailed area is, however, to take a limit with both x and n going to ∞, more precisely to take $x = nz$ with $z > m = \mathbb{E}X$. The main asymptotic result is that under some regularity conditions

$$\mathbb{P}(S_n > n(m + a)) \sim \frac{\text{constant}}{n^{1/2}} \exp\left\{ -n\kappa^*(m + a) \right\} \tag{1.6}$$

as $n \to \infty$ with $a > 0$ fixed and some suitable function κ^*. It is thus a result on behavior atypical of the LLN, and for this reason is often said to be of *large deviations* type (though this terminology could equally well refer to the case $z \to \infty$ with n fixed). It is discussed in more detail in Sect. 3.

2 Heavy Tails. Subexponential Distributions

2.1 Definition of Subexponentiality and Sufficient Conditions

We are only concerned with positive r.v.s X, i.e. with distributions F concentrated on $(0, \infty)$. The main cases of heavy tails are:

(a) distributions with a regularly varying tail, $\overline{F}(x) = L(x)/x^\alpha$ where $\alpha > 0$ and $L(x)$ is slowly varying, $L(tx)/L(x) \to 1$, $x \to \infty$, for all $t > 0$.[1] A main example is the Pareto, i.e. the distribution of $X = aY + b$ where $\mathbb{P}(Y > y) = 1/(1 + y)^\alpha$ and $a > 0, b \geq 0$;

(b) the lognormal distribution (the distribution of e^V where $V \sim \mathcal{N}(\mu, \sigma^2)$). The density and asymptotic tail (cf. Mill's ratio in Sect. A.7 in the Appendix) are given by

$$f(x) = \frac{1}{x\sqrt{2\pi\sigma^2}} e^{-(\log x - \mu)^2/2\sigma^2}, \tag{2.1}$$

$$\overline{F}(x) = \overline{\Phi}\big((\log x - \mu)/\sigma\big) \sim \frac{\sigma}{\log x \sqrt{2\pi}} e^{-(\log x - \mu)^2/2\sigma^2}; \tag{2.2}$$

(c) the Weibull distribution with decreasing hazard rate, $\overline{F}(x) = e^{-x^\beta}$ with $0 < \beta < 1$.

Here the regularly varying tails are the heaviest (and the heavier the smaller α is) and the Weibull the lightest (and the lighter the larger β is). The lognormal case is intermediate (here the heaviness increases with σ^2). This ordering is illustrated by

---

[1] Some main examples of slowly varying functions are: i) functions with a limit, $L(x) \to \ell$ where $0 < \ell < \infty$; ii) $L(x) \sim (\log x)^\beta$ with $-\infty < \beta < \infty$. But see Exercise IX.3.14 for an example with weird properties.

the behavior of the pth moments $\mathbb{E}X^p$. For regular variation these are finite only for $p < \alpha$ (and possibly for $p = \alpha$), and for the lognormal and Weibull for all p; however, for the lognormal $\mathbb{E}X^p = \exp\{p\mu + p^2\sigma^2/2\}$ grows so rapidly in p that the lognormal is not determined by its moments (cf. the celebrated *Stieltjes moment problem*).

The definition $\widehat{F}[s] = \infty$ for all $s > 0$ of heavy tails is too general to allow for many general non-trivial results, and instead we shall work within the class $\mathscr{S}$ of *subexponential* distributions. For the definition, we require that F is concentrated on $(0, \infty)$ and say then that F is *subexponential* ($F \in \mathscr{S}$) if

$$\frac{\overline{F^{*2}}(x)}{\overline{F}(x)} \to 2, \quad x \to \infty. \tag{2.3}$$

Here F^{*2} is the convolution square, that is, the distribution of independent r.v.s X_1, X_2 with distribution F. In terms of r.v.s, (2.3) then means $\mathbb{P}(X_1 + X_2 > x) \sim 2\mathbb{P}(X_1 > x)$.

To capture the intuition behind this definition, note first the following fact:

Proposition 2.1 *Let F be any distribution on $(0, \infty)$. Then:*

(a) $\mathbb{P}(\max(X_1, X_2) > x) \sim 2\overline{F}(x), \quad x \to \infty.$

(b) $\displaystyle\liminf_{x \to \infty} \frac{\overline{F^{*2}}(x)}{\overline{F}(x)} \geq 2.$

Proof By the inclusion-exclusion formula, $\mathbb{P}(\max(X_1, X_2) > x)$ equals

$$\mathbb{P}(X_1 > x) + \mathbb{P}(X_2 > x) - \mathbb{P}(X_1 > x, X_2 > x) = 2\overline{F}(x) - \overline{F}(x)^2 \sim 2\overline{F}(x),$$

proving (a). Since F is concentrated on $(0, \infty)$, we have $\{\max(X_1, X_2) > x\} \subseteq \{X_1 + X_2 > x\}$, and thus the lim inf in (b) is at least

$$\liminf \mathbb{P}(\max(X_1, X_2) > x)/\overline{F}(x) = 2. \qquad \square$$

Remark 2.2 The proof shows that the condition for $F \in \mathscr{S}$ is that the probability of the set $\{X_1 + X_2 > x\}$ is asymptotically the same as the probability of its subset $\{\max(X_1, X_2) > x\}$. That is, in the subexponential case *the only way $X_1 + X_2$ can get large is by one of the X_i becoming large*. This generalizes the study of the Pareto case in Example 1.5, cf. in particular (1.5), and should be compared to the different behavior in the light-tailed case in Examples 1.3–1.4.   ◊

Remark 2.3 As a generalization of (a), it is not difficult to show that if X_1, X_2 are independent with distribution F_i for X_i, then

$$\mathbb{P}\big(\max(X_1, X_2) > x\big) \sim \overline{F}_1(x) + \overline{F}_2(x). \qquad \qquad ◊$$

Regular variation is the simplest example of subexponentiality:

Proposition 2.4 *Any F with a regularly varying tail is subexponential. More generally, assume F_1, F_2 satisfy $\overline{F}_1(x) \sim c_1 \overline{F}_0(x)$, $\overline{F}_2(x) \sim c_2 \overline{F}_0(x)$ where $\overline{F}_0(x) = L(x)/x^\alpha$ for some slowly varying $L(x)$, some $\alpha > 0$ and some $c_1, c_2 \geq 0$ with $c = c_1 + c_2 > 0$. Then $F = F_1 * F_2$ satisfies $\overline{F}(x) \sim cL(x)/x^\alpha$.*

Proof It suffices to prove the second assertion, since the first is just the special case $c_1 = c_2 = 1$. Let $0 < \delta < 1/2$. Then the event $\{X_1 + X_2 > x\}$ is a subset of

$$\{X_1 > (1 - \delta)x\} \cup \{X_2 > (1 - \delta)x\} \cup \{X_1 > \delta x, X_2 > \delta x\}$$

(if say $X_2 \leq \delta x$, one must have $X_1 > (1 - \delta)x$ for $X_1 + X_2 > x$ to occur). Hence

$$\limsup_{x \to \infty} \frac{\overline{F}(x)}{\overline{F}_0(x)} = \limsup_{x \to \infty} \frac{\mathbb{P}(X_1 + X_2 > x)}{\overline{F}_0(x)}$$

$$\leq \limsup_{x \to \infty} \frac{\overline{F}_1((1-\delta)x) + \overline{F}_2((1-\delta)x)) + \overline{F}_1(\delta x)\overline{F}_2(\delta x)}{\overline{F}_0(x)}$$

$$= \limsup_{x \to \infty} \frac{cL((1-\delta)x)}{[(1-\delta)x]^\alpha L(x)/x^\alpha} + 0 = \frac{c}{(1-\delta)^\alpha}$$

(here we used $\overline{F}_0(\delta x) \sim \overline{F}_0(x)/\delta^\alpha$ to get $\overline{F}_1(\delta x)\overline{F}_2(\delta x)/\overline{F}_0(x) \to 0$). Letting $\delta \downarrow 0$, we get $\limsup \overline{F}(x)/\overline{F}_0(x) \leq c$, and combining with Remark 2.3 we get $\overline{F}(x)/\overline{F}_0(x) \to c$. □

For other types of distributions, a classical sufficient (and close to necessary) condition for subexponentiality is due to Pitman [142]. Recall that the hazard rate $\lambda(x)$ of a distribution F with density f is $\lambda(x) = f(x)/\overline{F}(x)$.

Proposition 2.5 *Let F have density f and hazard rate $\lambda(x)$ such that $\lambda(x)$ is decreasing for $x \geq x_0$ with limit 0 at ∞. Then $F \in \mathscr{S}$ provided*

$$\int_0^\infty e^{x\lambda(x)} f(x)\,\mathrm{d}x < \infty.$$

Proof We may assume that $\lambda(x)$ is everywhere decreasing (otherwise, replace F by a tail equivalent distribution with a failure rate which is everywhere decreasing, cf. Corollary 2.14 below). Define $\Lambda(x) = \int_0^x \lambda(y)\,\mathrm{d}y$. Then $\overline{F}(x) = e^{-\Lambda(x)}$ (standard but the proof is immediate by observing that $\lambda(x) = -\mathrm{d}/\mathrm{d}x \, \log \overline{F}(x)$), and we get

$$\frac{\overline{F^{*2}}(x)}{\overline{F}(x)} - 1 = \frac{\overline{F^{*2}}(x) - \overline{F}(x)}{\overline{F}(x)} = \frac{F(x) - F^{*2}(x)}{\overline{F}(x)}$$

$$= \int_0^x \frac{1 - F(x - y)}{\overline{F}(x)} f(y)\,\mathrm{d}y = \int_0^x \frac{\overline{F}(x - y)}{\overline{F}(x)} f(y)\,\mathrm{d}y$$

$$= \int_0^x e^{\Lambda(x)-\Lambda(x-y)-\Lambda(y)}\lambda(y)\,dy$$

$$= \int_0^{x/2} e^{\Lambda(x)-\Lambda(x-y)-\Lambda(y)}\lambda(y)\,dy \tag{2.4}$$

$$+ \int_0^{x/2} e^{\Lambda(x)-\Lambda(x-y)-\Lambda(y)}\lambda(x-y)\,dy. \tag{2.5}$$

For $y < x/2$,

$$\Lambda(x) - \Lambda(x-y) \le y\lambda(x-y) \le y\lambda(y).$$

The rightmost bound shows that the integrand in (2.4) is bounded by $e^{y\lambda(y)-\Lambda(y)}\lambda(y) = e^{y\lambda(y)}f(y)$, an integrable function by assumption. The middle bound shows that it converges to $f(y)$ for any fixed y since $\lambda(x-y) \to 0$. Thus by dominated convergence, (2.4) has limit 1. Since $\lambda(x-y) \le \lambda(y)$ for $y < x/2$, we can use the same domination for (2.5) but now the integrand has limit 0. Thus $\overline{F^{*2}}(x)/\overline{F}(x) - 1$ has limit $1 + 0$, proving $F \in \mathscr{S}$. $\qquad\square$

Example 2.6 Consider the DFR (decreasing failure rate) Weibull case $\overline{F}(x) = e^{-x^\beta}$ with $0 < \beta < 1$. Then $f(x) = \beta x^{\beta-1}e^{-x^\beta}$, $\lambda(x) = \beta x^{\beta-1}$. Thus $\lambda(x)$ is everywhere decreasing, and $e^{x\lambda(x)}f(x) = \beta x^{\beta-1}e^{-(1-\beta)x^\beta}$ is integrable. Thus, the DFR Weibull distribution is subexponential. $\qquad\diamond$

Example 2.7 In the lognormal distribution, we have by (2.1), (2.2) that $\lambda(x) \sim \log x/\sigma x$. This yields easily that $e^{x\lambda(x)}f(x)$ is integrable. Further, elementary but tedious calculations (which we omit) show that $\lambda(x)$ is ultimately decreasing. Thus, the lognormal distribution is subexponential. $\qquad\diamond$

2.2 Further Mathematical Properties

Proposition 2.8 *If $F \in \mathscr{S}$, then* $\dfrac{\overline{F}(x-y)}{\overline{F}(x)} \to 1$ *uniformly in $y \in [0, y_0]$ as $x \to \infty$.*

In terms of r.v.s: if $X \sim F \in \mathscr{S}$, then the overshoot $X - x | X > x$ converges in distribution to ∞. This follows since the probability of the overshoot to exceed y is $\overline{F}(x+y)/\overline{F}(x)$ which has limit 1. A distribution with the property $\overline{F}(x-y)/\overline{F}(x) \to 1$ is often called *long-tailed*.

Proof Consider first a fixed y. Here and in the following, we shall use several times the identity

$$\frac{\overline{F^{*(n+1)}}(x)}{\overline{F}(x)} = 1 + \frac{F(x) - F^{*(n+1)}(x)}{\overline{F}(x)} = 1 + \int_0^x \frac{1 - F^{*n}(x - z)}{\overline{F}(x)} F(dz)$$

$$= 1 + \int_0^x \frac{\overline{F^{*(n)}}(x - z)}{\overline{F}(x)} F(dz) . \tag{2.6}$$

Taking $n = 1$ and splitting the integral into two corresponding to the intervals $[0, y]$ and $(y, x]$, we get

$$\frac{\overline{F^{*2}}(x)}{\overline{F}(x)} \geq 1 + F(y) + \frac{\overline{F}(x - y)}{\overline{F}(x)}\big(F(x) - F(y)\big) .$$

If $\limsup \overline{F}(x - y)/\overline{F}(x) > 1$, we therefore get

$$\limsup \frac{\overline{F^{*2}}(x)}{\overline{F}(x)} > 1 + F(y) + 1 - F(y) = 2,$$

a contradiction. Finally, $\liminf \overline{F}(x - y)/\overline{F}(x) \geq 1$ since $y \geq 0$.

The uniformity now follows from what has been shown for $y = y_0$ and the obvious inequality

$$1 \leq \frac{\overline{F}(x - y)}{\overline{F}(x)} \leq \frac{\overline{F}(x - y_0)}{\overline{F}(x)}, \quad y \in [0, y_0] . \qquad \square$$

Corollary 2.9 *If X_1, X_2 are independent with distribution $F \in \mathscr{S}$, then the conditional distribution of $\min(X_1, X_2)$ given $X_1 + X_2 > x$ has limit F as $x \to \infty$, and the conditional probability that $\max(X_1, X_2) > x$ goes to 1.*

Proof The statement on the max was noted already in Remark 2.2. For the statement on the min, note first that

$$\mathbb{P}\big(\min(X_1, X_2) \leq z \mid X_1 + X_2 > x\big)$$

$$= \frac{1}{\mathbb{P}(X_1 + X_2 > x)} 2\,\mathbb{P}(X_1 \leq z, X_1 \leq X_2, X_1 + X_2 > x)$$

$$\sim \frac{1}{\overline{F}(x)} \int_0^z \overline{F}(y \vee (x - y))\, F(dy) . \tag{2.7}$$

Here $x - z \leq x - y \leq x$ in the integral for large x, and using the uniformity in Proposition 2.8 together with dominated convergence shows that (2.7) has limit $\int_0^z F(dy) = F(z)$. $\qquad \square$

Proposition 2.10 *Let A_1, A_2 be distributions on $(0, \infty)$ such that $\overline{A}_i(x) \sim a_i \overline{F}(x)$ for some $F \in \mathscr{S}$ and some constants a_1, a_2 with $a_1 + a_2 > 0$. Then $\overline{A_1 * A_2}(x) \sim (a_1 + a_2)\overline{F}(x)$.*

Proof Let X_1, X_2 be independent r.v.s such that X_i has distribution A_i. As in the proof of Proposition 2.1,

$$\liminf_{x \to \infty} \overline{A_1 * A_2}(x)/\overline{F}(x) \geq \liminf_{x \to \infty} \mathbb{P}\big(\max(X_1, X_2) > x\big)/\overline{F}(x)$$

$$= \liminf_{x \to \infty} \big(\overline{A}_1(x) + \overline{A}_1(x) - \overline{A}_1(x)\overline{A}_2(x)\big)/\overline{F}(x) = a_1 + a_2 - 0.$$

For the lim sup bound, define

$$C_1 = \{X_1 \leq v\}, \ C_2 = \{X_2 \leq v\},$$

$$C_3 = \{v < X_2 \leq x - v, X_1 > v\}, \ C_4 = \{X_2 > x - v, X_1 > v\}$$

cf. Fig. 1. Then

$$\mathbb{P}(X_1 + X_2 > x) \leq c_1 + c_2 + c_3 + c_4 \quad \text{where } c_i = \mathbb{P}(X_1 + X_2 > x; C_i)$$

(with equality if $x > 2v$). For any fixed v, Proposition 2.8 easily yields

$$c_1 = \int_0^v \overline{A}_2(x - y)A_1(dy) \sim a_2 \overline{F}(x)A_1(v) = a_2 \overline{F}(x)\big(1 + o_v(1)\big)$$

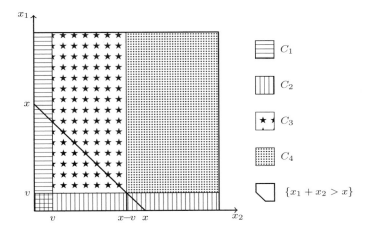

Fig. 1 The sets C_1, C_2, C_3, C_4

and similarly $c_2 = a_1 \overline{F}(x)\big(1 + o_v(1)\big)$. Also, $c_4 = \overline{F}(x - v)\overline{F}(v) = \overline{F}(x)o_v(1)$. Finally, letting $c_5 = \sup \overline{A}_1(z)/\overline{F}(z)$, $c_6 = \sup \overline{A}_2(z)/\overline{F}(z)$, we have

$$
\begin{aligned}
c_3 &= \int_v^{x-v} \overline{A}_1(x - y) A_2(dy) \ \le\ c_5 \int_v^{x-v} \overline{F}(x - y) A_2(dy) \\
&= c_5\Big[\overline{F}(x - v)\overline{A}_2(v) - \overline{A}_2(x - v)\overline{F}(v) + \int_v^{x-v} \overline{A}_2(x - y) F(dy)\Big] \\
&\le \overline{F}(x)o_v(1) + c_5 c_6 \mathbb{P}(X_1^* + X_2^* > x,\ v < X_2^* \le x - v) \ =\ \overline{F}(x)o_v(1)\,,
\end{aligned}
$$

where the third step follows by integration by parts, X_1^*, X_2^* are independent r.v.s with distribution F and the final step uses Corollary 2.9. Putting these estimates together gives

$$
\limsup_{x \to \infty} \overline{A_1 * A_2}(x)/\overline{F}(x) \ \le\ a_1 + a_2 + o_v(1)\,.
$$

The proof is completed by letting $v \uparrow \infty$. $\square$

Proposition 2.11 *If* $F \in \mathscr{S}$, *then for any* n $\overline{F^{*n}}(x)/\overline{F}(x) \ \to\ n$, $x \to \infty$.

This extension from the case $n = 2$ is often taken as the definition of the class $\mathscr{S}$; its intuitive content is the same as discussed in the case $n = 2$ above. Note that in the special case of regular variation, we can take $F_0 = F_1 = F$ and $F_2 = F^{*2}$ in Proposition 2.4 to get $\overline{F}^{*3}(x) \sim 3\overline{F}(x)$. Continuing in this manner, Proposition 2.11 follows for regular variation.

Proof The case $n = 2$ is just the definition, so taking $A_1 = F^{*2}$, $A_2 = F$ in Proposition 2.10 gives $\overline{F^{*3}}(x) \sim 3\overline{F}(x)$. Continuing in this manners gives the assertion. For an alternative self-contained proof, assume the proposition has been shown for n. Given $\epsilon > 0$, choose y such that $\big|\overline{F^{*n}}(x)/\overline{F}(x) - n\big| \le \epsilon$ for $x \ge y$. Then by (2.6),

$$
\frac{\overline{F^{*(n+1)}}(x)}{\overline{F}(x)} = 1 + \left(\int_0^{x-y} + \int_{x-y}^x\right) \frac{\overline{F^{*n}}(x - z)}{\overline{F}(x - z)} \frac{\overline{F}(x - z)}{\overline{F}(x)} F(dz)\,. \tag{2.8}
$$

Here the second integral can be bounded above by

$$
\sup_{v \ge 0} \frac{\overline{F^{*n}}(v)}{\overline{F}(v)} \frac{F(x) - F(x - y)}{\overline{F}(x)} = \sup_{v \ge 0} \frac{\overline{F^{*n}}(v)}{\overline{F}(v)} \frac{\overline{F}(x - y) - \overline{F}(x)}{\overline{F}(x)}\,, \tag{2.9}
$$

which converges to $n \cdot 0 = 0$ by Proposition 2.8 and the induction hypothesis. The first integral is bounded below and above by

$$(n \pm \epsilon) \int_0^{x-y} \frac{\overline{F}(x-z)}{\overline{F}(x)} F(\mathrm{d}z) = (n \pm \epsilon) \left(\int_0^x - \int_0^{x-y} \right) \frac{\overline{F}(x-z)}{\overline{F}(x)} F(\mathrm{d}z)$$

$$= (n \pm \epsilon) \left\{ \frac{\mathbb{P}(X_1 + X_2 > x, X_1 \le x)}{\overline{F}(x)} - \int_{x-y}^x \frac{\overline{F}(x-z)}{\overline{F}(x)} F(\mathrm{d}z) \right\}.$$

Here the first term in $\{\cdot\}$ converges to 1 by Corollary 2.9 and the second to 0 by the same argument as was used for (2.9). Combining these estimates and letting $\epsilon \downarrow 0$ shows that (2.8) has limit $n + 1$, completing the proof. $\square$

Corollary 2.12 *If $F \in \mathcal{S}$, then $\mathrm{e}^{\epsilon x} \overline{F}(x) \to \infty$, $\widehat{F}[\epsilon] = \infty$ for all $\epsilon > 0$.*

Proof For $0 < \delta < \epsilon$, we have by Proposition 2.8 that $\overline{F}(n) \ge \mathrm{e}^{-\delta} \overline{F}(n-1)$ for all large n so that $\overline{F}(n) \ge c_1 \mathrm{e}^{-\delta n}$ for all n. This implies $\overline{F}(x) \ge c_2 \mathrm{e}^{-\delta x}$ for all x, and this immediately yields the desired conclusions. $\square$

Corollary 2.13 *The class $\mathcal{S}$ is closed under tail-equivalence. That is, if $\overline{A}(x) \sim a\overline{F}(x)$ for some $F \in \mathcal{S}$ and some constant $a > 0$, then $A \in \mathcal{S}$.*

Proof Taking $A_1 = A_2 = A$, $a_1 = a_2 = a$ gives $\overline{A^{*2}}(x) \sim 2a\overline{F}(x) \sim 2a\overline{A}(x)$. $\square$

Corollary 2.14 *Let $F \in \mathcal{S}$ and let A be any distribution with a lighter tail, $\overline{A}(x) = \mathrm{o}\big(\overline{F}(x)\big)$. Then $A * F \in \mathcal{S}$ and $\overline{A * F}(x) \sim \overline{F}(x)$*

Proof Take $A_1 = A$, $A_2 = F$ so that $a_1 = 0$, $a_2 = 1$. $\square$

Exercises

2.1 Use Pitman's criterion to show that the Pareto distribution with tail $(1+x)^{-\alpha}$ is subexponential without referring to Proposition 2.5.

2.2 Show that the distribution with tail $\mathrm{e}^{-x/\log(x+\mathrm{e})}$ is subexponential.

2.3 Show that if $F \in \mathcal{S}$, then

$$\mathbb{P}(X_1 > x \mid X_1 + X_2 > x) \to \tfrac{1}{2}, \quad \mathbb{P}(X_1 \le y \mid X_1 + X_2 > x) \to \tfrac{1}{2} F(y).$$

That is, given $X_1 + X_2 > x$, the r.v. X_1 is w.p. $1/2$ 'typical' (with distribution F) and w.p. $1/2$ it has the distribution of $X_1 | X_1 > x$.

2.4 Show that if $X_1, X_2 \ge 0$ are independent and X_1 subexponential, then $\mathbb{P}(X_1 - X_2 > x) \sim \mathbb{P}(X_1 > x)$.

Notes

Good general references for subexponential distributions are Foss, Korshunov and Zachary [76] and Embrechts, Klüppelberg and Mikosch [66].

3 Large Deviations of Sums of Light-Tailed Random Variables

3.1 The Cumulant Function

Let X be an r.v. with distribution F. The *cumulant generating function* (c.g.f.) or just *cumulant function* $\kappa(\theta)$ is then defined as the logarithm of the moment generating function $\widehat{F}[\theta]$,

$$\kappa(\theta) \;=\; \log \widehat{F}[\theta] \;=\; \log \mathbb{E} e^{\theta X} \;=\; \log \int_{-\infty}^{\infty} e^{\theta x}\, F(\mathrm{d}x)\,. \tag{3.1}$$

Since $\widehat{F}[\theta]$ is well defined (possibly equal to ∞) for all θ, so is $\kappa(\theta)$. When needing to stress the dependence on X or F we write κ_X, κ_F.

Example 3.1 For the $\mathcal{N}(\mu, \sigma^2)$ case, $\kappa(\theta) = \theta\mu + \theta^2\sigma^2/2$. For X Poisson(λ), $\kappa(\theta) = \lambda(e^\theta - 1)$. For X exponential(λ), $\kappa(x) = \log\big(\lambda/(\lambda - \theta)\big)$; more generally, $\kappa(x) = \alpha \log\big(\lambda/(\lambda - \theta)\big)$ when X is gamma(α, λ). ◇

Assuming $\mathbb{E}X$, $\mathbb{V}\mathrm{ar}\, X$ to be well defined and finite, it follows immediately from $\widehat{F}'[0] = \mathbb{E}X$, $\widehat{F}''[0] = \mathbb{E}X^2$ that

$$\kappa'(0) \;=\; \mathbb{E}X, \quad \mathbb{V}\mathrm{ar}\, X \;=\; \kappa''(0)\,. \tag{3.2}$$

The term cumulant function stems from these first two and all higher-order derivatives $\kappa^{(k)}$ of κ being so-called *cumulants* or *invariants*, i.e. additive w.r.t. convolution. That is, if X_1, X_2 are independent then $\kappa^{(k)}_{X_1+X_2} = \kappa^{(k)}_{X_1} + \kappa^{(k)}_{X_2}$, as follows directly from the m.g.f. being multiplicative. The kth cumulant is defined as the kth derivative $\kappa^{(k)}(0)$ of κ at zero and is a polynomial in the first k moments. The third cumulant is just the centered third moment $\mathbb{E}\big[X - \mathbb{E}X\big]^3$, from which one could easily think from (3.2) that the kth cumulant is the centered kth moment $\mathbb{E}\big[X - \mathbb{E}X\big]^k$. However, this is not correct for $k \geq 4$, and the formulas expressing higher order cumulants in terms of moments are not particularly intuitive.

Proposition 3.2 *The c.g.f. $\kappa(\theta)$ is convex.*

Proof Let $\theta = \lambda\theta_1 + (1 - \lambda)\theta_2$ with $0 < \lambda < 1$. Writing $p = 1/\lambda$, $q = 1/(1 - \lambda)$, $Y_1 = e^{\theta_1 X/p}$, $Y_2 = e^{\theta_2 X/q}$, $Y_2 = e^{\theta_2 X/q}$, Hölder's inequality yields

$$\mathbb{E} e^{\theta X} \;=\; \mathbb{E}[Y_1 Y_2] \;\leq\; (\mathbb{E}Y_1^p)^{1/p}(\mathbb{E}Y_2^q)^{1/q} \;=\; (\mathbb{E} e^{\theta_1 X})^\lambda (\mathbb{E} e^{\theta_2 X})^{1-\lambda}\,.$$

Taking logarithms gives $\kappa(\theta) \leq \lambda\kappa(\theta_1) + (1 - \lambda)\kappa(\theta_2)$. □

It follows in particular that $\Theta = \{\theta : \kappa(\theta) < \infty\}$ is an interval.

3.2   The Legendre–Fenchel Transform

A general construction in the theory of convex functions is the *Legendre–Fenchel transform* or the *convex conjugate*, defined for $x \in \mathbb{R}$ by

$$\kappa^*(x) \; = \; \sup_{\theta \in \mathbb{R}} \big(\theta x - \kappa(\theta)\big). \qquad (3.3)$$

Since the function on the r.h.s. is 0 for $\theta = 0$, one has $\kappa^*(x) \geq 0$. Also, $\kappa^*(x)$ is convex as a supremum of affine (in particular convex) functions of x.

   An extremely important fact is that

$$x = \kappa'(\theta) \text{ with } \theta \in \Theta \quad \Rightarrow \quad \kappa^*(x) = \theta x - \kappa(\theta). \qquad (3.4)$$

See Fig. 2. This follows since for such a θ, the derivative of $\theta x - \kappa(\theta)$ is 0 so that θ must be the maximizer. (3.4) is the standard vehicle for computing $\kappa^*(x)$: simply solve $\kappa'(\theta(x)) = x$ for $\theta(x)$ and compute $\theta(x)x - \kappa(\theta(x))$ (complications may arise if x is one of the endpoints of the interval $\{\kappa'(\theta) : \theta \in \Theta\}$).

Example 3.3 For the standard normal case $\kappa(\theta) = \theta^2/2$, we get $\kappa'(\theta) = \theta$, so $\theta(x) = x$ and $\kappa^*(x) = x^2 - x^2/2 = x^2/2$. $\diamond$

Exercises

3.1 Verify that $\kappa^*(x) = \lambda x - 1 - \log(\lambda x)$ when X is exponential(λ).

3.2 Verify that $\kappa^*(x) = \lambda - x + x \log(x/\lambda)$ when X is Poisson(λ).

3.3 Show that if $\widetilde{X} = aX + b$, then $\widetilde{\theta}(x) = \theta(x)/a - b$. Compute thereby $\kappa^*(x)$ for the $\mathcal{N}(\mu, \sigma^2)$ distribution.

3.4 Verify that the cumulants of order 1,2,3 are $\mathbb{E}X$, $\mathbb{V}\text{ar}\, X$ and $\mathbb{E}\big[X - \mathbb{E}X\big]^3$.

3.5 Find $\kappa^*(x)$ and $\{x : \kappa^*(x) < \infty$ in the Bernoulli(p) distribution.

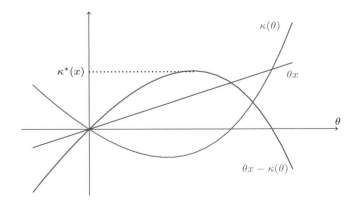

Fig. 2 $\kappa^*(x)$ as a supremum

3.3 Exponential Families and Change of Measure

Now let $X_1, X_2, \ldots$ be i.i.d. r.v.s with common distribution F and c.g.f. $\kappa(\cdot)$. For each $\theta \in \Theta = \{\theta \in \mathbb{R} : \kappa(\theta) < \infty\}$, we denote by F_θ the probability distribution with density $e^{\theta x - \kappa(\theta)}$ w.r.t. F. In standard statistical terminology, $(F_\theta)_{\theta \in \Theta}$ is the *exponential family generated by* F (it was encountered already in Sect. II.2). Similarly, $\mathbb{P}_\theta$ denotes the probability measure w.r.t. which $X_1, X_2, \ldots$ are i.i.d. with common distribution F_θ.

Proposition 3.4 *Let $\kappa_\theta(\alpha) = \log \mathbb{E}_\theta e^{\alpha X_1}$ be the c.g.f. of F_θ. Then*

$$\kappa_\theta(\alpha) = \kappa(\alpha + \theta) - \kappa(\theta), \quad \mathbb{E}_\theta X_1 = \kappa'(\theta), \quad \mathrm{Var}_\theta X_1 = \kappa''(\theta).$$

Proof The formula for $\kappa_\theta(\alpha)$ follows from

$$e^{\kappa_\theta(\alpha)} = \int_{-\infty}^{\infty} e^{\alpha x} F_\theta(dx) = \int_{-\infty}^{\infty} e^{(\alpha + \theta)x - \kappa(\theta)} F(dx) = e^{\kappa(\alpha + \theta) - \kappa(\theta)}.$$

We then get $\mathbb{E}_\theta X_1 = \kappa_\theta'(0) = \kappa'(\theta)$, $\mathrm{Var}_\theta X_1 = \kappa_\theta''(0) = \kappa''(\theta)$. □

Example 3.5 Let F be the normal distribution with mean μ and variance σ^2. Then $\Theta = \mathbb{R}$, $\kappa(\alpha) = \mu\alpha + \sigma^2\alpha^2/2$ so that

$$\kappa_\theta(\alpha) = \mu(\alpha + \theta) + \sigma^2(\alpha + \theta)^2/2 - \mu\theta - \sigma^2\theta^2/2 = (\mu + \theta\sigma^2)\alpha + \sigma^2\alpha^2/2,$$

which shows that F_θ is the normal distribution with mean $\mu + \theta\sigma^2$ and the same variance σ^2. ◇

Example 3.6 Let F be the exponential distribution with rate λ. Then $\Theta = (-\infty, \lambda)$, $\kappa(\alpha) = \log \lambda - \log(\lambda - \alpha)$, which gives $\kappa_\theta(\alpha) = \log(\lambda - \theta) - \log(\lambda - \theta - \alpha)$. That is, F_θ is the exponential distribution with rate $\lambda - \theta$. ◇

If $\mathbb{E}_\theta$ is the expectation operator corresponding to $\mathbb{P}_\theta$, then for any fixed n

$$\mathbb{E}_\theta f(X_1, \ldots, X_n) = \mathbb{E}\big[e^{\theta S_n - n\kappa(\theta)} f(X_1, \ldots, X_n)\big] \tag{3.5}$$

for all measurable $f : \mathbb{R}^n \to \mathbb{R}$ which are bounded or nonnegative, where $S_n = X_1 + \cdots + X_n$. This follows since the l.h.s. of (3.5) is

$$\int_{\mathbb{R}^n} f(x_1, \ldots, x_n) F_\theta(dx_1) \cdots F_\theta(dx_n)$$

$$= \int_{\mathbb{R}^n} f(x_1, \cdots, x_n) e^{\theta x_1 - \kappa(\theta)} F(dx_1) \ldots e^{\theta x_n - \kappa(\theta)} F(dx_n)$$

$$= \int_{\mathbb{R}^n} f(x_1, \ldots, x_n) e^{\theta S_n - n\kappa(\theta)} F(dx_1) \cdots F(dx_n),$$

which is the same as the r.h.s. (here $s_n = x_1 + \cdots + x_n$). Replacing first $f(x_1, \ldots, x_n)$ by $e^{-\theta s_n + n\kappa(\theta)} f(x_1, \ldots, x_n)$ and specializing next to an indicator function of a Borel set $A \subseteq \mathbb{R}^n$, we get

$$\mathbb{E} f(X_1, \ldots, X_n) = \mathbb{E}_\theta \big[e^{-\theta S_n + n\kappa(\theta)} f(X_1, \ldots, X_n) \big], \tag{3.6}$$

$$\mathbb{P}(A) = \mathbb{E}_\theta \big[e^{-\theta S_n + n\kappa(\theta)}; \, A \big]. \tag{3.7}$$

Thus we have expressed $\mathbb{P}(A)$ as an expectation corresponding to an i.i.d. sum with a changed increment distribution. At first sight this appears to be a complicated way of evaluating $\mathbb{P}(A)$. The point is that in a number of cases the F_θ-distribution has more convenient properties than F, as will be demonstrated by a number of applications throughout the chapter.

3.4 The Chernoff Bound and the Saddlepoint Approximation

Recall the definition of the Legendre–Fenchel transform $\kappa^*(\cdot)$ from Sect. 3.2. One has $\kappa^*(x) = \theta(x)x - \kappa\big(\theta(x)\big)$, where $\theta(x)$ is the solution of

$$x = \kappa'\big(\theta(x)\big) = \mathbb{E}_{\theta(x)} X_1 \tag{3.8}$$

whenever x is an interior point of the interval where $\kappa^*(x) < \infty$.

Theorem 3.7 *Let $x > \kappa'(0) = \mathbb{E} X_1$ and assume that (3.8) has a solution $\theta = \theta(x)$. Then*

$$\mathbb{P}(S_n > nx) \le e^{-n\kappa^*(x)}, \tag{3.9}$$

$$\frac{1}{n} \log \mathbb{P}(S_n > nx) \to -\kappa^*(x), \quad n \to \infty, \tag{3.10}$$

$$\mathbb{P}(S_n > nx) \sim \frac{1}{\theta \sqrt{2\pi \sigma_\theta^2 n}} e^{-n\kappa^*(x)}, \quad n \to \infty, \tag{3.11}$$

provided in addition for (3.10) that $\sigma_\theta^2 = \kappa''(\theta) < \infty$ and for (3.11) that $|\kappa'''(\theta)| < \infty$ and that F satisfies Cramér's condition (C).

Condition (C) is an analytical regularity condition on the characteristic function of F and states that $\lim_{s \to \pm\infty} \big| \mathbb{E} e^{\mathrm{i}s X} \big| = 0$. Before giving the proof, we add some remarks.

Remark 3.8 The inequality (3.9) goes under the name of the *Chernoff bound*, (3.11) is the *saddlepoint approximation*, and (3.10) (and some extensions to $\mathbb{P}(S_n \in A)$ for more general sets than half-lines) is associated with the name of Cramér. See further Sect. XIII.4.1.   ◇

Remark 3.9 The asymptotics in (3.10) are implied by those in (3.11) so that (3.10) is a weaker statement. The logarithmic form is typical for the large deviations area and essentially captures the rough first order. For example, (3.10) would be equally well compatible with the asymptotic form $18 \log x \exp\{-n\kappa^*(x) + n^{1/3}x^4\}$ of $\mathbb{P}(S_n > x)$ as with the correct answer (3.11). ◇

Proof of (3.9) Taking $A = \{S_n > nx\}$, $\theta = \theta(x)$ in (3.7), we get

$$\mathbb{P}(S_n > nx) = \mathrm{e}^{-n\kappa^*(x)} \mathbb{E}_\theta\big[\mathrm{e}^{-\theta(S_n - nx)}; \ S_n > nx\big]. \tag{3.12}$$

Since $\theta > 0$ and $S_n - nx > 0$ on A, (3.9) follows by bounding the indicator of A by 1. □

Proof of (3.10) When $\sigma_\theta^2 < \infty$, the CLT yields

$$\mathbb{P}_\theta\big(nx < S_n < nx + 1.96\sigma_\theta\sqrt{n}\big) \ \to \ \Phi(1.96) - \Phi(0) = 0.475.$$

Hence for large n,

$$\mathbb{P}(S_n/n > x) \geq \mathrm{e}^{-n\kappa^*(x)} \mathbb{E}_\theta\Big[\mathrm{e}^{-\theta(S_n - nx)}; \ nx < S_n < nx + 1.96\sigma_\theta\sqrt{n}\Big]$$

$$\geq \mathrm{e}^{-n\kappa^*(x)} \cdot 0.4\,\mathrm{e}^{-1.96\theta\sigma_\theta\sqrt{n}}.$$

Combining with (3.9) and taking logarithms, (3.10) follows. □

Proof of (3.11) That $|\kappa'''(\theta)| < \infty$ as required for (3.11) is equivalent to the third moment assumption $\mathbb{E}_\theta|X|^3 < \infty$. This together with (C) is needed in the proof to ensure a strong version of the local central limit theorem, more precisely that $\mathbb{E}_\theta\big[\mathrm{e}^{-\theta(S_n - nx)}; \ S_n > nx\big]$ can be calculated asymptotically as if the $\mathbb{P}_\theta$-distribution $(S_n - nx)/\sqrt{n}$ was exactly $\mathcal{N}(0, \sigma_\theta^2)$. That is,

$$\mathbb{E}_\theta\big[\mathrm{e}^{-\theta(S_n - nx)}; \ S_n > nx\big] \ \sim \ \frac{1}{\sqrt{2\pi n}} \int_0^\infty \mathrm{e}^{-\theta\sigma_\theta y}\mathrm{e}^{-y^2/2n}\,\mathrm{d}y. \tag{3.13}$$

For a more detailed explanation and references, see [6, p. 355–6]. The reason why the standard CLT does not suffice is that $\mathbb{E}_\theta\big[\mathrm{e}^{-\theta(S_n - nx)}; \ S_n > nx\big]$ cannot be written as a bounded continuous function of $(S_n - nx)/\sqrt{n}$.

Assuming (3.13), now just note that by dominated convergence the r.h.s. of (3.13) is asymptotically

$$\frac{1}{\sqrt{2\pi n}} \int_0^\infty \mathrm{e}^{-\theta\sigma_\theta y}\,\mathrm{d}y = \frac{(\theta\sigma_\theta)^{-1}}{\sqrt{2\pi n}}. \qquad\qquad □$$

Exercises

3.6 Identify the form of F_θ when F is binomial(n, p).

3.7 Identify the form of F_θ when F is Poisson(λ).

4 Tails of Sums of Light-Tailed Random Variables

We consider here the asymptotics of $\overline{F^{*n}}(x) = \mathbb{P}(S_n > x)$ and the corresponding density $f^{*n}(x)$ for an i.i.d. light-tailed sum as x goes to ∞ with n fixed. This is similar to the set-up for the subexponential limit theory developed in Sect. 2, but differs from the large deviations set-up of Sect. 3 where n and x went to ∞ at the same time. The theory which has been developed applies to distributions with a density or tail close to e^{-cx^β} where $\beta \geq 1$.

The cases $\beta = 1$ and $\beta > 1$ are intrinsically different (of course, $\beta < 1$ leads to a heavy tail). The difference may be understood by looking at two specific examples from Sect. 1 with $n = 2$. The first is the exponential distribution in Example 1.3 where $\beta = 1$, which shows that if $X_1 + X_2 > x$, then approximately X_1, X_2 are both uniform between 0 and x, with the joint distribution concentrating on the line $x_1 + x_2 = x$. On the other hand, for the normal distribution where $\beta = 2$, Example 1.4 shows that X_1, X_2 are both approximately equal to $x/2$ if $X_1 + X_2 > x$.

For ease of exposition, we make some assumptions that are simplifying but not all crucial. In particular, we take F to be concentrated on $(0, \infty)$ and having a density $f(x)$, and exemplify 'close to' by allowing a modifying regularly varying prefactor to e^{-cx^β}. For the density, this means

$$f(x) \sim L(x)x^{\alpha+\beta-1}e^{-cx^\beta}, \qquad x > 0, \tag{4.1}$$

with $L(\cdot)$ slowly varying. However, in the rigorous proofs we only take $n = 2$ and $L(x) \equiv d$ will be constant. We shall need the following simple lemma (cf. Exercise 4.1):

Lemma 4.1 *If* (4.1) *holds with* $\beta \geq 1$, *then* $\overline{F}(x) = \mathbb{P}(X > x) \sim L(x)x^\alpha e^{-cx^\beta}/c\beta$.

The case $\beta = 1$ of a close-to-exponential tail is easy:

Proposition 4.2 *Assume* $f(x) \sim L(x)x^{\alpha-1}e^{-cx}$ *with* $\alpha > 0$. *Then the density and the tail of an i.i.d. sum satisfy*

$$f^{*n}(x) \sim \frac{\Gamma(\alpha)^n}{\Gamma(n\alpha)}L(x)^n x^{n\alpha-1}e^{-cx}, \tag{4.2}$$

$$\overline{F^{*n}}(x) = \mathbb{P}(S_n > x) \sim \frac{\Gamma(\alpha)^n}{c\Gamma(n\alpha)}L(x)^n x^{n\alpha-1}e^{-cx}. \tag{4.3}$$

Proof By Lemma 4.1, it suffices to show (4.2). We give the proof for the case $n = 2$ only where

$$f^{*2}(x) = \int_0^x f(z)f(x-z)\,dz \tag{4.4}$$

and also take $L(x) \equiv d$ for simplicity. Denote by $\tilde{f}^{*2}(x)$ the rh.s. of (4.2) and write $\log_2 x = \log \log x$. Then

$$\mathbb{P}(X_1 > x - \log_2 x) \ \sim \ dx^{\alpha-1}(\log x)^c e^{-cx} \ = \ o\big(\tilde{f}^{*2}(x)\big)$$

by Lemma 4.1 since $\alpha > 0$, so the contribution to $f^{*2}(x)$ from the event $X_1 > x - \log_2 x$ can be neglected. The same is true for the event $X_1 < \log_2 x$ since if that occurs, we can only have $X_1 + X_2 = x$ if $X_2 > x - \log_2 x$. It thus suffices to show that

$$\int_{\log_2 x}^{x-\log_2 x} f(z)f(x-z)\,\mathrm{d}z \ \sim \ \tilde{f}^{*2}(x).$$

Since here $z \geq \log_2 x$ and $x - z \geq \log_2 x$ and $\log_2 x \to \infty$, the asymptotics of the l.h.s. is the same as that of

$$\int_{\log_2 x}^{x-\log_2 x} d^2 z^{\alpha-1} e^{-cz}(x-z)^{\alpha-1} e^{-c(x-z)}\,\mathrm{d}z.$$

But this equals

$$d^2 e^{-cx} x^{2\alpha-2} \int_{\log_2 x/x}^{1-\log_2 x/x} y^{\alpha-1}(1-y)^{\alpha-1} x\,\mathrm{d}y$$

$$\sim \ d^2 x^{2\alpha-1} e^{-cx}\,\mathrm{Beta}(\alpha,\alpha) \ = \ d^2 x^{2\alpha-1} e^{-cx}\frac{\Gamma(\alpha)^2}{\Gamma(2\alpha)} \ = \ \tilde{f}^{*2}(x).$$

$\square$

Tails that are lighter than exponential present more substantial difficulties. The main example is $\overline{F}(x)$ being close to a Weibull tail e^{-cx^β} with $\beta > 1$, and we give the rigorous proof of some results in that setting in Proposition 4.3 below. First, however, we give some intuition for a slightly more general case, $f(x) = h(x)e^{-\Lambda(x)}$, with h in some sense much less variable than Λ. Assume that Λ is such that the same principle applies as for the normal distribution, that if $X_1 + X_2 > x$, then X_1, X_2 are both close to $x/2$. Then the main contribution to the convolution integral (4.4) comes from a neighborhood of $z = x/2$ where

$$\Lambda(z) \ \approx \ \Lambda(x/2) + \Lambda'(x/2)(z - x/2) + \frac{\Lambda''(x/2)}{2}(z - x/2)^2,$$

$$\Lambda(x - z) \ \approx \ \Lambda(x/2) + \Lambda'(x/2)(x/2 - z) + \frac{\Lambda''(x/2)}{2}(x/2 - z)^2. \tag{4.5}$$

Also, the assumption on h indicates that h is roughly constant in this neighborhood. Noting that the $\Lambda'(x/2)$ terms in (4.5) cancel when adding, the expression (4.4) for

$f^{*2}(x)$ therefore approximately should equal

$$h(x/2)^2 \int \exp\left\{-2\Lambda(x/2) - \Lambda''(x/2)(z - x/2)^2\right\}$$

$$= h(x/2)^2 \exp\left\{-2\Lambda(x/2)\right\}\sqrt{\frac{\pi}{\Lambda''(x/2)}} \,.$$

Here is a precise result in that direction:

Proposition 4.3 *Assume that* $f(x) \sim L(x)x^{\alpha+\beta-1}e^{-cx^\beta}$ *with* $\beta > 1$. *Then the density and the tail of an i.i.d. sum satisfy*

$$f^{*n}(x) \sim L(x/n)^n k_n x^{\alpha_n+\beta-1}e^{-c_n x^\beta}, \tag{4.6}$$

$$\overline{F^{*n}}(x) = \mathbb{P}(S_n > x) \sim L(x/n)^n k_n x^{\alpha_n}e^{-c_n x^\beta}/c_n\beta, \tag{4.7}$$

where $c_n = c/n^{\beta-1}$, $\alpha_n = n\alpha + (n+1)\beta/2$ *and*

$$k_n = \left[\frac{\pi}{\beta(\beta - 1)c}\right]^{(n-1)/2} n^{-n\alpha-(n+1)\beta/2+1}. \tag{4.8}$$

Proof We give again the proof for the case $n = 2$ only and take $L(x) \equiv d$, $c = 1$. Lemma 4.1 gives in particular that $\overline{F}(x) \sim dx^\alpha e^{-x^\beta}/\beta$ and also shows that (4.7) follows from (4.6), so it suffices to show (4.6).
  Denote by

$$\tilde{f}^{*2}(x) = d^2\sqrt{\frac{\pi}{\beta(\beta - 1)}}\frac{1}{2^{2\alpha+3\beta/2-1}}x^{2\alpha+3\beta/2-1}e^{-x^\beta/2^{\beta-1}}$$

the rh.s. of (4.6) and choose $1/2 < a < 1$ such that $a^\beta > c_2 = 1/2^{\beta-1}$ (this is possible since $\beta > 1$). Then

$$\mathbb{P}(X_1 > ax) \sim d(ax)^\alpha e^{-a^\beta x^\beta}/\beta = o\left(\tilde{f}^{*2}(x)\right)$$

by Lemma 4.1. The same argument as in the proof of Proposition 4.2 therefore gives that the contribution to $f^{*2}(x)$ from the events $X_1 > ax$ and $X_1 < (1 - a)x$ can be neglected. It thus suffices to show that

$$\int_{(1-a)x}^{ax} f(z)f(x - z)\,dz \sim \tilde{f}^{*2}(x). \tag{4.9}$$

Since $z \geq (1-a)x$ and $x - z \geq ax$ in (4.9) are both bounded below by $(1-a)x$, which goes to ∞, the asymptotics of the l.h.s. is the same as that of

$$\int_{(1-a)x}^{ax} d^2 z^{\alpha+\beta-1}(x-z)^{\alpha+\beta-1} \exp\left\{-z^\beta - (x-z)^\beta\right\} dz. \tag{4.10}$$

We now use the substitution $z \to v$ given by

$$z = \frac{x}{2}(1 + v/x^{\beta/2}) \ \text{ or equivalently } x - z = \frac{x}{2}(1 - v/x^{\beta/2})$$

together with the expansion

$$(1+h)^\beta = 1 + h\beta + \frac{h^2}{2}\beta(\beta-1)\omega(h) = 1 + h\beta + \frac{h^2 2^\beta}{4\sigma^2}\omega(h),$$

where $1/\sigma^2 = 2^{1-\beta}\beta(\beta-1)$, $\omega(h) \to 1$ as $h \to 0$ and $\omega(h)$ has the form $(1+h^*)^{\beta-2}$, for some h^* between 0 and h. We can then write $z^\beta + (x-z)^\beta$ as

$$\frac{x^\beta}{2^\beta}\left(1 + \frac{v}{x^{\beta/2}} + \frac{v^2 2^\beta}{4\sigma^2 x^\beta}\omega(v/x^{\beta/2})\right) + \frac{x^\beta}{2^\beta}\left(1 - \frac{v}{x^{\beta/2}} + \frac{v^2 2^\beta}{4\sigma^2 x^\beta}\omega(v/x^{\beta/2})\right)$$

$$= \frac{x^\beta}{2^{\beta-1}} + \frac{v^2}{2\sigma^2}\delta(v/x^{\beta/2}),$$

where $\delta(h) = (\omega(h) + \omega(-h))/2$, so that (4.10) takes the form

$$\frac{d^2}{2^{2\alpha+2\beta-2}}x^{2\alpha+2\beta-2}\exp\left\{-\frac{x^\beta}{2^{\beta-1}}\right\}$$

$$\times \int_{(1/2-a)x^{\beta/2}}^{(a-1/2)x^{\beta/2}}\left(1 - \frac{v^2}{x^\beta}\right)^{\alpha+\beta-1}\exp\left\{-\frac{v^2}{2\sigma^2}\delta(v/x^{\beta/2})\right\}\frac{1}{2}x^{1-\beta/2}\,dv.$$

$$\tag{4.11}$$

When z varies between $(1-a)x$ and ax, $v/x^{\beta/2}$ varies between $1/2-a$ and $a-1/2$. But the choice $a < 1$ then ensures that $\delta(v/x^{\beta/2})$ is bounded in this range and goes to 1 as $x \to \infty$. Similar estimates for the powers allows us to use dominated convergence to conclude that the asymptotics of (4.11) and hence (4.10) is given by

$$\frac{d^2}{2^{2\alpha+2\beta-1}}x^{2\alpha+3\beta/2-1}\exp\left\{-\frac{x^\beta}{2^{\beta-1}}\right\}\int_{-\infty}^{\infty}\exp\left\{-\frac{v^2}{2\sigma^2}\right\}dv$$

$$= \frac{d^2\sqrt{2\pi\sigma^2}}{2^{2\alpha+2\beta-1}}x^{2\alpha+3\beta/2-1}\exp\left\{-\frac{x^\beta}{2^{\beta-1}}\right\}$$

$$= \frac{d^2}{2^{2\alpha+3\beta/2-1}}\left(\frac{\pi\,2^\beta}{\beta(\beta-1)}\right)^{1/2}x^{2\alpha+3\beta/2-1}\exp\left\{-\frac{x^\beta}{2^{\beta-1}}\right\},$$

which is the same as $\tilde{f}^{*2}(x)$.   □

Exercises

4.1 Prove Lemma 4.1.

4.2 Verify that Proposition 4.3 gives the correct result if F is standard normal.

Notes

The key reference for Proposition 4.3 is Balkema, Klüppelberg and Resnick [18]. A somewhat more elementary survey along the present lines is in Asmussen et al. [14]. The extension from $n = 2$ to $n > 2$ can be carried out by first deriving analogues of Propositions 4.2, 4.3 with different c, L, α, and then using induction. Keeping track of the constants is messy, as may be guessed from the complicated expressions (4.8)! The more advanced argument in [18] uses convex analysis and some non-standard central limit theory. Section 8 of [14] also contains some references for the more elementary Proposition 4.3.

5 Aggregate Claims and Compound Sums: Generalities

Let A be the total amount of claims to be covered by an insurance company in a given time period (say one year). The company will possess statistics to estimate basic quantities like $\mu_A = \mathbb{E}A$ and will have made sure that the premiums p received during the same time period exceed μ_A with some margin covering administration costs, profits to shareholders, etc. Nevertheless, due to the random nature of A, one is faced with an uncertainty as to whether in fact A is much larger than μ_A, and what is the probability that such unfortunate events occur. The topic of this and the following section is to present methods for quantitative evaluations of risks of this type.

Risk evaluation, as for any kind of probability calculation, requires a model. We will assume that $A = V_1 + \cdots + V_N$, where N is the total number of claims, V_n is the size of the nth claim, and that the V_n are i.i.d. with common distribution F and independent of N, an $\mathbb{N}$-valued r.v. with point probabilities $p_n = \mathbb{P}(N = n)$. The problem is to say something about $\mathbb{P}(A > x)$ for large x under these assumptions. Often the purpose of this is VaR (Value-at-Risk) calculations, that is, the evaluation of α-quantiles $q_\alpha(A)$, say for $\alpha = 95 , 97.5$ or 99%.

We will start with some elementary analytical formulas for transforms, moments, etc. For notation and terminology, see Sect. A.11 in the Appendix.

Proposition 5.1 *The m.g.f. of A is $\widehat{p}\big[\widehat{F}[\theta]\big]$, where $\widehat{p}[z] = \mathbb{E}z^N$ is the p.g.f. of N and $\widehat{F}[\theta] = \mathbb{E}e^{\theta V}$ the m.g.f. of F. Further,*

$$\mu_A = \mu_N \mu_V , \quad \sigma_A^2 = \sigma_N^2 \mu_V^2 + \mu_N \sigma_V^2 ,$$

where $\mu_A = \mathbb{E}A$, $\sigma_V^2 = \mathbb{V}\text{ar } V$, etc.

Proof The expression for the m.g.f. follows from

$$\mathbb{E}e^{\theta A} = \sum_{n=0}^{\infty} \mathbb{P}(N = n)\mathbb{E}\big[e^{\theta A} \mid N = n\big] = \sum_{n=0}^{\infty} p_n \mathbb{E}e^{\theta(V_1 + \cdots + V_n)}$$

$$= \sum_{n=0}^{\infty} p_n \widehat{F}[\theta]^n = \widehat{p}\big[\widehat{F}[\theta]\big].$$

Differentiating, we obtain the first derivative of the m.g.f. as $\widehat{p}'\big[\widehat{F}[\theta]\big]\widehat{F}'[\theta]$, and letting $\theta = 0$ we get

$$\mu_A = \widehat{p}'\big[\widehat{F}[0]\big]\widehat{F}'[0] = \widehat{p}'[1]\mu_V = \mu_N \mu_V.$$

The second moment (and thereby $\mathbb{V}\mathrm{ar}\ A$) can be found in the same way by one more differentiation. Alternatively,

$$\mathbb{V}\mathrm{ar}\ A = \mathbb{V}\mathrm{ar}\big(\mathbb{E}[A \mid N]\big) + \mathbb{E}\big(\mathbb{V}\mathrm{ar}[A \mid N]\big) = \mathbb{V}\mathrm{ar}(N\mu_V) + \mathbb{E}(N\sigma_V^2)$$

$$= \sigma_N^2 \mu_V^2 + \mu_N \sigma_V^2.$$

$$\square$$

Exercises
5.1 Find the central third moment $\mathbb{E}(A - \mathbb{E}A)^3$ in terms of the moments of N and the V_i.

5.1 Poisson Compounding

In Chap. IV, we will meet geometric compounding, i.e. $p_n = (1 - \rho)\rho^n$, as a technical tool in ruin theory. However, by far the most important case from the point of view of aggregate claims is Poisson compounding, $p_n = e^{-\lambda}\lambda^n/n!$. As explained in Sect. I.5, the Poisson assumption can be motivated in a number of different, though often related, ways. For example:

1. Assume that the number of policy holders is M, and that the mth produces claims according to a Poisson process with rate (intensity) λ_m. In view of the interpretation of the Poisson process as a model for events which occur in an 'unpredictable' and time-homogeneous way, this is a reasonable assumption for example in car insurance or other types of accident insurance. If further the policy holders behave independently, then the total claims process (the superposition of the individual claim processes) is again Poisson, now with rate $\lambda = \lambda_1 + \cdots + \lambda_M$. In particular, if N is the number of claims in a time period of unit length, then N has a Poisson(λ) distribution.

2. Assume again that the number of policy holders is M, and that the number of claims produced by each during the given time period is 0 or 1, w.p. p for 1. If p is small and M is large, the law of small numbers (Theorem 5.1) therefore implies that N has an approximate Poisson(λ) distribution where $\lambda = Mp$.

In the Poisson case, $\widehat{p}[z] = e^{\lambda(z-1)}$, and therefore the c.g.f. of A is (replace z by $\widehat{F}[\theta]$, cf. Proposition I.5.1, and take logarithms)

$$\kappa_A(\theta) = \lambda(\widehat{F}[\theta] - 1). \qquad (5.1)$$

Further, Proposition 5.1 yields

$$\mu_A = \lambda\mu_V, \quad \sigma_A^2 = \lambda\sigma_V^2 + \lambda\mu_V^2 = \lambda\mathbb{E}V^2. \qquad (5.2)$$

As a first application of (5.1), we shall show:

Proposition 5.2 *As $\lambda \to \infty$ with F fixed, $(A - \mu_A)/\sigma_A \xrightarrow{D} \mathcal{N}(0, 1)$.*

Proof We must show that the c.g.f. of $(A - \mu_A)/\sigma_A$ converges to the c.g.f. of the standard normal distribution, i.e. that

$$\log \mathbb{E}\exp\{\theta(A - \mu_A)/\sigma_A\} \to \theta^2/2. \qquad (5.3)$$

Using

$$\widehat{F}[\alpha] = 1 + \alpha\mu_V + \alpha^2\mathbb{E}V^2/2 + o(\alpha^2), \quad \alpha \to 0,$$

$\theta/\sigma_A = O(\lambda^{-1/2})$, and (5.3), the l.h.s. of (5.3) becomes

$$\lambda(\widehat{F}[\theta/\sigma_A] - 1) - \theta\mu_A/\sigma_A$$
$$= \lambda(\mu_V\theta/\sigma_A + \mathbb{E}V^2\theta^2/2\sigma_A^2 + o(1/\lambda)) - \theta\mu_A/\sigma_A.$$

Inserting (5.2), the r.h.s. reduces to $\theta^2/2 + o(1)$.   □

Proposition 5.2 suggests the following approximations:

$$\mathbb{P}(A > x) \approx 1 - \Phi((x - \mu_A)/\sigma_A), \qquad (5.4)$$

$$q_\alpha(A) \approx \mu_A + \sigma_A q_{1-\alpha}(\Phi), \qquad (5.5)$$

where $q_{1-\alpha}(\Phi)$ is the $1 - \alpha$-quantile in the normal distribution. To arrive at (5.5) from (5.4), set the r.h.s. of (5.4) equal to α and solve for $x = q_\alpha(A)$. In Sect. 7.3, we will derive refinements of these approximations based upon higher order expansions in the CLT.

Remark 5.3 A word of warning should be said right away: the CLT can only be expected to provide a good fit in the center of the distribution. Thus, it is quite questionable to use (5.4) for the case of main interest, large x, and (5.5) for α close to 1. Similar remarks apply to the refinements of Sect. 7.3. $\Diamond$

Explicit distributions of compound Poisson sums are basically only known in the following case:

Example 5.4 Assume that V is exponential with rate μ. Then $V_1 + \cdots + V_n$ is gamma(n, μ) and so the density of A on $(0, \infty)$ is

$$f_A(x) = \sum_{n=1}^{\infty} e^{-\lambda} \frac{\lambda^n}{n!} \frac{\mu^n x^{n-1}}{(n-1)!} e^{-\mu x} . \tag{5.6}$$

Recall that the *modified Bessel function of order* $k \in \mathbb{N}$ is defined by

$$I_k(z) = (z/2)^k \sum_{n=0}^{\infty} \frac{(z^2/4)^n}{n!(n+k)!} .$$

Letting $z^2/4 = \lambda \mu x$, it follows that (5.6) can be written as

$$\frac{\lambda \mu}{x} e^{-\lambda - \mu x} \sum_{n=0}^{\infty} \frac{(z^2/4)^n}{n!(n+1)!} = \frac{\lambda \mu}{xz/2} e^{-\lambda - \mu x} I_1(z)$$

$$= \frac{\lambda \mu}{x^{3/2}\sqrt{\lambda \mu}} e^{-\lambda - \mu x} I_1\left(2\sqrt{\lambda \mu x}\right) .$$

$\Diamond$

6   Panjer's Recursion

Consider $A = \sum_{i=1}^{N} V_i$, let $p_n = \mathbb{P}(N = n)$, and assume that there exist constants a, b such that

$$p_n = \left(a + \frac{b}{n}\right) p_{n-1}, \quad n = 1, 2, \ldots . \tag{6.1}$$

For example, this holds with $a = 0$, $b = \beta$ for the Poisson distribution with rate β since

$$p_n = e^{-\beta} \frac{\beta^n}{n!} = \frac{\beta}{n} e^{-\beta} \frac{\beta^{n-1}}{(n-1)!} = \frac{\beta}{n} p_{n-1} .$$

Proposition 6.1 *Assume that F is concentrated on $\{0, 1, 2, \ldots\}$ and write $f_j = \mathbb{P}(V_i = j)$, $j = 1, 2, \ldots$, $g_j = \mathbb{P}(A = j)$, $j = 0, 1, \ldots$. Then $g_0 = \sum_0^\infty f_0^n p_n$ and*

$$g_j = \frac{1}{1 - a f_0} \sum_{k=1}^{j} \left(a + b\frac{k}{j} \right) f_k g_{j-k}, \quad j = 1, 2, \ldots. \tag{6.2}$$

In particular, if $f_0 = 0$, then

$$g_0 = p_0, \quad g_j = \sum_{k=1}^{j} \left(a + b\frac{k}{j} \right) f_k g_{j-k}, \quad j = 1, 2, \ldots. \tag{6.3}$$

Remark 6.2 The crux of Proposition 6.1 is that the algorithm is much faster than the naive method, which would consist in noting that

$$g_j = \sum_0^\infty p_n f_j^{*n}, \tag{6.4}$$

where f^{*n} is the nth convolution power of f. In particular, in the case $f_0 = 0$

$$g_j = \sum_{n=1}^{j} p_n f_j^{*n} \tag{6.5}$$

(this is valid not only for $j > 0$ but also for $j = 0$ when $f_0 = 0$, since then the sum is empty and hence 0 by convention), and one then calculates the f_j^{*n} recursively by $f_j^{*1} = f_j$,

$$f_j^{*n} = \sum_{k=1}^{j-1} f_k^{*(n-1)} f_{j-k}. \tag{6.6}$$

It follows that the complexity (number of arithmetic operations required) is $O(j^3)$ for (6.5), (6.6) but only $O(j^2)$ for Proposition 6.1. $\diamond$

Proof of Proposition 6.1 The expression for g_0 is obvious. By symmetry (exchangeability; see Proposition A.5 for a formal proof),

$$\mathbb{E}\left[a + b\frac{V_i}{j} \;\Big|\; \sum_{i=1}^{n} V_i = j \right] \tag{6.7}$$

is independent of $i = 1, \ldots, n$. Since the sum over i is $na + b$, the value of (6.7) is therefore $a + b/n$. Hence by (6.1), (6.5) we get for $j > 0$ that

$$
\begin{aligned}
g_j &= \sum_{n=1}^{\infty} \left(a + \frac{b}{n} \right) p_{n-1} f_j^{*n} \\
&= \sum_{n=1}^{\infty} \mathbb{E} \left[a + b \frac{V_1}{j} \,\Big|\, \sum_{i=1}^{n} V_i = j \right] p_{n-1} f_j^{*n} \\
&= \sum_{n=1}^{\infty} \mathbb{E} \left[a + b \frac{V_1}{j} \,;\, \sum_{i=1}^{n} V_i = j \right] p_{n-1} \\
&= \sum_{n=1}^{\infty} \sum_{k=0}^{j} \left(a + b \frac{k}{j} \right) f_k f_{j-k}^{*(n-1)} p_{n-1} \\
&= \sum_{k=0}^{j} \left(a + b \frac{k}{j} \right) f_k \sum_{n=0}^{\infty} f_{j-k}^{*n} p_n = \sum_{k=0}^{j} \left(a + b \frac{k}{j} \right) f_k g_{j-k} \\
&= a f_0 g_j + \sum_{k=1}^{j} \left(a + b \frac{k}{j} \right) f_k g_{j-k} \,,
\end{aligned}
$$

and (6.2) follows. (6.3) is a trivial special case.   □

If the distribution F of the V_i is non-lattice, it is natural to use a discrete approximation. To this end, let $V_{i,+}^{(h)}$, $V_{i,-}^{(h)}$ be V_i rounded upwards, resp. downwards, to the nearest multiple of h and let $A_{\pm}^{(h)} = \sum_{1}^{N} V_{i,\pm}^{(h)}$. An obvious modification of Proposition 6.1 applies to evaluate the distribution $F_{\pm}^{(h)}$ of $A_{\pm}^{(h)}$, letting $g_{j,\pm}^{(h)} = \mathbb{P}(A_{\pm}^{(h)} = jh)$ and

$$
f_{k,-}^{(h)} = \mathbb{P}\left(V_{i,-}^{(h)} = kh \right) = F((k+1)h) - F(kh)\,, \quad k = 0, 1, 2, \ldots,
$$

$$
f_{k,+}^{(h)} = \mathbb{P}\left(V_{i,+}^{(h)} = kh \right) = F(kh) - F((k-1)h)\,, \quad k = 1, 2, \ldots.
$$

Then the error on the tail probabilities (which can be taken arbitrarily small by choosing h small enough) can be evaluated by

$$
\sum_{j=\lfloor x/h \rfloor}^{\infty} g_{j,-}^{(h)} \leq \mathbb{P}(A \geq x) \leq \sum_{j=\lceil x/h \rceil}^{\infty} g_{j,+}^{(h)} \,.
$$

Notes

Further examples beyond the Poisson (and in fact the only ones, cf. Sundt and Jewell [168]) where (6.1) holds are the binomial distribution and the negative binomial (in particular, geometric) distribution. The geometric case is of particular importance in ruin theory, and we touch upon the relevance of Panjer's recursion in that setting in Chap. IV.

For a number of recursion schemes more or less related to Panjer's recursion, see Chapters 4–5 of Dickson [61]. Further relevant references include Sundt and Jewell [168], Sundt [167] and Ch. 6 of Albrecher, Beirlant and Teugels [3].

7 Tails of Compound Sums

7.1 Heavy Tails: The Subexponential Approximation

We recall the definition of a subexponential distribution: $B \in \mathscr{S}$ if $\overline{B}^{*n}(x) \sim n\overline{B}(x)$ for $n = 2$ (and then for all $n \geq 2$). The main results on tail asymptotics of compound subexponential sums is:

Theorem 7.1 *Consider a compound sum* $A = V_1 + \cdots + V_N$ *where* $\mathbb{E}z^N < \infty$ *for some* $z > 1$ *and the* V_n *are i.i.d. with common distribution* $F \in \mathscr{S}$. *Then* $\mathbb{P}(A > x) \sim \mathbb{E}N \cdot \overline{F}(x)$.

The proof requires the following technical result:

Lemma 7.2 *If* $B \in \mathscr{S}$, $\epsilon > 0$, *then there exists a constant* $K = K_\epsilon$ *such that* $\overline{B^{*n}}(x) \leq K(1 + \epsilon)^n \overline{B}(x)$ *for all* n *and* x.

Proof Define $\delta > 0$ by $(1 + \delta)^2 = 1 + \epsilon$, choose T such that $(\overline{B}(x) - \overline{B^{*2}}(x))/\overline{B}(x) \leq 1 + \delta$ for $x \geq T$ and let $A = 1/\overline{B}(T)$, $\alpha_n = \sup_{x \geq 0} \overline{B^{*n}}(x)/\overline{B}(x)$. Then by (2.6),

$$\alpha_{n+1}$$

$$\leq 1 + \sup_{x \leq T} \int_0^x \frac{\overline{B^{*n}}(x - z)}{\overline{B}(x)} B(dz) + \sup_{x > T} \int_0^x \frac{\overline{B^{*n}}(x - z)}{\overline{B}(x - z)} \frac{\overline{B}(x - z)}{\overline{B}(x)} B(dz)$$

$$\leq 1 + A + \alpha_n \sup_{x > T} \int_0^x \frac{\overline{B}(x - z)}{\overline{B}(x)} B(dz) \leq 1 + A + \alpha_n(1 + \delta).$$

The truth of this for all n together with $\alpha_1 = 1$ implies $\alpha_n \leq K(1 + \delta)^{2n}$, where $K = (1 + A)/\epsilon$. $\qquad\square$

Proof of Theorem 7.1 Recall from above that $\overline{F^{*n}}(x) \sim n\overline{F}(x)$, $x \to \infty$, and that for each $z > 1$ there is a $D < \infty$ such that $\overline{F^{*n}}(x) \leq \overline{F}(x)Dz^n$ for all x. We get

$$\frac{\mathbb{P}(V_1 + \cdots + V_N > x)}{\overline{F}(x)} = \sum_{n=0}^{\infty} \mathbb{P}(N = n) \frac{\overline{F^{*n}}(x)}{\overline{F}(x)} \to \sum_{n=0}^{\infty} \mathbb{P}(N = n) \cdot n = \mathbb{E}N,$$

using dominated convergence with $\sum \mathbb{P}(N = n) Dz^n$ as majorant. $\qquad\square$

Notes
Lemma 7.2 is due to Kesten (who actually never published this result himself). One should
be warned that the numerical accuracy of the approximation $\mathbb{P}(A > x) \sim \mathbb{E}N \cdot \overline{F}(x)$ is not
outstanding, see Ch. 6 of [3]!

7.2 Light Tails: The Saddlepoint Approximation

We consider the case of a Poisson N where $\mathbb{E}e^{\alpha A} = e^{\kappa(\alpha)}$ with $\kappa(\alpha) = \lambda(\widehat{F}[\alpha] - 1)$.
The exponential family generated by A is given by

$$\mathbb{P}_\theta(A \in dx) = \mathbb{E}\left[e^{\theta A - \kappa(\theta)}; A \in dx\right].$$

In particular,

$$\kappa_\theta(\alpha) = \log \mathbb{E}_\theta e^{\alpha A} = \kappa(\alpha + \theta) - \kappa(\theta) = \lambda_\theta(\widehat{F}_\theta[\alpha] - 1),$$

where $\lambda_\theta = \lambda\widehat{F}[\theta]$ and F_θ is the distribution given by

$$F_\theta(dx) = \frac{e^{\theta x}}{\widehat{F}[\theta]} F(dx).$$

This shows that the $\mathbb{P}_\theta$-distribution of A has a similar compound Poisson form as
the $\mathbb{P}$-distribution, only with λ replaced by λ_θ and F by F_θ.

Following the lines of Sect. 3.4, we shall derive a tail approximation by
exponential tilting. For a given x, we define the saddlepoint $\theta = \theta(x)$ by $\mathbb{E}_\theta A = x$,
i.e. $\kappa'_\theta(0) = \kappa'(\theta) = x$.

Proposition 7.3 *Assume that* $\lim_{r \uparrow r^*} \widehat{F}''[r] = \infty$,

$$\lim_{r \uparrow r^*} \frac{\widehat{F}'''[r]}{\left(\widehat{F}''[r]\right)^{3/2}} = 0, \tag{7.1}$$

where $r^* = \sup\{r : \widehat{F}[r] < \infty\}$. *Then as* $x \to \infty$,

$$\mathbb{P}(A > x) \sim \frac{e^{-\theta x + \kappa(\theta)}}{\theta\sqrt{2\pi \lambda \widehat{F}''[\theta]}}. \tag{7.2}$$

Proof Since $\mathbb{E}_\theta A = x$, $\text{Var}_\theta(A) = \kappa''(\theta) = \lambda\widehat{F}''[\theta]$, (7.1) implies that the limiting
$\mathbb{P}_\theta$-distribution of $(A - x)/\sqrt{\lambda\widehat{F}''[\theta]}$ is standard normal. Hence

$$\mathbb{P}(A > x) = \mathbb{E}_\theta\left[e^{-\theta A + \kappa(\theta)}; A > x\right] = e^{-\theta x + \kappa(\theta)}\mathbb{E}_\theta\left[e^{-\theta(A-x)}; A > x\right]$$

$$\sim e^{-\theta x + \kappa(\theta)} \int_0^\infty e^{-\theta\sqrt{\lambda\widehat{F}''[\theta]}y}\frac{1}{\sqrt{2\pi}}e^{-y^2/2}\,dy$$

$$= \frac{e^{-\theta x + \kappa(\theta)}}{\theta \sqrt{2\pi \, \lambda \widehat{F}''[\theta]}} \int_0^\infty e^{-z} e^{-z^2/(2\theta^2 \lambda \widehat{F}''[\theta])} \, dz$$

$$\sim \frac{e^{-\theta x + \kappa(\theta)}}{\theta \sqrt{2\pi \, \lambda \widehat{F}''[\theta]}} \int_0^\infty e^{-z} \, dz = \frac{e^{-\theta x + \kappa(\theta)}}{\theta \sqrt{2\pi \, \lambda \widehat{F}''[\theta]}}.$$

□

Remark 7.4 The proof is somewhat heuristical in the CLT steps. To make it a rigorous proof, some regularity of the density $f(x)$ of F is required. In particular, either of the following is sufficient:

A. f is gamma-like, i.e. bounded with $f(x) \sim c_1 x^{\alpha-1} e^{-\delta x}$.
B. f is log-concave, or, more generally, $f(x) = q(x) e^{-h(x)}$, where $q(x)$ is bounded away from 0 and ∞ and $h(x)$ is convex on an interval of the form $[x_0, x^*)$ where $x^* = \sup\{x : f(x) > 0\}$. Furthermore $\int_0^\infty f(x)^\zeta \, dx < \infty$ for some $\zeta \in (1, 2)$.

For example, **A** covers the exponential distribution and phase-type distributions, **B** covers distributions with finite support or with a density not too far from e^{-x^α} with $\alpha > 1$.   ◇

Notes
Proposition 7.3 goes all the way back to Esscher [70], and (7.2) is often referred to as the *Esscher approximation*. Some alternatives are discussed in Ch. 6 of [3]. For further details, see Embrechts et al. [65], Jensen [95] and references therein.

7.3   The NP Approximation

The (first-order) *Edgeworth expansion* states that if the characteristic function $\widehat{g}(u) = \mathbb{E} e^{iuY}$ of an r.v. Y satisfies

$$\widehat{g}(u) \approx e^{-u^2/2}(1 + i\delta u^3), \tag{7.3}$$

where δ is a small parameter, then

$$\mathbb{P}(Y \le y) \approx \Phi(y) - \delta(1 - y^2)\varphi(y) \tag{7.4}$$

[note as a pitfall of this approximation that the r.h.s. of (7.4) may be negative and is not necessarily an increasing function of y for $|y|$ large].

Heuristically, (7.4) is obtained by noting that by Fourier inversion, the density of Y is

$$g(y) = \frac{1}{2\pi} \int_{-\infty}^{\infty} e^{-iuy} \hat{g}(u)\, du \approx \frac{1}{2\pi} \int_{-\infty}^{\infty} e^{-iuy} e^{-u^2/2}(1 + i\delta u^3)\, du$$

$$= \varphi(y) - \delta(y^3 - 3y)\varphi(y),$$

and from this (7.4) follows by integration.

In concrete examples, the CLT for $Y = Y_\delta$ is usually derived via expanding the ch.f. as

$$\hat{g}(u) = \mathbb{E}e^{iuY} = \exp\left\{ iu\kappa_1 - \frac{u^2}{2}\kappa_2 - i\frac{u^3}{3}\kappa_3 + \frac{u^4}{4!}\kappa_4 + \cdots \right\},$$

where $\kappa_1, \kappa_2, \ldots$ are the cumulants; in particular,

$$\kappa_1 = \mathbb{E}Y, \quad \kappa_2 = \mathrm{Var}(Y), \quad \kappa_3 = \mathbb{E}(Y - \mathbb{E}Y)^3.$$

Thus if $\mathbb{E}Y = 0$, $\mathrm{Var}(Y) = 1$ as above, one needs to show that $\kappa_3, \kappa_4, \ldots$ are small. If this holds, one expects the u^3 term to dominate the terms of order $u^4, u^5, \ldots$ so that

$$\hat{f}(u) \approx \exp\left\{ -\frac{u^2}{2} - i\frac{u^3}{3}\kappa_3 \right\} \approx \exp\left\{ -\frac{u^2}{2} \right\} \left(1 - i\frac{u^3}{6}\kappa_3 \right),$$

so that we should take $\delta = -\kappa_3/6$ in (7.4).

Rather than with the tail probabilities $\mathbb{P}(A > x)$, the *NP (normal power) approximation* deals with the quantile $a_{1-\epsilon}$, defined as the solution of $\mathbb{P}(A \le a_{1-\epsilon}) = 1 - \epsilon$ (in finance, $a_{1-\epsilon}$ is the VaR). A particular case is $a_{0.975}$, which is the most common case for VaR calculations in insurance.

Let $Y = (A - \mathbb{E}A)/\sqrt{\mathrm{Var}(A)}$ and let $y_{1-\epsilon}, z_{1-\epsilon}$ be the $1 - \epsilon$–quantile in the distribution of Y, resp. the standard normal distribution. If the distribution of Y is close to $\mathcal{N}(0, 1)$, $y_{1-\epsilon}$ should be close to $z_{1-\epsilon}$ (cf., however, Remark 5.3!), and so as a first approximation we obtain

$$a_{1-\epsilon} = \mathbb{E}A + y_{1-\epsilon}\sqrt{\mathrm{Var}(A)} \approx \mathbb{E}A + z_{1-\epsilon}\sqrt{\mathrm{Var}(A)}. \tag{7.5}$$

A correction term may be computed from (7.4) by noting that the $\Phi(y)$ terms dominate the $\delta(1 - y^2)\varphi(y)$ term. This leads to

$$1 - \epsilon \approx \Phi(y_{1-\epsilon}) - \delta(1 - y_{1-\epsilon}^2)\varphi(y_{1-\epsilon}) \approx \Phi(y_{1-\epsilon}) - \delta(1 - z_{1-\epsilon}^2)\varphi(z_{1-\epsilon})$$

$$\approx \Phi(z_{1-\epsilon}) + (y_{1-\epsilon} - z_{1-\epsilon})\varphi(z_{1-\epsilon}) - \delta(1 - z_{1-\epsilon}^2)\varphi(z_{1-\epsilon})$$

$$= 1 - \epsilon + (y_{1-\epsilon} - z_{1-\epsilon})\varphi(z_{1-\epsilon}) - \delta(1 - z_{1-\epsilon}^2)\varphi(z_{1-\epsilon})$$

which combined with $\delta = -\mathbb{E}Y^3/6$ yields

$$y_{1-\epsilon} = z_{1-\epsilon} + \frac{1}{6}(z_{1-\epsilon}^2 - 1)\mathbb{E}Y^3.$$

Using $Y = (A - \mathbb{E}A)/\sqrt{\mathbb{V}\mathrm{ar}(A)}$, this yields the NP approximation

$$a_{1-\epsilon} = \mathbb{E}A + z_{1-\epsilon}(\mathbb{V}\mathrm{ar}(A))^{1/2} + \frac{1}{6}(z_{1-\epsilon}^2 - 1)\frac{\mathbb{E}(A - \mathbb{E}A)^3}{\mathbb{V}\mathrm{ar}(A)}. \tag{7.6}$$

Under the Poisson assumption, the kth cumulant of A is $\lambda\mu_V^{(k)}$ and so $\kappa_k = \lambda\mu_V^{(k)}/(\lambda\mu_V^{(2)})^{k/2}$. In particular, κ_3 is small for large λ but dominates $\kappa_4, \kappa_5, \ldots$ as required. We can rewrite (7.6) as

$$a_{1-\epsilon} = \lambda\mu_V + z_{1-\epsilon}(\lambda\mu_V^{(2)})^{1/2} + \frac{1}{6}(z_{1-\epsilon}^2 - 1)\frac{\mu_V^{(3)}}{\mu_V^{(2)}}. \tag{7.7}$$

Notes
We have followed largely Sundt [166]. Another main reference is Daykin et al. [54]. Note, however, that [54] distinguishes between the NP and Edgeworth approximations, and that our terminology for the Edgeworth expansion may not be standard.

Chapter IV: Ruin Theory

1 The Cramér–Lundberg Model

Let $R(t)$ be the reserve of the insurance company at time t and $u = R(0)$. Ruin occurs if the reserve goes negative, and so the time of ruin is

$$\tau(u) = \inf\{t > 0 : R(t) < 0\},$$

with the usual convention that the inf of an empty subset of $\mathbb{R}$ is ∞ so that $\tau(u) = \infty$ when $R(t) \geq 0$ for all t. Ruin theory is concerned with studying the probability

$$\psi(u) = \mathbb{P}\big(\tau(u) < \infty\big)$$

that ruin ever occurs (the *infinite horizon ruin probability*) and the probability

$$\psi(u, T) = \mathbb{P}\big(\tau(u) \leq T\big)$$

that ruin occurs before time T (the *finite horizon ruin probability*).

The prototype model is the *Cramér–Lundberg model* or the *Cramér–Lundberg risk process*

$$R(t) = u + ct - \sum_{k=1}^{N(t)} V_k,$$

where c is the rate of premium inflow, N is a counting process such that the number of claims in $[0, t]$ is $N(t)$ and $V_1, V_2, \ldots$ are the claim sizes. The assumptions are that N is Poisson with rate denoted by β, that V_k has distribution B, and that all r.v.s $N, V_1, V_2, \ldots$ are independent.

S. Asmussen, M. Steffensen, *Risk and Insurance*, Probability Theory and Stochastic Modelling 96, https://doi.org/10.1007/978-3-030-35176-2_4

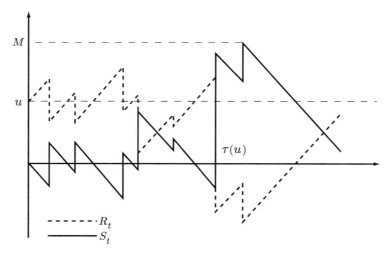

Fig. 1 The processes R and S

Mathematically, it will be convenient to work with the *claim surplus process S* defined by

$$S(t) = u - R(t) = \sum_{k=1}^{N(t)} V_k - ct$$

rather than with R itself. The sample path connection between R and S is illustrated in Fig. 1, where $M = \sup_{t \geq 0} S(t)$.

In terms of S, we have the initial condition $S(0) = 0$ and

$$\tau(u) = \inf\{t > 0 : S(t) > u\},$$
$$\psi(u) = \mathbb{P}\big(\tau(u) < \infty\big) = \mathbb{P}(M > u),$$
$$\psi(u, T) = \mathbb{P}\big(\tau(u) \leq T\big).$$

Define

$$\mu_B = \mathbb{E}V, \quad \mu_S = \mathbb{E}S(1) = \beta\mu_B - c, \quad \rho = \frac{\beta\mu_B}{c}, \quad \eta = -\frac{\mu_S}{\beta\mu_B} = \frac{1}{\rho} - 1.$$

Here η is the relative amount by which the premiums exceed the expected claims and for this reason denoted the *safety loading*. The insurance company will typically ensure that the premiums are sufficient to ensure a positive loading, i.e. that one of the equivalent conditions $\mu_S < 0$, $\rho < 1$ or $\eta > 0$ hold. It is sometimes argued that typical values of η are 10%–25%.

Note that for fixed t, the term $\sum_{k=1}^{N(t)} V_k$ occurring in R and S is a compound sum of the type studied in Sect. III.5.1. We have seen there that its c.g.f. is $\beta(\widehat{B}[\alpha] - 1)$ where $\widehat{B}[\cdot]$ is the m.g.f. of B, and therefore the c.g.f. of $S(1)$ is

$$\kappa(\alpha) = \log \mathbb{E}e^{\alpha S(1)} = \beta(\widehat{B}[\alpha] - 1) - \alpha c .$$

Similarly,

$$\log \mathbb{E}e^{\alpha S(t)} = \beta t(\widehat{B}[\alpha] - 1) - \alpha c t = t\kappa(\alpha), \tag{1.1}$$

$$\mathbb{E}S(t) = \mu_S t = (\beta\mu_B - c)t , \tag{1.2}$$

$$\mathbb{V}\mathrm{ar}\, S(t) = \beta\mu_B^{(2)}t \quad \text{where } \mu_B^{(2)} = \mathbb{E}V^2 . \tag{1.3}$$

Proposition 1.1 $S(t)/t \overset{\text{a.s.}}{\to} \beta\mu_B - c$ or, equivalently, $R(t)/t \overset{\text{a.s.}}{\to} c - \beta\mu_B$. In particular, if $\eta \leq 0$, then $\psi(u) = 1$ for all $u \geq 0$, whereas $\psi(u) < 1$ for all $u \geq 0$ if $\eta > 0$.

Proof Let T_n denote the nth interclaim time and $\sigma_n = T_1 + \cdots + T_n$ the time of the nth claim. Then the T_n are i.i.d. exponential(β) r.v.s, so $\sigma_n/n \overset{\text{a.s.}}{\to} 1/\beta$. Further,

$$V_1 + \cdots + V_n - c\sigma_{n+1} \leq S(t) \leq V_1 + \cdots + V_n - c\sigma_n$$

when $\sigma_n \leq t < \sigma_{n+1}$. Such a t is of order n/β as are σ_n, σ_{n+1}, and hence

$$\lim_{t\to\infty} \frac{S(t)}{t} = \lim_{n\to\infty} \frac{V_1 + \cdots + V_n}{n/\beta} - c = \beta\mu_B - c ,$$

as claimed.

If $\eta < 0$, $S(t)/t$ thus has a strictly positive limit which implies $S(t) \overset{\text{a.s.}}{\to} \infty$ and $M = \infty$ a.s. so that $\psi(u) = \mathbb{P}(M > u) = 1$. Similarly, $S(t) \overset{\text{a.s.}}{\to} -\infty$ and $M < \infty$ a.s. when $\eta > 0$. It is then easy to see that one cannot have $\mathbb{P}(M > u) = 1$ for any $u \geq 0$.

The case $\eta = 0$ is slightly more intricate and omitted. □

Exercises

1.1 Let $\psi_{c,\beta}(u)$, $\psi_{c,\beta}(u, T)$ denote the ruin probabilities with premium rate c and Poisson rate β, and fixed claim size distribution B. Show by considering the risk process $\widetilde{R}(t) = R(t/c)$ that

$$\psi_{c,\beta}(u) = \psi_{1,\beta/c}(u), \quad \psi_{c,\beta}(u, T) = \psi_{1,\beta/c}(u, cT).$$

Notes
Ruin theory is classical and some texts with particular emphasis on the topic are (in chronological order) Gerber [79], Grandell [84], Dickson [61], Asmussen and Albrecher [8], Schmidli [156], and Konstantinides [103]. Virtually all material of this chapter can be found in these sources.

The area is closely related to queueing theory. In fact, if $c = 1$ the claim surplus process $S(t)$ reflected at 0 is the same as the M/G/1 workload process $V(t)$, from which it can be deduced that $\psi(u) = \mathbb{P}(V > u)$, where V is the limit in distribution of $V(t)$ as $t \to \infty$ (a proper r.v. when $\rho < 1$). See, e.g., [8, III.2]. Here ρ becomes the traffic intensity of the queue. In view of Exercise 1.1, it is actually no loss of generality to assume $c = 1$. This is convenient for translating queueing results to risk theory and vice versa, and is done in much of the literature, including the first author's earlier book [8].

2   First Results: Martingale Techniques

Define $M(t) = e^{\alpha S(t) - t\kappa(\alpha)}$.

Proposition 2.1 *For any α such that $\widehat{B}[\alpha] < \infty$, $M = \{M(t)\}_{t \geq 0}$ is a martingale w.r.t. the filtration given by $\mathscr{F}(t) = \sigma\big(S(v) : 0 \leq v \leq t\big)$.*

M commonly goes under the name of the *Wald martingale*.

Proof Consider $0 \leq t < t + v$ and define

$$\widetilde{S}(v) = S(t + v) - S(t) = \sum_{k=N(t)+1}^{N(t+v)} V_k - cv.$$

Properties of Poisson process increments imply that $\widetilde{S}(v)$ is independent of $\mathscr{F}(t)$, and that $\widetilde{S}(v) \overset{\mathscr{D}}{=} S(v)$. Hence

$$\mathbb{E}\big[M(t + v) \,\big|\, \mathscr{F}(t)\big] = \mathbb{E}\big[\exp\{\alpha S(t) + \alpha \widetilde{S}(v) - \kappa(\alpha)(t + v)\} \,\big|\, \mathscr{F}(t)\big]$$

$$= \exp\{\alpha S(t) - \kappa(\alpha)t\}\mathbb{E}\big[\exp\{\alpha \widetilde{S}(v) - \kappa(\alpha)v\} \,\big|\, \mathscr{F}(t)\big]$$

$$= M(t)\mathbb{E}\exp\{\alpha S(v) - \kappa(\alpha)v\} = M(t),$$

where we used (1.1) in the last step.   □

The *Cramér condition* states that $\eta > 0$ and there exists a $\gamma > 0$ such that $\kappa(\gamma) = 0$ and $\kappa'(\gamma) < \infty$. It typically holds when the tail of B is light. In the insurance risk literature, γ goes under the name of the *adjustment coefficient* and is most often denoted by R.

The situation is illustrated in Fig. 2. Note that we always have $\kappa(0) = 0$ and that $\kappa'(0) = \mathbb{E}S(1) = \beta\mu_B - c < 0$ because of the assumption $\eta > 0$. In general, γ is not explicitly available even if $\widehat{B}[\alpha]$ is so (but see Corollary 2.4 and Exercise 2.1 below).

At the time $\tau(u)$ of ruin, S upcrosses level u by making a jump, and we let $\xi(u) = S\big(\tau(u)\big) - u$ denote the overshoot (defined on $\{\tau(u) < \infty\}$ only).

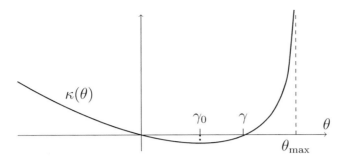

Fig. 2 The κ function

Proposition 2.2 *Under the Cramér condition,*

$$\psi(u) = \frac{e^{-\gamma u}}{\mathbb{E}\big[e^{\gamma \xi(u)} \mid \tau(u) < \infty\big]}. \tag{2.1}$$

Proof The optional stopping theorem asserts that $\mathbb{E}M(\tau \wedge t) = \mathbb{E}M(0) = 1$ for any stopping time $\tau \le \infty$ and any $t < \infty$. We take $\alpha = \gamma$ in Proposition 2.1, which has the appealing feature that $M(t)$ simplifies to $e^{\gamma S(t)}$. Thus

$$1 = \mathbb{E}e^{\gamma S(\tau(u) \wedge t)} = \mathbb{E}\big[e^{\gamma S(\tau(u))}; \tau(u) \le t\big] + \mathbb{E}\big[e^{\gamma S(t)}; \tau(u) > t\big]. \tag{2.2}$$

As $t \to \infty$, the second term converges to 0 since $[\cdot]$ is bounded by $e^{\gamma u}$ and converges to 0 because of $\gamma > 0$ and $S(t) \to -\infty$ (cf. the $\eta > 0$ part of Proposition 1.1). Hence in the limit, (2.2) takes the form

$$1 = \mathbb{E}\big[e^{\gamma S(\tau(u))}; \tau(u) < \infty\big] = e^{\gamma u}\mathbb{E}\big[e^{\gamma \xi(u)}; \tau(u) < \infty\big]$$
$$= e^{\gamma u}\mathbb{E}\big[e^{\gamma \xi(u)} \mid \tau(u) < \infty\big]\psi(u).$$

$\square$

Corollary 2.3 (Lundberg's Inequality) *Under the Cramér condition,* $\psi(u) \le e^{-\gamma u}$ *for all* $u \ge 0$.

Proof Just note that $\xi(u) \ge 0$.

$\square$

Corollary 2.4 *If B is exponential(δ) with $\delta > \beta$, then $\psi(u) = \rho e^{-\gamma u}$ for all u where $\gamma = \delta - \beta/c$, $\rho = \beta/(\delta c)$.*

Proof That $\gamma = \delta - \beta/c$ is immediately seen by taking $\alpha = \delta - \beta/c$ in

$$\kappa(\alpha) = \beta(\widehat{B}[\alpha] - 1) - \alpha c = \beta\Big(\frac{\delta}{\delta - \alpha} - 1\Big) - \alpha c.$$

The available information on the jump at the time of ruin is that the distribution given $\tau(u) = t$ and $S(t-) = x \leq u$ is that of a claim size V given $V > u - x$. Thus by the memoryless property of the exponential distribution, the conditional distribution of the overshoot $\xi(u) = V - u + x$ is again exponential(δ). Thus

$$\mathbb{E}\big[e^{\gamma\xi(u)} \,\big|\, \tau(u) < \infty\big] = \int_0^\infty e^{\gamma v}\delta e^{-\delta v}\,dv = \int_0^\infty \delta e^{-\beta/c\cdot v}\,dv = \frac{\delta}{\beta/c} = \frac{1}{\rho},$$

using $\mu_B = 1/\delta$ in the last step.   □

One of the most celebrated results in ruin theory states that $\psi(u)$ is asymptotically exponential with rate γ:

Theorem 2.5 (The Cramér–Lundberg Approximation) *Under the Cramér condition,* $\psi(u) \sim Ce^{-\gamma u}$ *as* $u \to \infty$, *where*

$$C = \frac{1-\rho}{\kappa'(\gamma)} = \frac{1-\rho}{\beta\widehat{B}'[\gamma] - c} = \frac{1 - \beta\mu_B/c}{\beta\mathbb{E}[Ve^{\gamma V}] - c}.$$

The proof is, however, more difficult. It needs tools beyond martingales and is given in Sect. 4.

Note that Lundberg's inequality implies that one must have $C \leq 1$, and that the Cramér-lundberg approximation is exact for exponential claims.

Exercises
2.1 Compute γ and C for the Cramér–Lundberg model with $c = 1$, $\beta = 3$ and B being the distribution with density

$$\frac{1}{2}3e^{-3x} + \frac{1}{2}7e^{-7x}.$$

3   Ladder Heights. Heavy Tails

The first *ladder height* is defined as the first positive value of the claim surplus process $\{S(t)\}$, that is, as the distribution of $S(\tau_+)$ where $\tau_+ = \tau_+(1) = \tau(0)$. The *ladder height distribution* G_+ is given by

$$G_+(x) = \mathbb{P}\big(S(\tau_+) \leq x, \tau_+ < \infty\big).$$

Then G_+ is concentrated on $(0, \infty)$, i.e., has no mass on $(-\infty, 0]$, and is defective, i.e.

$$\|G_+\| = G_+(\infty) = \mathbb{P}(\tau_+ < \infty) = \psi(0) < 1,$$

when $\eta > 0$ (there is positive probability that $\{S(t)\}$ will never rise above level 0).

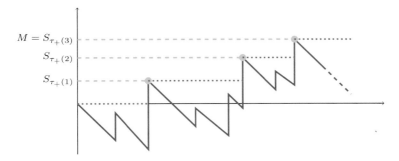

Fig. 3 Ladder heights

The term *ladder height* is motivated by the shape of the process $\{M(t)\}$ of relative maxima, see Fig. 3 where the ladder points are the green balls and the red-dotted lines are the ladder steps. The first ladder height is precisely $S(\tau_+)$, and the maximum M is the total height of the ladder, i.e. the sum of all the ladder heights (if $\eta > 0$, there are only finitely many). In Fig. 3, the second ladder point is $S(\tau_+(2))$ where

$$\tau_+(2) = \inf\{t > \tau_+(1) : S(t) > S(\tau_+(1))\}$$

is the time of the next relative maximum after $\tau_+(1) = \tau_+$, the second ladder height (step) is $S(\tau_+(2)) - S(\tau_+(1))$ and so on. Because of the simple compound Poisson structure, the ladder heights are i.i.d., a fact which turns out to be extremely useful.

Remarkably, there is an explicit formula for G_+. Recall that $\overline{B}(x) = 1 - B(x)$ denotes the tail of B and let the *integrated tail distribution* B_0 be defined by having density $b_0(x) = \overline{B}(x)/\mu_B$.

Theorem 3.1 *For the Cramér–Lundberg model with $\rho = \beta\mu_B/c < 1$, G_+ is given by the defective density $g_+(x) = \beta\overline{B}(x)/c = \rho b_0(x)$ on $(0, \infty)$. Equivalently, $G_+ = \rho B_0$.*

Before giving the proof, we state some important consequences.

Corollary 3.2 *The infinite horizon ruin probability $\psi(u)$ with zero initial reserve $u = 0$ is given by $\psi(0) = \rho$. More generally,*

$$\psi(u) = (1 - \rho) \sum_{n=1}^{\infty} \rho^n \overline{B}_0^{*n}(u) \quad \text{for all } u \geq 0.$$

Proof By Theorem 3.1, $\psi(0) = \mathbb{P}(\tau_+ < \infty) = \|G_+\| = \rho$. Therefore in Fig. 3, the number K of ladder heights is geometric on $\mathbb{N}$ with success parameter $1 - \rho$, i.e. $\mathbb{P}(K \geq n) = \rho^n$ (note that the conditional probability of one more ladder step given there are n already equals $\mathbb{P}(\tau_+ < \infty) = \rho$). Further, by Theorem 3.1 the conditional distribution of a ladder step given that the corresponding ladder epoch is finite is B_0. Thus, we obtain a representation of the maximal claims surplus as

$$M = \sup_{t \geq 0} S(t) \stackrel{\mathscr{D}}{=} \sum_{n=0}^{K} Y_n, \tag{3.1}$$

where the Y_n are i.i.d. with distribution B_0 and independent of K. It follows that

$$\psi(u) = \mathbb{P}(M > u) = \sum_{n=0}^{\infty} \mathbb{P}(K = n)\mathbb{P}(Y_1 + \cdots + Y_n > u)$$

$$= (1 - \rho) \sum_{n=1}^{\infty} \rho^n \overline{B}_0^{*n}(u).$$

□

Note that the sum in (3.1) is a compound sum of just the same type as discussed in Chap. III. In particular, Panjer's recursion (Sect. III.6) becomes available as a numerical tool since the geometric case is one of the three special cases allowed for.

Corollary 3.3 *Assume that B_0 is subexponential and that $\rho < 1$. Then*

$$\psi(u) \sim \frac{\rho}{1 - \rho} \overline{B}_0(u), \quad u \to \infty.$$

Proof The proof is similar to the proof of Theorem III.7.1, with the small difference that the compound sum there was Poisson whereas it is geometric here. Since $\overline{B}_0^{*n}(u) \sim n\overline{B}_0(u)$ by subexponentiality, we get formally that

$$\psi(u) \sim (1 - \rho) \sum_{n=1}^{\infty} \rho^n n\overline{B}_0(u) = \frac{\rho}{1 - \rho} \overline{B}_0(u).$$

The justification is a dominated argument similar to the one in the proof of Theorem III.7.1 and based upon Lemma III.7.2, which states that given $\epsilon > 0$, there exists a constant $k = k_\epsilon$ such that $\overline{B}_0^{*n}(x) \leq k(1 + \epsilon)^n \overline{B}_0(x)$ for all n and x. Here ϵ should be chosen with $\rho(1 + \epsilon) < 1$. □

We will give a very short proof of Theorem 3.1. It may be slightly on the heuristical side, but serves to understand why the result is true.

Proof of Theorem 3.1 The argument is illustrated in Fig. 4.

Assume B has a density b and let $g(u)$ be the density of G_+ at $u > 0$. Note that if there is a claim arrival before dt (case (a) in the figure), then $S(\tau_+) \in [u, du)$ occurs precisely when the claim has size u. Hence the contribution to $g(u)$ from this event is $b(u)\beta\, dt\, du$. If there are no claim arrivals before dt as in (b) and (c), consider the

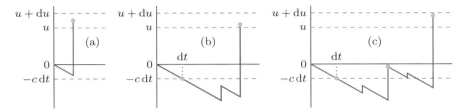

Fig. 4 Determining the ladder height distribution

process $\{\widetilde{S}(t)\}_{t\geq 0}$ where $\widetilde{S}(t) = S(t+\mathrm{d}t) - S(\mathrm{d}t) = S(t+\mathrm{d}t) - c\,\mathrm{d}t$. For $S(\tau_+) \in [u, \mathrm{d}u)$ to occur, $\widetilde{S}$ must either have its first ladder height in $[u+c\,\mathrm{d}t,\ u+c\,\mathrm{d}t+\mathrm{d}u)$ (case (b)), or the first ladder height must be in $[0, c\,\mathrm{d}t)$ and the second ladder height be in $[u, u+c\,\mathrm{d}t+\mathrm{d}u)$ (case (c)). It follows that

$$
\begin{aligned}
g(u)\,\mathrm{d}u &= b(u)\beta\,\mathrm{d}t\,\mathrm{d}u + (1-\beta\,\mathrm{d}t)\big(g(u+c\,\mathrm{d}t)\,\mathrm{d}u + g(0)\,c\,\mathrm{d}t\,g(u)\,\mathrm{d}u\big) \\
&= b(u)\beta\,\mathrm{d}t\,\mathrm{d}u + (1-\beta\,\mathrm{d}t)\big(g(u)\mathrm{d}u + g'(u)c\,\mathrm{d}t\,\mathrm{d}u + g(0)c\,\mathrm{d}t\,g(u)\,\mathrm{d}u\big) \\
&= g(u)\,\mathrm{d}u + \mathrm{d}t\,\mathrm{d}u\big(-\beta g(u) + g'(u)c + g(0)g(u)c + \beta b(u)\big), \\
g'(u) &= \big(\beta/c - g(0)\big)g(u) - \beta b(u)/c. \tag{3.2}
\end{aligned}
$$

Integrating from 0 to x gives

$$
g(x) = g(0) + \big(\beta/c - g(0)\big)\mathbb{P}(S_{\tau_+} \leq x, \tau_+ < \infty) - \beta B(x)/c.
$$

Letting $x \to \infty$ and assuming (heuristical but reasonable!) that $g(x) \to 0$, we get

$$
0 = g(0) + \big(\beta/c - g(0)\big)\mathbb{P}(\tau_+ < \infty) - \beta/c = \big(g(0) - \beta/c\big)\mathbb{P}(\tau_+ = \infty).
$$

Since $\mathbb{P}(\tau_+ = \infty) > 0$ because of the assumption $\eta > 0$, we therefore have $g(0) = \beta/c$. Thus (3.2) simply means $g'(u) = -\beta b(u)/c$, and the solution satisfying $g(0) = \beta/c$ is $\beta\overline{B}(u)/c = \rho b_0(u)$. □

4 Proof of the Cramér–Lundberg Approximation

Recall that the Cramér–Lundberg approximation is $\psi(u) \sim Ce^{-\gamma u}$ as $u \to \infty$, where $C = (1-\rho)/(\beta\widehat{B}'[\gamma] - c)$. There is no elementary proof of the result in its full generality. The original proof by Cramér used complex analysis, whereas most later ones are based on renewal theory, and this is the approach we take here.

A *renewal equation* is an integral equation of the form

$$
Z(u) = z(u) + \int_0^u Z(u-y)\,F(\mathrm{d}y) \tag{4.1}
$$

where the unknown is the function $Z(u)$, F is a non-negative measure and $z(u)$ a known function. It is called *proper* if F is a probability measure, i.e. of total mass $\|F\| = 1$, whereas it is *defective* if $\|F\| < 1$. The *key renewal theorem* gives the asymptotics of $Z(u)$ (under conditions to which we return below) and will allow a very short proof of the Cramér–Lundberg approximation given the following result:

Corollary 4.1 *The ruin probability $\psi(u)$ satisfies the defective renewal equation*

$$\psi(u) = \frac{\beta}{c} \int_u^\infty \overline{B}(y)\,\mathrm{d}y + \int_0^u \psi(u-y)\frac{\beta}{c}\overline{B}(y)\,\mathrm{d}y. \tag{4.2}$$

Proof Since $G_+(\mathrm{d}y) = \beta\overline{B}(y)/c\,\mathrm{d}y$, the assertion is equivalent to

$$\psi(u) = \overline{G}_+(u) + \int_0^u \psi(u-y)G_+(\mathrm{d}y). \tag{4.3}$$

This follows directly from Theorem 3.1 and Fig. 3: $\overline{G}_+(u)$ is the probability that there is a first ladder step and that it leads to ruin. If ruin it not attained in the first ladder step, at least there needs to be such a first step, say at y where necessarily $y \in [0, u]$, after which the process $\{S(\tau_+ + t) - y\}$ needs to exceed $u - y$. Since the conditional distribution of this process given $S(\tau_+) = y$ is the same as the distribution of $\{S(t)\}$, the probability of this is just $\psi(u - y)$. This gives the result.   □

Before applying the key renewal theorem, we make two more elementary observations: first that Corollary 4.1 immediately gives the Laplace transform of $\psi(u)$ (Corollary 4.2) and next (Remark 4.3) that this implies that the value of C stated above is the only possible one. Recall that the m.g.f.s of B, B_0 are denoted $\widehat{B}[s] = \mathbb{E}\mathrm{e}^{sV}$, $\widehat{B}_0[s]$ and so the Laplace transforms are $\widehat{B}[-s]$, $\widehat{B}_0[-s]$.

Corollary 4.2 *For $\Re(s) > -\gamma$, the Laplace transform of the ruin probability is*

$$\int_0^\infty \mathrm{e}^{-su}\psi(u)\,\mathrm{d}u = \frac{\beta - \beta\mathbb{E}\mathrm{e}^{-sV} - \rho s}{s(\beta - s - \beta\mathbb{E}\mathrm{e}^{-sV})} = \frac{\beta - \beta\widehat{B}[-s] - \rho s}{s(\beta - s - \beta\widehat{B}[-s])}. \tag{4.4}$$

Proof In convolution notation, the renewal equation (4.1) can be recast as $Z = z + F * Z$. Letting $\widehat{Z}[s] = \int_0^\infty \mathrm{e}^{-sx}Z(x)\,\mathrm{d}x$, $\widehat{z}[s] = \int_0^\infty \mathrm{e}^{-sx}z(x)\,\mathrm{d}x$, this gives

$$\widehat{Z}[s] = \widehat{z}[s] + \widehat{F}[s]\widehat{Z}[s], \quad \widehat{Z}[s] = \frac{\widehat{z}[s]}{1 - \widehat{F}[s]}. \tag{4.5}$$

In (4.2), we have $\widehat{F}[s] = \widehat{G}_+[s] = \rho\widehat{B}_0[s]$ where

$$\widehat{B}_0[s] = \frac{1}{\mu_B}\int_0^\infty \mathrm{e}^{su}\overline{B}(u)\,\mathrm{d}u = \int_0^\infty \frac{\mathrm{e}^{su} - 1}{s\mu_B}B(\mathrm{d}u) = \frac{\widehat{B}[s] - 1}{s\mu_B}. \tag{4.6}$$

Similarly,

$$\widehat{z}[s] = \int_0^\infty e^{su}\overline{G}_+(u)\,du = \rho\int_0^\infty e^{su}\overline{B}_0(u)\,du = \frac{\rho(\widehat{B}_0[s]-1)}{s}. \qquad (4.7)$$

Combining (4.5)–(4.7) and using $\rho = \beta\mu_B$, easy algebra then gives

$$\int_0^\infty e^{su}\psi(u)\,du = \frac{\beta - \beta\widehat{B}[s] + \rho s}{s(\beta + s - \beta\widehat{B}[s])} = \frac{-\kappa(s)+(\rho-1)s}{-s\kappa(s)}. \qquad (4.8)$$

For (4.4), now just replace s by $-s$. The condition $\Re(s) > -\gamma$ is needed because of the singularity $\kappa(\gamma) = 0$.   □

Remark 4.3 Assume it shown that $\psi(u) \sim \widetilde{C}e^{-\widetilde{\gamma}u}$ for some $\widetilde{\gamma}, \widetilde{C} > 0$. Since $|\kappa(s)| < \infty$ for $s < \gamma$ and $\kappa(\gamma) = 0$, (4.8) then gives $\widetilde{\gamma} = \gamma$. Further, the asymptotics of (4.8) as $s \uparrow \gamma$ is

$$\int_0^\infty e^{su}\psi(u)\,du \sim \frac{(\rho-1)\gamma}{-\gamma\kappa'(\gamma)(\gamma - s)} = \frac{C}{\gamma - s}. \qquad (4.9)$$

On the other hand, $\psi(u) \sim \widetilde{C}e^{-\gamma u}$ gives

$$\int_0^\infty e^{su}\psi(u)\,du \sim \frac{\widetilde{C}}{\gamma - s} \qquad (4.10)$$

and so we conclude that $\widetilde{C} = C$.

The argument looks almost like a complete proof. However, the problem is that (4.9) only implies $\psi(u) \sim Ce^{-\gamma u}$ under additional conditions that may appear weak but nevertheless are difficult to verify in the present context (the precise formulation goes under the name of *Tauberian theorems*, cf. [25]).   ◊

Proof of the Cramér–Lundberg approximation Multiplying (4.2) by $e^{\gamma u}$ on both sides, it takes the form (4.1) with

$$Z(u) = e^{\gamma u}\psi(u), \quad z(u) = e^{\gamma u}\overline{G}_+(u), \quad F(dy) = e^{\gamma y}\beta\overline{B}(y)/c\,dy = \rho e^{\gamma y}B_0(dy).$$

This is a proper renewal equation since (4.6) and $\kappa(\gamma) = 0$ implies

$$\|F\| = \rho\widehat{B}_0[\gamma] = \frac{\beta(\widehat{B}[\gamma]-1)}{\gamma c} = 1.$$

The key renewal theorem asserts [6, V.4] that the solution $Z(u)$ of (4.1) satisfies

$$Z(u) \sim \frac{1}{\mu_F}\int_0^\infty z(x)\,dx \quad \text{as } x \to \infty$$

under a suitable regularity conditions. Of these conditions, the one on F is automatic since F has a density. For the one on z, assume e.g. that $\widehat{B}[\gamma + \epsilon] < \infty$ for some $\epsilon > 0$ and check that then $z(u)$ has a majorant of the type $De^{-\delta x}$. □

Remark 4.4 An alternative proof uses an exponential change of measure similar to the techniques in Sect. III.3, replacing $\mathbb{P}$ by some $\mathbb{P}_\theta$. The likelihood ratio on $[0, T]$ is $\exp\{\theta S(T) - T\kappa(\theta)\}$ which taking $\theta = \gamma$ just becomes $e^{\gamma S(T)}$. Here T can be replaced by a stopping time τ, and taking $\tau = \tau(u)$, $\theta = \gamma$ we get

$$\psi(u) = \mathbb{P}(\tau(u) < \infty) = \mathbb{E}_\gamma\left[e^{-\gamma S(\tau(u))}; \tau(u) < \infty\right].$$

However, $\kappa'_\gamma(0) = \kappa'(\gamma) > 0$ implies $\mathbb{P}_\gamma(\tau(u) < \infty) = 1$. Recalling that $\xi(u) = S(\tau(u)) - u$ is the overshoot, we therefore have $\psi(u) = e^{-\gamma(u)}\mathbb{E}_\gamma e^{-\gamma\xi(u)}$. Here the process $\{\xi(u)\}_{u \geq 0}$ starts from scratch (regenerates) at each ladder point $u = u_k = S(\tau_+(k))$ where the left limit is 0. From the theory of regenerative process, it is then easy to see that $\xi(u)$ converges in $\mathbb{P}_\gamma$-distribution to some $\xi(\infty)$, and thus $\psi(u) \sim Ce^{-\gamma u}$ where $C = \mathbb{E}e^{-\gamma\xi(\infty)}$. Some analytic manipulations finally give that $\mathbb{E}e^{-\gamma\xi(\infty)} = \beta\widehat{B}'[\gamma] - c$ as it should be.

This proof is, however, not more elementary than the one above. The substantial piece of theory used is the convergence theorem for regenerative processes, and this result and the key renewal theorem turn out to be easy consequences of each other, cf. [6, Ch. VI]. ◊

Exercises
4.1 Show that the survival probability $Z(u) = 1 - \psi(u)$ satisfies the defective renewal equation

$$Z(u) = 1 - \rho + \int_0^u Z(u - y)\beta\overline{B}(y)/c\,dy. \tag{4.11}$$

4.2 Rederive the expression (4.4) for the Laplace transform of $\psi(u)$ by using the geometric sum representation $\psi(u) = (1 - \rho)\sum_1^\infty \rho^n \overline{B}_0^{*n}(u)$ in Corollary 3.2 instead of the renewal equation,

4.3 Fill in the details in the proof that $\psi(u) \sim \widetilde{C}e^{-\gamma u}$ implies (4.10).

5   Finite Time Ruin Probabilities

Obviously, the time horizon faced by an insurance company is not infinite — the frame within which risk assessments are relevant may rather be of order $T = 1$–5 years. So, it may be of no concern that $\psi(u)$ is large provided $\psi(u, T)$ is not so. Giving some estimates of the order of $\psi(u, T)$ is the topic of this section. An equivalent problem is the analysis of the distribution of the ruin time $\tau(u)$.

The problem is easy if the loading is negative so the company makes an overall loss:

Proposition 5.1 *Assume* $\eta < 0$. *Then* $\tau(u) < \infty$ *for all* u. *More precisely,* $\tau(u)/u \overset{\text{a.s.}}{\to} 1/\mu_S$ *as* $u \to \infty$.

Proof As observed in Proposition 1.1, $S(t)/t \overset{\text{a.s.}}{\to} \mu_S = \beta\mu_B - c > 0$ as $t \to \infty$. Now just appeal to Exercise 5.1. $\qquad\square$

It remains to determine what happens subject to the more relevant condition $\eta > 0$ (or equivalently $\mu_S < 0$), assumed in the following. This can be split into two subcases: given ruin occurs, what is the typical order of $\tau(u)$? And what happens within a shorter horizon?

5.1 Finite Horizon Ruin with Light Tails

We assume here the Cramér condition from Sect. 2 and let γ be defined as there. Recall that $\kappa(\alpha) = \beta\big(\widehat{B}[\alpha] - 1\big) - \alpha c$ so that

$$\kappa'(\alpha) = \beta\widehat{B}'[\alpha] - c = \beta\mathbb{E}[Ve^{\alpha V}] - c.$$

Proposition 5.2 *Assume* $\eta > 0$. *Then* $\tau(u)/u \to 1/\kappa'(\gamma)$ *in* $\mathbb{P}(\cdot \mid \tau(u) < \infty)$-*probability as* $u \to \infty$. *That is,*

$$\mathbb{P}\big(|\tau(u)/u - 1/\kappa'(\gamma)| > \epsilon \mid \tau(u) < \infty\big) \to 0 \ \text{for all } \epsilon > 0.$$

Proof We use an exponential change of measure $\mathbb{P}_\gamma$ as in Remark 4.4. Since $\mathbb{E}_\gamma S(1) = \kappa'(\gamma) > 0$, Proposition 5.1 implies $\mathbb{P}_\gamma A(u, \epsilon) \to 0$ where $A(u, \epsilon)$ is the event $\{|\tau(u)/u - 1/\kappa'(\gamma)| > \epsilon\}$. Using $S_{\tau(u)} \geq u$ therefore gives

$$\mathbb{P}A(u, \epsilon) = \mathbb{E}_\gamma\big[e^{-\gamma S(\tau(u))}; A(u, \epsilon)\big] \leq e^{-\gamma u}\mathbb{P}_\gamma A(u, \epsilon) = o(e^{-\gamma u}),$$

$$\mathbb{P}\big[A(u, \epsilon) \mid \tau(u) < \infty\big] = \frac{\mathbb{P}A(u, \epsilon)}{\psi(u)} \sim \frac{\mathbb{P}A(u, \epsilon)}{Ce^{-\gamma u}} \to 0.$$

$\qquad\square$

The implication of Proposition 5.2 is that the typical order of $\tau(u)$ is $u/\kappa'(\gamma)$. For a shorter time horizon T, we can write $T = yu$ with $y < 1/\kappa'(\gamma)$ and define α_y, γ_y by

$$\kappa'(\alpha_y) = \frac{1}{y}, \quad \gamma_y = \alpha_y - y\kappa(\alpha_y) = y\kappa^*(1/y), \tag{5.1}$$

where κ^* is the convex conjugate (Legendre–Fenchel transform) of κ, cf. Sect. III.3.2. Note that $\alpha_y > \gamma_0$ and that $\gamma_y > \gamma$ since $y < 1/\kappa'(\gamma)$.

Proposition 5.3 $\psi(u, yu) \le e^{-\gamma_y u}$, $\quad y < \dfrac{1}{\kappa'(\gamma)}$.

Proof Since $\kappa(\alpha_y) > 0$, we get

$$\psi(u, yu) = e^{-\alpha_y u}\mathbb{E}_{\alpha_y}\big[e^{-\alpha_y \xi(u) + \tau(u)\kappa(\alpha_y)}; \ \tau(u) \le yu\big]$$

$$\le e^{-\alpha_y u}\mathbb{E}_{\alpha_y}\big[e^{\tau(u)\kappa(\alpha_y)}; \ \tau(u) \le yu\big] \le e^{-\alpha_y u + yu\kappa(\alpha_y)}.$$

$\square$

Remark 5.4 It may appear that the proof uses considerably less information on α_y than is inherent in the definition (5.1). However, the point is that we want to select an α which produces the largest possible exponent in the inequalities. From the proof it is seen that this amounts to α maximizing $\alpha - y\kappa(\alpha)$. Differentiating w.r.t. α, we arrive at the expression in (5.1). This is basically the same situation as for the Chernoff bound III.(3.9). $\Diamond$

5.2 Finite Horizon Ruin with Heavy Tails

With heavy tails, the intuition is that of one big jump. In the ruin setting, this means that the claim surplus process S should develop in its typical way up to the time of ruin, that is, with drift $\mu_S = \beta\mu_B - c < 0$, and then to go above u as result of a single big claim. If the time of the jump is t, the size of the jump should thus exceed $u - \mu_S t = u + |\mu_S|t$. This leads to assessing the infinite horizon ruin probability $\psi(u)$ by

$$\psi(u) \approx \int_0^\infty \beta\overline{B}\big(u + |\mu_S|t\big)\,dt = \frac{\beta}{|\mu_S|}\int_u^\infty \overline{B}(z)\,dz = \frac{\rho}{1-\rho}\overline{B}_0(u),$$

$$(5.2)$$

where B_0 is the integrated tail distribution. This is indeed what was proved rigorously in Corollary 3.3.

The same heuristics applies to finding the conditional distribution of the ruin time given it is finite. The result is the following, which has an intrinsically different form from the light-tailed case. There the conditional distribution of τ (when properly scaled) concentrates around one point, here it spreads out over the entire positive half-axis.

Proposition 5.5

(i) *Assume that B is regularly varying with tail* $\overline{B}(x) = L(x)/x^\alpha$. *Then* $\tau(u)/u \to$ *W in* $\mathbb{P}(\cdot \mid \tau(u) < \infty)$-*probability as* $u \to \infty$, *where W is the Pareto r.v. with* $\mathbb{P}(W > y) = 1/\big(1 + (c - \beta\mu_B)y\big)^{\alpha-1}$.

(ii) *Assume that B is lognormal*(μ, σ^2) *and let* $e(x) = x/\log x$. *Then* $\tau(u)/e(u) \to V$ *in* $\mathbb{P}(\cdot \mid \tau(u) < \infty)$-*probability as* $u \to \infty$, *where V is exponential.*

(iii) *Assume that B is Weibull with tail* $\overline{B}(x) = e^{-x^\delta}$ *where* $\delta < 1$ *and let* $e(x) = x^{1-\delta}$. *Then* $\tau(u)/u \to V$ *in* $\mathbb{P}(\cdot \mid \tau(u) < \infty)$-*probability as* $u \to \infty$, *where V is exponential.*

Proof For (i), define

$$A(\epsilon, t) = \big\{ |S(v) - \mu_S v| < \epsilon v + K \text{ for all } v < t \big\}$$

with $K = K(\epsilon)$ chosen such that $\mathbb{P}A(\epsilon, t) > 1 - \epsilon$ for all t (this is possible according to the LLN in Proposition 1.1). Then as above, we get

$$\mathbb{P}\big(y < \tau(u)/u < \infty\big) = \int_{yu}^\infty \mathbb{P}(\tau(u) \in dt) \geq \int_{yu}^\infty \mathbb{P}\big(\tau(u) \in dt; A(\epsilon, t)\big)$$

$$\geq \int_{yu}^\infty \mathbb{P}A(\epsilon, t) \, \beta \overline{B}\big(u + K + (|\mu_S| + \epsilon)t\big) \, dt$$

$$\geq \frac{\beta(1 - \epsilon)}{|\mu_S| + \epsilon} \int_{u(1+(|\mu_S|+\epsilon)y)+K}^\infty \overline{B}(z) \, dz$$

$$= \frac{\rho(1 - \epsilon)}{1 - \rho + \epsilon/c} \overline{B}_0\big(u[1 + (|\mu_S| + \epsilon)y] + K\big).$$

Using the asymptotics of $\psi(u)$ in (5.2) gives

$$\liminf_{u\to\infty} \mathbb{P}\big(\tau(u)/u > y \mid \tau(u) < \infty\big)$$

$$\geq \frac{(1 - \epsilon)(1 - \rho)}{1 - \rho + \epsilon/c} \liminf_{u\to\infty} \frac{\overline{B}_0\big(u[1 + (|\mu_S| + \epsilon)y] + K\big)}{\overline{B}_0(u)}$$

$$= \frac{(1 - \epsilon)(1 - \rho)}{1 - \rho + \epsilon/c} \cdot \frac{1}{\big(1 + (|\mu_S| + \epsilon)y\big)^{\alpha-1}}.$$

Letting $\epsilon \downarrow 0$ gives

$$\liminf_{u\to\infty} \mathbb{P}\big(\tau(u)/u > y \mid \tau(u) < \infty\big) \geq \frac{1}{\big(1 + |\mu_S|y\big)^{\alpha-1}} = \mathbb{P}(W > y).$$

For the lim sup part, we first take $y = 0$ to obtain

$$\limsup_{u \to \infty} \frac{1}{\psi(u)} \int_0^\infty \mathbb{P}\big(\tau(u) \in dt;\, A(\epsilon, t)^c\big)$$

$$= \limsup_{u \to \infty} \frac{1}{\psi(u)} \left[\int_0^\infty \mathbb{P}\big(\tau(u) \in dt\big) - \int_0^\infty \mathbb{P}\big(\tau(u) \in dt;\, A(\epsilon, t)\big) \right]$$

$$= 1 - \liminf_{u \to \infty} \frac{1}{\psi(u)} \int_0^\infty \mathbb{P}\big(\tau(u) \in dt;\, A(\epsilon, t)\big)$$

$$\le d(\epsilon) = 1 - \frac{(1 - \epsilon)(1 - \rho)}{1 - \rho + \epsilon/c}.$$

Further, similar calculations as above give

$$\int_{yu}^\infty \mathbb{P}\big(\tau(u) \in dt;\, A(\epsilon, t)\big) \le \frac{1 - \rho}{1 - \rho - \epsilon/c} \cdot \overline{B}_0\big(u[1 + (|\mu_S| - \epsilon)y] + K\big).$$

Combining gives

$$\limsup_{u \to \infty} \mathbb{P}\big(\tau(u)/u > y \mid \tau(u) < \infty\big)$$

$$\le d(\epsilon) + \frac{1 - \rho}{1 - \rho - \epsilon/c} \limsup_{u \to \infty} \frac{\overline{B}_0\big(u[1 + (|\mu_S| - \epsilon)y] + K\big)}{\overline{B}_0(u)}$$

$$= d(\epsilon) + \frac{1 - \rho}{1 - \rho - \epsilon/c} \frac{1}{\big(1 + (|\mu_S| - \epsilon)y\big)^{\alpha - 1}},$$

which has limit $0 + 1/(1 + |\mu_S|y)^{\alpha - 1} = \mathbb{P}(W > y)$ as $\epsilon \downarrow 0$. This proves part (i); (ii), (iii) are left as Exercise 5.2.   □

Finally, consider the ruin probability in shorter time horizons than those given by Proposition 5.5.

Proposition 5.6 *Assume that B is regularly varying with tail $\overline{B}(x) = L(x)/x^\alpha$ and that $T(u)/u \to 0$ as $u \to \infty$. Then $\psi\big(u, T(u)\big) \sim \beta \overline{B}(u)$ as $u \to \infty$*

Proof Arguing as above, we have

$$\psi\big(u, T(u)\big) \approx \int_0^{T(u)} \beta \overline{B}\big(u + |\mu_S|t\big) \, dt.$$

An upper bound for the integrand is $\beta \overline{B}(u)$, whereas an asymptotic lower bound is

$$\beta \overline{B}\big(u + |\mu_S|T(u)\big) = \beta \overline{B}\big(u(1 + o(1))\big) \sim \beta \overline{B}(u).$$

□

Exercises

5.1 Let $f : [0, \infty) \to \mathbb{R}$ be a (deterministic) function such that $f(t)/t \to m$ as $t \to \infty$ for some $m > 0$ and that $\sup_{t \le A} f(t) < \infty$ for all A. Show that $\tau(u)/u \to 1/m$ as $u \to \infty$ where $\tau(u) = \inf\{t > 0 : f(t) \ge u\}$.

5.2 Prove parts (ii), (iii) of Proposition 5.5 and identify the rate parameters. A heuristic calculation along the lines of (5.2) will do!

5.3 Formulate and prove the analogues of Proposition 5.6 for a lognormal or Weibull B.

5.4 In the setting of Proposition 5.6, what is the asymptotic distribution of $\tau(u)$ given $\tau(u) \le T(u)$?

Notes
A broad survey of finite horizon ruin probabilities for the Cramér–Lundberg model can be found in Asmussen and Albrecher [8, Ch. V]. The present proof of Proposition 5.5 is conceptually substantially simpler than the one given in [8, X.4].

6 Markov Regime Switching

Obviously, the Cramér–Lundberg model involves a number of simplifying assumptions. Historically, the most popular generalization has been the so-called *Sparre–Andersen* or *renewal model* where one replaces the Poisson arrival process by a renewal process, i.e., interclaim times $T_1, T_2, \ldots$ are allowed to have a distribution G more general than exponential(β). However, in our view the motivation for this extension is weak: the only realistic example beyond the exponential we can think of is that G is a one point distribution, say that claims in a time period of length h (a month or a year, etc.) are pooled to aggregate claims arriving at times $h, 2h, 3h, \ldots$ We therefore do not go into the renewal model, but give instead a brief treatment of a better motivated generalization, in financial terminology going under the name of *Markov regime switching* and in queueing theory having a parallel in what is most commonly called *Markov-modulation* there. The next section then goes into another extension of the Cramér–Lundberg model, in a different direction.

In Markov regime switching, there is a background Markov process $J = \{J(t)\}_{0 \le t < \infty}$ with a finite state space E and intensity matrix $\mathbf{\Lambda} = (\lambda_{ij})_{i,j \in E}$. The arrival intensity is β_i when $J(t) = i$ and claims arriving when $J(t) = i$ have distribution B_i. Thus, $\{J(t)\}$ describes the environmental conditions (regime) for the risk process. As usual, $\{S(t)\}$ denotes the claim surplus process. Consistent with stochastic analysis notation and representations to be used for the life insurance models in Chaps. V–VII, we can write $\mathrm{d}S(t) = \mathrm{d}S^{J(t)}(t)$ where $\{S^i(t)\}$ is a standard Cramér–Lundberg process with parameters c_i, β_i, B_i. Further $\tau(u) = \inf\{t \ge 0 : S(t) > u\}$ is the ruin time and the ruin probabilities with initial environment i are

$$\psi_i(u) \ = \ \mathbb{P}_i\big(\tau(u) < \infty\big), \qquad \psi_i(u,T) \ = \ \mathbb{P}_i\big(\tau(u) \le T\big),$$

where as usual $\mathbb{P}_i$ refers to the case $J(0) = i$.

Examples of how such a mechanism could be relevant in risk theory follow.

Example 6.1 Consider car insurance, and assume that weather conditions play a major role for the occurrence of accidents. For example, we could distinguish between normal and icy road conditions, leading to E having two states n, i and corresponding arrival intensities β_n, β_i and claim size distributions B_n, B_i; one expects that $\beta_i > \beta_n$ and presumably also that $B_n \ne B_i$, meaning that accidents occurring during icy road conditions lead to claim amounts which are different from the normal ones.

In this two-state setting, the duration of normal and icy periods have exponential distributions. One can, however, incorporate more general cases in the model by using so-called *phase-type* distributions, defined as lifetimes of terminating Markov processes, but we shall not go into this here. ◊

Example 6.2 In health insurance for a disease occurring in epidemics, one could take $E = \{n, s, p, d\}$ with n for normal (no epidemic) conditions, s for the initial stage of the epidemic, p for its peak stage and d for its decline stage. Again, this gives exponential distributions of the sojourn times in the different stages, and again, this can be remedied by using phase-type distributions. ◊

The stationary limiting distribution of $\{J(t)\}$ is represented as an E-row vector π. It exists whenever Λ is irreducible which is assumed throughout, and can be computed as the positive solution of $\pi \Lambda = 0$, $\pi \mathbf{1} = 1$ where $\mathbf{1}$ is the E-column-vector of ones.

6.1   The Averaged Model

It is fundamental for much of the intuition on the model to note that to each risk process in a Markovian environment with $J(t)$ irreducible, one can associate in a natural way a standard Cramér-Lundberg process by averaging over the environment. More precisely, this has premium rate c^*, Poisson rate β^* and claim size distribution B^* given by

$$c^* \ = \ \sum_{i \in E} \pi_i c_i, \quad \beta^* \ = \ \sum_{i \in E} \pi_i \beta_i, \quad B^* \ = \ \sum_{i \in E} \frac{\pi_i \beta_i}{\beta^*} B_i.$$

For example, for the point process $N(t)$ of arrivals of claims one has:

Proposition 6.3 *As $t \to \infty$,* $\dfrac{N(t)}{t} \overset{\text{a.s.}}{\to} \beta^* \ \mathbb{P}_i\text{-a.s. for all } i \in E.$

Proof Let $\ell_i(t)$ denote the time spent by $\{J(t)\}$ in state i before t. By the LLN for Markov processes, $\ell_i(t)/t \to \pi_i$ $\mathbb{P}_i$-a.s. With N^i the Poisson arrival process for the Cramér–Lundberg process S^i above, we therefore get

$$\frac{N(t)}{t} \overset{\mathscr{D}}{=} \sum_{i \in E} \frac{N^i(\ell_i(t))}{t} = \sum_{i \in E} \frac{\ell_i(t)}{t} \frac{N^i(\ell_i(t))}{\ell_i(t)} \to \sum_{i \in E} \pi_i \beta_i = \beta^*. \qquad \Box$$

Similar arguments identify B^* as the average distribution of the claims over a large time horizon and give

$$\lim_{t \to \infty} \frac{S(t)}{t} = \sum_{i \in E} \pi_i (\beta_i \mu_i - c_i) \tag{6.1}$$

with $\mu_i = \widehat{B}_i'[0]$ the mean of B_i. Letting μ_S denote the r.h.s. of (6.1), we get in a similar way as in Proposition 1.1 that:

Proposition 6.4 $\psi_i(u) < 1$ for all i when $\mu_S < 0$, and $\psi_i(u) = 1$ for all i when $\mu_S > 0$.

In view of this discussion, one can view the regime switching model as the averaged model with some randomization added. From the general principle that typically added variation increases the risk, one therefore expects larger ruin probabilities for the regime switching model, at least asymptotically. This was indeed shown under some additional conditions in Asmussen et al. [13].

6.2 The Matrix m.g.f.

As a generalization of the m.g.f., consider the matrix $\widehat{F}_t[\alpha]$ with ijth element $\mathbb{E}_i\big[e^{\alpha S(t)}; J(t) = j\big]$.

Let as above $\{S^i(t)\}$ be a standard Cramér–Lundberg process with parameters c_i, β_i, B_i and $\kappa^i(\alpha) = \log \mathbb{E}e^{\alpha S^i(1)}$.

Proposition 6.5 $\widehat{F}_t[\alpha] = e^{t K[\alpha]}$ where

$$K[\alpha] = \Lambda + \big(\kappa^i(\alpha)\big)_{\text{diag}} = \Lambda + \big(\beta_i(\widehat{B}_i[\alpha] - 1)\big)_{\text{diag}}.$$

Proof Conditioning on $J(t) = k$ gives that up to o(h) terms,

$$\mathbb{E}_i\big[e^{\alpha S(t+h)}; J(t+h) = j\big]$$

$$= (1 + \lambda_{jj} h)\mathbb{E}_i\big[e^{\alpha S(t)}; J(t) = j\big]\mathbb{E}_j e^{\alpha S_h^{(j)}} + \sum_{k \neq j} \lambda_{kj} h \mathbb{E}_i\big[e^{\alpha S(t)}; J(t) = k\big]$$

$$= \mathbb{E}_i\big[e^{\alpha S(t)}; J(t) = j\big]\big(1 + h\kappa^j(\alpha)\big) + h \sum_{k \in E} \mathbb{E}_i\big[e^{\alpha S(t)}; J(t) = k\big]\lambda_{kj}.$$

In matrix formulation, this means that

$$\widehat{\boldsymbol{F}}_{t+h}[\alpha] = \widehat{\boldsymbol{F}}_t[\alpha]\big(\boldsymbol{I} + h\kappa^{(j)}(\alpha)\big)_{\text{diag}} + h\widehat{\boldsymbol{F}}_t[\alpha]\boldsymbol{\Lambda}$$

so that $\widehat{\boldsymbol{F}}_t'[\alpha] = \widehat{\boldsymbol{F}}_t[\alpha]\boldsymbol{K}$. In conjunction with $\widehat{\boldsymbol{F}}_0[\alpha] = \boldsymbol{I}$, this implies $\widehat{\boldsymbol{F}}_t[\alpha] = e^{t\boldsymbol{K}[\alpha]}$ according to the standard solution formula for systems of linear differential equations. □

The assumption that $\boldsymbol{\Lambda}$ is irreducible implies that $\boldsymbol{K}[\alpha]$ is so. By Perron–Frobenius theory (cf. [6, Sects. I.6, XI.2c]), we infer that $\boldsymbol{K}[\alpha]$ has a real eigenvalue $\kappa(\alpha)$ with maximal real part. The corresponding left and right eigenvectors $\boldsymbol{v}^{(\alpha)}$, $\boldsymbol{h}^{(\alpha)}$ may be chosen with strictly positive components. Since $\boldsymbol{v}^{(\alpha)}$, $\boldsymbol{h}^{(\alpha)}$ are only given up to constants, we are free to impose two normalizations, and we shall take

$$\boldsymbol{v}^{(\alpha)}\boldsymbol{h}^{(\alpha)} = 1, \quad \boldsymbol{\pi}\boldsymbol{h}^{(\alpha)} = 1,$$

where $\boldsymbol{\pi} = \boldsymbol{v}^{(0)}$ is the stationary distribution. Then $\boldsymbol{h}^{(0)} = \boldsymbol{1}$.

The function $\kappa(\alpha)$ plays in many respects the same role as for the simple Cramér–Lundberg case, as will be seen in the following. In particular, appropriate generalizations of the Wald martingale can be defined in terms of $\kappa(\alpha)$ (and $\boldsymbol{h}^{(\alpha)}$, cf. Proposition 6.7), $\kappa'(0)$ is the overall drift of $\{S(t)\}$ (Proposition 6.9), and we get analogues of Lundberg's inequality (Corollary 6.10) and the Cramér–Lundberg approximation (Theorem 6.11).

Corollary 6.6 $\mathbb{E}_i\big[e^{\alpha S(t)}; J(t) = j\big] \sim h_i^{(\alpha)}v_j^{(\alpha)}e^{t\kappa(\alpha)}.$

Proof By Perron–Frobenius theory. □

We also get an analogue of the Wald martingale for random walks:

Proposition 6.7 $\mathbb{E}_i e^{\alpha S(t)}h_{J(t)}^{(\alpha)} = h_i^{(\alpha)}e^{t\kappa(\alpha)}.$ *Furthermore,*

$$\big\{e^{\alpha S(t)-t\kappa(\alpha)}h_{J(t)}^{(\alpha)}\big\}_{t\geq 0}$$

is a martingale.

Proof For the first assertion, just note that

$$\mathbb{E}_i e^{\alpha S(t)}h_{J(t)}^{(\alpha)} = \boldsymbol{1}_i^{\mathsf{T}}\widehat{\boldsymbol{F}}_t[\alpha]\boldsymbol{h}^{(\alpha)} = \boldsymbol{1}_i^{\mathsf{T}}e^{t\boldsymbol{K}[\alpha]}\boldsymbol{h}^{(\alpha)} = \boldsymbol{1}_i^{\mathsf{T}}e^{t\kappa(\alpha)}\boldsymbol{h}^{(\alpha)} = e^{t\kappa(\alpha)}h_i^{(\alpha)}.$$

It then follows that

$$\mathbb{E}\big[e^{\alpha S(t+v)-(t+v)\kappa(\alpha)}h_{J(t+v)}^{(\alpha)} \,\big|\, \mathscr{F}_t\big]$$

$$= e^{\alpha S(t)-t\kappa(\alpha)}\mathbb{E}\big[e^{\alpha(S(t+v)-S(t))-v\kappa(\alpha)}h_{J(t+v)}^{(\alpha)} \,\big|\, \mathscr{F}_t\big]$$

$$= e^{\alpha S(t)-t\kappa(\alpha)}\mathbb{E}_{J(t)}\big[e^{\alpha S(v)-v\kappa(\alpha)}h_{J(v)}^{(\alpha)}\big] = e^{\alpha S(t)-t\kappa(\alpha)}h_{J(t)}^{(\alpha)}.$$

□

Let $k^{(\alpha)}$ denote the derivative of $h^{(\alpha)}$ w.r.t. α, and write $k = k^{(0)}$.

Corollary 6.8 $\mathbb{E}_i S(t) = t\kappa'(0) + k_i - \mathbb{E}_i k_{J(t)} = t\kappa'(0) + k_i - \mathbf{1}_i^\top e^{\Lambda t} k.$

Proof By differentiation in Proposition 6.7,

$$\mathbb{E}_i \left[S(t) e^{\alpha S(t)} h^{(\alpha)}_{J(t)} + e^{\alpha S(t)} k^{(\alpha)}_{J(t)} \right] = e^{t\kappa(\alpha)} \left(k^{(\alpha)}_i + t\kappa'(\alpha) h^{(\alpha)}_i \right). \tag{6.2}$$

Let $\alpha = 0$ and recall that $h^{(0)} = \mathbf{1}$ so that $h^{(0)}_i = h^{(0)}_{J(t)} = 1.$ □

The next result represents the a.s. limit of $S(t)/t$ in terms of $\kappa(\cdot)$, in a similar vein as the expression $\mathbb{E}X = \kappa'_X(0)$ for the mean of an r.v. X in terms of its c.g.f. κ_X:

Proposition 6.9 $\kappa'(0) = \sum_{i \in E} \pi_i (\beta_i \mu_i - c_i).$

Proof Note that

$$K[\alpha] = \Lambda + \alpha (\beta_i \mu_i - c)_{\text{diag}} + o(\alpha) \quad \text{so that } K'[0] = (\beta_i \mu_i - c)_{\text{diag}}. \tag{6.3}$$

Differentiating $K[\alpha] h^{(\alpha)} = \kappa(\alpha) h^{(\alpha)}$ w.r.t. α gives

$$K'[\alpha] h^{(\alpha)} + K[\alpha] k^{(\alpha)} = \kappa'(\alpha) h^{(\alpha)} + \kappa(\alpha) k^{(\alpha)}.$$

Letting $\alpha = 0$, multiplying by $v^{(0)} = \pi$ to the left and using (6.3) together with $h^{(0)} = \mathbf{1}$, $\pi K[0] = \mathbf{0}$ and $\kappa(\alpha) = 0$ gives

$$\pi (\beta_i \mu_i - c_i)_{\text{diag}} h^{(\alpha)} + 0 = \kappa'(0) \pi \mathbf{1} + 0 = \kappa'(0). \quad \square$$

6.3   Cramér–Lundberg Theory

Since the definition of $\kappa(s)$ is a direct extension of the definition for the Cramér–Lundberg model, the Lundberg equation is $\kappa(\gamma) = 0$. We assume that a solution $\gamma > 0$ exists and use notation like $\mathbb{P}_{L;i}$ instead of $\mathbb{P}_{\gamma;i}$; also, for brevity we write $h = h^{(\gamma)}$ and $v = v^{(\gamma)}$.

Corollary 6.10 (Lundberg's Inequality) $\psi_i(u) \leq \dfrac{h_i}{\min_{j \in E} h_j} e^{-\gamma u}.$

Proof Use the martingale in Proposition 6.7 with $\alpha = \gamma$ together with arguments similar to those in the proofs of Corollary 2.3 and Proposition 2.2. □

Assuming it has been shown that $C = \lim_{u\to\infty} \mathbb{E}_{L;i}[e^{-\gamma\xi(u)}/h_{J_{\tau(u)}}]$ exists and is independent of i, it also follows immediately that:

Theorem 6.11 (The Cramér–Lundberg Approximation)   *In the light-tailed case, it holds for some constant C that $\psi_i(u) \sim h_i C e^{-\gamma u}$ for all i.*

However, the calculation of C is non-trivial.

Exercises
6.1 In continuation of the operational time argument in Exercise 1.1, define

$$\theta(T) = \int_0^T c_{J(t)}\, dt, \quad \widetilde{J}(t) = J_{\theta^{-1}(t)}, \quad \widetilde{S}(t) = S_{\theta^{-1}(t)}.$$

Show that $\{\widetilde{S}(t)\}$ is a Markov regime switching risk process with constant premium rate $\widetilde{c} \equiv 1$ and parameters $\widetilde{\lambda}_{ij} = \lambda_{ij}/c_i$, $\widetilde{\beta}_i = \beta_i/c_i$, and that the time change does not affect the infinite horizon ruin probabilities, i.e. $\widetilde{\psi}_i(u) = \psi_i(u)$.

6.2 Fill in the details in the proof of Lundberg's inequality in Corollary 6.10.

Notes
An extensive treatment of the model is given in Asmussen and Albrecher [8, Ch. VII]. An important extension [6, Ch. XI] is to allow some extra claims arriving with distribution, say, B_{ij} at epochs of state transitions $i \to j$ of $J(t)$. This has the appealing feature of making the claim surplus process dense in $D[0, \infty)$ and also includes renewal arrivals with phase-type interarrival time distributions.

7   Level-Dependent Premiums

We now consider the model where the company's income rate is not fixed at c (so far interpreted as the premium rate) but is a function $c(u)$ of the current reserve u. That is, the model is

$$R(t) = u + \int_0^t c(R(s))\, ds - \sum_{i=1}^{N(t)} V_i. \tag{7.1}$$

Examples:

a) The company adjusts its premium to the reserve, such that the premium is smaller for large u where the risk is smaller and larger for small u where the ruin probability $\psi(u)$ is non-negligible.
b) The company charges a constant premium rate p but invests its money at interest rate r. Then $c(u) = p + ru$.
c) The company invests at rate r as in b) when $R(t) > 0$ but borrows the deficit in the bank when the reserve goes negative, say at interest rate r^*. Thus at deficit $x > 0$ (meaning $R(t) = -x$), the payout rate of interest is r^*x and *absolute ruin*

occurs when this exceeds the premium inflow p, i.e. when $x > p/r^*$, rather than when the reserve itself becomes negative. In this situation, we can put $\widetilde{R}(t) = R(t) + p/r^*$,

$$\widetilde{c}(u) = \begin{cases} p + r(u - p/r^*) & u > p/r^*, \\ p - r^*(p/r^* - u) & 0 \le u \le p/r^*. \end{cases}$$

Then the ruin problem for $\{\widetilde{R}(t)\}$ is of the type defined above, and the probability of absolute ruin with initial reserve $R(0) = u \in [-p/r^*, \infty)$ is given by $\widetilde{\psi}(u + p/r^*)$.

The main general result about the ruin probability $\psi(u)$ is an integral equation for its derivative $\psi'(u)$:

Proposition 7.1 *For the model* (7.1), $g(u) = -\psi'(u)$ *satisfies*

$$c(u)g(u) = (1 - \psi(0))\beta\overline{B}(u) + \beta \int_0^u \overline{B}(u - y)g(y)\,\mathrm{d}y. \tag{7.2}$$

Proof For some small $h > 0$, we write $\psi(u) = \psi_0(u) + \psi_{1+}(u)$, where $\psi_0(u), \psi_{1+}(u)$ are the contributions from the events $N(h) = 0$, resp. $N(h) \ge 1$. Neglecting $o(h)$ terms here and in the following, the reserve will have grown to $u + c(u)h$ at time h if $N(h) = 0$ so that

$$\psi_0(u) = (1 - \beta h)\psi(u + c(u)h) = \psi(u) - \beta\psi(u)h + \psi'(u)c(u)h.$$

If $N(h) \ge 1$, only the event $N(h) = 1$ contributes and ruin will occur immediately if the claim V_1 exceeds u. Otherwise, the reserve has value $u - y \in [0, u]$ right after the claim and so

$$\psi_{1+}(u) = \beta h\left[\overline{B}(u) + \int_0^u \psi(u - y) B(\mathrm{d}y)\right]$$

$$= \beta h\left[\overline{B}(u) - \psi(0)\overline{B}(u) + \psi(u) - \int_0^u \psi'(u - y)\overline{B}(y)\,\mathrm{d}y\right]$$

$$= \beta h\left[\overline{B}(u)(1 - \psi(0)) + \psi(u) - \int_0^u \psi'(u)\overline{B}(u - y)\,\mathrm{d}y\right],$$

where we used integration by parts in the second step. Thus

$$0 = \psi_0(u) + \psi_{1+}(u) - \psi(u)$$

$$= h\Big[-\beta\psi(u) + \psi'(u)c(u)$$

$$+ \beta\overline{B}(u)(1 - \psi(0)) + \beta\psi(u) - \beta \int_0^u \psi'(u)\overline{B}(u - y)\,\mathrm{d}y\Big].$$

Here the $\beta\psi(u)$ terms cancel and after division by h, we arrive at an expression equivalent to (7.2).   □

Equation (7.2) has the form

$$g(u) = f(u) + \int_0^u K(u, y)g(y)\,dy.$$

This is a *Volterra equation*, and the general form of the solution is an infinite sum $\sum_0^\infty K^n f$ of the iterates K^n of the integral kernel K. This is obviously far from explicit. However, for B exponential one can reduce it to an ODE to get the following explicit result. Define

$$\omega(x) = \int_0^x \frac{1}{c(t)}\,dt.$$

Then $\omega(x)$ is the time it takes for the reserve to reach level x provided it starts with $R_0 = 0$ and no claims arrive. Note that it may happen that $\omega(x) = \infty$ for all $x > 0$, say if $c(x)$ goes to 0 at rate x or faster as $x \downarrow 0$.

Corollary 7.2 *Assume that B is exponential with rate δ, $\overline{B}(x) = e^{-\delta x}$ and that $\omega(x) < \infty$ for all $x > 0$. Then for $u > 0$, the ruin probability is $\psi(u) = \int_0^\infty g(y)\,dy$, where*

$$g(x) = \frac{(1 - \psi(0))\beta}{c(x)} \exp\{\beta\omega(x) - \delta x\}. \tag{7.3}$$

Further $\psi(0) = \dfrac{J}{1 + J}$, where $J = \displaystyle\int_0^\infty \frac{\beta}{c(x)} \exp\{\beta\omega(x) - \delta x\}dx$.

Proof We may rewrite (7.2) as

$$g(x) = \frac{1}{c(x)}\left\{(1 - \psi(0))\beta e^{-\delta x} + \beta e^{-\delta x}\int_0^x e^{\delta y}g(y)\,dy\right\} = \frac{\beta}{c(x)}e^{-\delta x}h(x),$$

where $h(x) = (1 - \psi(0)) + \int_0^x e^{\delta y}g(y)\,dy$, so that

$$h'(x) = e^{\delta x}g(x) = \frac{\beta}{c(x)}h(x).$$

Thus

$$\log h(x) = \log h(0) + \int_0^x \frac{\beta}{c(t)}dt = \log h(0) + \beta\omega(x),$$

$$h(x) = h(0)e^{\beta\omega(x)} = (1 - \psi(0))e^{\beta\omega(x)},$$

$$g(x) = e^{-\delta x}h'(x) = e^{-\delta x}(1 - \psi(0))\beta\omega'(x)e^{\beta\omega(x)},$$

which is the same as the expression in (7.3). That $\psi(0)$ has the asserted value follows from

$$\psi(0) = \int_0^\infty g(y)\,dy = (1 - \psi(0))J\,. \qquad\qquad \square$$

Notes
Ruin theory with level-dependent premiums $c(x)$ is surveyed in Asmussen and Albrecher [8, Ch. VIII]. It is notable that taking $c(x)$ random, as occurs for example by investing the reserve in a risky asset, makes the ruin probability substantially larger, cf. [8, VIII.6].

8 The Diffusion Approximation

The diffusion approximation of a risk process is close to the classical invariance principle for a random walk S_n, stating in a functional limit formulation that

$$\left\{\left(S_{\lfloor nt\rfloor} - \lfloor nt\rfloor m\right)/n^{1/2}\right\}_{t\ge 0} \xrightarrow{\mathcal{D}} W_{0,\sigma^2}\,, \tag{8.1}$$

where $m = \mathbb{E}S_1$, $\sigma^2 = \operatorname{Var}S_1$, $\xrightarrow{\mathcal{D}}$ indicates convergence in distribution in the Skorokhod space $D[0,\infty)$ and $W_{\mu,\sigma^2} = \{W_{\mu,\sigma^2}(t)\}_{t\ge 0}$ is Brownian motion with drift μ and variance constant σ^2. The connection is that the surplus process $\{S(t)\}$ can be viewed as a continuous time random walk since it has independent stationary increments (is a Lévy process).

The positive loading of the premiums is a key feature of insurance risk modelling and makes the centering by the mean in (8.1) somewhat inconvenient. Instead, suitable assumptions on the loading give a functional limit theorem where the limit is Brownian motion with non-zero drift. This has several applications. One, to be given later in Chap. XI dealing with stochastic control problems, is to provide a justification for using Brownian motion as a model for the reserve rather than the more detailed Cramér–Lundberg dynamics. Another is to use the continuous mapping theorem to get approximations for the Cramér–Lundberg ruin probabilities in terms of these for Brownian motion. This, together with the basics of the functional limit theorem, is the focus of this section.

We will need families of claim surplus processes. The representation we use is

$$S^c(t) = \sum_{i=1}^{N(t)} V_i - ct\,,$$

where c varies but the Poisson rate β of N and the distribution B of the V_i are fixed. Positive loading then means $c > c_0 = \beta\mu_B$, and the key functional limit theorem to be derived is:

Theorem 8.1 *As $c \downarrow c_0$,*

$$\left\{(c - c_0)S^c\big(\lfloor t/(c - \rho)^2\rfloor\big)\right\}_{t\ge 0} \xrightarrow{\mathcal{D}} W_{-1,\sigma^2} \quad \textit{where } \sigma^2 = \beta\mu_B^{(2)}\,.$$

From this, the continuous mapping theorem easily gives an approximation for the finite horizon ruin probabilities:

Corollary 8.2 *As $c \downarrow c_0$, it holds for any fixed $T < \infty$ that*

$$\psi^c\big(x/(c - c_0), \, T/(c - c_0)^2\big) \;\to\; \mathbb{P}\Big(\sup_{t \leq T} W_{-1,\sigma^2}(t) > x \Big).$$

Note that the r.h.s. is an inverse Gaussian tail and hence explicitly computable as

$$1 - \Phi\Big(\frac{x}{\sqrt{T\sigma^2}} + \sqrt{T\sigma^2} \Big) + e^{-2x/\sigma^2} \Phi\Big(-\frac{x}{\sqrt{T\sigma^2}} + \sqrt{T\sigma^2} \Big).$$

Here weak convergence theory only tells that Corollary 8.2 is accurate when T is bounded away from 0 and ∞. However, an additional argument gives the result also for $T = \infty$:

Theorem 8.3 *As $c \downarrow c_0$,*

$$\psi^c\big(x/(c - c_0)\big) \;\to\; \mathbb{P}\Big(\sup_{t \geq 0} W_{-1,\sigma^2}(t) > x \Big) \;=\; e^{-2x/\sigma^2}.$$

The implication of these results is to substitute $u = x/(c - c_0)$ to get the approximations

$$\psi^c\big(u, \, T/(c - c_0)^2\big) \;\approx\; \mathbb{P}\Big(\sup_{t \leq T} W_{-1,\sigma^2}(t) > (c - c_0)u \Big),$$

$$\psi^c(u) \;\approx\; e^{-2(c-c_0)u/\sigma^2}.$$

Whereas the proofs of Theorem 8.1 and Corollary 8.2 require the machinery of weak convergence in function spaces, there is a more elementary proof of Theorem 8.3. We start with this and proceed then to the more advanced material.

Lemma 8.4 *Define $M(c, \beta) = \sup_{t \geq 0} S^c(t)$. Then $(c - c_0)M(c, \beta)$ converges in distribution as $c \downarrow c_0$ to the exponential distribution with rate $2/\sigma^2$.*

Proof We use the continuity theorem for the Laplace transform. By (3.1), we have

$$\mathbb{E}e^{\alpha M(c,\beta)} \;=\; \mathbb{E}e^{\alpha M(1,\beta/c)} \;=\; \frac{1 - \beta\mu_B/c}{1 - \beta\mu_B/c\widehat{B}_0[\alpha]} \;=\; \frac{c - c_0}{c - c_0\widehat{B}_0[\alpha]}.$$

Therefore

$$\mathbb{E}e^{s(c-c_0)M(c,\beta)} \;\approx\; \frac{c - c_0}{c - c_0\big(1 + \mu_{B_0}s(c - c_0)\big)} \;=\; \frac{1}{1 - c_0\mu_{B_0}s}.$$

The result then follows since $c_0\mu_{B_0} = \beta\mu_B\mu_B^{(2)}/2\mu_B = \sigma^2/2$. $\square$

For the proof of Theorem 8.3, now just note that

$$\psi^c\big(x/(c-c_0)\big) = \mathbb{P}\big((c-c_0)M(c,\beta) > x\big).$$

Proof of Theorem 8.1 and Corollary 8.2 We shall need the standards facts that if $\chi, \eta, \eta_n, \xi_n$ are random elements of $D[0,\infty)$, then

a) $\xi_n = \{\xi_n(t)\}_{t\geq 0} \xrightarrow{\mathcal{D}} \xi = \{\xi(t)\}_{t\geq 0}$ in $D[0,\infty)$ if and only if $\{\xi_n(t)\}_{0\leq t\leq T}$
$\xrightarrow{\mathcal{D}} \{\xi(t)\}_{0\leq t\leq T}$ in $D[0,T]$ for any fixed $0 < T < \infty$,

b) if in addition $\|\xi_n - \eta_n\|_T \xrightarrow{\mathbb{P}} 0$ for all T where $\|\chi\|_T = \sup_{t\leq T}|\chi(t)|$, then also $\eta_n \xrightarrow{\mathcal{D}} \xi$ in $D[0,\infty)$.

Write $S^c(t) = C(t) - (c-c_0)t$, where $C(t) = \sum_1^{N(t)} V_i - c_0 t$. The key step is to apply the random walk invariance principle (8.1) and b) to ξ_n, η_n where $\xi_n(t) = C(\lfloor nt \rfloor)/n^{1/2}$, $\eta_n(t) = C(t)/n^{1/2}$. Since $C(1), C(2), \dots$ is a mean zero random walk with $\mathbb{V}\mathrm{ar}\, C(1) = \sigma^2 = \beta\mu_B^{(2)}$, (8.1) immediately gives $\xi_n \xrightarrow{\mathcal{D}} \xi$, where $\xi = W_{0,\sigma^2}$. To verify the condition of b), note that by sample path inspection $C(k+1) - c_0 \leq C(t) \leq C(k) + c_0$ for $k \leq t \leq k+1$ and therefore $\|\xi_n - \eta_n\|_T \leq M_{T,n}/n^{1/2}$, where $M_{T,n} = \max_{k<nT}|C(k+1) - C(k)|$. But the $C(k+1) - C(k)$ are i.i.d. and centered with finite second moment, and so

$$\mathbb{P}(M_{T,n} > \epsilon n^{1/2}) \leq nT\mathbb{P}(C(1) > \epsilon n^{1/2}) \leq T\frac{\mathbb{E}\big[C(1);\, C(1) > \epsilon n^{1/2}\big]}{\epsilon^2} \to 0$$

for any $\epsilon > 0$. This establishes $\|\xi_n - \eta_n\|_T \xrightarrow{\mathbb{P}} 0$ and $\eta_n \xrightarrow{\mathcal{D}} W_{0,\sigma^2}$.

Now Theorem 8.1 will follow if for any sequence c_n satisfying $n^{1/2}(c_n - c_0) \to 1$ it holds that $\chi_n \to W_{-1,\sigma^2}$, where

$$\chi_n(t) = \frac{1}{n^{1/2}}S^{c_n}(nt) = \eta_n(t) - \frac{c_n - c_0}{n^{1/2}}.$$

But this is obvious from $\eta_n \xrightarrow{\mathcal{D}} W_{0,\sigma^2}$ and $n^{1/2}(c_n - c_0) \to 1$. $\square$

Notes
In queueing theory, results of the type of the present section go under the name of *heavy traffic limit theorems*. The term heavy traffic also has an obvious interpretation in risk theory: on average, the premiums exceed only slightly the expected claims. That is, heavy traffic conditions mean that the safety loading η is positive but small, or equivalently that β is only slightly smaller than $\beta_{\max} = c/\mu_B$.

Chapter V: Markov Models in Life Insurance

1 The Contract Payments and the Probability Model

We consider an insurance policy issued at time 0 and terminating at a fixed finite time n (not necessarily an integer!). There is a finite set of states of the policy, $\mathcal{J} = \{0, \ldots, J\}$, and $Z(t)$ denotes the state of the policy at time $t \in [0, n]$. By convention, 0 is the initial state at time 0. As in Sect. A.3 of the Appendix, Z is taken as a time-inhomogeneous Markov process (RCLL, i.e. with paths in $D[0, \infty)$), the history of the policy up to and including time t is represented by the σ-algebra $\mathcal{F}^Z(t) = \sigma(Z(s), s \in [0, t])$, and its development is given by the filtration $\mathscr{F}^Z = (\mathcal{F}^Z(t))_{t \in [0,n]}$. Further, $N^k(t)$ denotes the number of transitions into state k before t, and we write $N = (N^k)_{k \in \mathcal{J}}$ for the associated J-dimensional and RCLL counting process.

Let $B(t)$ denote the *total amount of contractual benefits less premiums* payable during the time interval $[0, t]$. We assume that it develops in accordance with the dynamics

$$\mathrm{d}B(t) = \mathrm{d}B^{Z(t)}(t) + \sum_{k:\, k \neq Z(t-)} b^{Z(t-)k}(t)\, \mathrm{d}N^k(t), \qquad (1.1)$$

where B^j is a deterministic and sufficiently regular function specifying payments due during sojourns in state j, and b^{jk} is a deterministic and sufficiently regular function specifying payments due upon transition from state j to state k. We assume that each B^j decomposes into an absolutely continuous part and a discrete part, i.e.

$$\mathrm{d}B^{Z(t)}(t) = b^{Z(t)}(t)\, \mathrm{d}t + \Delta B^{Z(t)}(t), \qquad (1.2)$$

S. Asmussen, M. Steffensen, *Risk and Insurance*, Probability Theory and Stochastic Modelling 96, https://doi.org/10.1007/978-3-030-35176-2_5

where $\Delta B^j(t) = B^j(t) - B^j(t-)$, when different from 0, is a jump representing a lump sum payable at time t if the policy is then in state j. The set of time points with jumps in $(B^j)_{j \in \mathcal{J}}$ is denoted $\mathcal{T} = \{t_0, t_1, \ldots, t_q\}$, where $t_0 = 0$ and $t_q = n$.

We note that the distribution of the time-continuous Markov process Z is fully described by the intensities μ_{ij}. The intensities fully characterize the transition probabilities and, for a Markov process, the transition probabilities fully characterize the finite-dimensional distributions which, again, characterizes the distribution of the Markov process. The transition probabilities are characterized by the intensities via Kolmogorov's differential equations given in Theorem A.1 of the Appendix. The states of the policy relate to the state of an insured's life, or possibly the states of two or more. These states are decisive to the payment stream specified in the contract. Throughout, we show examples of state models, payment streams and valuation results. We start out here by showing the canonical cases of a 2-, 3-, and 4-state model, respectively, in order give the reader some applications to have in mind when we develop the mathematics further below.

2   Canonical Models

2.1   Mortality Modelling

Originally, the distribution of the lifetime of an individual was described by a so-called mortality- or life table. A mortality table is a table of the number l_x of survivors at age x from a cohort consisting of l_0 newborns. As an example, the following numbers are extracted from the official Danish 2015–2016 mortality table for women, with cohort size $10^5 = 100{,}000$.

x	10	20	30	40	50	60	70	80	90	99
l_x	99,621	99,533	99,349	98,934	97,876	94,025	86,643	68,740	31,163	3,494

In a mortality table it is not spelled out that l_x is thought of as the expected number of survivors as one is not interested in quantifying the risk behind the numbers in such a table. Working with probabilistic terms it is more appropriate to work with the probability distribution, density, and similar terms of the non-negative stochastic variable describing the age of death T of an individual. We introduce the distribution function $F(t) = \mathbb{P}(T \le t)$ and the density $f(t) = F'(t) = \frac{\mathrm{d}}{\mathrm{d}t} F(t)$. In this case, it is natural to speak of $\overline{F}(t) = 1 - F(t)$ as the survival function. All these functions describe, of course, the distribution of T. So does the mortality rate (or force of mortality) $\mu(t)$ which we define below. Outside the domain of human lifetime modelling, e.g. in engineering where the lifetime of a system or device is modeled, the mortality rate is called the hazard rate or the failure rate. The mortality rate is defined by

$$\mu(t) = \frac{f(t)}{\overline{F}(t)}.$$

Intuitively, it is the rate of the death probability conditional on survival. We can namely write, sloppily,

$$\mu(t)dt = \frac{f(t)dt}{\overline{F}(t)} \approx \frac{F(t+dt) - F(t)}{\overline{F}(t)} = \mathbb{P}(T \le t + dt \mid T > t).$$

Note carefully that μ is not a probability in itself. Only after multiplying it by a (short, actually infinitesimally short) interval length does it become a (small, actually infinitesimally small) probability.

Since the chain rule gives that

$$\frac{d}{dt} \log \overline{F}(t) = -\frac{f(t)}{\overline{F}(t)} = -\mu(t),$$

we have a differential equation for the function $\log \overline{F}(t)$. Since $\overline{F}(0) = 1$, the side condition to the differential equation is $\log \overline{F}(0) = 0$, such that $\log \overline{F}(t) = -\int_0^t \mu(s)\, ds$, or

$$\overline{F}(t) = e^{-\int_0^t \mu(s)\, ds}.$$

The probability distribution and the density can therefore be expressed in terms of μ,

$$F(t) = 1 - e^{-\int_0^t \mu(s)\, ds},$$

$$f(t) = e^{-\int_0^t \mu(s)\, ds} \mu(t).$$

One can, in general, connect mortality tables with the probabilistic approach by the link

$$\overline{F}(t) = \frac{l_t}{l_0}.$$

The functions F, $\overline{F}$, f, and μ are equally good at formalizing the distribution of T. However, for two reasons, it is natural to think of μ as the fundamental function from where the other functions are derived. Firstly, we are primarily interested in probabilities and densities conditional upon survival until a given age and it turns out that these quantities do not change structure, expressed in terms of μ, compared with the unconditional ones. This is convenient, at least from a notational point of view. E.g., we can write the probability of surviving further t years, conditional upon surviving until age x as

$$\mathbb{P}(T > x + t \mid T > x) = \frac{\mathbb{P}(T > x + t)}{\mathbb{P}(T > x)}$$

$$= \frac{e^{-\int_0^{x+t} \mu(s)\,ds}}{e^{-\int_0^x \mu(s)\,ds}} = e^{-\int_x^{x+t} \mu(s)\,ds} = e^{-\int_0^t \mu(x+s)\,ds}.$$

Also, we can calculate the density in the conditional distribution as

$$f(t \mid x) = -\frac{d}{dt}\mathbb{P}(T > x + t \mid T > x) = -\frac{d}{dt}\frac{\overline{F}(x+t)}{\overline{F}(x)} = -\frac{d}{dt}e^{-\int_x^{x+t} \mu(s)\,ds}$$

$$= e^{-\int_x^{x+t} \mu(s)\,ds}\mu(x+t) = e^{-\int_0^t \mu(x+s)\,ds}\mu(x+t).$$

We see that the structure remains across conditioning upon survival. Only the ages at which μ is evaluated under the integral change to the residual ages to survive, after having survived age x.

Another argument for considering the mortality rate as fundamental, is that relatively simple functions μ fit human mortality fairly well. Often a parametric model is used and it is a parametric statistical exercise to fit the parameters to the data observed. The first attempt to model human mortality by a probability distribution was made by *de Moivre* in 1724, who assumed a maximal life span of ω and F to be uniform on $(0, \omega)$. De Moivre's suggestion was to take $\omega = 86$ but obviously, the uniform distribution is very crude. In modern life insurance mathematics, a popular choice is the *Gompertz–Makeham* distribution where

$$\mu(t) = a + be^{ct}, \qquad \int_0^t \mu(s)\,ds = at + \frac{b}{c}(e^{ct} - 1).$$

Gompertz suggested the age-dependent exponential term of μ in 1825 to formalize the acceleration of mortality with age. *Makeham* added the age-independent constant in 1860 to account for mortality which is not related to ageing, e.g. accidents. An example is the set of parameters in the so-called G82 base, which was used by Danish insurance companies for several decades. They are

$$a = 0.0005, \quad b = 0.000075858, \quad c = \log(1.09144).$$

The hazard rate and the density of the G82 base are given in Fig. 1. The shape of the density is obviously close to what one expects for human mortality.

Some points worth noting in mortality modelling:

1) Whether a parametric model such as G82 exactly matches the mortality in the population is not that essential because of later regulation by bonus, see Sects. VI.5–7 below.

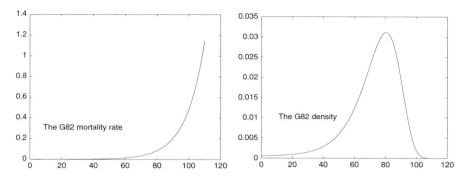

Fig. 1 Hazard rate and density for G82

2) A point to be taken seriously in life insurance is the so-called *select mortality*, meaning that the mortality of the insured is often different from the mortality in the population as a whole.

3) A crucial topical problem is the change in the mortality pattern (so-called *longevity*): people now get significantly older, so that $\mu(x)$ cannot be assumed to be the same now as 20 years ago or in 20 years from now. In different terms, a cohort born 25 years ago does not have the same mortality pattern as one born 60 years ago. A major difficulty is that this development is very hard to predict quantitatively. Again, bonus mitigates this.

2.2 The Survival Model

The simplest possible example is the two-state survival model shown in Fig. 2.

In terms of the distribution of the lifetime T of the insured, $\mu_{01}(t)$ is simply the hazard rate. Note that 'lifetime' should be understood as the time from the contract is signed until death and not the total lifetime. Thus, we always condition upon survival until the age at which the contract is signed. From the comments in Sect. 2.1 it follows that this can be handled by replacing $\mu(t)$ by $\mu(x + t)$ in all formulas below. We disregard this emphasis on the age x upon initiation of the contract and use the abbreviated form $\mu(t)$ from now on.

Within the survival model, we can formalize simple examples of insurance and pension contracts via specification of the payment coefficients in B. The contract

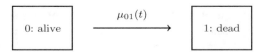

Fig. 2 Survival model

is not formulated in terms of differentials of payment streams but in plain words that should be understood by non-experts. Therefore, a part of the exercise is the translation of the contract formulation into mathematics. Throughout, think of $n \leq \infty$ as the time of the contract termination and of $m \leq n$ as the retirement time.

Example 2.1 (Pure Endowment)

- Contract formulation: the contract pays out 1 unit upon survival until termination $n = m$.
- Mathematical formulation: All payment coefficients are zero except for

$$\Delta B^0(n) = 1.$$

This product is bought by an individual who wants a lump sum benefit payment upon retirement to support consumption during retirement.   ◇

Example 2.2 (Term Insurance)

- Contract formulation: The contract pays out 1 unit at the time of death if death occurs before termination n. A whole-life insurance is specified through $n = \infty$.
- Mathematical formulation: All payment coefficients are zero except for

$$b^{01}(t) = \mathbb{1}_{t<n}.$$

This product is bought by an individual who has dependants (e.g. spouse or children) and wishes to support their consumption in case he dies. The dependants receive a lump sum upon death to support consumption after death. The benefit makes up for the consumption opportunities that are naturally lost when he dies. They are closely related to income during years at the labor market. It is therefore natural to stop the death benefit at the time of retirement m. But there is in principle nothing wrong with having a term insurance that also covers post-retirement ages.   ◇

Example 2.3 (Deferred Temporary Life Annuity)

- Contract formulation: The contract pays out 1 unit per time unit upon survival from time $m < n$ until termination n. A non-deferred life annuity is specified through $m = 0$. A non-temporary life annuity is specified through $n = \infty$.
- Mathematical formulation: All payment coefficients are zero except for

$$b^0(t) = \mathbb{1}_{m<t<n}.$$

This product is bought by an individual who wants a benefit payment stream during retirement until death to support consumption during that period. The motivation for this product is similar to the motivation for the pure endowment. However, in contrast to the pure endowment, the total amount which is the accumulated benefits paid from m to n or death, whatever occurs first, is a stochastic amount. The uncertainty of the time horizon of the life annuity is meaningful from the insured's point of view. Namely, his need for consumption also has the same uncertain time horizon. Therefore the life annuity actually takes away the uncertainty in the amount the insured can consume until he dies. ◇

By the three contracts above, we have specified the three elementary types of common payment streams. The pure endowment and the term insurance are both so-called lump sum payments with the difference that one pays upon survival to a deterministic time and the other pays upon death at a stochastic time. Both the pure endowment and the temporary life annuity are paid upon survival with the difference that one pays out a lump sum and the other is paid out continuously.

The payment process B is introduced to encompass both benefits and premiums. Thus the payment processes specified above never come alone in a contract since there has to be a premium side as well. All three benefits can be paid for by a single premium. Then the contract is fully paid for by a lump sum paid at time 0. This means that $\Delta B^0(0)$ is a negative (so far unspecified) amount that has to be settled in accordance with some fairness criterion.

More often, the contract is paid for by a so-called level premium, i.e. a premium rate paid from the policy holder to the insurance company until death or a certain age, say m, whatever occurs first. This premium payment has the same profile as the non-deferred temporary life annuity, temporary until time m, but with a negative sign since it is a premium or a negative benefit. Let us denote the premium rate by π (Greek p abbreviates *premium*), such that

$$b^0(t) = -\pi \, \mathbb{1}_{0 < t < m} \,.$$

What is now left is to determine a fair π. This is postponed until the next section.

Within the survival model, we can characterize the transition probabilities by the corresponding Kolmogorov equations. In Fig. 2, $\mu_{01}(t)$ above the arrow indicates that the transition from alive to dead occurs with that intensity. Obviously, if we just have the survival model in mind, talking about states and Markov processes is unnecessarily complicated. The model consists of one single stochastic variable, namely the stochastic lifetime T, the distribution of which we represent by its hazard rate $\mu(t)$ as discussed in Sect. 2.1. There are good reasons for doing so. Often we are interested in the survival probability and the density conditional upon survival until a certain point in time. This is so, because both at initiation of the contract and for valuation at later points in time, we want to take into account that the policyholder survived that far—or not, if he didn't.

The formulas in Sect. 2.1 can, of course, be obtained as solutions to Kolmogorov's differential equation in the special case of the survival model. Only one intensity $\mu_{01}(t)$ is in play here, and we write $\mu(t) = \mu_{01}(t)$. Consider first the conditional survival probability $\mathbb{P}(T > s \mid T > t) = p_{00}(t, s)$. Since $p_{10}(t, s) = 0$ (which is obvious, but can also itself be calculated from the corresponding ODE): the backward ODE (A.3.5) for p_{00} reduces to

$$\frac{\partial}{\partial t} p_{00}(t, s) = -\mu(t) p_{00}(t, s); \quad p_{00}(s, s) = 1,$$

with solution $e^{-\int_t^s \mu} = e^{-\int_t^s \mu(v)\,dv}$. Consider next

$$\mathbb{P}(T \le s \mid T > t) \;=\; p_{01}(t, s) \;=\; 1 - e^{-\int_t^s \mu}. \tag{2.1}$$

The final expression in (2.1) can be seen directly from $p_{00}(t, s) + p_{01}(t, s) = 1$ but can also be obtained otherwise. Since $p_{11}(t, s) = 1$ (which is obvious, but can also itself be calculated from the corresponding ODE), the backward ODE for p_{01} reduces to

$$\frac{\partial}{\partial t} p_{01}(t, s) \;=\; \mu(t) p_{01}(t, s) - \mu(t); \qquad p_{01}(s, s) = 0,$$

with solution

$$p_{01}(t, s) = \int_t^s e^{-\int_t^u \mu} \mu(u)\,du, \tag{2.2}$$

which gives (2.1). Although the expression (2.1) is more compact, the integral expression (2.2) allows a different heuristic interpretation that is helpful later. For the policyholder to die over the interval t to s, he first has to survive until an intermediary time point u and he does so w.p. $e^{-\int_t^u \mu}$. Then he has to die in $[u, u + du)$ and, conditional on surviving that far, he does so w.p. $\mu(u)\,du$. Now, finally the integral sums over all time points where death can occur.

The survival model is trivially a Markov model. This is most easily understood by realizing that, at any point in time, either the future or the past is fully known. Therefore, conditional on the present, the future and the past are independent and, thus, the process is Markov.

Exercises
2.1 Let e_x denote the life expectancy of an x-year-old. Show that

$$e_x \;=\; \int_0^\infty t e^{-\int_0^t \mu} \mu(t)\,dt \;=\; \int_0^\infty e^{-\int_0^t \mu}\,dt.$$

2.3   The Disability Model

The canonical example of a three-state model is the so-called disability model shown in Fig. 3.

Within this model we can, for example, specify the disability annuity that pays 1 unit per time unit as long as the policyholder is disabled until the contract ends. The mathematical formulation is $b^1(t) = 1,\, t \le n$, and if the contract is paid for by a level premium as long as the policy holder is active, we add that $b^0(t) = -\pi,\, t \le n$. An alternative way to cover the disability risk is to specify a disability sum paid out

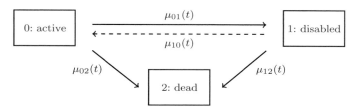

Fig. 3 Disability model with mortality, disability, and possible recovery

upon disability with the mathematical formulation $b^{01}(t) = 1, t \le n$. The products are bought by an individual who needs a benefit income to support his consumption during disability as a compensation for the labor income that has fallen away. Since time spent in disability is uncertain, the time span of the need for consumption is also uncertain. Therefore, the disability annuity does a better job in leveling out consumption opportunities during the course of life than the disability sum does. Nevertheless, the disability sum serves well to cover single costs from restructuring living condition upon disability like, for example, making a home handicap friendly.

Within the disability model we can characterize the transition probabilities by the corresponding Kolmogorov equations,

$$\frac{\partial}{\partial t} p_{00}(t, s) = \big(\mu_{01}(t) + \mu_{02}(t)\big) p_{00}(t, s) - \mu_{01}(t) p_{10}(t, s); \quad p_{00}(s, s) = 1,$$

$$\frac{\partial}{\partial t} p_{01}(t, s) = \big(\mu_{01}(t) + \mu_{02}(t)\big) p_{01}(t, s) - \mu_{01}(t) p_{11}(t, s); \quad p_{01}(s, s) = 0,$$

$$\frac{\partial}{\partial t} p_{10}(t, s) = \big(\mu_{10}(t) + \mu_{12}(t)\big) p_{10}(t, s) - \mu_{10}(t) p_{00}(t, s); \quad p_{10}(s, s) = 0,$$

$$\frac{\partial}{\partial t} p_{11}(t, s) = \big(\mu_{10}(t) + \mu_{12}(t)\big) p_{11}(t, s) - \mu_{10}(t) p_{01}(t, s); \quad p_{11}(s, s) = 1.$$

Obviously, once these probabilities are known, p_{02} can be determined by $p_{00} + p_{01} + p_{02} = 1$ and p_{12} can be determined by $p_{10} + p_{11} + p_{12} = 1$. We see that the 4-dimensional system of differential equations characterizing $(p_{00}, p_{01}, p_{10}, p_{11})$ is essentially two 2-dimensional systems of differential equations, namely one system covering (p_{00}, p_{10}) and another system covering (p_{01}, p_{11}). We can present the solution to p_{00} in the form (A.3.7) given in the Appendix, which yields

$$p_{00}(t, s) = \int_{t}^{s} e^{-\int_{t}^{u}(\mu_{01}+\mu_{02})} \mu_{01}(u) p_{10}(u, s) du + e^{-\int_{t}^{s}(\mu_{01}+\mu_{02})}. \tag{2.3}$$

This is, however, non-constructive since a similar expression for p_{10} contains p_{00} itself. Still, it allows for the following interpretation: what does it take to go from 0 to 0 over (t, s)? Well, either we never leave state 0 and the probability of this is accounted for in the last term of (2.3). Or, we stay in state 0 until an intermediary time point u (w.p. $e^{-\int_t^u (\mu_{01}+\mu_{02})}$), become disabled over the next infinitesimal time interval (with conditional probability $\mu_{01}(u)du$), and then, from the disability state, go back to active over the residual time interval (u, s), possibly turning disabled more times during that time interval (w.p. $p_{10}(u, s)$). Since this first occurrence of disability can happen at any point in time over (t, s) we add up, finally, via the integral. In spite of this interpretation, there is no way around simultaneous calculation of p_{00} and p_{10}, e.g. via numerical solution of the 2-dimensional differential equation. However, if we set the recovery rate to zero, $\mu_{10}(t) \equiv 0$, p_{00} drops out of the equation for p_{10} and p_{01} drops out of the equation for p_{11} such that the two 2-dimensional systems are reduced to four 1-dimensional differential equations. Here, p_{10} collapses to zero, and p_{00} reduces to the last term of (2.3), since the policy holder can only go from active to active over (s, t) by staying there uninterruptedly. We get $p_{11}(t, s) = e^{-\int_t^s \mu_{12}}$, which can be plugged directly into the solution for p_{01} to get

$$p_{01}(t, s) = \int_t^s e^{-\int_t^u (\mu_{01}+\mu_{02})} \mu_{01}(u) e^{-\int_u^s \mu_{12}} \, du.$$

This is a constructive presentation of a solution with an interpretation similar to the one mentioned above for p_{00}.

In the case of a positive reactivation intensity, $\mu_{10} > 0$, apart from solving the 2-dimensional equation for p_{00} and p_{10} simultaneously, there are different numerical ideas to approximate the solution to the 2-dimensional system by (a series of) 1-dimensional differential equations. One idea is to set $p_{10}^{(1)}$ to zero (wrongly, but in accordance with the terminal condition) and produce an approximate $p_{00}^{(1)}$ based on that. Then produce a $p_{10}^{(2)}$ based on $p_{00}^{(1)}$ and a $p_{00}^{(2)}$ based on $p_{10}^{(2)}$. Continue until the $p_{00}^{(n)}$ and $p_{10}^{(n)}$ is satisfactorily close to $p_{00}^{(n-1)}$ and $p_{10}^{(n-1)}$. Doing so, we have to solve a series of equations but since these are all 1-dimensional, this may be much easier than solving one single 2-dimensional system.

Another idea is based on keeping track of the number of times the active state and the disability state have been (re-)entered. The idea is to have an approximate feed-forward model which can be solved by computing simple integrals recursively, cf. Sect. A.3.1 of the Appendix. In Fig. 4 we illustrate the model with reactivation

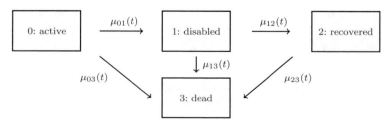

Fig. 4 Disability model with mortality, disability, and possible recovery

but where disability cannot occur more than once. If we now want to approximate p_{00} in the original model, we can calculate $p_{00} + p_{02}$ in the approximate model. In the approximate model p_{00} is simple. The probability p_{02} is now obtained by first, simply, calculating p_{22}, plugging it into the 1-dimensional equation for p_{12}, and then finally plugging this into the 1-dimensional equation for p_{02}. Obviously the approximate model can be extended with further numbers of occurrence of disability and reactivation. If we model potentially k disability occurrences, the original p_{00} is approximated by $p_{00} + p_{02} + \cdots + p_{0(k+1)}$. The more possible occurrences we take into account, the better we approximate the probabilities in the original model.

Exercises

2.2 Assume that $\mu_{02} = \mu_{12}$. Show that

$$p_{00}(t, n) + p_{01}(t, n) = e^{-\int_t^n \mu_{02}}.$$

Interpret the assumption and the result.

2.4 The Spouse Model

The canonical example of a four-state model is the so-called spouse model shown in Fig. 5.

Within that model we can, for example, specify the spouse annuity that pays 1 unit per time unit to the wife after the death of her husband until she dies or the contract ends, whatever occurs first. The mathematical formulation is $b^2(t) = 1, t < n$, and if the contract is paid for by a level premium paid as long as both members of the couple are alive, we add $b^0(t) = -\pi, t < n$. An alternative way for the wife to

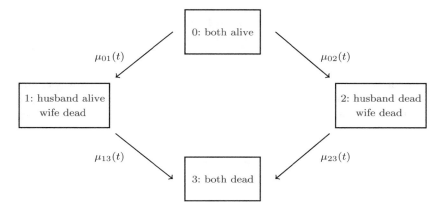

Fig. 5 Spouse model of two lives

cover the death risk of her husband is to specify a death sum paid out upon his death, conditional on the wife being alive, with the mathematical formulation $b^{02}(t) = 1$, $t < n$. This is obviously different from a plain death coverage on the husband's life since no payments are made if the wife dies first.

As we did for the disability model, we can now specify the various transition probabilities in the spouse model. Although the number of states is higher, the case is simpler since the model has, already from the beginning, a feed-forward (hierarchical) structure. Proceeding as for the disability model without recovery or as in Sect. A.3.1 of the Appendix, we first compute the $p_{ii}(t, s) = e^{-\int_t^s \mu_i}$. Using the backward equation, formula (A.3.6) in the Appendix, then gives $p_{01}(t, s) = \int_t^s e^{-\int_t^u \mu_i} \mu_{01}(u) p_{11}(u, s)\, du$ and similar expressions for p_{02}, p_{13}, p_{23}. Then p_{03} is finally obtained by $p_{00} + p_{01} + p_{02} + p_{03} = 1$.

The spouse model is an illustrative tool to discuss dependence between two positive stochastic variables, in this case containing the lifetimes of the husband and the wife, respectively. Intuitively, there are three 'levels' of dependence that can occur. First take $\mu_{01} = \mu_{23}$ and $\mu_{02} = \mu_{13}$. Then we have, as a starting point, independence between the two lives since the mortality intensity of one member of the married couple is not influenced by the death of the other. However, even in that model we can introduce a weak 'level' of dependence by letting the intensities themselves be stochastic. It should be stressed that then Z is not a Markov process anymore so we are a bit beyond the basic theory here. Only conditionally upon the μ_{ij} is Z Markov. This means that the two lives are dependent because they belong to the same generation and therefore are influenced by the same (uncertain) demographic evolution of that generation. It makes no difference whether the two lives are married or not.

A stronger 'level' of dependence occurs in a so-called contagion model, where we let (leaving the idea of stochastic intensities) μ_{23} differ from μ_{01} and μ_{13} differ from μ_{02}. If, for instance, we take $\mu_{23} > \mu_{01}$ and $\mu_{13} > \mu_{02}$ we speak of positive contagion in the sense that the death of one life accelerates the death of the other. There seems to be statistical evidence of positive contagion from the wife to the husband ($\mu_{13} > \mu_{02}$) whereas there is no significant contagion from the husband to the wife ($\mu_{23} \approx \mu_{01}$). One can speculate about the sociological or psychological explanation of that difference. A widely accepted hypothesis is that the wife (on average) is the more active socializer and upon her death, her husband essentially dies in loneliness and isolation, whereas this is not the case the other way around.

The strongest 'level' of dependence occurs if we add a transition intensity from state 0 to state 3, $\mu_{03} > 0$. This means that the same event kills both lives and can be realistic in the spouse model since the two lives tend to drive the same car and take the same flight, etc. In that respect, it may be, in particular, the accidental deaths that dominate μ_{03}, assuming this to be independent of age. This is the 'joint death' model. Note that, setting all other intensities than μ_{03} to zero, we have that the two lifetimes are equal.

Macro-dependence, contagion and joint death models are easy to understand in the context of life insurance but are used even more in the sister field, credit risk. There, the two members of the couple are two firms that are alive as long as they are solvent or a similar measure of creditworthiness. One can again imagine the lifetimes of the two firms being dependent just because they are influenced by the same (uncertain) macro-economic developments, corresponding to stochastic intensities. But one can also argue for a stronger dependence. If one firm delivers some goods which are needed in the production of the other firm, we may have positive contagion for both firms. Conversely, the two firms may be competing in the same market, such that the death of one firm opens up market opportunities for the other. This may lead to negative contagion, formalized by $\mu_{23} < \mu_{01}$ and $\mu_{13} < \mu_{02}$, meaning that the insolvency of one firm decelerates the insolvency of the other. Finally, the 'joint death' intensity may be costive if the two firms are part of the same group.

Exercises

2.3 Write down the expressions for p_{02}, p_{13}, and p_{23}. Assume now that $\mu_{01} = \mu_{23}$ and $\mu_{02} = \mu_{13}$. Show that

$$p_{02}(t, n) = e^{-\int_t^n \mu_{01}} \int_t^n e^{-\int_t^s \mu_{02}} \mu_{02}(s) \, ds \, .$$

Interpret the assumption and interpret the result.

2.4 Sketch a spouse model where each member can become disabled before death. Discuss the dependency between the two courses of life one can imagine in this situation.

3 Valuation of the Payments

We assume that the investment portfolio earns return on investment by a deterministic interest rate $r(t)$. As before, we use the short-hand notation $\int_t^s r = \int_t^s r(\tau) \, d\tau$ (equalling $r(s - t)$ if $r(t) \equiv r$) throughout.

The insurer is interested in evaluation of the future liabilities with respect to the contract. The canonical approach to this quantity is to think of the insurer having issued a large number of similar contracts with payment functions linked to independent lives. The law of large numbers then leaves the insurer with a liability per insured which tends to the expected present value of future benefits less premiums, given the past history of the policy, as the number of policy holders tends to infinity. We say that the valuation technique is based on diversification of risk. The conditional expected present value is called the *reserve* and appears on the liability side of the insurer's balance scheme. It is given by

$$V(t) = \mathbb{E}\left[\int_t^n e^{-\int_t^s r} \, dB(s) \, \middle| \, \mathcal{F}(t)\right]. \tag{3.1}$$

We can now plug in the payment function B to get

$$V(t) = \mathbb{E}\left[\int_t^n e^{-\int_t^s r}\left(dB^{Z(s)}(s) + \sum_{\ell: \ell \neq Z(s-)} b^{Z(s-)\ell}(s)\, dN^\ell(s)\right) \Big| \mathcal{F}(t)\right].$$

In the basic framework, there are two reasons why we can replace conditioning upon $\mathcal{F}(t)$ by conditioning upon $Z(t)$. Part of the explanation is that Z is Markov, but this is clearly not enough. It is also crucial that the future payments are, exclusively, dependent on the future positions of Z. This has nothing to do with Z being Markov or not. Both conditions need to be fulfilled and each one of them can destroy the idea of the reserve fully specified by Z. The process Z may not be Markov while, still, future payments depend exclusively on future positions of Z. This is, for example, the case in the disability model if the reactivation intensity depends on the duration of the disability. Alternatively, Z may still be Markov, while the payments depend on the history of Z. This is the case, for instance, in the disability model if the disability annuity depends on the duration of sojourn in the disability state. Anyway, as long as Z is Markov and all payments depend on the current state of Z, we can introduce the reserve function

$$V^{Z(t)}(t) = \mathbb{E}\left[\int_t^n e^{-\int_t^s r}\left(dB^{Z(s)}(s) + \sum_{\ell: \ell \neq Z(s-)} b^{Z(s-)\ell}(s)\, dN^\ell(s)\right) \Big| Z(t)\right],$$

(3.2)

which simply takes the current state of Z as argument.

Calculation of V has various purposes. It is a 'best estimate of the present value of future payments', which is a fundamental entry on the liability side of an insurance company's balance scheme. The reserve is also used for pricing in the sense that we formulate a fairness criterion in terms of V. We say that the contract is actuarially fair if

$$V(0-) = 0. \tag{3.3}$$

This means that prior to initiation of the contract, the benefits (positive increments of B) and premiums (negative increments of B) must fulfill this single equation. In a concrete situation there are typically several payment coefficients in B. All these payment coefficients, except for one, have to be specified in the contract by 'someone else', that is the policy holder, his labor union, his employer or his wife. Then the last payment coefficient which is not already specified by 'someone else' is determined by the insurance company as the single unknown in the single equivalence equation (3.3). The procedure is often referred to as the *equivalence principle*. Later in this chapter, we present examples of how this works. For now, we concentrate on the calculation of V.

There are different routes to take in order to calculate V. One approach is the direct representation via transition probabilities.

Proposition 3.1 *The reserve V can be written in the form*

$$V^j(t) = \int_t^n e^{-\int_t^s r} \sum_k p_{jk}(t,s)\left(dB^k(s) + \sum_{\ell:\ell\neq k} \mu_{k\ell}(s)b^{k\ell}(s)\,ds\right). \qquad (3.4)$$

Proof Taking the conditional expectation under the integral and under the (with respect to Z) constant discount factor, we get

$$V^j(t) = \int_t^n e^{-\int_t^s r} \mathbb{E}\big[dB^{Z(s)}(s) \,\big|\, Z(t) = j\big] \qquad (3.5)$$

$$+ \int_t^n e^{-\int_t^s r} \sum_{\ell:\ell\neq Z(s-)} \mathbb{E}\big[b^{Z(s-)\ell}(s)\,dN^\ell(s) \,\big|\, Z(t) = j\big]. \qquad (3.6)$$

For (3.5) we have that

$$\mathbb{E}\big[dB^{Z(s)}(s) \,\big|\, Z(t) = j\big] = \sum_k p_{jk}(t,s)\,dB^k(s),$$

while for (3.6)

$$\mathbb{E}\left[\sum_{\ell:\ell\neq Z(s-)} b^{Z(s-)\ell}(s)\,dN^\ell(s) \,\bigg|\, Z(t) = j\right]$$

$$= \mathbb{E}\left[\mathbb{E}\left[\sum_{\ell:\ell\neq Z(s-)} b^{Z(s-)\ell}(s)\,dN^\ell(s) \,\bigg|\, Z(t) = j, Z(s-)\right] \,\bigg|\, Z(t) = j\right]$$

$$= \mathbb{E}\left[\sum_{\ell:\ell\neq Z(s-)} b^{Z(s-)\ell}(s)\mathbb{E}\big[dN^\ell(s) \,\big|\, Z(t) = j, Z(s-)\big] \,\bigg|\, Z(t) = j\right]$$

$$= \sum_{\ell:\ell\neq Z(s-)} \mathbb{E}\big[b^{Z(s-)\ell}(s)\mu_{Z(s-)\ell}(s)\,ds \,\big|\, Z(t) = j\big]$$

$$= \sum_k \sum_{\ell:\ell\neq k} \mathbb{E}\big[\mathbb{1}\big(Z(s-) = k\big)b^{k\ell}(s)\mu_{k\ell}(s)\,ds \,\big|\, Z(t) = j\big]$$

$$= \sum_k \sum_{\ell:\ell\neq k} p_{jk}(t,s)\mu_{kl}(s)b^{k\ell}(s)\,ds$$

$$= \sum_k p_{jk}(t,s) \sum_{\ell:\ell\neq k} \mu_{k\ell}(s)b^{k\ell}(s)\,ds.$$

Inserting these two results into (3.5)–(3.6) gives (3.4). $\qquad\qquad\square$

We write the reserve more densely as

$$V^j(t) = \int_t^n e^{-\int_t^s r} \, dC_{t,j}(s),$$ (3.7)

with the (conditional expected) cash flow payment process $C_{t,j}$ defined by

$$dC_{t,j}(s) = \sum_k p_{jk}(t,s)\left(dB^k(s) + \sum_{\ell:\ell\neq k} \mu_{k\ell}(s)b^{k\ell}(s)\,ds\right).$$

We can calculate the transition probabilities $p_{jk}(t,s)$, plug them into $dC_{t,j}(s)$ and sum them up via (3.7). In that case, note that we need the transition probabilities $p_{ij}(t,s)$ for a fixed t but for many different times s. For this it is more efficient to work with the forward Kolmogorov equations than the backward equations. They give all the probabilities we need in one single numerical run of the (multi-dimensional) system. The route via the transition probabilities is particularly convenient if we want to calculate the reserve for different interest rates, e.g. if we are interested in the interest rate sensitivity of the reserve, $\frac{\partial}{\partial r} V(t,r)$. In that case, we only need to calculate the burdensome cash flow once while multiple aggregations in (3.7) are simple. If we are only interested in calculating the reserve once, and for other reasons that we come back to, we can embed the Kolmogorov equation for the probabilities into a system of differential equation characterizing the reserve itself. This is the celebrated Thiele differential equation. We show different routes to take.

Proposition 3.2 *The reserve V^j fulfills the system of ordinary differential equations such that at points t of differentiability, i.e. with $\Delta B^j(t) = 0$,*

$$\frac{d}{dt}V^j(t) = r(t)V^j(t) - b^j(t) - \sum_{k:k\neq j} \mu_{jk}(t)\left(b^{jk}(t) + V^k(t) - V^j(t)\right),$$ (3.8)

$$V^j(n) = 0.$$

When $\Delta B^j(t) \neq 0$, we have the gluing condition

$$V^j(t-) = V^j(t) + \Delta B^j(t).$$

Proof First we consider time points of differentiability. Using the formula

$$\frac{d}{dt}\int_t^n g(t,s)\,ds = -g(t,t) + \int_t^n \frac{d}{dt}g(t,s)\,ds$$

(cf. (A.1.3) in the Appendix) we get by direct differentiation that

$$
\begin{aligned}
\frac{d}{dt} V^j(t) &= \frac{d}{dt} \int_t^n e^{-\int_t^s r} \left(\sum_k p_{jk}(t,s) \left[dB^k(s) + \sum_{\ell:\ell\neq k} \mu_{k\ell}(s) b^{k\ell}(s)\, ds \right] \right) \\
&= -e^{-\int_t^t r} \left(\sum_k p_{jk}(t,t) \left[b^k(t) + \sum_{\ell:\ell\neq k} \mu_{k\ell}(t) b^{k\ell}(t) \right] \right) \\
&\quad + \int_t^n \frac{d}{dt} e^{-\int_t^s r} \left(\sum_k p_{jk}(t,s) \left[dB^k(s) + \sum_{\ell:\ell\neq k} \mu_{k\ell}(s) b^{k\ell}(s) \right] \right) ds \\
&\quad + \int_t^n e^{-\int_t^s r} \left(\sum_k \frac{d}{dt} p_{jk}(t,s) \left[dB^k(s) + \sum_{\ell:\ell\neq k} \mu_{k\ell}(s) b^{k\ell}(s) \right] \right) ds \\
&= -\left(b^j(t) + \sum_{k:k\neq j} \mu_{jk}(t) b^{jk}(t) \right) + r(t) V^j(t) \\
&\quad + \int_t^n e^{-\int_t^s r} \left(\sum_k \frac{d}{dt} p_{jk}(t,s) \left[dB^k(s) + \sum_{\ell:\ell\neq k} \mu_{k\ell}(s) b^{k\ell}(s) \right] \right) ds \\
&= -\left(b^j(t) + \sum_{\ell:\ell\neq j} \mu_{j\ell}(t) b^{j\ell}(t) \right) + r(t) V^j(t) \\
&\quad + \int_t^n e^{-\int_t^s r} \left(\sum_k \mu_{j\cdot}(t) p_{jk}(t,s) \left[dB^k(s) + \sum_{\ell:\ell\neq k} \mu_{k\ell}(s) b^{k\ell}(s) \right] \right) ds \\
&\quad - \int_t^n e^{-\int_t^s r} \left(\sum_k \sum_{i:i\neq j} \mu_{ji}(t) p_{ik}(t,s) \left[dB^k(s) + \sum_{\ell:\ell\neq k} \mu_{k\ell}(s) b^{k\ell}(s) \right] \right) ds \\
&= -\left(b^j(t) + \sum_{\ell:\ell\neq j} \mu_{j\ell}(t) b^{j\ell}(t) \right) + \left(r(t) + \mu_{j\cdot}(t) \right) V^j(t) - \sum_{i:i\neq j} \mu_{ji}(t) V^i(t) \\
&= -\left(b^j(t) + \sum_{k:k\neq j} \mu_{jk}(t) b^{jk}(t) \right) + \left(r(t) + \mu_{j\cdot}(t) \right) V^j(t) - \sum_{k:k\neq j} \mu_{jk}(t) V^k(t)
\end{aligned}
$$

where $\mu_{j\cdot}(t) = \sum_{j\neq k} \mu_{jk}(t)$ and the fourth step uses Kolmogorov's differential equations.

Second we consider time points where V^j is not differentiable. At these points we have that

$$ V^j(t-) = V^j(t) + \Delta B^j(t). \qquad \square $$

The gluing condition specifies how V^j jumps at deterministic time points where $\Delta B^j \neq 0$. The terminal condition for the reserve is $V^j(n) = 0$, which together with the gluing condition gives that

$$ V^j(n-) = \Delta B^j(n), $$

which may just be zero if there is no lump sump payment at the termination of the contract.

The term $R^{jk}(t) = b^{jk}(t) + V^k(t) - V^j(t)$ is spoken of as the 'sum at risk'.

The terms in the Thiele differential equation (3.8) have a nice interpretation. The statewise reserve evolves, in parts, as a bank account earning interest, $r(t)V^j(t)$, and with money flowing in or out, $-b^j(t)$. Recall that if b^j is positive, this should be thought of as the benefit rate paid from the insurance company to the policy holder. If b^j is negative, the term $-b^j(t)$ is positive and should be thought of as the premium rate paid from the policy holder to the insurance company. In addition to the bank account type movements, the policy holder has to pay a risk premium $\mu_{jk}(t)\big(b^{jk}(t) + V^k(t) - V^j(t)\big)$ for each occurrence risk there is from state j at time t, therefore the sum $\sum_{k:k\neq j}$. The rate of event k risk is $\mu_{jk}(t)$ and if that happens, the loss is the sum at risk R^{jk} since then the company has to pay out the transition payment b^{jk} and set up the new reserve corresponding to state k, V^k. On the other hand they do not need to set aside the old reserve anymore, so the net effect on the reserve is $V^k - V^j$.

Speaking of (3.8) as the dynamics of the reserve is a bit confusing. Thiele's differential equation is a backward system of ODEs characterizing V for all time points and all states. Once we have solved this, we plug in $(t, Z(t))$ to set up the actual reserve. However, since $Z(t)$ is constant between the jumps of Z we have that the differential equation also forms a forward differential equation,

$$dV(t, Z(t))\big|_{Z(t-)=Z(t)=j} = \frac{d}{dt}V^j(t)\, dt\,.$$

The interpretation above of the reserve equation being an account that is written forward by interest rates, payment rates and risk premiums paid out should be seen in that light. It is crucial to understand that this coincidence (to interpret piecewise the backward characterizing equation as a forward equation) only holds because Z is piecewise constant.

In the proof of Proposition 3.2, we simply differentiated the integral expression for V and made direct use of Kolmogorov's backward differential equations for the transition probabilities. There is also a more direct way to construct the differential equation. Therefore we give an alternative proof.

Proof First we consider

$$m(t) = \mathbb{E}\left[\int_{0-}^{n} e^{-\int_0^s r}\, d\big(-B(s)\big)\,\Big|\, \mathcal{F}(t)\right]$$

$$= \int_{0-}^{t} e^{-\int_0^s r}\, d\big(-B(s)\big) - e^{-\int_0^t r} V^{Z(t)}(t)\,,$$

which is a martingale since $\mathbb{E}[X \mid \mathcal{F}(t)]$ is so for any integrable X. We are now interested in deriving the dynamics of this martingale. First we calculate

$$e^{\int_0^t r} \, d\left(\int_{0-}^t e^{-\int_0^s r} \, d(-B(s))\right) = d(-B(t))$$

$$= \left(-b^{Z(t)}(t) \, dt - \Delta B^{Z(t)}(t) - \sum_{k:k \neq Z(t-)} b^{Z(t-)k}(t) \, dN^k(t)\right)$$

$$= \left(-b^{Z(t)}(t) \, dt - \Delta B^{Z(t)}(t) - \sum_{k:k \neq Z(t)} \mu_{Z(t)k}(t) b^{Z(t)k}(t) \, dt\right)$$

$$- \sum_{k:k \neq Z(t-)} b^{Z(t-)k}(t) \, dM^k(t),$$

where M^k is the compensated version of the counting process N^k, cf. (A.3.8) in the Appendix, and hence is a martingale. Assuming sufficient regularity of $V^{Z(t)}(t)$ we can calculate the dynamics of the reserve,

$$e^{\int_0^t r} \, d\left(e^{-\int_0^t r} V^{Z(t)}(t)\right) \tag{3.9}$$

$$= -r(t) V^{Z(t)}(t) \, dt + V_t^{Z(t)}(t) \, dt + V^{Z(t)}(t) - V^{Z(t)}(t-)$$

$$+ V^{Z(t)}(t) - V^{Z(t-)}(t)$$

$$= \left(-r(t) V^{Z(t)}(t) + V_t^{Z(t)}(t)\right) dt + V^{Z(t)}(t) - V^{Z(t)}(t-)$$

$$+ \sum_{k:k \neq Z(t-)} \left(V^k(t) - V^{Z(t-)}(t)\right) dN^k(t)$$

$$= \left(-r(t) V^{Z(t)}(t) + V_t^{Z(t)}(t) + \sum_{k:k \neq Z(t)} \mu_{Z(t)k}(t)\left(V^k(t) - V^{Z(t-)}(t)\right)\right) dt$$

$$+ V^{Z(t)}(t) - V^{Z(t)}(t-) + \sum_{k:k \neq Z(t-)} \left(V^k(t) - V^{Z(t-)}(t)\right) dM^k(t).$$

Now, we plug in these dynamics into $dm(t)$ and separate the martingale term away from the systematic increment, which is just the rest. We get that

$$e^{\int_0^t r} \, dm(t)$$

$$= \left(r(t) V^{Z(t)}(t) - b^{Z(t)}(t) - V_t^{Z(t)}(t) - \sum_{k:k \neq Z(t)} \mu_{Z(t)k}(t) R^{Z(t)k}(t)\right) dt$$

$$- R^{Z(t)}(t) - \sum_{k:k \neq Z(t-)} R^{Z(t-)k}(t) \, dM^k(t),$$

where we use the notation $R^{jk}(t) = b^{jk}(t) + V^k(t) - V^j(t)$ and $R^j(t) = \Delta B^j(t) + V^j(t) - V^j(t-)$.

Since we have a martingale on the r.h.s., the systematic increment on the r.h.s. must be zero. Since this has to hold for any realization of Z,

$$r(t)V^j(t) - b^j(t) - V_t^j(t) - \sum_{k:k\neq j} \mu_{jk}(t)R^{jk}(t) = 0,$$

which is just Thiele's differential equation, and $R^j(t) = 0$, which is the gluing condition.   □

The martingale methodology demonstrated here should be followed by a few remarks. Firstly, we have assumed sufficient regularity of V for the arguments to hold. This is not a big problem here and connects to our assumption on sufficient regularity of payment functions and transition intensities. Secondly, the martingale method in itself just teaches us that (3.8) holds for V but not up-front whether (3.8) characterizes V uniquely. We speak of (3.8) as a necessary condition on V. Only by concluding from (3.8) that it has a structure such that the solution is unique can we take this final analytical step. In contrast, the methodology demonstrated below proves directly that any function solving (3.8) indeed equals the reserve defined in (3.1). By such a result, we can also speak of (3.8) as a sufficient condition on V. We conclude our general treatment by establishing such a verification result.

Proposition 3.3 *Assume that there exists a function V^j solving*

$$\frac{d}{dt}V^j(t) = r(t)V^j(t) - b^j(t) - \sum_{k:k\neq j} \mu_{jk}(t)R^{jk}(t) \text{ for } t \notin \mathcal{T}, \qquad (3.10)$$

$$V^j(t-) = \Delta B^j(t) + V^j(t) \text{ for } t \in \mathcal{T}, \quad V^j(n) = 0.$$

Then this function is indeed equal to the reserve defined in (3.2).

Proof For simplicity, we take $r(t) \equiv r$ constant. Given a function $V^j(t)$, we have by Itô's formula (Sect. A.4 in the Appendix) that

$$d(e^{-rt}V^{Z(t)}(t))$$

$$= e^{-rt}\left(-rV^{Z(t)}(t) + dV^{Z(t)}(t)\right)$$

$$= e^{-rt}\left(-rV^{Z(t)}(t)\,dt + V_t^{Z(t)}(t)\,dt + \left(V^k(t) - V^{Z(t-)}(t-)\right)dN^k(t)\right.$$

$$\left. + V^{Z(t)}(t) - V^{Z(t)}(t-)\right).$$

By integrating over (t, n) we get

$$
e^{-rt} V^{Z(t)}(t) = -\int_t^n d\big(e^{-rs} V^{Z(s)}(s)\big) + e^{-rn} V^{Z(n)}(n)
$$

$$
= -\int_t^n e^{-rs}\Big(-r V^{Z(s)}(s)\,ds + V_s^{Z(s)}(s)\,ds + \big(V^k(s) - V^{Z(s-)}(s-)\big)\,dN^k(s)
$$

$$
+ V^{Z(s)}(s) - V^{Z(s)}(s-)\Big) + e^{-rn} V^{Z(n)}(n).
$$

We know by (3.10) that

$$
V_t^{Z(t)}(t) - r V^{Z(t)}(t)
$$

$$
= -b^{Z(t)}(t) - \sum_{k:\, k \neq Z(t)} \mu_{Z(t)k}(t)\big(b^{Z(t)k}(t) + V^k(t) - V^{Z(t)}(t)\big),
$$

and

$$
V^{Z(t)}(t) - V^{Z(t)}(t-) = \Delta B^{Z(t)}(t),
$$

so that

$$
e^{-rt} V^{Z(t)}(t)
$$

$$
= -\int_t^n e^{-rs}\Big(-r V^{Z(s)}(s)\,ds + V_s^{Z(s)}(s)\,ds + \big(V^k(s) - V^{Z(s-)}(s-)\big)\,dN^k(s)
$$

$$
+ V^{Z(s)}(s) - V^{Z(s)}(s-)\Big) + e^{-rn} V^{Z(n)}(n)
$$

$$
= \int_t^n e^{-rs}\Big(dB(s) - \sum_{k:\, k \neq Z(s-)} R^{Z(s-)k}(s)\,dM^k(s)\Big) + e^{-rn} V^{Z(n)}(n).
$$

Replacing $V^{Z(n)}(n)$ by 0 in accordance with the terminal condition, taking conditional expectation on both sides, and multiplying by e^{rt} yields

$$
V^j(t) = \mathbb{E}\Big[\int_t^n e^{-r(s-t)}\,dB(s)\,\Big|\,Z(t) = j\Big]. \qquad \square
$$

Exercises
3.1 Consider a payment stream that pays out 1 unit at time n conditional on sojourn in state k at that time point. Argue that, if we take $r = 0$, then the statewise reserve $V^j(t)$ coincides with the transition probability $p_{jk}(t, n)$, i.e.

$$
V^j(t) = p_{jk}(t, n).
$$

Show that the Thiele differential equation system characterizing V^j in that case boils down to Kolmogorov's differential equation. Does the Thiele differential equation generalize the Kolmogorov forward or backward equation?

3.2 We are interested in calculating the expected time spent in state k during the time interval $(t, n]$ conditional on being in state j at time t. Argue that, if we take $r = 0$, then this quantity can be calculated as a special case of the reserve by considering the payment stream specified by $b^k(t) = 1$ and all other payments equal to zero. Write down the Thiele differential equation, including side conditions, that characterizes the quantity.

4   Valuation in Canonical Models

Throughout this section, ε^s denotes the Dirac measure at s, and we write $I^j(t) = \mathbb{1}_{Z(t)=j}$.

4.1   The Survival Model

We are now going to evaluate a contract example in the survival model to learn how the machinery works. We assume that the contract combines all standard benefits that we introduced in Sect. 2. The life insurance part of the contract specifies that a constant insurance sum b^{01} is paid out upon death before time m. The savings part of the contract specifies both a lump sum $\Delta B^0(m)$ paid out upon survival until time m and a temporary life annuity at constant rate b^0 starting at time m and running until time $n \geq m$. The contract is paid for through a combination of a single premium π_0 and a level premium π paid until time m or death, whatever occurs first. The whole payment process of the contract can now be formalized by

$$dB(t) = -\pi_0 I^0(t)\, d\varepsilon^0(t) - \pi\, \mathbb{1}_{0<t<m} I^0(t)\, dt + b^{01}\mathbb{1}_{0<t<m}\, dN^1(t)$$
$$+ \Delta B^0 I^0(t)\, d\varepsilon^m(t) + b^0 \mathbb{1}_{m<t<n} I^0(t)\, dt \,.$$

Here we have specified the elements of the payment process in chronological order. Of course one could change order for different logical decompositions. E.g. one could gather terms with dt or terms with $I^0(t)$ with different interpretations.

We can write down the reserve in terms of transition probabilities in accordance with (3.4),

$$V^0(t) = \int_t^n e^{-\int_t^s r} \left[p_{00}(t, s)\big(dB^0(s) + \mu_{01}(s)b^{01}(s)\, ds\big) \right].$$

Then we just need to calculate $p_{00}(t, \cdot)$ in order to obtain the cash flow $C_{t,0}(s)$ and the reserve can be calculated.

Alternatively, we can now write down the Thiele differential equation for this contract. From m to n the differential equation for V^0 reads

$$\frac{d}{dt} V^0(t) = r(t) V^0(t) - b^0 - \mu_{01}(t)\big(- V^0(t)\big) = \big(r(t) + \mu_{01}(t)\big) V^0(t) - b^0,$$

with terminal condition

$$V^0\big(n-\big) = V^0(n) + \Delta B^0(n) = 0.$$

For $m \le t \le n$, it follows by Sect. A.2 in the Appendix that the solution is

$$V^0(t) = b^0 \int_t^n e^{-\int_t^s (r + \mu_{01})} \, ds.$$

At time m we have the gluing condition

$$V^0(m-) = V^0(m) + \Delta B^0(m) = b^0 \int_m^n e^{-\int_m^s (r + \mu_{01})} \, dt + \Delta B^0(m),$$

which now forms the terminal condition for a new segment of differentiability of V where

$$\frac{d}{dt} V^0(t) = r(t) V^0(t) + \pi - \mu_{01}(t)\big(b^{01} - V^0(t)\big).$$

For $0 \le t < m$, this has solution

$$V^0(t) = \int_t^m e^{-\int_0^s (r + \mu_{01})} \big(\mu_{01}(s) b^{01} - \pi\big) ds$$

$$+ b^0 \int_m^n e^{-\int_t^s (r + \mu_{01})} \, ds + e^{-\int_t^m (r + \mu_{01})} \Delta B^0(m).$$

At time 0 we have again a gluing condition

$$V^0(0-) = V^0(0) + \Delta B^0(0)$$

$$= \int_0^m e^{-\int_0^s (r + \mu_{01})} \big(\mu_{01}(s) b^{01} - \pi\big) ds$$

$$+ b^0 \int_m^n e^{-\int_0^s (r + \mu_{01})} \, ds + e^{-\int_0^m (r + \mu_{01})} \Delta B^0(m) - \pi_0.$$

For reserving in the accounting balance scheme we are done since we know $V^0(t)$ for all t. For pricing, we need to settle the single unknown payment coefficient after all others have been settled by 'someone else' in accordance with the equation $V^0(0-) = 0$. We can, for example, say that $(\pi_0, \pi, b^{01}, b^0)$ are all specified and the insurance company now determines $\Delta B^0(m)$ by isolating it in $V^0(0-) = 0$,

$$\Delta B^0(m) = -\int_0^m e^{\int_s^m (r+\mu_{01})} \left(\mu_{01}(s)b^{01} - \pi\right) ds$$
$$- b^0 \int_m^n e^{-\int_m^s (r+\mu_{01})} ds + e^{\int_0^m (r+\mu_{01})} \pi_0 .$$

We call this the 'equivalence pure endowment sum' as the pure endowment payment is the free unknown coefficient that is finally settled in accordance with the equivalence principle (3.3). Taking $\Delta B^0(m)$ as the unknown is not unrealistic. Often one specifies $\pi_0 = 0$ whereas (π, b^{01}, b^0) are all specified as proportions of an annual salary rate a, e.g. $\pi = 0.2a$, $b^{01} = 3a$, and $b^0 = 0.6a$. Hereby it is specified that one can afford to pay 20% of the salary to the insurance scheme. One has a need for three years' salary as a life insurance sum and a life annuity of 60% of the income. This corresponds to a so-called replacement ratio, consumption as retiree divided by consumption during saving, of 75%, since

$$\frac{b^0}{a - \pi} = \frac{0.6a}{a - 0.2a} = 0.75.$$

The residual $\Delta B^0(m)$ serves as an extra saving that allows for extra consumption immediately after retirement or additional consumption during retirement to increase the replacement ratio.

4.2 The Disability Model

We now turn to the canonical three state model. Let us consider a so-called disability annuity paying out a constant disability annuity rate b^1 during periods of disability. Let us add a constant disability sum b^{01} paid out upon occurrence of disability. The contract is paid for during periods of labor market activity by a level premium of rate π. All payments stop at retirement age $m = n$, and all other payment coefficients are set to zero. The payment stream is now formalized by

$$dB(t) = -\pi I^0(t) dt + b^{01} dN^{01}(t) + b^1 I^1(t) dt \quad \text{for } 0 < t < n .$$

Again we can write down the reserve in terms of transition probabilities in accordance with (3.4). If e.g. $Z(t) = 0$, we have

$$V^0(t) = \int_t^n e^{-\int_t^s r} \, dC_{t,0}(s),$$

$$dC_{t,0}(s) = \left(p_{00}(t, s)\left(-\pi + \mu_{01}(s)b^{01} \right) + p_{01}(t, s)b^1 \right) ds.$$

Then we just need to calculate $p_{00}(t, \cdot)$ and $p_{01}(t, \cdot)$ in order to obtain the cash flow $C_{t,0}(s)$ and the reserve can be calculated.

Alternatively, we can now write down the Thiele differential equation for this contract. From 0 to n the differential equation reads

$$\frac{d}{dt}V^0(t) = r(t)V^0(t) + \pi - \mu_{01}(t)\left(b^{01} + V^1(t) - V^0(t)\right) + \mu_{02}(t)V^0(t),$$

$$\frac{d}{dt}V^1(t) = r(t)V^1(t) - b^1 - \mu_{10}(t)\left(V^0(t) - V^1(t)\right) + \mu_{12}(t)V^1(t).$$

The terminal conditions become

$$V^0(n-) = V^0(n) + \Delta B^0(n) = 0, \quad V^1(n-) = V^1(n) + \Delta B^1(n) = 0.$$

Now, this is a system of equations where V^1 appears in the ordinary differential equation for V^0 and vice versa. So, both equations have to be solved numerically and simultaneously. Only in the special simple case of no reactivation can the two equations be solved one at a time. If $\mu^{10} = 0$, V^1 is solved by

$$\frac{d}{dt}V^1(t) = r(t)V^1(t) - b^1 + \mu_{12}(t)V^1(t); \quad V^1(n-) = 0,$$

with solution

$$V^1(t) = \int_t^n e^{-\int_t^s (r+\mu_{12})} b^1 \, ds.$$

Knowing V^1 we can solve the differential equation for V^0 as

$$V^0(t) = \int_t^n e^{-\int_t^s (r+\mu_{01}+\mu_{02})} \left[\mu^{01}(s)\left(b^{01} + V^1(s)\right) - \pi\right] ds.$$

The gluing condition at time 0 is just $V^0(0-) = V^0(0) + \Delta B^0(0) = V^0(0)$. For reserving in the accounting balance scheme we are done. For pricing, we need to settle the one unknown payment coefficients after all others have been settled by 'someone else' in accordance with the equation $V^0(0-) = V^0(0) = 0$. For example, we can say that (b^{01}, b^0) are both specified and the insurance company now determines π by isolating it in $V^0(0-) = 0$,

$$
\pi = \frac{\int_0^n e^{-\int_0^s (r+\mu_{01}+\mu_{02})} \mu_{01}(s)\left(b^{01} + V^1(s)\right) ds}{\int_0^n e^{-\int_0^s (r+\mu_{01}+\mu_{02})} ds}.
$$

The natural idea is again to specify b^1 in terms of the salary rate, e.g. $b^1 = 0.8a$, and b^{01} allows then for extra consumption upon disability.

Exercises

4.1 Consider the elementary pure endowment contract described in Example 2.1, paid for by a level premium of rate π until retirement m or death, whatever occurs first. Argue that the differential equation characterizing the reserve is

$$
\frac{d}{dt} V^0(t) = (r + \mu(t)) V^0(t) - \pi, \qquad V^0(m-) = 1.
$$

Show, by direct differentiation, that the differential equation is solved by the expression

$$
V^0(t) = e^{-\int_t^m r+\mu} - \pi \int_t^m e^{-\int_t^s r+\mu} ds.
$$

(Hint: This is a special case of the contract studied in this subsection.)

4.2 Consider the elementary term insurance contract described in Example 2.2, paid for by a level premium of rate π until retirement m or death, whatever occurs first. Write down the differential equation characterizing the reserve and its solution. (Hint: This is a special case of the contract studied in this subsection.)

Assume now that μ is constant. Show that $\pi = \mu$ fulfills the equivalence principle and specify the reserve function for that case. Interpret the model and the contract.

4.3 Consider the elementary deferred temporary life annuity contract described in Example 2.3, paid for by a level premium of rate π until retirement m or death, whatever occurs first. Write down the differential equation characterizing the reserve and its solution. Consider an individual who earns at rate a until retirement m or death, whatever occurs first. He wishes to buy a life annuity such that he has a constant consumption opportunity (salary less premium rate before retirement and annuity benefit after retirement) until death. Calculate the proportion of the income a he should pay as premium.

4.4 The disability model can be used to price a so-called premium waiver. Consider, for example, a savings contract that pays out, say, 1 as lump sum upon survival until retirement m, conditional on being alive (the elementary pure endowment). However, the level premium rate π is paid only as long as the policy holder is active in the labor market. This feature is called a premium waiver. Specify the payment stream B and the cash flow C. Calculate the excess level premium one has to pay for a pure endowment with premium waiver compared to a pure endowment without premium waiver. The difference can, naturally, be spoken of as the price for the premium waiver.

4.5 Consider the spouse annuity described in Sect. 2.4, i.e. the model in Fig. 5 and payments given by $b^2(t) = 1$, $t < n$, and $b^0(t) = -\pi$, $t < n$. Write down the payment process B and the cash flow C. Write down the Thiele differential equation characterizing the reserves (V^0, V^1, V^2). Calculate the equivalence premium rate π.

Notes

Earlier, life insurance mathematics was built up around mortality tables without much concern about probability theory. Sverdrup [169] and Hickman [87] suggested the probabilistic approach in works that contain special cases of the general Markov process formulation. Hoem [88] suggested to use the full formulation of a continuous-time finite-state Markov chains and after two decades and a number of publications on the topic, Hoem [90] presented an overview at a congress. Since the end of the 80s the tool has been widely accepted by both academics and practitioners as a relevant and powerful formalization of classical risk in a life insurance contract.

The mathematics of life insurance was often formulated in discrete time, with one-period transition probabilities being the fundamentals of the probability model and with lump sum payments at the end of periods, e.g. monthly or yearly. Although payments definitely occur in discrete time and transitions therefore can be assumed to do the same for valuation purposes, the continuous-time mathematics appears much more elegant and reader friendly, and time discretization can appropriately be postponed to the computation of numbers. However, still there are geographical differences in textbooks on the subject. Scandinavia, in particular, and large parts of continental Europe, in general, are determined to work out the formulas in continuous-time, theoretically and practically.

The Thiele differential equation was formulated for the survival model in (unpublished) works by Danish astronomer and insurance mathematician T.N. Thiele, in (generally cited) 1875; see Lauritzen [105] for further historical aspects of Thiele's work. The works by Hoem [88], [90], and others a hundred years later, generalized the tool to the multi-state case. Further generalizations occur in publications from the last 40 years, some of which appear in the following chapters of this book.

Chapter VI: Financial Mathematics in Life Insurance

In this chapter, we study products offered by insurance companies that, unlike the pure insurance policies in Chap. V, give investors both insurance and investment under a single integrated plan. The prototypes of such products are so-called unit-linked[1] insurance plans where the benefits are somehow connected to the development of some asset price process $S(t)$. Thereby the pricing involves financial considerations and not just the Markov process machinery of Chap. V. Also for with-profit products, the payments are connected to asset price processes, though in a more complicated way than for the unit-linked plan.

At the probability level, this leads into stochastic calculus, where martingales play a main role. The process $Y(t)$ we encounter typically has dynamics of the type

$$dY(t) = a(t) \, dt + b(t) \, dW(t), \qquad (0.1)$$

where W is standard Brownian motion and $a(t), b(t)$ are adapted processes. An argument that is often encountered is that $Y(t)$ can only be a martingale if $a(t) \equiv 0$, another is that taking expectations gives

$$\mathbb{E}Y(t) = Y(0) + \int_0^t \mathbb{E}a(s) \, ds, \quad \mathbb{E}Y(t) = \mathbb{E}Y(n) - \int_t^n \mathbb{E}a(s) \, ds$$

(assuming for the first identity that $Y(0)$ is deterministic). Similar identities hold for many conditional expectations.

---

[1] Also called equity-linked

© Springer Nature Switzerland AG 2020
S. Asmussen, M. Steffensen, *Risk and Insurance*, Probability Theory
and Stochastic Modelling 96, https://doi.org/10.1007/978-3-030-35176-2_6

1 Background and Simple Claims

We consider a financial market consisting of one asset with price process S^0 with dynamics given by

$$dS^0(t) = r(t)S^0(t)\,dt$$

for a deterministic interest rate r. This is actually the money market account that allowed us to discount future payment whenever we formed a 'present value' in Chap. V. However, we are now going to consider an additional process S to be part of that financial market in the sense that we can write contracts on that process but, at the moment, not in the sense that the process itself forms the price process of an asset. We assume that the process S is a time-continuous Markov process with continuous paths. Furthermore, we assume that the dynamics of S are given by the stochastic differential equation (SDE)

$$dS(t) = \alpha(t)S(t)\,dt + \sigma(t)S(t)\,dW(t), \tag{1.1}$$

$S(0) = s_0$, where W is Brownian motion, and α and σ are adapted processes.

Note from the formulation above that a financial market here is taken to consist of a set of traded assets (S^0 in this case) and a set of further state processes (S above) that specify the states of the world in which we are interested in the sense that we want to write financial contracts in terms of these states of the world and to discuss their prices. State processes that are themselves traded asset prices, as well as state processes that are not, may in general show up in all coefficients of the market. For example, one could have that r is a function of S above. Here, we just assume that it is not. In that case, one could think of saying that S is not part of the market, since it has no relation to market prices. However, in our definition of a financial market we keep it there because we are interested in learning about prices of contracts derived from S. Later on, but not until then, we are going to include S in the set of traded asset prices. This does not change the states of the world of our interest but it enlarges the set of potential investment objects from $\{S^0\}$ to $\{S^0, S\}$.

A financial contract derived from S is a specified payment stream fully determined by the outcome of the process S. We consider first a so-called simple claim which consists of one single lump sum payment at time n, which is determined by the value of S at that time point only. Thus, we consider the claim

$$\Phi\big(S(n)\big).$$

The question is now, what can we say about the price of this claim? One criterion to work with is the idea of no arbitrage. If we introduce the claim to the market without introducing the possibility of an arbitrage, we say that the price of the claim is consistent with the market prices. Arbitrage is defined in terms of a portfolio h which is a vector process of positions, adapted to the states of the world, in the available assets. The portfolio value is a process of scalar products of the portfolio and the asset prices. As long as the market consists of S^0 only, a portfolio is just a number $h^0(t)$ of S^0 held with time t-value $h^0(t)S^0(t)$. When we add the claim Φ to the market, the portfolio is $h = (h^0, h^\Phi)$ and the value at time t is $V^h(t) = h^0(t)S^0(t) + h^\Phi(t)V^\Phi(t)$, where V^Φ is the value of the claim (to be determined later). We shall only consider portfolios which are *self-financing* or *self-balancing*, meaning that they can only change value as a consequence of changes in S^0, S. Mathematically,

$$dV^h(t) = h^0(t)\, dS^0(t) + h^\Phi(t)\, dV^\Phi(t) = r(t)h^0(t)S^0(t)\, dt + h^\Phi(t)\, dV^\Phi(t).$$
$$(1.2)$$

An arbitrage means that we can construct a portfolio with value V^h such that

$$V^h(0) = 0, \quad \mathbb{P}(V^h(n) \geq 0) = 1, \quad \mathbb{P}(V^h(n) > 0) > 0. \qquad (1.3)$$

Let us assume that there exists a martingale measure, meaning that there exists a measure $\mathbb{Q}$, equivalent to $\mathbb{P}$, such that all discounted prices are martingales. Discounting means dividing by S^0, and we start by realizing that S^0 is a martingale by construction, since $S^0/S^0 = 1$ is a martingale under any measure. But the existence of a martingale measure in the market means that the martingale condition holds for all other discounted prices as well. For instance, if the claim Φ is included in the market, then also

$$e^{-\int_0^t r}V^\Phi(t) = \mathbb{E}^Q\left[e^{-\int_0^n r}V^\Phi(n)\,\middle|\,\mathcal{F}(t)\right].$$

This implies that the discounted portfolio value is also a $\mathbb{Q}$-martingale,

$$e^{-\int_0^t r}V^h(t) = \mathbb{E}^Q\left[e^{-\int_0^n r}V^h(n)\,\middle|\,\mathcal{F}(t)\right]. \qquad (1.4)$$

To see this, note that

$$d\left[-e^{-\int_0^t r}V^h(t)\right] = r(t)e^{-\int_0^t r}V^h(t)\, dt - e^{-\int_0^t r}\, dV^h(t),$$

which by inserting (1.2) and performing some simple calculations gives

$$d\left[e^{-\int_0^t r}V^h(t)\right] = h^\Phi(t) \cdot d\left[e^{-\int_0^t r}V^\Phi(t)\right]. \qquad (1.5)$$

The martingale property of $e^{-\int_0^t r}V^\Phi(t)$ and Itô integrals then gives the martingale property of $e^{-\int_0^t r}V^h(t)$.

From (1.4) it follows that

$$0 = \mathbb{E}^Q\left[e^{-\int_0^n r}V^h(n)\right] \qquad (1.6)$$

for any portfolio value V^h obtained from a position in S^0 and V^Φ and normalized by a residual position in S^0 at time 0 such that $V^h(0) = 0$. Since $\mathbb{Q}$ and $\mathbb{P}$ are equivalent, arbitrage is equivalent to

$$\mathbb{Q}(V^h(n) \geq 0) = 1, \quad \mathbb{Q}(V^h(n) > 0) > 0. \tag{1.7}$$

But (1.7) and (1.6) cannot both be true and therefore the existence of the equivalent measure rules out the arbitrage possibility.

We have here **indicated but not proved** the strong link between the existence of a martingale measure and the no arbitrage condition, which is known as the *first fundamental theorem of asset pricing*. That theorem states that a market model is arbitrage free if and only if there exists a martingale measure. Based on this insight we are now going to derive a price representation of the claims. Namely, as above we have that

$$V^\Phi(t) = \mathbb{E}^\mathbb{Q}\big[e^{-\int_t^n r} V^\Phi(n) \,\big|\, \mathcal{F}(t)\big],$$

But since holding the contract at time n and receiving $\Phi(S(n))$ are equivalent, it must be true that

$$V^\Phi(n) = \Phi(S(n)).$$

Thus, we have the representation of the price of the claim in the following form

$$V^\Phi(t) = \mathbb{E}^\mathbb{Q}\big[e^{-\int_t^n r} \Phi(S(n)) \,\big|\, \mathcal{F}(t)\big]. \tag{1.8}$$

for some measure $\mathbb{Q}$ equivalent to $\mathbb{P}$. We do not know for which martingale measure this is true. We just know that there exists a martingale measure for which it is true, because if there were not, access to the claim would lead to arbitrage opportunities. We stress that up-front we do not know much about $\mathbb{Q}$ except for the fact that it is equivalent to $\mathbb{P}$. Not knowing much about the equivalent measure $\mathbb{Q}$ is essentially equivalent to not knowing much about the value V. But the representation as a conditional expected value in (1.8) is still valid. It may just not be unique.

That there exists claims that cannot be priced uniquely does not mean that no claims can be priced uniquely. There may exist claims for which the price representation is independent of the martingale measure $\mathbb{Q}$, and for such a claim, the representation gives a unique price. Here, there is such a claim, namely the (essentially only, boring, and trivial) example given by

$$\Phi(S(n)) = 1, \quad \text{where } V^\Phi(t) = e^{-\int_t^n r}.$$

Of course, introducing S and the set of martingale measures is redundant if the purpose is to price the claim 1 in a market with the money market account S^0. Yet, the price is derived above to illustrate that, even if not all claims are uniquely priced, some may be.

A claim is called *hedgeable* if there exists a portfolio h such that $V^h(n) = \Phi(S(n))$ and, in that case, the portfolio h is said to hedge the claim $\Phi(S(n))$. Since all discounted price processes are martingales under the martingale measures we have for any portfolio that

$$e^{-\int_0^t r} V^h(t) = \mathbb{E}^\mathbb{Q}\!\left[e^{-\int_0^n r} V^h(n) \mid \mathcal{F}(t)\right].$$

If, in particular, the portfolio hedges $\Phi(S(n))$, we use $V^h(n) = \Phi(S(n))$ to write

$$V^h(t) = \mathbb{E}^\mathbb{Q}\!\left[e^{-\int_t^n r} \Phi(S(n)) \mid \mathcal{F}(t)\right]. \tag{1.9}$$

For hedgeable claims, this is true for any martingale measure and for any hedging portfolio. In our specific case, at the moment not many claims are hedgeable, though. However, there is such a claim, namely the (essentially only, boring, and trivial) example

$$\Phi(S(n)) = 1.$$

The trivial hedging strategy is a portfolio consisting of investing, at time t, the full portfolio V^h in the money market account, such that $V^h(t)/S^0(t)$ is the 'number' of money market accounts held. Consequently,

$$dV^h(t) = \frac{V^h(t)}{S^0(t)} dS^0(t) = rV^h(t)\, dt, \tag{1.10}$$

which can easily fulfill the terminal condition $V^h(n) = \Phi(S(n)) = 1$ by letting

$$V^h(t) = e^{-\int_t^n r}.$$

By comparing (1.8) and (1.9), we see that the representation of a price is equivalent to the representation of a hedging portfolio. For a hedging portfolio to exist this representation has to be unique, so a further connection is that the existence of a unique price is equivalent to the existence of a hedging portfolio and the price is, in that case, exactly equal to the value of the portfolio, $V^\Phi(t) = V^h(t)$. In that case, this value is unique and can be represented as

$$V^\Phi(t) = V^h(t) = \mathbb{E}^\mathbb{Q}\!\left[e^{-\int_t^n r} \Phi(S(n)) \mid \mathcal{F}(t)\right]. \tag{1.11}$$

We discussed above that the claim $\Phi(S(n)) = 1$ is the one (essentially only, boring, and trivial) example of such a claim and we even specified the hedging strategy.

A market is called *complete* if all claims are hedgeable. Given the representations in (1.8) and (1.9) it is tempting to conclude that all claims are hedgeable if the martingale measure is unique because, in that case, the right-hand sides of (1.8) and

(1.9) are unique and represent the unique price and the unique hedging portfolio value, respectively. We have here **indicated but not proved** the strong link between the uniqueness of the martingale measure and completeness, which is known as the *second fundamental theorem of asset pricing*. That theorem states that (under the assumption of no arbitrage or, equivalently, the existence of a martingale measure) a market model is complete if and only if the martingale measure is unique.

It is clear that the market consisting of the asset S^0 and the non-traded state process S is not complete. The portfolio (1.10) is the only portfolio we can construct and for such a portfolio we have no chance to hedge, say, $S(n)$. However, we now enrich the market by assuming that S is actually itself a price process of an asset in which we can invest. This completely changes the picture above. In that case, we know that, in order to avoid arbitrage in the market $\{S^0, S\}$, we have that the set of martingale measures is restricted such that S/S^0 is also a martingale. By the Itô formula we can calculate

$$
\begin{aligned}
d\left(\frac{S(t)}{S^0(t)}\right) &= \frac{dS(t)}{S^0(t)} - \frac{S(t)}{S^0(t)^2} dS^0(t) \\
&= \left(\alpha(t) - r\right)\frac{S(t)}{S^0(t)} dt + \sigma(t)\frac{S(t)}{S^0(t)} dW(t).
\end{aligned}
\tag{1.12}
$$

Now for an adapted process ϕ, the process $W^{\mathbb{Q}}$ defined by

$$
dW^{\mathbb{Q}}(t) = dW(t) - \phi(t)\, dt
$$

is Brownian motion under the measure $\mathbb{Q}$, which is equivalent to $\mathbb{P}$ with Radon–Nikodym derivative $d\mathbb{Q}/d\mathbb{P}$ given by Girsanov's theorem. We see that if we choose

$$
\phi(t) = \frac{r - \alpha(t)}{\sigma(t)},
$$

then (1.12) becomes

$$
d\left(\frac{S(t)}{S^0(t)}\right) = \frac{dS(t)}{S^0(t)} - \frac{S(t)}{S^0(t)^2} dS^0(t) = \sigma(t)\frac{S(t)}{S^0(t)} dW^{\mathbb{Q}}(t),
$$

so that the measure $\mathbb{Q}$ defined by $\phi = (r - \alpha)/\sigma$ is indeed a martingale measure. Also, we can represent the dynamics of S in the form

$$
\begin{aligned}
dS(t) &= \alpha(t)S(t)\, dt + \sigma(t)S(t)\left(dW^{\mathbb{Q}}(t) + \phi(t)\, dt\right) \\
&= rS(t)\, dt + \sigma(t)S(t)\, dW^{\mathbb{Q}}(t).
\end{aligned}
\tag{1.13}
$$

This is sometimes spoken of as $\mathbb{Q}$-dynamics. It is a bit misleading to speak of the dynamics as being a specific kind of dynamics since the dynamics are what they are and should therefore just be called dynamics. However, here these dynamics then have two different representations. One in terms of W, which is Brownian motion under the objective measure $\mathbb{P}$, (1.1), and therefore carelessly spoken of as $\mathbb{P}$-dynamics, and another one in terms of W^Q, which is Brownian motion under the martingale measure $\mathbb{Q}$, (1.13), and therefore carelessly spoken of as $\mathbb{Q}$-dynamics.

We can now conclude that there exists only one martingale measure. This follows from the fact (related to the martingale representation) that an equivalent change of measure of Brownian motion can only be done by changing the drift and that setting $\phi(t) = (r - \alpha(t))/\sigma(t)$ is the only possible choice above. We learned above that this means that all claims are hedgeable and that all claims have a unique price which equals the value of the hedging portfolio and which, in the case of a simple claim Φ, can be represented as

$$\mathbb{E}^Q\left[e^{-\int_t^n r}\Phi\big(S(n)\big)\,\middle|\,\mathcal{F}(t)\right].$$

The hedging portfolio can be constructed as the portfolio (h^0, h) that solves

$$dV^\Phi(t) = dV^h(t) = h^0(t)\,dS^0(t) + h(t)\,dS(t). \tag{1.14}$$

In general, however, $dV^\Phi(t)$ is not easy to find. Indeed, from the martingale representation theorem we only know that the martingale

$$m(t) = e^{-\int_0^t r}V^\Phi(t) = \mathbb{E}^Q\left[e^{-\int_0^n r}\Phi\big(S(n)\big)\,\middle|\,\mathcal{F}(t)\right] \tag{1.15}$$

has the representation

$$dm(t) = \eta(t)\,dW^Q(t) \tag{1.16}$$

for some adapted process η, but we get no constructive tool for finding η.

In the Markovian case, however, life is much simpler. If $\sigma(t) = \sigma\big(t, S(t)\big)$, using here sloppily the same letter σ for both the volatility process and the volatility function, we have from (1.13) that S is Markovian under the measure $\mathbb{Q}$. In particular, the conditional distribution of $S(n)$ given $\mathcal{F}(t)$ depends only on $S(t)$ so that the r.h.s. of (1.11) is a function of t and $S(t)$ only. That is, with a similar notational convention as used for σ we can write

$$V^\Phi(t) = V^\Phi\big(t, S(t)\big) = \mathbb{E}^Q\left[e^{-\int_t^n r}\Phi\big(S(n)\big)\,\middle|\,S(t)\right]. \tag{1.17}$$

Note that it is actually not a problem if α is not a function of S only since any non-Markovian feature of S only hidden in α disappeared in the measure transformation. Only a non-Markovian feature of S also hidden in σ is troublesome since this is preserved under the measure transformation. In the Markovian case we have from

Itô's formula that

$$
e^{\int_0^t r} d\left(e^{-\int_0^t r} V^\Phi(t, S(t))\right) \tag{1.18}
$$

$$
= -r \, dt \, V^\Phi(t, S(t)) + V_t^\Phi(t, S(t)) \, dt + V_s^\Phi(t, S(t)) \, dS(t)
$$

$$
+ \frac{1}{2} V_{ss}^\Phi(t, S(t)) \sigma^2(t, S(t)) S^2(t) \, dt
$$

$$
= \left(V_t^\Phi(t, S(t)) + V_s^\Phi(t, S(t)) r S(t) - r V^\Phi(t, S(t))\right) dt
$$

$$
+ \frac{1}{2} V_{ss}^\Phi(t, S(t)) \sigma^2(t, S(t)) S^2(t) \, dt
$$

$$
+ V_s^\Phi(t, S(t)) \sigma(t, S(t)) S(t) \, dW^Q(t) .
$$

Comparing with (1.14) and (1.16) we can derive three conclusions:

- The number h of assets S in the hedge portfolio is obtained by matching the diffusive term on the r.h.s. of (1.14) with the diffusive term on the r.h.s. of (1.18). This means

$$
h(t) \sigma(t, S(t)) S(t) = V_s^\Phi(t, S(t)) \sigma(t, S(t)) S(t) ,
$$

implying

$$
h(t) = h(t, S(t)) = V_s^\Phi(t, S(t)) . \tag{1.19}
$$

- The amount $h^0 S^0$ invested in S^0 in the hedge portfolio can then be derived from

$$
V^\Phi(t, S(t)) = h^0(t, S(t)) S^0(t) + h(t, S(t)) S(t) .
$$

- The price of the claim which is equal to the value of the hedging portfolio is characterized by

$$
V_t^\Phi(t, s) = r V^\Phi(t, s) - V_s^\Phi(t, s) r s - \frac{1}{2} V_{ss}^\Phi(t, s) \sigma^2(t, s) s^2 , \tag{1.20}
$$

$$
V^\Phi(n, s) = \Phi(s).
$$

This is so because the drift in (1.18) has to be zero for any realization of $(t, S(t))$.

In the introduction of the market, we assumed in (1.1) that σ was an adapted process. To obtain the valuation formula (1.17) we assume that σ is a function of time and stock price and we, sloppily, use the same letter σ for the process $\sigma(t)$ and the function $\sigma(t, S(t))$. The function σ appears in the differential equation (1.20). We now focus on the special case where σ is constant. This is the celebrated *Black–Scholes* market. Independence of the stock price is the crucial assumption here, and we could easily have allowed σ to be a deterministic function of time. From (1.13) we have the dynamics of S in the form

$$dS(t) = rS(t)dt + \sigma S(t)\, dW^Q(t),$$

and under these dynamics we can price simple claims by calculating the expectation in (1.17) or, equivalently, solve the PDE (1.20), with $\sigma(t, s) = \sigma$,

$$V_t^\Phi(t, s) = rV^\Phi(t, s) - V_s^\Phi(t, s)rs - \frac{1}{2}V_{ss}^\Phi(t, s)\sigma^2 s^2, \qquad (1.21)$$

$$V^\Phi(n, s) = \Phi(s).$$

The PDE (1.21) is called the *Black–Scholes differential equation*.

A special case of the solution is obtained if $\Phi(s) = (s - K)^+$, namely the so-called *Black–Scholes formula*. Since the SDE for S is solved by

$$S(n) = S(t)\exp\left\{(r - \sigma^2/2)(n - t) + \sigma\left(W^Q(n) - W^Q(t)\right)\right\},$$

the expectation in (1.17) is simply an integral with respect to the density of the normal distribution,

$V(t, S(t))$

$$= e^{-r(n-t)}\int_{-\infty}^\infty \left[S(t)\exp\{(r - \sigma^2/2)(n - t) + \sigma\sqrt{n - t}x\} - K\right]^+ \varphi(x)\, dx$$

$$= S(t)\Phi\left(d_1(t, S(t))\right) - e^{-r(n-t)}K\Phi\left(d_2(t, S(t))\right),$$

where $\varphi(x) = e^{-x^2/2}/\sqrt{2\pi}$ is the density of the standard normal distribution, Φ its c.d.f., and

$$d_1(t, S(t)) = \frac{1}{\sigma\sqrt{n - t}}\log(S(t)/K) + \frac{1}{\sigma}\left(r + \frac{\sigma^2}{2}\right)\sqrt{n - t},$$

$$d_2(t, S(t)) = \frac{1}{\sigma\sqrt{n - t}}\log(S(t)/K) + \frac{1}{\sigma}\left(r - \frac{\sigma^2}{2}\right)\sqrt{n - t}$$

$$= d_1(t, S(t)) - \sigma\sqrt{n - t}.$$

To complete the picture, we verify that this solution actually also solves the Black–Scholes PDE (1.21). To perform the differentiations needed is a time-consuming, but worthwhile exercise. Using

$$\log(s/K) = \sigma\sqrt{n-t}\,d_1(t,s) - (n-t)(r+\sigma^2/2)$$
$$= \sigma\sqrt{n-t}\,d_2(t,s) - (n-t)(r-\sigma^2/2),$$

we can calculate the partial derivatives as

$$V_t(t,s) = s\frac{e^{-d_1^2/2}}{\sqrt{2\pi}}\left(\frac{1}{2}\frac{1}{n-t}d_1(t,s) - \frac{1}{\sigma}(r+\sigma^2/2)\frac{1}{\sqrt{n-t}}\right)$$
$$- re^{-r(n-t)}K\Phi(d_2(t,s))$$
$$- e^{-r(n-t)}K\frac{e^{-d_2^2/2}}{\sqrt{2\pi}}\left(\frac{1}{2}\frac{1}{n-t}d_2(t,s) - \frac{1}{\sigma\sqrt{n-t}}(r+\sigma^2/2)\right),$$

$$V_s(t,s) = \Phi(d_1(t,s)) + \frac{e^{-d_1^2/2}}{\sqrt{2\pi}}\frac{1}{\sigma\sqrt{n-t}} - e^{-r(n-t)}K\frac{e^{-d_2^2/2}}{\sqrt{2\pi}}\frac{1}{\sigma\sqrt{n-t}}\frac{1}{s},$$

$$V_{ss}(t,s) = \frac{1}{s}\frac{e^{-d_1^2/2}}{\sqrt{2\pi}}\left(\frac{1}{\sigma\sqrt{n-t}} - d_1(t,s)\frac{1}{\sigma^2(n-t)}\right)$$
$$+ \frac{1}{s^2}e^{-r(n-t)}K\frac{e^{-d_2^2/2}}{\sqrt{2\pi}}\left(d_2(t,s)\frac{1}{\sigma^2(n-t)} + \frac{1}{\sigma\sqrt{n-t}}\right).$$

Plugging these derivatives into the Black–Scholes equation, we can verify that V is a solution. Since also $V(n,s) = (s-K)^+$, we have found a solution to the Black–Scholes PDE (1.21) that represents the price.

Exercises

1.1 Consider the model where σ is constant and the claim is $\Phi(s) = s^b$.

(i) calculate the price by taking plain expectation

$$V^\Phi(t,S(t)) = \mathbb{E}^Q\left[e^{-\int_t^n r}\Phi(S(n))\,\Big|\,S(t)\right]$$

with

$$dS(t) = rS(t)\,dt + \sigma S(t)\,dW^Q(t).$$

(Hint: First write $S(n)$ as $S(t)$ multiplied by the stochastic accumulation factor.)

(ii) Verify your result by plugging it into the PDE (1.20).
(iii) Specify the hedging strategy.
(iv) Comment on all three results in the special cases $b = 0$ and $b = 1$.

2 Payment Streams

If the claim is not a simple claim but rather a stream of payments linked to the process S, we need to adapt the insight from Sect. 1 above. We stay in the Black–Scholes market represented by the two price processes

$$dS^0(t) = r S^0(t) \, dt \,,$$

$$dS(t) = \alpha S(t) \, dt + \sigma S(t) \, dW(t) = r S(t) \, dt + \sigma S(t) \, dW^Q(t) \,,$$

where W is Brownian motion under the physical measure $\mathbb{P}$, W^Q is Brownian motion under the unique martingale measure $\mathbb{Q}$ and r, α, σ are constants rather than functions depending on t.

We consider a financial contract issued at time 0 and terminating at a fixed finite time n. Let $B(t)$ denote the total amount of contractual payments during the time interval $[0, t]$. We assume that it develops in accordance with the dynamics

$$dB(t) = b(t, S(t)) \, dt + \Delta B(t, S(t)) \,, \tag{2.1}$$

where $b(t, s)$ and $\Delta B(t, s)$ are deterministic and sufficiently regular functions specifying payments if the stock value is s at time t. The decomposition of B into an absolutely continuous part and a discrete part conforms with formulas V.(1.1)–(1.2). Again, we denote the set of time points with jumps in B by $\mathcal{T} = \{t_0, t_1, \ldots, t_q\}$ where $t_0 = 0$ and $t_q = n$.

We are now interested in calculating the value of the future payments of the contract. By arguments similar to the ones in the previous section, we have that the reserve (the value of future payments) equals

$$V(t, S(t)) = \mathbb{E}^Q \left[\int_t^n e^{-\int_t^v r} dB(v) \,\Big|\, S(t) \right]. \tag{2.2}$$

In fact, a simple way to arrive at (2.2) is to think of B as a sum (integral) of simple contracts maturing at time points v between t and n. Some of these contracts specify payments of $b(v, S(v)) \, dv$ whereas others specify payments of $\Delta B(v, S(v))$. Now we can value each contribution to the integral separately and add up via the integral. Since the valuation operator is just the $\mathbb{Q}$-expectation, we can use its linearity to obtain (2.2).

The simple claim studied in the previous section is specified by

$$b(t, s) = 0, \quad \Delta B(n, s) = \Phi(s), \quad \Delta B(t, s) = 0 \text{ for } t < n,$$

and for this claim (2.2), reduces to (1.17).

.

Proposition 2.1 *The reserve $V(t, s)$ fulfills the system of ODEs such that at points t of differentiability $(\Delta B(t, s) = 0)$,*

$$V_t(t, s) = rV(t, s) - b(t, s) - V_s(t, s)rs - \frac{1}{2}V_{ss}(t, s)\sigma^2 s^2, \qquad (2.3a)$$

$$V(n, s) = 0. \qquad (2.3b)$$

Outside points t of differentiability $(\Delta B(t, s) \neq 0)$, we have the gluing condition

$$V(t-, s) = V(t, s) + \Delta B(t, s).$$

Proof First we consider the $\mathbb{Q}$-martingale defined by

$$m(t) = \mathbb{E}^{\mathbb{Q}}\left[\int_{0-}^{n} e^{-\int_0^v r}\, d(-B(v)) \,\Big|\, \mathcal{F}(t)\right]$$

$$= \int_{0-}^{t} e^{-\int_0^v r}\, d(-B(v)) - e^{-\int_0^t r}V(t, S(t)),$$

where $\mathcal{F}$ is the natural filtration of S. We are now interested in deriving the dynamics of this martingale. First we calculate

$$e^{\int_0^t r}\, d\left(\int_{0-}^{t} e^{-\int_0^v r}\, d(-B(v))\right)$$

$$= d(-B(t)) = -b(t, S(t))\, dt - \Delta B(t, S(t)).$$

Assuming sufficient regularity of $V(t, S(t))$, we can calculate the dynamics of the reserve. This corresponds to the derivation in (1.18):

$$e^{\int_0^t r}\, d\left(e^{-\int_0^t r}V(t, S(t))\right)$$

$$= \left(V_t(t, S(t)) + V_s(t, S(t))rS(t) - rV(t, S(t))\right)dt$$

$$+ \frac{1}{2}V_{ss}(t, S(t))\sigma^2 S^2(t)\, dt + V_s(t, S(t))\sigma S(t)\, dW^{\mathbb{Q}}(t)$$

$$+ V(t, S(t)) - V(t-, S(t)).$$

Now, we plug in these dynamics in $dm(t)$ and separate the martingale term from the systematic increment, which is just the rest. We get that

$$e^{\int_0^t r}\, dm(t)$$

$$= \left(rV(t, S(t)) - b(t, S(t)) - V_t(t, S(t)) - V_s(t, S(t))rS(t)\right)dt$$

$$- \frac{1}{2} V_{ss}(t, S(t)) \sigma^2 S^2(t) \, dt - R(t, S(t))$$

$$- V_s(t, S(t)) \sigma S(t) \, dW^Q(t) \, ,$$

where we use the notation $R(t, s) = \Delta B(t, s) + V(t, s) - V(t-, s)$.

Multiplying both sides by $e^{-\int_0^t r}$, the martingale property of m implies that the systematic increment on the r.h.s. (the dt term) must be zero. Since this has to hold for any realization of S, even $S(t) = s$, we conclude that

$$r V(t, s) - b(t, s) - V_t(t, s) - V_s(t, s) r s - \frac{1}{2} V_{ss}(t, s) \sigma^2 s^2 = 0 \, ,$$

and that $R(t, s) = 0$, which is the gluing condition. $\qquad\square$

As in Sect. V.3, the martingale methodology demonstrated here provides a necessary condition on V under the assumption of regularity. We now present the verification argument stating that any function solving (2.3) indeed equals the contract value given by (2.2), so that (2.3) is indeed sufficient for characterizing V.

Proposition 2.2 *Assume that there exists a function V solving*

$$V_t(t, s) = r V(t, s) - b(t, s) - V_s(t, s) r s - \frac{1}{2} V_{ss}(t, s) \sigma^2 s^2 \, , \tag{2.4}$$

$$V(t-, s) = \Delta B(t, s) + V(t, s) \, , \tag{2.5}$$

$$V(n, s) = 0 \, .$$

Then this function is indeed equal to the reserve defined in (2.2).

Proof Given a function $V(t, s)$, we have by Itô's lemma that

$$e^{rt} \, d\left[e^{-rt} V(t, S(t)) \right]$$

$$= \left[-r V(t, S(t)) + V_t(t, S(t)) + V_s(t, S(t)) r S(t) \right.$$

$$\left. + \frac{1}{2} V_{ss}(t, S(t)) \sigma^2 S^2(t) \right] dt$$

$$+ V_s(t, S(t)) \sigma S(t) \, dW^Q(t) \, + \, V(t, S(t)) - V(t-, S(t)) \, .$$

By integrating over (t, n) and using (2.5) we get

$$- e^{-rt} V(t, S(t)) = \int_t^n d\left[e^{-rv} V(v, S(v)) \right] \tag{2.6}$$

$$= \int_t^n e^{-rv} \left[-r V(v, S(v)) + V_t(v, S(v)) + V_s(v, S(v)) r S(v) \right.$$

$$+ \frac{1}{2} V_{ss}(v, S(v))\sigma^2 S^2(v)\Big] dv$$

$$+ \int_t^n e^{-rv} V_s(v, S(v))\sigma S(v)\, dW^Q(v)$$

$$+ \sum_{t \le v \le n} e^{-rv}\big(V(v, S(v)) - V(v-, S(v))\big),$$

where for the last term we used (2.5). We know by (2.4) and (2.5) that

$$V_t(v, S(v)) \;=\; rV(v, S(v)) - b(v, S(v)) - V_s(v, S(v))rS(v)$$
$$- \frac{1}{2} V_{ss}(v, S(v))\sigma^2 S(v)^2$$

and

$$V(v-, S(v)) = \Delta B(v, S(v)) + V(v, S(v)).$$

Thus (2.6) becomes

$$e^{-rt} V(t, S(t)) \;=\; \int_t^n e^{-rv}\big[dB(v) - V_s(v, S(v))\sigma S(v)\, dW^Q(v)\big].$$

Taking $\mathbb{Q}$-expectation on both sides and multiplying by e^{rt} yields the desired conclusion

$$V(t, s) \;=\; \mathbb{E}^{\mathbb{Q}}\Big[\int_t^n e^{-r(v-t)}\, dB(v) \,\Big|\, S(t) = s\Big]. \qquad\qquad \square$$

Remark 2.3 That the reserve V is differentiable between jump points is certainly obvious in the Markov setting of Sect. V.3, and there the verification step given in Proposition V.3.3 is not really necessary since one may just use the uniqueness result for simple systems of ODEs between jumps and glue together at jumps. However, in later sections and chapters with more complicated models, this becomes less obvious. The results are therefore formulated in terms of verification theorems. ◇

Let us derive the hedging strategy. We can derive the dynamics of V by inserting the differential equation and the gluing condition characterizing V to achieve the second equality below,

$$dV(t, S(t)) \;=\; V_t(t, S(t))\, dt + V_s(t, S(t))\, dS(t) + \frac{1}{2} V_{ss}(t, S(t))\sigma^2 S^2(t)\, dt$$
$$+ V(t, S(t)) - V(t-, S(t))$$
$$= rV(t, S(t))\, dt - dB(t, S(t)) + V_s(t, S(t))\,(\alpha - r)\, S(t)\, dt$$
$$+ V_s(t, S(t))\sigma S(t)\, dW(t).$$

Now we can find the hedging strategy in stocks by matching the dW term in dV with the dW term in the hedging portfolio. The nominal stock position hS is deduced from the relation

$$h(t)S(t)\sigma \;=\; V_s(t, S(t))S(t)\sigma\,,$$

implying

$$h(t) \;=\; V_s(t, S(t))\,. \tag{2.7}$$

Thus, the hedge has the same form as in (1.19) but the investment strategy is of course more general since we allow for a stream of payments rather than a payment at a single time. The rest of the value, $V(t, S(t)) - h(t)S(t)$, is the amount invested in the riskfree asset.

Exercises

2.1 Consider the claim that pays out the stock price both at termination $t_2 = n$ and at an intermediary time point $t_1 < t_2$. Show by using Proposition 2.2 that $V(t, s) = 2s$ for $0 < t < t_1$. Determine the hedging strategy for $0 < t < t_1$ and for $t_1 < t < n$.

3 Unit-Link Insurance

We start out by recalling from Chap. V the dynamics of the reserve for an insurance contract with payment stream given by

$$dA(t) \;=\; dA^{Z(t)}(t) + \sum_{k:k\neq Z(t-)} a^{Z(t-)k}(t)\,dN^k(t)\,,$$

$$dA^{Z(t)}(t) \;=\; a^{Z(t)}(t)\,dt + \Delta A^{Z(t)}(t)\,.$$

The letter A is just chosen because we reserve the letter B for a different payment stream to be introduced below. The dynamics of the reserve are given by

$$dV^{Z(t)}(t) \;=\; rV^{Z(t)}(t)\,dt - dA^{Z(t)}(t) - \sum_{k:k\neq Z(t)} \mu_{Z(t)k}(t)R^{Z(t)k}(t)\,dt$$

$$+ \sum_{k:k\neq Z(t-)} \left(V^k(t) - V^{Z(t-)}(t)\right)dN^k(t)\,, \tag{3.1}$$

where

$$R^{jk}(t) \;=\; a^{ij}(t) + V^k(t) - V^j(t)$$

are the sums at risk. The reserve fulfills by definition the terminal condition $V^{Z(n)}(n) = 0$ and, as such, it actually forms a so-called backward stochastic differential equation. In Sect. V.3 we calculated V^j on the basis of the payment coefficients such that $V^{Z(t)}(t) = \sum_j I^j(t)V^j(t)$. Since $\Delta V^{Z(n)}(n) = \Delta A^{Z(n)}(n)$, the terminal condition $V^{Z(n)}(n) = 0$ is equivalent to the side condition $V^{Z(n)}(n-) = \Delta A^{Z(n)}(n)$, which is fulfilled since we have from the terminal condition for the reserve that $V^j(n-) = \Delta A^j(n)$. If we now want the contract to fulfill the equivalence principle, we need to settle all payment coefficients to obey

$$V^0(0-) = 0. \tag{3.2}$$

As preparation for the design of a unit-link contract, it is useful to turn the prospective view around to a retrospective one by skipping the terminal condition and considering a process V defined as the solution to the forward stochastic differential equation with initial condition (3.2). This V is not (necessarily) the conditional expected value of future payments anymore, since the payment coefficients are no longer tied together once we have given up the terminal condition $V^{Z(n)}(n-) = \Delta A^{Z(n)}(n)$. Let us introduce a different name for the forward version of the process, namely X. Also, let us allow the payment coefficients of A to depend on X itself. Thus, X is a forward stochastic differential equation with initial condition $X(0-) = 0$ and dynamics inherited from (3.1), i.e., letting

$$\varrho^{jk}(t, x) = a^{jk}(t, x) + \chi^k(t, x) - x, \tag{3.3}$$

(for the role of χ, see Remark 3.1 below), X is governed by

$$dX(t) = rX(t)\,dt - dA^{Z(t)}(t, X(t-)) - \sum_{k:k\neq Z(t)} \mu_{Z(t)k}(t)\varrho^{Z(t)k}(t, X(t))\,dt$$

$$+ \sum_{k:k\neq Z(t-)} \left(\chi^k(t, X(t-)) - X(t-)\right) dN^k(t).$$

Before starting to derive a payment stream B based on X, we allow the account to be invested not only in the risk free asset but also partly in the risky asset. More precisely, we invest the proportion $\pi(t, X(t))$ of X in stocks in the Black–Scholes market. Then the dynamics of X generalize to

$$dX(t) = X(t)\left((r + \pi(t, X(t))(\alpha - r))\,dt + \sigma\pi(t, X(t))\,dW(t)\right)$$

$$- dA^{Z(t)}(t, X(t-)) - \sum_{k:k\neq Z(t} \mu_{Z(t)k}(t)\varrho^{Z(t)k}(t, X(t))\,dt$$

$$+ \sum_{k:k\neq Z(t-)} \left(\chi^k(t, X(t-)) - X(t-)\right) dN^k(t). \tag{3.4}$$

The payment stream B is then defined by

$$dB(t) = dB^{Z(t)}(t, X(t-)) + \sum_{k: k \neq Z(t-)} b^{Z(t-)k}(t, X(t-)) \, dN^k(t),$$

where

$$dB^{Z(t)}(t, X(t-)) = b^{Z(t)}(t, X(t)) \, dt + \Delta B^{Z(t)}(t, X(t-)).$$

Remark 3.1 We can consider X as a savings account into which the insured deposits and from which the company withdraws statewise payments and risk premia between the jumps of Z. Upon a jump in Z to k at time t, X jumps to $\chi^k(t, X(t-))$ where $\chi^k(t, x)$ is a function being specified somehow, directly or indirectly, in the contract. The traditional choice is $\chi^k(t, x) = 0$ for the state k corresponding to death. One could simply have $\chi^j(t, x) = x$ for all other states j, but the present formulation opens up many other possibilities.   ◇

Example 3.2 By far the most important special case of the unit-link contract is the pure unit-link contract, specified by choosing

$$b^j = a^j, \quad b^{jk} = a^{jk}, \quad \Delta B^j = \Delta A^j.$$

One simple product design is to set

$$\Delta B^j(n, x) = \Delta A^j(n, x) = x.$$

That is, at termination all the savings go back to the insured.   ◇

We are interested in evaluating the payment stream B. The question is now under which measure should we take the expectation to form the expected value of future payments. In the payment process, there are types of risk that are fundamentally different. One difference between risks is of course the fact that some of the risk is of jump type through the state process Z while another part of the risk is diffusive risk through investment in the diffusive asset. However, this is not the key difference when discussing under which measure we should take an expectation in order to calculate something we can reasonably speak of as a financial value. The key difference is that we think of the jump risk as something that relates to a single individual policy holder or, perhaps, one single household. We also think of a portfolio of policy holders, or households, as living independent lives. Thus, when forming a large portfolio of policies, one can argue by diversification that a plain expectation under the physical probability measure is a meaningful operation to form a financial value. This is in sharp contrast to the diffusive risk, which is common for all policies in the portfolio. There, no version of the LLN helps us to consider a plain expectation as relevant for pricing. However, we are then offered something else instead, namely the financial market. The financial market teaches us how to evaluate market risk and this is exactly what the diffusive risk is.

There are rich possibilities for building up theories for the argument but, hopefully, with this reflection, the reader finds it reasonable to evaluate the payment stream under a product measure, denoted by $\mathbb{P} \otimes \mathbb{Q}$, splitting up the risks into diversifiable ones and market risks. The product measure evaluates financial risk under the $\mathbb{Q}$ measure uniquely determined by the market price of the stock whereas it evaluates diversifiable risk under the objective measure $\mathbb{P}$. The process (X, Z) is Markov and all payments in B depend on the current position or transition of (X, Z). Therefore we are interested in calculating

$$V^{Z(t)}(t, X(t)) = \mathbb{E}^{\mathbb{P} \otimes \mathbb{Q}} \left[\int_t^n e^{-\int_t^s r} \, dB(s) \,\Big|\, X(t), Z(t) \right], \tag{3.5}$$

where, under $\mathbb{P} \otimes \mathbb{Q}$, the intensity of N^k is not changed away from the given one $\mu_{Z(t)k}(t)$ and

$$dW^{\mathcal{Q}}(t) = dW(t) - \frac{r - \alpha}{\sigma} \, dt$$

is Brownian motion, in terms of which we can rewrite (3.4) as

$$\begin{aligned}
dX(t) = {}& rX(t) \, dt + \sigma \pi(t, X(t)) X(t) \, dW^{\mathcal{Q}}(t) \\
& - dA^{Z(t)}(t, X(t-)) - \sum_{k:\, k \neq Z(t)} \mu_{Z(t)k}(t) \varrho^{Z(t)k}(t, X(t)) \, dt \\
& + \sum_{k:\, k \neq Z(t-)} \left(\chi^k(t, X(t-)) - X(t-) \right) dN^k(t) \, .
\end{aligned}$$

When conditioning on the specific values (x, j) of $(X(t), Z(t))$, we write $V^j(t, x)$. Introduce

$$\begin{aligned}
\mathcal{D}_x V^j(t, x) = {}& V_x^j(t, x) \left(rx - a^j(t, x) - \sum_{k:\, k \neq j} \mu_{jk}(t) \varrho^{jk}(t, x) \right) \\
& + \frac{1}{2} V_{xx}^j(t, x) \pi^2(t, x) \sigma^2 x^2, \\
R^{jk}(t, x) = {}& b^{jk}(t, x) + V^k(t, \chi^k(t, x)) - V^j(t, x) \, . \tag{3.6}
\end{aligned}$$

We then have:

Proposition 3.3 *Assuming that $V^j(t, x)$ is sufficiently differentiable, the following set of PDEs is fulfilled:*

$$\begin{aligned}
V_t^j(t, x) = {}& rV^j(t, x) - b^j(t, x) \\
& - \sum_{k:\, k \neq j} \mu_{jk}(t) R^{jk}(t, x) - \mathcal{D}_x V^j(t, x) \, , \tag{3.7a}
\end{aligned}$$

$$V^j(t-, x) = \Delta B^j(t, x) + V^j\big(t, x - \Delta A^j(t, x)\big), \tag{3.7b}$$

$$V^j(n, x) = 0. \tag{3.7c}$$

Conversely, if there exists a set of functions $V^j(t, x)$ such that (3.7) holds, then $V^{Z(t)}(t, X(t))$ is indeed equal to the reserve defined in (3.5).

Proof Given a function $V^j(t, x)$, we have by Itô that

$$
\begin{aligned}
e^{rs}\mathrm{d}\big(e^{-rs}V^{Z(s)}(s, X(s))\big) &= -r V^{Z(s)}(s, X(s))\,\mathrm{d}s + \mathrm{d}V^{Z(s)}(t, X(s)) \\
&= -r V^{Z(s)}(s, X(s))\,\mathrm{d}s + V_s^{Z(s)}(s, X(s))\,\mathrm{d}s + \mathcal{D}_x V^{Z(s)}(s, X(s))\,\mathrm{d}s \\
&\quad + V_x^{Z(t)}(s, X(s))\,\sigma\,\pi(s, X(s))X(s)\,\mathrm{d}W^{\mathbb{Q}}(s) \\
&\quad + \sum_k \Big[V^k\big(s, \chi^k(s, X(s-))\big) - V^{Z(s-)}(s-, X(s-))\Big]\mathrm{d}N^k(s) \\
&\quad + V^{Z(s)}\big(s, X(s-) - \Delta A^{Z(s)}(s, X(s-))\big) - V^{Z(s)}(s-, X(s-)).
\end{aligned}
$$

By integrating over (t, n) and using the terminal condition (3.7c) we get

$$
\begin{aligned}
e^{-rt}V^{Z(t)}(t, X(t)) &= -\int_t^n \mathrm{d}\big(e^{-rs}V^{Z(s)}(s, X(s))\big) \\
&= -\int_t^n e^{-rs}\Big[-r V^{Z(s)}(s, X(s)) + V_s^{Z(s)}(s, X(s)) + \mathcal{D}_x V^{Z(s)}(s, X(s))\Big]\mathrm{d}s \\
&\quad -\int_t^n e^{-rs}V_x^{Z(t)}(s, X(s))\,\sigma\,\pi(s, X(s))X(s)\,\mathrm{d}W^{\mathbb{Q}}(s) \\
&\quad -\int_t^n e^{-rs}\sum_k \Big[V^k\big(s, \chi^k(s, X(s-))\big) - V^{Z(s-)}(s-, X(s-))\Big]\mathrm{d}N^k(s) \\
&\quad -\sum_{t \le s \le n} e^{-rs}V^{Z(s)}\big(s, X(s-) - \Delta A^{Z(s)}(s, X(s-))\big) - V^{Z(s)}(s-, X(s-)).
\end{aligned}
$$

We know by (3.7a) that

$$V_t^j(t, x) - r V^j(t, x) + \mathcal{D}_x V^j(t, x) = -b^j(t, x) - \sum_{k: k \ne j} \mu_{jk}(t) R^{jk}(t, x)$$

and

$$
\begin{aligned}
V^{Z(t)}\big(t, X(t-) - \Delta A^{Z(t)}(t, X(t-))\big) &- V^{Z(t)}(t-, X(t-)) \\
&= -\Delta B^{Z(t)}(t, X(t-)),
\end{aligned}
$$

so that

$$
\mathrm{e}^{-rt} V^{Z(t)}(t, X(t)) - \int_t^n \mathrm{e}^{-rs}\, \mathrm{d}B(s)
$$

$$
= -\int_t^n \mathrm{e}^{-rs} \sum_{k: k \neq Z(s-)} R^{Z(s-)k}(s, X(s-))\, \mathrm{d}M^k(s)
$$

$$
- \int_t^n \mathrm{e}^{-rs} V_x^{Z(t)}(s, X(s))\, \sigma\, \pi(s, X(s)) X(s)\, \mathrm{d}W^{\mathbb{Q}}(s)
$$

$$
+ \int_t^n \mathrm{e}^{-rs} b^{Z(s-)k}(s, X(s))\, \mathrm{d}N^k(s)
$$

$$
- \sum_{t \leq s \leq n} \mathrm{e}^{-rs} V^{Z(s)}\big(s, X(s-) - \Delta A^{Z(s)}(s, X(s-))\big) - V^{Z(s)}(s-, X(s-))
$$

$$
= \int_t^n \mathrm{e}^{-rs} \Big(\mathrm{d}B(s) - \sum_{k: k \neq Z(s-)} R^{Z(s-)k}(s, X(s-))\, \mathrm{d}M^k(s)
$$

$$
- V_x(s, X(s)) \pi(s, X(s)) \sigma X(s)\, \mathrm{d}W^{\mathbb{Q}}(s) \Big).
$$

Taking expectation under the product measure $\mathbb{P} \otimes \mathbb{Q}$ on both sides, and multiplying by e^{rt} yields

$$
V^j(t, x) = \mathbb{E}^{\mathbb{P} \otimes \mathbb{Q}}\Big[\int_t^n \mathrm{e}^{-r(s-t)}\, \mathrm{d}B(s) \,\Big|\, Z(t) = j, X(t) = x \Big]. \qquad \square
$$

Before discussing different constructions via different specifications of payment coefficients, let us turn to the issue of hedging. Note that the investment strategy parametrized by π is not meant to be a hedging strategy but an investment strategy driving X forward. And X is just the index which determines the payment coefficients of B. We can derive the dynamics of V by inserting the differential equation and the gluing condition characterizing V to achieve the second equality below,

$$
\mathrm{d}V^{Z(t)}(t, X(t))
$$

$$
= V_t^{Z(t)}(t, X(t))\, \mathrm{d}t + V_x^{Z(t)}(t, X(t))\mathrm{d}X(t)
$$

$$
+ \frac{1}{2} V_{xx}^{Z(t)}(t, X(t)) \sigma^2 \pi^2(t, X(t)) X(t)^2\, \mathrm{d}t
$$

$$
+ \sum_k \big[V^k\big(t, \chi^k(t, X(t-))\big) - V^{Z(t-)}(t-, X(t-)) \big]\, \mathrm{d}N^k(t)
$$

$$
+ V^{Z(t)}\big(t, X(t-) - \Delta A^{Z(t)}(t, X(t-))\big) - V^{Z(t)}(t-, X(t-))
$$

$$= rV^{Z(t)}(t, X(t))\, dt + V_x^{Z(t)}(t, X(t))X(t)\pi(t, X(t))(\alpha - r)\, dt$$

$$+ V_x^{Z(t)}(t, X(t))X(t)\sigma\pi(t, X(t))\, dW(t)$$

$$- dB(t) + \sum_{k:k\neq Z(t-)} R^{Z(t-)}(t, X(t-))\, dM^k(t).$$

Now we can find the hedging strategy in stocks by matching the dW term in dV with the dW term in the hedging portfolio. The nominal stock position hS is deduced from the relation

$$h(t)S(t)\sigma\, dW(t)\ =\ V_x^{Z(t)}(t, X(t))X(t)\sigma\pi(t, X(t))\, dW(t)\,,$$

implying

$$h(t)S(t)\ =\ V_x^{Z(t)}(t, X(t))X(t)\pi(t, X(t))\,. \tag{3.8}$$

The rest of the value, $V^{Z(t)}(t, X(t)) - h(t)S(t)$, is then invested in the riskfree asset.

It is, however, extremely important to understand that what we speak of above as a hedging strategy does not really hedge the cash flow of payment in B. They are not hedgeable. Actually, what is hedged is some specific conditionally expected payments in B. The hedge fails by the martingale term

$$\sum_{k:k\neq Z(t-)} R^{Z(t-)}(t, X(t-))\, dM^k(t)\,,$$

which appears in dV but has no investment counterpart to offset it. On the other hand, the LLN comes to help here.

Example 3.4 As in Example 3.2, consider the pure unit-link contract specified by

$$b^j = a^j,\ \ b^{jk} = a^{jk},\ \ \Delta B^j(t) = \Delta A^j(t),\ t < n\ \ \Delta B^j(n, x) = \Delta A^j(n, x) = x\,.$$

For this specification of the payment coefficients in B, we have that, for $t < n$,

$$V^j(t, x) = x. \tag{3.9}$$

We need to check that our candidate function (3.9) fulfills the system specified in Proposition 3.3. The terminal condition (3.7c) is fulfilled by definition of (3.5). What is left is to check (3.7a) and (3.7b). Firstly, we calculate for (3.9)

$$V_t^j(t, x) = 0,\ \ V_x^j(t, x) = 1,\ \ V_{xx}^j(t, x) = 0\,.$$

Then we note that equation (3.7a) reads

$$0 = rx - b^j(t, x) - \sum_{k:k\neq j} \mu_{jk}(t)\left(b^{jk}(t, x) + \chi^k(t, x) - x\right)$$

$$- \left(rx - a^j(t, x) - \sum_{k:k\neq j} \mu_{jk}(t)\left(a^{jk}(t, x) + \chi^k(t, x) - x\right)\right),$$

which is true since the payment coefficients of A and B match. Further, we have that (3.7b) reads

$$x = \Delta B^j(t, x) + x - \Delta A^j(t, x),$$

which is true since the payment coefficients of A and B match. Thus, our candidate function fulfills the differential equation, and we conclude that (3.9) is true.

One important conclusion to draw is that this contract, by construction, fulfills the equivalence principle

$$V^0(0-, X(0-)) = X(0-) = 0,$$

where the first equality is the candidate function (3.9) and the second equality is the initial condition for X.

The hedging strategy can be calculated from (3.8) as

$$h(t)S(t) = X(t)\pi(t, X(t)),$$

and we see that the amount $h(t)S(t)$ that should be invested in the risky asset equals the amount invested in risky assets in the index process $X(t)\pi(t, X(t))$. Thus, in this particular case, π is actually the hedging strategy itself.   ◊

Example 3.5 The standard insurance contract type we studied in Chap. V is a special case of our general construction of the unit-link contract. We let

$$b^j = a^j, \quad b^{jk} = a^{jk}, \quad \Delta B^j = \Delta A^j,$$

and note that this means that there are no payments in B that depend on X. Then the reserve V does not depend on X either and (3.7) reduces to the standard differential system,

$$V_t^j(t) = rV^j(t) - b^j(t) - \sum_{k:k\neq j} \mu_{jk}(t)R^{jk}(t),$$

$$V^j(t-) = \Delta B(t) + V^j(t), \quad V^j(n, x) = 0.$$

Since V does not depend on X, the hedging position in stocks is zero and everything is invested in the riskfree asset.   ◊

Example 3.6 The contract type we studied in Sect. 2 is another special case of our general construction of the unit-link contract. We let

$$\Delta A(0) = s_0, \quad \chi^k(t, x) = x, \quad \pi(t, x) = 1,$$

and all other payment coefficients of A be zero. Then the dynamics of X reduces to

$$dX(t) = X(t)\big(\alpha\, dt + \sigma\, dW(t)\big), \quad X(0) = s_0,$$

which means that $X = S$. Further, we let

$$b^{jk}(t, x) = 0,$$

and let $dB^j(t, x)$ be independent of j. Then no payments depend on Z. Then the reserve V does not depend on Z either, and (3.7) reduces to

$$V_t(t, x) = rV(t, x) - b(t, x) - V_x(t, x)rx - \frac{1}{2}V_{xx}(t, x)\sigma^2 x^2,$$

$$V(t-, x) = \Delta B(t, x) + V(t, x), \quad V(n, x) = 0.$$

The hedging strategy becomes, using that $X = S$,

$$h(t)S(t) = V_s(t, S(t))S(t) \quad \Rightarrow \quad h(t) = V_s(t, S(t)).$$

As expected, we find again here the hedging strategy presented in (1.19) and (2.7) and in this particular case of the unit-link contract, this is really a hedging strategy, since all risk is hedgeable as there is no jump risk in the payment stream.   ◇

Example 3.7 The most important special case of the general contract is the specification within the survival model and where $\chi^1(t, x) = 0$, meaning that the process X is set to zero upon death. Then

$$dX(t) = X(t)\big((r + \pi(t, X(t))(\alpha - r)\big)\, dt + \sigma\pi(t, X(t))X(t)\, dW(t)$$
$$- dA^0(t, X(t)) - \mu(t)\big(a^{01}(t, X(t)) - X(t)\big)\, dt - X(t-)\, dN(t),$$

and

$$V_t(t, x) = rV(t, x) - b(t, x) - \mu(t)\big(b^{01}(t, x) - V(t, x)\big) - \mathcal{D}_x V(t, x), \tag{3.10}$$

$$V(t-, x) = \Delta B(t, x) + V\big(t, x - \Delta A^0(t, x)\big),$$

$$V(n, x) = 0.$$

Different contracts are now designed through different specifications of the payment coefficients of B. A standard unit-link contract with guarantee specifies that

$$\Delta B(n, x) = \max(x, G) \qquad (3.11)$$

for some guarantee G. Recall that for the pure unit-link case where $\Delta B(n, x) = x$, we have that $V(0-, X(0-)) = X(0-) = 0$. Now, we are instead cutting off $X(n)$ by the guarantee G in the terminal payoff. This has to be paid for in one way or another. One possibility is to let the guarantee be paid for by a distinction between a^0 and b^0, where we think of them as (negative) premium payments in the saving phase. The policy holder pays in the premium rate $-b^0$ but only $-a^0 < -b^0$ is added to X. Then

$$-b^0 + a^0 > 0,$$

can be thought of as the price that the policy holder has to pay for the terminal guarantee. If $-a^0(t, x) = -b^0(t) - \delta x$, we say that the premium for the guarantee is paid as a marginal δ in the return on X. This is a somewhat standard way of pricing maturity guarantees in unit-link insurance. The guarantee G is naturally connected to the payment stream B, e.g. through a money-back-guarantee,

$$G = -\int_0^n b^0(t) \, dt.$$

The death sum b^{01} is often linked to x but the practical implementation varies a lot. We just highlight two canonical examples, namely the non-linear form

$$b^{01}(t, x) = \max(x, G(t)),$$

and the linear form

$$b^{01}(t, x) = F(t) + G(t)x,$$

where $F(t) \equiv 0$ and $G(t) \not\equiv 1$, and $F(t) > 0$ and $G(t) \equiv 1$, resp., are standard combinations. The linearity of b^{01} in x is inherited by the risk premium $\mu(t)(b^{01}(t, x) - V(t, x))$ in (3.10) with potential computational benefits. The idea of considering the difference between A and B as the payment for possible option structures in payments such as (3.11) can be generalized to multistate models with countless possible contract designs. Also in that case, one can achieve computational benefits by letting not only the transition payment like b^{01} above, but also χ^k be linear in x.

A simple and practical way of getting around pricing issues is to design the investment strategy π such that the price is negligible. This is done in certain types of life-cycle investment contracts where π tends to 0 as t tends to n. If this happens sufficiently early and fast, the price of the guarantee becomes correspondingly small and may be neglected.   ◇

Exercises
3.1 The class of variable annuities can be formalized as a special case of the contract studied in this section. Consider the survival model and product specification in Example 3.7. Assume a time point of retirement m such that after m, the death benefit is zero and we set

$$a^0(t, x) = b^0(t, x)) = \frac{x}{\int_t^n e^{-\int_t^s r^* + \mu^*} ds},$$

where we introduce a so-called calculation basis (r^*, μ^*). We can now simply set $\Delta B(n, x) = x$ without really influencing the payments (why? Hint: study the dynamics of X as t approaches n) and with that terminal payment, we conclude that $V^0(t, x) = x$ (why?).

Determine the dynamics of $b^0(t, X(t))$ when the policy holder is still alive. (Hint: recognize a geometric Brownian motion). Find the relation between (r^*, μ^*) and (r, α, π, μ) that determines whether b is increasing or decreasing.

As for the terminal payment in (3.11), we can also add guarantees to the design such that, for example,

$$b^0(t, x) = \max\left(\frac{x}{\int_t^n e^{-\int_t^s r^* + \mu^*} ds}, G\right).$$

Can we still set $\Delta B(n, x) = x$ without influencing the payments and conclude that $V^0(t, x) = x$? Suggest a way to charge the policy holder for the guarantee G. (Note: in practice the variation of product and guarantee is enormous and computational issues have been studied in the literature over the last few decades.)

4 With-Profit Insurance and the Dynamics of the Surplus

The unit-link contract is an alternative to an original product design called *with-profit insurance*, developed from the classical insurance mathematics from Chap. V. The with-profit contract takes as its starting point the long term uncertainty nature of pension saving. Life insurance contracts are typically long-term contracts with time horizons up to half a century or more. Calculation of reserves is based on assumptions on interest rates and transition intensities until termination. Two difficulties arise in this connection. First, these are quantities which are difficult to predict even on a short-term basis. Second, the policy holder may be interested in participating in returns on risky assets rather than riskfree assets.

The basic set-up in the following is the Markov model from Chap. V, and reserves are calculated as there. Thus the investment features considered in Sect. 3 are not present, but investment will emerge in a different way in connection with the *surplus* defined below.

In with-profit insurance, the insurer makes a first prudent guess on the future interest rates and transition intensities in order to be able to set up a reserve, knowing quite well that realized returns and transitions differ. This first guess on interest rates and transition intensities, here denoted by (r^*, μ^*), is called the *first-order basis* (or the *technical basis*), and gives rise to the *first-order reserve* V^*. The set of payments B settled under the first-order basis is called the *first-order payments*. By settled we mean that B fulfills the equivalence principle under the technical first-order assumption,

$$V^*(0-) = \mathbb{E}^*\left[\int_{0-}^{n} e^{-\int_0^t r^*}\, dB(t)\right] = 0,$$

where $\mathbb{E}^*$ denotes expectation with respect to the measure under which the transition intensities are μ_{ij}^*. This measure is not necessarily equivalent to the objective measure. Further, also during the course of the policy one works with an artificial technical prospective evaluation

$$V^{j*}(t) = \mathbb{E}^*\left[\int_t^n e^{-\int_t^s r^*}\, dB(s)\,\Big|\, Z(t) = j\right]. \tag{4.1}$$

Example 4.1 Consider whole life insurance with premium π paid as a lump sum at time 0 and let T be the lifetime of the insured. Calculating π according to the equivalence principle and using the parameters $r, \mu(t)$ for the interest rate and the mortality rate then gives

$$\pi = \mathbb{E}\left[e^{-rT}\right] = \int_0^\infty e^{-rt}\mu(t)e^{-\int_0^t \mu}\, dt = 1 - r\int_0^\infty e^{-rt}e^{-\int_0^t \mu}\, dt\,.$$

This expression is obviously decreasing in r and increasing in μ. Being prudent, the company therefore sets the first-order basis with r^* somewhat smaller than what one expects and $\mu^*(t)$ somewhat larger. $\diamond$

However, the insurer and the policy holder agree, typically through the legislation regulating the contract, that the realized returns and transitions, different from the artificial assumption (r^*, μ^*), should be reflected in a realized payment stream different from the guaranteed one. For this reason the insurer adds to the first-order payments a *dividend payment stream*. We denote this payment stream by D and assume that its structure corresponds to the structure of B, i.e.

$$dD(t) = dD^{Z(t)}(t) + \sum_{k:\,k\neq Z(t-)} \delta^{Z(t-)k}(t)\, dN^k(t), \tag{4.2}$$

$$dD^{Z(t)}(t) = \delta^{Z(t)}(t)\, dt + \Delta D^{Z(t)}(t)\,.$$

Here, however, the elements of D are not assumed to be deterministic. In contrast, the dividends should reflect realized returns and transitions relative to the first-order basis assumptions. Thus, the policy holder experiences the payment stream $B + D$.

From that introduction, we can now categorize essentially all types of life and pension insurance by their specification of D. Such a specification includes possible constraints on D, the way D is settled, and the way in which D materializes into payments for the policy holder or others. We do not give a thorough exposition of the various types of life insurance existing here but just give a few hints as to what we mean by categorization. One important message here is that we enter more deeply into a complex area of insurance law, since the way the performance of the contract materializes in dividends is often written in the law that regulates the contract rather than in the contract itself.

First, we distinguish between D being constrained or not. When dividends are constrained to the benefit of the policy holder, i.e. D is positive and increasing, we speak of *participating* or *with-profit life insurance*. We speak in that case of B as the *guaranteed payments*. In with-profit insurance, the positive dividend payments belong to the policy holder. In a different product notion, so-called *pension funding*, there is no constraint on D. In some periods dividends may be negative, i.e. D may be decreasing and D can go negative. In that case, it makes no sense to speak of B as guaranteed. In pension funding, however, the insured himself is typically not affected by dividends. Instead, there is an employer who pays or receives dividends.

Second, we distinguish between whether D is directly paid out to the policy holder or employer as cash or whether it is converted into an adjustment of the first-order payments. The idea here is that the policy holder already expressed, in part, a demand for a certain payment profile through the first-order payments B. It appears natural that the insurer uses that information when returning dD to the policy holder. One possibility is that dividends are returned as a discount (positive or negative, in general) in future *first-order premiums*, i.e. as a correction to the future decrements of B, whereas *first-order benefits* are fixed. We speak in that case of *defined benefits*, where 'defined' is read as 'unaffected by dividend payments'. Actually, the case where dividends are paid out as cash during periods of premium payment can also be spoken of as defined benefits because the cash dividends are experienced as a discount in the premium payment. Another possibility is that dividends are returned as a *bonus* (positive or negative, in general) in future *first-order benefits*, i.e. as a correction to the future increments of B, whereas *first-order premiums* are fixed. We speak in that case of *defined contributions*, where 'defined' is read as 'unaffected by dividend payments' and 'contributions' is just a synonym for our premiums. In the following sections, we formalize in a dense manner the dividend and bonus schemes explained in this paragraph.

Before we go back to the mathematics, we finish this digression to practical product design and regulation by linking the different distinctions above. Typically, the pension funding contract is handled as a defined benefits contract. The benefits are then often linked to the salary development of the policy holder, and the regulation of premiums through dividends affect the contributions (premiums) from the employer to the pension fund. The pension relation is closely linked to the employer relation and the pension benefit obligation is an obligation of the employer. However, a fund manager monitors these obligations and regulate the premiums through dividend payments as a reaction to the performance. These contracts dominated the English speaking markets (UK, US, Aus) in the last century but, over the last 30–50 years, are gradually being replaced by the unit-link contract. One of the challenges of the construction is that the employer plays the role as a pension fund with real long-term obligations and thereby potential risk. So, the transition from pension funding to unit-link should be seen as a transition from the employer holding pension obligations and related risk to pension funds holding pension obligations and pension funds or policy holders holding risk.

Typically, the with-profit contract is handled as a defined contributions contract. Then the regulation of benefits is always to the advantage of the policy holders, which explains why we introduced the word bonus for that extra benefit payment on top of what was guaranteed. It may still be the employer that pays in the premiums or contributions but after that the employer is free of all obligations. Then it is up to the pension fund to live up to the guaranteed payments. Dividends are simply kept in the pension fund as a source of paying out the bonus benefits. These contracts dominated the continental European markets in the last century. After the introduction of unit-link insurance to English speaking markets, some analysts expected a fast entry of that product type into continental Europe. However, the pace has been slower and much more diverse than expected. One of the reasons is that the with-profit contract did not suffer from the problematic employer obligations that the unit-link contract was introduced to cure in the pension funding contract. Nevertheless, many continental European countries *did* gradually introduce the unit-link contract as an alternative to the with-profit contract.

For all types of contracts, independently of constraints, etc., the question remains: How should dividends reflect the realized returns and transitions?

A natural measure of realized performance is the surplus given by the excess of assets over liabilities. Assuming that the company invests payments in a self-financing portfolio with value process G and that liabilities are measured by the first-order reserve, we get the *surplus*

$$Y(t) = Y_0(t) - V^{Z(t)*}(t) \quad \text{where} \quad Y_0(t) = \int_{0-}^{t} \frac{G(t)}{G(s)} \mathrm{d}\big(-(B+D)(s)\big),$$

(4.3)

where $Y_0(t)$ is the total payments in the past accumulated with capital gains from investing in G.

We now assume that a proportion $q(t)$ of G, written in the form

$$q(t) = \frac{\pi\big(t, Y(t)\big) Y(t)}{Y_0(t)},$$

(4.4)

is invested in a risky asset modeled as in Sect. 1. Then the dynamics of G are given by

$$dG(t) = rG(t)\, dt + \sigma q(t)G(t)\, dW^Q(t).$$

Note that we choose the specification directly in terms of W^Q, Brownian motion under the valuation measure, and that r is the realized interest rate, not the r^* in the technical base. Introduce

$$R^{Z(t)k*}(t) = b^{Z(t)k}(t) + V^{k*}(t) - V^{Z(t)*}(t).$$

Proposition 4.2 *The dynamics of the surplus Y are given by*

$$dY(t) = d\big(- (B+D)(t)\big) + \big(r\, dt + \sigma q(t)\, dW^Q(t)\big)Y_0(t) - dV^{Z(t)*}(t)$$
$$= rY(t)\, dt + \sigma \pi(t, Y(t))Y(t)\, dW^Q(t) + d(C-D)(t),$$

*where C is the **surplus contribution process** defined by*

$$dC(t) = c^{Z(t)}(t)\, dt - \sum_{k:k\neq Z(t-)} R^{Z(t-)k*}\, dM^k(t),$$

$$c^j(t) = (r - r^*)V^{j*}(t) + \sum_{k:k\neq j} \big(\mu_{jk}^*(t) - \mu_{jk}(t)\big)R^{jk*}(t).$$

Proof The first expression follows from

$$dY_0(t) = d\big(- (B+D)(t)\big) + dG(t)\int_0^t \frac{1}{G(s)} d\big(- (B+D)(s)\big)$$
$$= d\big(- (B+D)(t)\big) + \frac{dG(t)}{G(t)}Y_0(t).$$

Inserting the dynamics of $V^{Z(t)*}(t)$ as given in formula V.(3.9) then gives

$$dY(t) = d\big(- (B+D)(t)\big)$$
$$+ r\big(Y_0(t) - V^{Z(t)*}(t)\big)\, dt + \sigma \pi(t, Y(t))Y(t)\, dW^Q(t)$$
$$- V_t^{Z(t)*}(t)\, dt - \sum_{k:k\neq Z(t-)} \Big[\big(V^{k*}(t) - V^{Z(t-)*}(t-)\big)\, dN^k(t)\Big]$$
$$+ rV^{Z(t)*}(t)\, dt - \big(V^{Z(t)*}(t) - V^{Z(t)*}(t-)\big)$$
$$= rY(t)\, dt + \sigma \pi(t, Y(t))Y(t)\, dW^Q(t)$$
$$+ d\big(- (B+D)(t)\big) + rV^{Z(t)*}(t)\, dt$$

$$- \left(r^* V^{Z(t)*}(t) - b^{Z(t)}(t) - \sum_{k:\,k \neq Z(t)} \mu^*_{Z(t)k}(t) R^{Z(t)k*}(t) \right) dt$$

$$- \sum_{k:\,k \neq Z(t-)} \left[\left(V^{k*}(t) - V^{Z(t-)*}(t-) \right) dN^k(t) \right]$$

$$- \left(V^{Z(t)*}(t) - V^{Z(t)*}(t-) \right)$$

$$= rY(t)\,dt + \sigma \pi(t, Y(t)) Y(t)\,dW^Q(t) - dD(t)$$

$$+ rV^{Z(t)*}(t)\,dt - \left(r^* V^{Z(t)*}(t) - \sum_{k:\,k \neq Z(t)} \mu^*_{Z(t)k}(t) R^{Z(t)k*}(t) \right) dt$$

$$- \sum_{k:\,k \neq Z(t-)} \left[\left(b^{Z(t-)k}(t) + V^{k*}(t) - V^{Z(t-)*}(t-) \right) dN^k(t) \right]$$

$$= rY(t)\,dt + \sigma \pi(t, Y(t)) Y(t)\,dW^Q(t) + d(C - D)(t). \qquad \square$$

Note that we have written the payment process C in a slightly different way than usual. Namely, we define the coefficient c^j as the residual from the martingale term rather than just a jump term. The idea is that the insurance policy holder is not meant to participate in the performance of diversifiable risk, only in systematic risk and systematic contributions to the surplus. Thus, we throw away—or rather, leave it to the owners of the pension fund to cover—the martingale term from our notion of surplus to form the index underlying dividend payments. Also, when we proceed, we are going to allow lump sum dividend payments upon termination only. Then a revised version of Y is

$$dY(t) = rY(t)\,dt + \sigma \pi(t, Y(t)) Y(t)\,dW^Q(t)$$

$$+ \left(c^{Z(t)}(t) - \delta^{Z(t)}(t) \right) dt - \Delta D^{Z(t)}(t)\,d\varepsilon^n(t). \qquad (4.5)$$

The dynamics of Y show that π is actually the proportion of the surplus invested in the risky asset. This explains why we chose to start out, in the specification of G, with the proportion $q(t)$ given by (4.4).

The statewise surplus contribution rate c^j plays a crucial role in the product design. It unveils how the surplus is built up by the systematic mismatch between the realized and assumed interest rate, $r - r^*$, weighted with the technical reserve $V^{j*}(t)$, and the assumed and realized transition intensities, $\mu^*_{jk}(t) - \mu_{jk}(t)$, weighted with the sums at risk, $R^{jk*}(t)$. One can now analyze how to prepare for systematic positive surplus contributions by choosing, at time 0, the first-order basis (r^*, μ^*) appropriately. Deciding on a first-order basis before selling the contract is the first important risk management decision to make within the with-profit product design.

After time 0 the remaining decisions concern the investment strategy π and the dividend distribution strategy δ. We have already assumed that the investment strategy is specified in terms of Y, and we are now going to assume the same construction for the dividend rate, namely that $\delta^{Z(t)}(t)$ is specified through the function $\delta^j(t, x)$ such that

$$\delta^{Z(t)}(t) \; = \; \delta^{Z(t)}(t, Y(t)).$$

This means that, at any point in time, the dividends are fully determined in terms of the state process (Z, Y). Thus, the surplus Y is taken to be the underlying index driving all decisions about investment and redistribution.

Exercises

4.1 Revisit the statewise surplus contribution $c^j(t)$ in Proposition 4.2 and the comment in the second last paragraph above. Consider a survival model.

For the term insurance contract paid by a single premium, specify $R^{01*}(t)$. Find the relation between r^* and r and between μ^* and μ such that both the interest rate component and the mortality rate component of $c^0(t)$ are non-negative. Does your result change if the insurance contract is instead paid by a level premium as long as the policy holder is alive? Interpret the result you found for μ^* and μ.

For the deferred life annuity contract paid by a single premium, specify $R^{01*}(t)$. Find the relation between r^* and r and between μ^* and μ such that both the interest rate component and the mortality rate component of $c^0(t)$ are non-negative. Does your result change if the insurance contract is instead paid by a level premium in the deferment period as long as the policy holder is alive? Interpret the result you found for μ^* and μ.

5   Cash Dividends and Market Reserve

In this section, we consider market consistent valuation of payments within the with-profit product design, corresponding to the value studied in Sect. 3. We consider the case where the dividends are paid out in cash. More precisely, we have the value corresponding to (3.5), which in this section reads

$$V^{Z(t)}(t, Y(t)) \; = \; \mathbb{E}^{\mathbb{P}\otimes\mathbb{Q}}\left[\int_t^n e^{-\int_t^s r}\, d(B + D)(s)\,\Big|\, Z(t), Y(t)\right]. \tag{5.1}$$

The first thing to observe is that the value can be naturally decomposed into the value of guaranteed payments B and the value of dividend payments D. The market value of the guaranteed payments is

$$V^{Z(t)}(t) \; = \; \mathbb{E}^{\mathbb{P}}\left[\int_t^n e^{-\int_t^s r}\, dB(s)\,\Big|\, Z(t)\right].$$

We recall the dynamics of Y, B, C, and D,

$$dY(t) = rY(t)\,dt + \sigma\pi(t, Y(t))Y(t)\,dW^Q(t) + \left(c^{Z(t)}(t) - \delta^{Z(t)}(t)\right)dt - \Delta D^{Z(t)}(t)\,d\varepsilon^n(t),$$

$$dB(t) = dB^{Z(t)}(t) + \sum_{k:k\neq Z(t-)} b^{Z(t-)k}(t)\,dN^k(t),$$

$$dC(t) = c^{Z(t)}(t)\,dt, \quad dD(t) = \delta^{Z(t)}(t)\,dt + \Delta D^{Z(t)}(t)\,d\varepsilon^n(t).$$

From here we see that our set-up actually fits into our general construction of a unit-link contract. The underlying state process here is a special case of the state process of the unit-link contract with

$$dA(t) = dD(t) - dC(t), \quad \chi^k(t, y) = y,$$

such that, noting that A has no payments at jumps of Z,

$$\varrho^{jk}(t, y) = 0 + \chi^k(t, y) - y = 0,$$

where ϱ^{jk} was introduced in (3.3). The payment process from the unit-link set-up which is linked to the state process Y is here the total payment process $B + D$, noting that only the coefficients of D actually depend on Y.

With this translation, we can now specify directly a differential equation characterizing V by reference to Proposition 3.3. Introduce

$$\mathcal{D}_y V^j(t, y) = V_y^j(t, y)\left(ry + c^j(t) - \delta^j(t, y)\right) + \frac{1}{2}V_{yy}^j(t, y)\pi^2(t, y)\sigma^2 y^2,$$

$$\tag{5.2}$$

$$R^{jk}(t, y) = b^{jk}(t) + V^k(t, y) - V^j(t, y).$$

Proposition 5.1 *Assume that there exists a set of functions $V^j(t, y)$ such that*

$$V_t^j(t, y) = rV^j(t, y) - b^j(t) - \delta^j(t, y)$$

$$- \sum_{k:k\neq j} \mu_{jk}(t)R^{jk}(t, y) - \mathcal{D}_y V^j(t, y), \tag{5.3}$$

$$V^j(t-, y) = \Delta B^j(t) + \Delta D^j(t, y) + V^j\left(t, y - \Delta D^j(t, y)\right) \tag{5.4}$$

and $V^j(n, y) = 0$. Then this function is indeed equal to the reserve defined in (5.1).

So far, we have just given the differential system characterizing the reserve. We have not discussed which functional dependence of dividends on Y might be relevant. For such a discussion, we need to know the insurer's and the policy holder's agreement on how performance is reflected in dividends. In practice, dividends are always increasing in Y. Then a good performance is shared between the two parties by the insurer paying back part of the surplus as positive dividends. A bad performance can be shared between the two parties by the insurer collecting part of the deficit as negative dividends. Since there may be constraints on D, e.g. D increasing, these qualitative estimates are not necessarily strict, though. There are only a few examples of a functional dependence that allow for more explicit calculations.

Exercises

5.1 The surplus dynamics in (4.5) is a construction although it appears to be a conclusion. We can construct it slightly differently by subtracting from the dynamics a payment to the equity holders of the pension fund. Letting ψ be the statewise rate of payments to equity, we simply subtract $\psi^{Z(t)}(t, Y)) \, dt$ from the dynamics specified in (4.5). How does this influence the differential equation (5.3)?

6   The Pure Case of Cash Dividends

A canonical special case occurs if the surplus is fully emptied at termination. Having allowed for a lump sum dividend upon termination, this is easily formalized by

$$\Delta D^j (n, y) = y.$$

In this case we can establish a simple solution to (5.3) and (5.4). We can guess the solution

$$V^j (t, y) = f^j (t) + y, \tag{6.1}$$

and plug it into (5.3) and (5.4). The linearity of ΔD fits with the linearity of V, so that it all boils down to an ODE for f, reading

$$f_t^j (t) = r f^j (t) - b^j (t) - c^j (t) - \sum_{k : k \neq j} \mu_{jk}(t) \left(b^{jk}(t) + f^k (t) - f^j (t) \right), \tag{6.2}$$

$$f^j (t-) = \Delta B^j (t) + f^j (t), \qquad f^j (n) = 0.$$

This differential equation can be recognized as a differential equation characterizing the present value of future payments in B and C,

$$f^j (t) = \mathbb{E}^{\mathbb{P}} \left[\int_t^n e^{-\int_t^s r} \, d(B + C)(s) \, \middle| \, Z(t) = j \right] = V_B^j (t) + V_C^j (t). \tag{6.3}$$

But since we also have that

$$V^j(t, y) = V_B^j(t) + V_D^j(t, y), \qquad (6.4)$$

we can write from (6.1), (6.3), and (6.4),

$$V_D^j(t, y) = V_C^j(t) + y. \qquad (6.5)$$

We have even one further insight to make, namely

$$f^j(t) = V^{j*}(t). \qquad (6.6)$$

Consider it as a guess and plug it into (6.2) to arrive at

$$V_t^{j*}(t) = r^* V^{j*}(t) - b^j(t) - \sum_{k:k \neq j} \mu_{jk}^*(t)\left(b^{jk}(t) + V^{k*}(t) - V^{j*}(t)\right),$$

$$V^{j*}(t-) = \Delta B^j(t) + V^{j*}(t), \qquad V^{j*}(n) = 0.$$

But this is just the Thiele differential equation applied under the first-order basis (r^*, μ^*).

During the previous century, when market valuation was not well-established as a valuation principle, the presentation of the total reserve as

$$V^j(t, y) = V^{j*}(t) + y$$

was standard in accounting schemes. During this century, where the market valuation has become well-established as a valuation principle, the presentation of the total reserve as

$$V^j(t, y) = V_B^j(t) + V_D^j(t, y)$$

has become standard in accounting schemes. One may still be interested in visualizing V^{j*} by decomposing V_D^j according to (6.5), so that

$$V^j(t, y) = V_B^j(t) + V_C^j(t) + y.$$

Here, $V^{j*}(t)$ can be obtained immediately as $V_B^j(t) + V_C^j(t)$, due to (6.1) and (6.6).

The term V_B is called *Guaranteed Benefits* (or just *GB*) and V_D is called *Future Discretionary Benefits* (or just *FDB*). Only if we have $\Delta D^j(n, y) = y$ is it really that easy to calculate future discretionary benefits and thereby the total reserve residually, based on the retrospective account Y. However, Y does not necessarily have to be emptied as a lump sum upon termination. Of course, it can be emptied before by appropriate specification of δ, e.g. $\delta^j(t, y) = q^j(t)y$ and then $q^j(t) \to \infty$

as $t \to n$. We can then still write $\Delta D^j(t, y) = y$ but since $Y(n-) = 0$ almost surely, this leads to no lump sum dividend upon termination.

Exercises
6.1 Argue that $Y(0-) = 0$, $f^{0-}(0) = 0$, and $V^0(0-, 0) = 0$, and conclude that the contract is fair. If we want to be sure to pay out a positive dividend upon termination, we could set $\Delta D^j(n, y) = y^+$. In that case, the contract is not necessarily fair anymore. Construct a situation where it actually is, i.e. a situation where $Y(n) \geq 0$ a.s., such that the 'positive part'-sign is redundant. Explain and formalize how you could, in general, use the payments to equity holders, introduced in Exercise 7.1, to make the contract fair if it is not.

7   Bonus Payments and Market Reserve

In this section, we consider again a market valuation of future payments corresponding to (3.5). Now, we assume the scheme where only a part of the dividends are paid out in cash as shown in the previous section. The rest of the dividend payments are used to change some of the future guaranteed payments. When we write some of the future guaranteed payments, the point is that a part of the future payments is fixed whereas another part is floating as dividends are paid out. The idea is that we have two basic payment streams B_1 and B_2. Both are standard payment streams with deterministic payment coefficients. Thus, for $i = 1, 2$,

$$dB_i(t) = dB_i^{Z(t)}(t) + \sum_{k:k \neq Z(t-)} b_i^{Z(t-)k}(t) \, dN^k(t).$$

Correspondingly, we have the technical reserves

$$V_i^{j*}(t) = \mathbb{E}^* \left[\int_t^n e^{-\int_t^s r^*} \, dB_i(s) \,\Big|\, Z(t) = j \right],$$

for $i = 1, 2$ with reserve dynamics

$$dV_i^{Z(t)*}(t) = r^* V_i^{Z(t)*}(t) \, dt - dB_i^{Z(t)}(t) - \sum_{k:k \neq Z(t)} \mu_{Z(t)k}^*(t) R_i^{Z(t)k} \, dt$$

$$+ \sum_{k:k \neq Z(t-)} \left(V_i^{k*}(t) - V_i^{Z(t-)*}(t) \right) dN^k(t),$$

where

$$R_1^{jk}(t) = b_1^{jk}(t) + V_1^{k*}(t) - V_1^{j*}(t).$$

At time 0, the payment process $B_1 + B_2$ is considered as guaranteed and the coefficients of B_1 and B_2 are constrained by the equivalence principle under the technical basis

$$V_1^{Z(0)*}(0-) + V_2^{Z(0)*}(0-) = 0. \tag{7.1}$$

The fixed part and the floating part can be any decomposition of the guaranteed payment stream at time 0. However, to fix ideas, we think of the fixed part as the premiums and the floating part as the benefits, i.e. B_1 is decreasing and B_2 is increasing. The payment process B_1 is fixed and part of the payment process realized by the policy holder. The payment process B_2 is also fixed but it only serves as the profile of the benefit stream. At time zero, the policy holder is, indeed, guaranteed exactly one times the payment stream B_2. But during the course of the contract, additional portions of benefit streams of profile B_2 are added. They are bought for the dividends allocated to increasing benefits, D_2, whereas the rest of the dividends $D_1 = D - D_2$ are paid out in cash. The infinitesimal dividend payout $dD_2(t) = \delta_2(t)\, dt$ is an infinitesimal single premium for which we buy an infinitesimal number of streams of future payments in B_2. If this infinitesimal investment in an additional contract is priced under the first-order basis and we introduce $Q(t)$ for the number of payment processes B_2 the policy holder holds at time t, we have by the equivalence principle that

$$dD_2(t) = dQ(t)V_2^{Z(t)*}(t).$$

The interpretation is that the single premium is $dD_2(t)$ and the payment profile we want to buy costs $V_2^{Z(t)*}(t)$. This means that we can afford $dQ(t)$ of them. Letting $Q(0-) = 1$, the policy holder now experiences in total the payment process with dynamics

$$dB(t) + dD_1(t), \tag{7.2}$$

where

$$dB(t) = dB_1(t) + Q(t-)\, dB_2(t).$$

The first part $dB_1(t)$ is the fixed part of the guaranteed payments from time $0-$. The second part $Q(t-)\, dB_2(t)$ is the floating payments that have the profile $dB_2(t)$ and of which we have bought the number $Q(t)$ in the past. The third part dD_1 is the part of the dividends payouts for which we do not buy additional contracts. One may think of all dividends going to increase the floating guarantees, i.e. $D = D_2$ and $D_1 = 0$. However, for the sake of generality we allow for cash dividends even in this section. Also, this allows us to formalize the pure case in the next section more easily. Note that we do not allow for lump sum dividends in D_2 at time n as at that time point there are no further future payments in B_2 to increase. However, we do allow for a lump sum dividend in D_1 at time n and in the pure case we use this to empty the surplus upon termination.

The total technical reserve at time t is the technical value of all future payments guaranteed at time t, i.e.

$$\mathbb{E}^*\left[\int_t^n e^{-\int_t^s r^*}\,dB_1(s)\,\Big|\,\mathcal{F}(t)\right] + \mathbb{E}^*\left[\int_t^n e^{-\int_t^s r^*}\,Q(t)\,dB_2(s)\,\Big|\,\mathcal{F}(t)\right]$$
$$= V_1^{Z(t)*}(t) + Q(t)V_2^{Z(t)*}(t).$$

We denote this value by X and have the following relation between X and Q,

$$X(t) = V_1^{Z(t)*}(t) + Q(t)V_2^{Z(t)*}(t)\,, \quad \text{where } Q(t) = \frac{X(t) - V_1^{Z(t)*}(t)}{V_2^{Z(t)*}(t)}. \tag{7.3}$$

Note that, by the equivalence principle (7.1) and $Q(0-) = 1$, X fulfills the initial condition $X(0-) = 0$.

The relation between X and Q allows us to consider the coefficients of the payment stream B as functions of Q or X. If we spell out these ingredients we have

$$dB(t) = dB_1(t) + Q(t-)\,dB_2(t)$$
$$= dB_1(t) + \frac{X(t-) - V_1^{Z(t-)*}(t-)}{V_2^{Z(t-)*}(t-)}\,dB_2(t)$$
$$= dB^{Z(t)}(t, X(t)) + \sum_k b^{Z(t-)k}(t, X(t-))\,,$$

where

$$dB^j(t, x) = dB_1^j(t) + \frac{x - V_1^{j*}(t-)}{V_2^{j*}(t-)}\,dB_2^j(t)\,,$$

$$b^{jk}(t, x) = b_1^{jk}(t) + \frac{x - V_1^{j*}(t)}{V_2^{j*}(t)}b_2^{jk}(t)\,.$$

Note that from

$$V_2^{j*}(n-) = \Delta B_2^j(n)\,, \qquad V_1^{j*}(n-) = \Delta B_1^j(n)\,,$$

we have that

$$\Delta B^j(n, x) = \Delta B^j_1(n) + \frac{x - V^{j*}_1(n-)}{V^{j*}_2(n-)} \Delta B^j_2(n) = x. \qquad (7.4)$$

We can derive the dynamics of X and get, noting that Q is continuous by continuity of D_2,

$$
\begin{aligned}
dX(t) &= dV^{Z(t)*}_1(t) + dQ(t) V^{Z(t)*}_2(t) + Q(t) dV^{Z(t)*}_2(t) \\
&= r^*\big(V^{Z(t)*}_1(t) + Q(t) V^{Z(t)*}_2(t)\big) dt - dB^{Z(t)}_1(t) - Q(t) dB^{Z(t)}_2(t) \\
&\quad - \sum_{k:k\neq Z(t)} \mu^*_{Z(t)k}(t)\big(R^{Z(t)k}_1(t) + Q(t) R^{Z(t)k}_2(t)\big) dt \\
&\quad + \sum_{k:k\neq Z(t-)} \big(V^{k*}_1(t) + Q(t) V^{k*}_2(t) - V^{Z(t-)*}_1(t) \\
&\quad - Q(t) V^{Z(t-)}_2(t)\big) dN^k(t) + dD_2(t) \\
&= r^* X(t) dt - dB^{Z(t)}(t, X(t)) - \sum_{k:k\neq Z(t)} \mu^*_{Z(t)k}(t) R^{Z(t)k*}(t, X(t)) dt \\
&\quad + \sum_{k:k\neq Z(t-)} \big(\chi^{Z(t-)k}(t, X(t-)) - X(t-)\big) dN^k(t) + dD_2(t),
\end{aligned}
$$

where

$$R^{jk}_2(t) = b^{jk}_2(t) + V^{k*}_2(t) - V^{j*}_2(t),$$

$$R^{jk*}(t, x) = b^{jk}(t, x) + \chi^{jk}(t, x) - x,$$

$$\chi^{jk}(t, x) = V^{k*}_1(t) + \frac{x - V^{j*}_1(t)}{V^{j*}_2(t)} V^{k*}_2(t).$$

We can now define the surplus of the accumulated contracts as

$$Y(t) = -\int_0^t \frac{G(t)}{G(s)} d(B + D_1)(s) - X(t).$$

At first glance it looks fundamentally different from (4.3) but the ingredients are the same. The integral is the value of all past payments accumulated with capital gains, including payments that have been added to B by increasing Q in the past. The process X is the technical value of the future payment stream guaranteed from time t, see (7.3). This also includes all increments of Q in the past. Assuming that a proportion of G given by the quotient $q(t) = \pi\big(t, X(t), Y(t)\big) Y(t)/\big(Y(t) + X(t)\big)$, see (4.3), is invested in a risky asset modeled as in Sect. 4, the dynamics of G are given by

$$dG(t) = rG(t)\,dt + \sigma\,q(t)\,G(t)\,dW^Q(t).$$

We can now derive the dynamics of Y,

$$
\begin{aligned}
dY(t) &= d(-B - D_1)(t) - dX(t) \\
&\quad + \left(r\,dt + \sigma\frac{\pi(t, X(t), Y(t))Y(t)}{Y(t) + X(t)}\,dW^Q(t)\right)\!\big(Y(t) + X(t)\big) \\
&= d(-B - D_1)(t) + r\big(Y(t) + X(t)\big)\,dt + \sigma\pi(t, X(t), Y(t))Y(t)\,dW^Q(t) \\
&\quad - \left(r^* X(t)\,dt - dB^{Z(t)}(t) - \sum_{k:k\neq Z(t)} \mu^*_{Z(t)k}(t)R^{Z(t)k*}(t, X(t))\,dt\right) \\
&\quad - \sum_{k:k\neq Z(t-)} \big(\chi^{Z(t-)k}(t, X(t-)) - X(t-)\big)\,dN^k(t) - \delta^2(t)\,dt \\
&= rY(t)\,dt + \sigma\pi(t, X(t), Y(t))Y(t)\,dW^Q(t) + d(C - D)(t) \\
&\quad - \sum_{k:k\neq Z(t-)} R^{Z(t-)k*}(t, X(t-))\,dM^k(t),
\end{aligned}
$$

with

$$dC(t) = c^{Z(t)}(t, X(t))\,dt, \tag{7.5}$$

$$c^j(t, x) = (r - r^*)x + \sum_{k:k\neq j} \big(\mu^*_{jk}(t) - \mu_{jk}(t)\big)R^{jk*}(t, x).$$

Again, we throw away the non-systematic part of the surplus, namely the martingale term

$$\sum_{k:k\neq Z(t-)} R^{Z(t-)k*}(t, X(t-))\,dM^k(t),$$

and redefine Y by the dynamics

$$dY(t) = rY(t)\,dt + \sigma\pi(t, X(t), Y(t))Y(t)\,dW^Q(t) + d(C - D)(t). \tag{7.6}$$

Having established the dynamics of X and the dynamics of Y, we now formalize the dividend processes D_1 and D_2 by coefficients as functions of the state process $(t, X(t), Y(t), Z(t))$. Thus, we assume that we have functions $\delta_1^j(t, x, y)$,

$\Delta D_1^j(t, x, y)$, and $\delta_2^j(t, x, y)$ such that

$$dD_1(t) = \delta_1^{Z(t)}\big(t, X(t), Y(t)\big)\, dt + \Delta D_1^{Z(t)}\big(t, X(t-), Y(t-)\big)\, d\varepsilon^n(t),$$

$$dD_2(t) = \delta_2^{Z(t)}\big(t, X(t), Y(t)\big)\, dt.$$

The market value of all future payments including future regulations from time t and onwards is given by

$$V^j(t, x, y)$$

$$= \mathbb{E}^{\mathbb{P}\otimes\mathbb{Q}}\left[\int_t^n e^{-r(s-t)}\, d(B + D_1)(s)\,\Big|\, Z(t) = j, X(t) = x, Y(t) = y\right].$$

$$(7.7)$$

Note from the coefficients of X and Y that the process $\big(Z(t), X(t), Y(t)\big)$ is Markovian and the future payment coefficients of $d(B + D_1)$ depend on this process only. Actually, if $D^1 = 0$, the future payment coefficients of $d(B + D_1)$ do not depend on Y, but they depend on X and X contains Y in its coefficients, so even in that case we need the full state process $\big(Z(t), X(t), Y(t)\big)$. We can now form a PDE to characterize $V^j(t, x, y)$. We leave the proof aside and refer to the structure of the proof of similar results above. Introduce

$$\mathcal{D}_y V^j(t, x, y) = V_y^j(t, x, y)\big(ry + c^j(t, x) - \delta^j(t, x, y)\big)$$

$$+ \frac{1}{2}V_{yy}^j(t, x, y)\pi^2(t, y)\sigma^2 y^2,$$

$$\mathcal{D}_x V^j(t, x, y) = V_x^j(t, x, y)\Big[r^*x - b^j(t, x) + \delta_2^j(t, x, y)$$

$$- \sum_{k:k\neq j} \mu_{jk}^*(t) R^{jk*}(t, x)\Big],$$

$$R^{jk}(t, x, y) = b^{jk}(t, x) + V^k\big(t, \chi^{jk}(t, x), y\big) - V^j(t, x, y).$$

Proposition 7.1 *Assume that there exists a function $V^j(t, x, y)$ such that*

$$V_t^j(t, x, y) = r V^j(t, x, y) - b^j(t, x) - \delta_1^j(t, x, y)$$

$$- \sum_{k:k\neq j} \mu_{jk}(t) R^{jk}(t, x, y) - \mathcal{D}_y V^j(t, x, y) - \mathcal{D}_x V^j(t, x, y),$$

$$(7.8)$$

$$V^j(t-, x, y) \;=\; \Delta B^j(t, x) + \Delta D_1^j(t, x, y)$$
$$+ \; V^j\big(t, x - \Delta B^j(t, x), y - \Delta D^j(t, x, y)\big), \qquad (7.9)$$

$$V^j(n, x, y) \;=\; 0.$$

Then this function is indeed equal to the reserve defined in (7.7).

Exercises

7.1 In this exercise, we generalize the impact of the fee to the equity holders from Exercise 5.1 from the dividend case studied in that section to the bonus case studied in this section. The surplus dynamics in (7.6) is a construction although it appears a conclusion. We can construct it slightly differently by subtracting from the dynamics a payment to the equity holders of the pension fund. Letting ψ be the statewise rate of payments to equity, we simply subtract $\psi^{Z(t)}(t, X(t), Y(t))\, dt$ from the dynamics specified in (7.6). How does this influence the differential equation (7.8)?

8   The Pure Case of Bonus Payments

Again, a canonical special case occurs if the surplus is fully emptied at termination and we now formalize this situation by

$$\Delta D_1^j(n, x, y) = y. \qquad (8.1)$$

In this case we can establish a simple solution to (7.8). We can guess the solution

$$V^j(t, x, y) = x + y, \qquad (8.2)$$

and plug it into (7.8). With this guess, we have that

$$\mathcal{D}_y V^j(t, x, y) \;=\; ry + c^j(t, x) - \delta^j(t, x, y),$$

$$\mathcal{D}_x V^j(t, x, y) \;=\; r^* x - b^j(t, x) + \delta_2^j(t, x, y) - \sum_{k:k\neq j} \mu_{jk}^*(t) R^{jk*}(t, x),$$

$$R^{jk*}(t, x, y) \;=\; b^{jk}(t, x) + \chi^{jk}(t, x) - x.$$

The l.h.s. of (7.8) is zero and the r.h.s. is

$$r(x + y) - b^j(t, x) - \delta_1^j(t, x, y) - \sum_{k:k\neq j} \mu_{jk}(t) R^{jk}(t, x, y)$$

$$- \big(ry + c^j(t, x) - \delta^j(t, x, y)\big)$$

$$- \Big(r^* x - b^j(t, x) + \delta_2^j(t, x, y) - \sum_{k:k\neq j} \mu_{jk}^*(t) R^{jk}(t, x, y)\Big),$$

which is also zero by the definition of $c^j(t, x)$ in (7.5). The updating equation (7.9) gives $x + y$ on both sides and, finally,

$$V^j(n-, x, y) = \Delta B^j(n, x) + \Delta D_1^j(n, x, y) = x + y,$$

which follows from (7.4) and (8.1), such that the terminal condition is also fulfilled.

During the previous century where market valuation was not well-established as a valuation principle, the presentation of the total reserve as

$$V^j(t, x, y) = x + y$$

was standard in accounting schemes. During this century, where the market valuation has become well-established as a valuation principle, a different decomposition has become standard. Introduce

$$V_B(t)^{Z(t)} = V_1^{Z(t)}(t) + Q(t)V_2^{Z(t)}(t) \quad \text{where } Q(t) = \frac{X(t) - V_1^{Z(t)*}(t)}{V_2^{Z(t)*}(t)}.$$

Here $V_1^{Z(t)}(t)$, $V_2^{Z(t)}(t)$ are the market values of the basic payment processes B_1 and B_2 as given by

$$V_i^{Z(t)}(t) = \mathbb{E}^{\mathbb{P}\otimes\mathbb{Q}}\left[\int_t^n e^{-\int_t^s r}\, dB_i(s) \,\Big|\, Z(t)\right].$$

This means that V_B is the market value of all future payments guaranteed at time t. Note carefully the difference between V_B and

$$\mathbb{E}^{\mathbb{P}\otimes\mathbb{Q}}\left[\int_t^n e^{-\int_t^s r}\, dB(s) \,\Big|\, Z(t), X(t), Y(t)\right],$$

which is the market value of future payments in B where, for example, $dB(s)$ includes numbers of $dB_2(s)$ added via increments of Q over (t, s). Now, decompose $V^{Z(t)}(t, X(t), Y(t))$ into

$$V^{Z(t)}(t, X(t), Y(t)) = V_B^{Z(t)}(t) + Y(t) + X(t) - V_B^{Z(t)}(t). \tag{8.3}$$

V_B is called the *Guaranteed Benefits* (or just *GB*) and the residual term $V^{Z(t)}(t, X(t), Y(t)) - V_B^{Z(t)}(t)$ is called the *Future Discretionary Benefits* (or just *FDB*). We only write this residual term as the sum of $Y(t)$ and $X(t) - (V_1^j(t) + Q(t)V_2^j(t))$ in order to visualize the dependence on the retrospective values $X(t)$ and $Y(t)$ in the accounting scheme. Only if we have $\Delta D_1^j(t, x, y) = y$ is it really that easy to calculate future discretionary benefits and the reserve residually, based on the retrospective accounts X and Y. As in the case with only cash dividends, Y

does not necessarily have to be emptied as a lump sum. If it is emptied before in one way or another, then we can keep $\Delta D_1^j(n, x, y) = y$ such that the calculations above still hold. But since $Y(n-) = 0$, this leads to no lump sum dividends upon termination after all. This means that we can think of a situation where no dividends are paid out in cash prior to termination. Upon termination, Y is emptied by $\Delta D_1^j(n, x, y) = y$. But D_2 is designed to fully empty Y as we approach n, so that $Y(n-) = 0$ and $D_1 = 0$, i.e. no cash dividends are paid out at all.

Exercises

8.1 In this exercise, we generalize the points about fairness from Exercise 6.1 from the dividend case studied in that section to the bonus case studied in this section. Argue that $X(0-) = 0$, $Y(0-) = 0$, and $V^0(0-, 0, 0) = 0$, and conclude that the contract is fair. If we want to be sure to pay out a positive dividend upon termination, we could set

$$\Delta D^j (n, x, y) = y^+.$$

In that case, the contract is not necessarily fair anymore. Construct a situation where it actually is, i.e. a situation where $Y(n) \geq 0$ a.s., such that the 'positive part'-sign is redundant. Explain and formalize how you could, in general, use the payments to equity holders, introduced in Exercise 7.1, to make the contract fair if it is not.

9 Comparison of Products

In this chapter we have formalized unit-link insurance and with-profit insurance under the title financial mathematics in life insurance. In both cases the analysis is based on the policy-specific Markov chain model for the insurance risk that we developed in Chap. V. In both cases we modeled the financial risk with a Black–Scholes market. We did not do this because we believe that the Black–Scholes market is a good fit to reality over the time horizon of a pension contract. We did this because this is the simplest and most well-known multi-asset market that allows for no-arbitrage valuation of contracts that exchange financial risk, and because it is good enough for making a series of important points. Finally, in both cases there were only models for diversifiable risk and for hedgeable risk. No risk was left undiversifiable and unhedgeable. Finally, in both cases we modeled a single policy rather than a portfolio of policies. All these common underlying assumptions made the unit-link policy and the with-profit policy look somewhat the same. We even made a virtue out of that by realizing that the formalization of the with-profit Sect. 5 was essentially a special case of the unit-link contract from Sect. 3 with the surplus Y playing the role as the retrospective account X. This allowed us to just refer to Proposition 3.3 when establishing Proposition 5.1 The with-profit contract got slightly more involved by using dividends to buy additional payment streams in Sect. 7. But it is still much alike the unit-link contract except that the process that underlies the payments is now two-dimensional (X, Y). The surplus Y generates dividends that are paid into the account X from which the payments, including bonus payments, are paid to the policy holders. The general resemblance calls on this separate section about the similarities and dissimilarities between the contract designs. The reader is urged to take note of the strong hints to solving some of the exercises in this chapter that we provide here.

For both contracts we paid quite some attention to the pure cases. The pure cases were covered by Example 3.2 where the retrospective unit-link account is ultimately fully emptied to the policy holder, and by Sects. 6 and 8, respectively, where the with-profit surplus Y is ultimately emptied to the policy holder, directly or indirectly via X, respectively. In all these cases the prospective market valuation becomes particularly simple since only retrospective accounts are needed for prospective valuation. Formulas (3.8), (6.1), and (8.2) essentially all show how the prospective valuation are just linear combinations of a deterministic function (V^* in (6.1)) and the state processes (X in (3.8) and (6.1) and X and Y in (8.2)). This is remarkable for one particular reason of major importance. What we have indicated but not proved is that prospective valuations for these contract designs are model-free in the sense that they do not rely on a particular model for the financial market.

For the with-profit case where the payments in B are thought of as guarantees, it is natural to ask the simple question whether it is even possible to implement a strategy such that $Y(n-)$ is non-negative. Otherwise the terminal dividend payment leads to an ultimate negative dividend payment which will be realized by the policy holder as a shortfall on the guaranteed benefit $\Delta B(n)$. The answer is intuitive and simple: We can in principle protect Y downwards by the following design of π: Let π go to zero when Y goes to zero. Then V^* and X, respectively, are protected against shortfalls. And if we do not need to protect the technical values but instead the market values, we can modify the investment strategy: In the dividend case, let π go to zero when Y goes to $-V_C$. Then V_B is protected against a shortfall. In the bonus case, let π go to zero when Y goes to $-X + V_B$. Then V_B is protected against a shortfall. Note carefully that whereas access to the retrospective formulas is model-free, the discussion about investment strategies that secure a non-negative terminal dividend payment is not. Sufficiency of the designs of π explained above relies, for example, on continuous asset prices. If asset prices contain jumps, these strategies should be modified (if jumps are sufficiently bounded) or will not exist (if jumps are not sufficiently bounded). Also, whether these investment strategies can be implemented at all depends on investment restrictions of the pension fund, since regulatory and/or large investor issues may arise. If any of these issues arise or if, simply, the pension fund follows a strategy which does not deliver the downward protection, model-dependence enters, generally, prospective valuation and, specifically, the calculation of a fair payment to equity holders in balance with protection of the guarantee formalized through the positive part of Y, see Exercise 6.1.

In a sense, the invention of the unit-link contract came as a pleasant relief from model-dependent discussions about protection of guarantees and the existence of investment strategies that deliver such protection. Nothing has to be protected in (2.7), all capital gains are directly absorbed in the account X and the policy holder is not protected downwards. With the introduction of guarantees in the unit-link case, formalized in (3.10), we are, however, back to model-dependence in the prospective valuation and model-dependent calculation of, say, a fair balance between the guarantee G and the marginal return on X one should pay for it.

The model-independence in the pure cases means that we can even introduce stochastic interest rates (which, of course, relates to model-independence with respect to the financial market model) and stochastic transition rates (which relates to model-independence beyond the financial market model). This is just a matter of replacing the deterministic interest and transition rates driving the state processes by the realized ones. We do not prove this here but just mention it as an extremely important feature shared across product designs among all the pure cases. Stochastic interest rates are typically introduced along with a bond market for trading interest rate sensitive assets like bonds, and if such assets are exploited by the fund, this is reflected in the capital gains. This is very different from the non-pure cases where stochastic models for interest and transition rates heavily influence the prospective valuation. It is here that an important difference between the (pure) unit-link product and the with-profit product shows up: If V_B should be protected against a shortfall by strategic investments as explained above, even the pure with-profit products involve prospective valuation. The decomposition of V into V_B and $V - V_B$ that took place in both (6.1) and (8.3) is relevant for both asset management and for market-based reporting and, even in the pure case, that decomposition is certainly not model-free. This paragraph bridges parts of this chapter to parts of Chap. VII, where impact from stochastic interest and transition rates are studied. The obvious application is to the calculation of V_B in this section. Note, however, that this dissimilarity between unit-link and with-profit products only holds as long as the unit-link contract does not contain any guarantees. Once guarantees are introduced in unit-link products, impact from stochastic interest and transition rates inevitably enters valuation in their product design as well.

Until now, the dissimilarities between unit-link and with-profit products have been subtle. However, one profound dissimilarity has been hidden in the formulas and, partly, in the text up until now. Whereas the account X in unit-link products is fully individual, the surplus Y in with-profit products is shared within a group of individuals. Thus, to mimic real-life accounting schemes for the bonus case, we would in principle have m individual accounts X but only one surplus process Y to which all individuals contribute and from which dividend payments to all individuals are paid out. However, if no systematic redistribution of capital between individuals is allowed, our set-up is adequate for calculation of fair balances between guarantees and fees and for calculation of the portfolio accounting values which are just the sums of corresponding individual accounting values. One may need one generalization, though. If a policy holder enters a group which already has a surplus $Y > 0$ upon entrance, this should be reflected in an initial Y value of y_0. In that case, though, we just replace the fairness criterion $V^0(0-, 0) = 0$ by $V^0(0-, y_0) = 0$. The intuition behind the fairness arising from that criterion is that the policy holder leaves behind a surplus which has the same financial value as the surplus he inherited upon initiation. Then, again, systematic redistribution is avoided. This intuition does not depend on whether y_0 is inherited from previous policy holders (in case of which there may be issues with systematic redistribution, just not involving the policy holder we consider) or is sponsored by equity holders. If one allows for some level of systematic redistribution, the "window of fairness" widens but the set-up here is still the cornerstone from which all calculations must begin.

Two further points on the idea of collectivism (through sharing Y) in with-profit products are relevant. Our set-up is formally rich enough to cover even the real situation. Just consider the state space of Z as being generated from a portfolio of m policy holders. If we are in a survival model, this leads to 2^m states in Z, corresponding to all combinations of survivors. If the contracts are the same, it can be reduced to m states, corresponding to the number of deaths, starting from state 0 and jumping from state q to state $q + 1$ upon the first death after the q deaths already occurred. Rarely, these ideas are implemented in the literature on with-profit contracts. More often, insurance risk is ignored by projecting the portfolio of contracts to a space without insurance risk. The basic idea is to replace all payments by their expectations conditional on financial risk. One has to be careful, though. It turns out that blindly replacing payments by conditionally expected payments does not, in general, lead to portfolio accounts X and Y that are just the expected individual accounts, which is what one could have hoped for. Non-linearities in the system may be harmful and issues about the Markov property may arise. There is still room for research on the balance between sufficient precision, as in close to being correct, and implementability, as in being legal and computationally tractable, of the projections.

Finally, we want to stress that the overall resemblance between product designs makes a point in itself. Sometimes the contracts are considered fundamentally different, even on aspects where they are not. It is time to remind ourselves that the very purpose of the contracts are the same: To exchange payments streams between an individual and a financial institution, linked to the development of that individual's personal life history and the development of capital gains obtained in the financial market. The exchange of payment streams serves the noble purpose of giving access to a relatively stable consumption over the lifetime in spite of earning money over a shorter time period of active participation in the labor market. The capital gains give access to a higher (expected) consumption level than if the money were just kept under the pillow of the pension fund. Further, the participation in the financial market to some extent hedges uncertainty about future consumer prices so that it is actually the goods consumed rather than the money spent which is relatively stable. This is why the pension and insurance market is there at all, and this holds for all pension contract designs.

Notes

Financial mathematics is a really successful application of mathematics and stochastics to the industry over the last decades, kicked off by Black and Scholes [27] in 1973 and Merton [119] in 1976. Generalization of claims, markets, and objects of study in mathematical finance is overwhelming and we return to some of them in the dimension of interest rate models and bond market theory in Chap. VII. Of course, over the time horizon of a pension contract of 40 years or more, the Black–Scholes model appears somewhat oversimplified. Nevertheless, we have performed the integration of financial mathematics in pension contracts with simple diffusion models here. This is the natural starting point, and generalizations from there are easier after having understood the role of financial risk in the insurance market in a simple market model.

Already shortly after the seminal papers from 1973, Brennan and Schwartz [34] in 1976 realized the application to pricing and valuation of insurance contracts. Later expositions and generalizations by Aase and Persson [1] and Steffensen [160] are more in line with the exposition in this book. Many market generalizations and ideas originating from financial mathematics have spin-off publications in the insurance literature, showing applications in insurance markets. Apart from the examples mentioned above, we also highlight Møller [127] as a financial approach to incomplete market pricing and hedging. We mention here that tools and techniques from finance also found their way into non-life insurance with Delbaen and Haezendonck [55] as an original example and the introduction of capital gains in classical risk models as a general trend. However, asset management plays a much more important role in pensions and the integration between insurance and finance is much more distinctive and necessary there.

The formalization of surplus, dividends, and bonus builds on ideas and formalism dating back to Ramlau–Hansen [147] in 1991 and Norberg [133] in 1999. Modern financial mathematics was integrated into these concepts from insurance by Steffensen [164]. The formalization in this book builds on these ideas. Current developments in market-based valuation in accounting and solvency place demands and needs for rereading and rewriting these insights with modern formalism and modern insight. This is also what is being done in the present exposition.

Chapter VII: Special Studies in Life Insurance

1 Duration-Dependent Intensities and Payments

In this section, we study two extensions to the general Markov framework studied in Chap. V. The two extensions draw to some extent on the same mathematical generalization, but arise from two practically very different sources. One extension is an extension to duration-dependent intensities. This is a canonical generalization of the probability model and has nothing to do with generalization of the contract specification. Another extension is an extension to duration-dependent payments. This is a canonical generalization of the contract specification and has nothing to do with generalization of the probability model. The reason why the mathematics of the two extensions coincide is that they are both handled by extension of the state space from Z to (Z, U), where U is a stochastic process describing, at any point in time, the duration of sojourn in the current state Z.

The generalization of the probability model is formalized by intensities μ_{ij} taking two time arguments instead of just one, such that $\mu_{ij}(t, u)$ is the intensity of jumping to state j at time t, given that Z has been in state i, uninterruptedly, during a time period of length u. Then

$$M^k(t) = N^k(t) - \int_0^t \mu_{Z(s)k}(s, U(s)) \, ds$$

forms a martingale, where U increases linearly at rate 1 from 0 during sojourns in any state, but is reset to 0 at a transition to a different state. Formally, the dynamics are

$$dU(t) = dt - U(t-) \sum_k dN^k(t), \quad U(0) = u_0.$$

© Springer Nature Switzerland AG 2020
S. Asmussen, M. Steffensen, *Risk and Insurance*, Probability Theory
and Stochastic Modelling 96, https://doi.org/10.1007/978-3-030-35176-2_7

Such generalizations of Markov processes play an important role in stochastic process theory and are called *semi-Markov processes*.

A concrete motivation for such duration-dependence is easily obtained by considering the disability model, see Fig. V.3. Let us think about the disability state covering different types of diseases that are, however, not observed or, at least, the information is not used in reserving. We just observe that the policy holder is disabled. An example is two types of disabling diseases. Either the policy holder suffers from an innocent broken leg, in which case recovery happens fast and mortality is not accelerated, or the policy holder suffers from a lethal cancer, in which case recovery is uncertain and mortality is accelerated. Now, intuitively, the duration of the sojourn in the disability state contains relevant information about the tendency to recover and die, respectively, at a given age. If the policy holder jumped, recently, to the disability state, we do not really know much about which disease he suffers from. If, alternatively, the policy holder has spent a longer time as disabled (at the same given age) we tend to believe that cancer is the reason, since, otherwise, he would already have recovered. As consequence, the death intensity is higher and the recovery intensity is lower due to the high death intensity and the low recovery intensity if suffering from cancer. This intuition can be made formal by the idea of partial observation, where we have an underlying Markov process including information about the type of disability. If instead some states are aggregated and we only observe the aggregate process, then this is not (necessarily) Markovian any more. The aggregate process can become a semi-Markov process as the disability example hints at. We do not pursue the mathematical theory of aggregate processes here but just take the intuition from the disability example as a motivating example.

The generalization of the contract specification is formalized by the payment process

$$dB(t) = dB^{Z(t)}(t, U(t)) + \sum_{k: k \neq Z(t-)} b^{Z(t-)k}(t, U(t-)) dN^k(t),$$

where

$$dB^{Z(t)}(t, U(t)) = b^{Z(t)}(t, U(t)) dt + \Delta B^{Z(t)}(t, U(t)).$$

The functions $b^{jk}(t, u)$, $b^j(t, u)$, and $\Delta B^j(t, u)$ are deterministic and sufficiently regular, and specify the contractual payments.

Also for this generalization, we provide two simple and realistic motivating examples. In the survival model illustrated in Fig. V.2, one may specify a k-year *annuity certain* starting upon survival until time m. By annuity certain we mean that the annuity starts upon survival until time m and runs, in that case, certainly, i.e. unconditionally on survival until time $m + k$. An elementary version can be formalized by

$$b^0(t, U(t)) = \mathbb{1}_{m < t < m+k}, \quad b^1(t, U(t)) = \mathbb{1}_{m < t < m+k} \mathbb{1}_{U(t) < t-m}.$$

This coverage resembles in one sense the pure endowment as all payments are fully determined by survival until a specific time point. But in a different sense it resembles a life annuity since the benefits are paid out continuously after reaching age m.

Another annuity certain could start upon death before time m and then run certainly for k years thereafter. This can be formalized by

$$b^1(t, U(t)) = \mathbb{1}_{k > U(t) > t - m} .$$

This coverage resembles in one sense the term insurance as all payments are fully determined by death before a specific time point. But in a different sense, it resembles a spouse's annuity since the benefits are paid out continuously after death before time m.

Both examples in the survival model serve as motivating examples only, since the general structure analyzed below is not really needed there. Simple tricks lead to simplifications of the reservation problems. Our last motivating example is more involved. Here we consider a disability annuity paying out a disability benefit at rate 1 during periods of disability. However, we now add the contractual feature of a waiting period k, saying that the disability does not start until the policy holder has been disabled, uninterruptedly, for a period of length k, k in practice typically being 3 or 6 months. Within the disability model illustrated by Fig. V.3, this coverage can be formalized by

$$b^1(t, U(t)) = \mathbb{1}_{U(t) > k} .$$

In the general framework of duration-dependent intensities and duration-dependent payments we are interested in evaluating the reserve defined generally in formula V.(3.1). It is clear that with duration-dependence, Z is insufficient as an argument in V.(3.2). However, (Z, U) is sufficient and we form

$$V^{Z(t)}(t, U(t)) = \mathbb{E}\left[\int_t^n e^{-\int_t^s r} \, dB(s) \,\Big|\, Z(t), U(t) \right]. \tag{1.1}$$

We now turn to the general characterization of V.

Proposition 1.1 *Assume that there exists a function $V^j(t, u)$ solving*

$$V_t^j(t, u) = r(t) V^j(t, u) - b^j(t, u)$$

$$- \sum_{k: k \neq j} \mu_{jk}(t, u)\big(b^{jk}(t, u) + V^k(t, 0) - V^j(t, u)\big) - V_u^j(t, u),$$

$$V^j(t-, u) = V^j(t, u) + \Delta B^j(t, u), \tag{1.2a}$$

$$V^j(n, u) = 0. \tag{1.2b}$$

Then this function is indeed equal to the reserve defined in (1.1).

The set-up adding a state process to Z and developing a deterministic differential equation to characterize the reserve was already done in Chap. VI. Actually, the state process X introduced for the unit-link insurance contract was much more involved than our extra state process U. In fact, U can be obtained as a special case of X by specifying

$$a^{jk}(t, x) = x, \quad a^j(t, x) = -1 + rx, \quad \Delta A^j(t, x) = 0,$$

$$\chi^k(t, x) = 0, \quad \pi(t, x) = 0.$$

In $a^j(t, x)$, the term -1 corresponds to counting time. The somewhat odd term rx in $a^j(t, x)$ serves to kill the interest rate in the dynamics of X. One could have proposed, instead, to just put $r = 0$. However, we do need an interest rate different from zero to represent discounting in (1.1). So, the specification of $a^j(t, x)$ is exactly what it takes to generate the right dynamics of X while keeping the discount factor in (1.1). The term $\chi^k(t, x) = 0$ resets the duration process to zero at every jump time and $a^{jk}(t, x)$ is set such that $\varrho^{jk}(t, x) = 0$. Thus, we have that $X = U$. With this specification of X, we have that

$$\mathcal{D}_x V^j(t, x) = V_x^j(t, x), \quad R^{jk}(t, x) = b^{jk}(t, x) + V^k(t, 0) - V^j(t, x).$$

In Chap. VI, the process X was only introduced to drive the payment coefficients, not the transition intensities. However, the proof of Proposition VI.3.3 can be directly extended to allow for X dependence also in μ. In Chap. VI, this would have been an awkward idea, allowing for transition intensities to depend on the state of the financial process X. However, here it is exactly what it takes to allow μ to depend on U with the very special case of X formalized above. Apart from that extension, Proposition VI.3.3 collapses into Proposition V.1.1.

Exercises

1.1 Consider a disability annuity in a disability model without recovery. The disability annuity pays out an elementary annuity rate 1 during disability until time n, provided that disability occurred before time m. Such a product is called a disability with temporary risk because the risk that triggers the initiation of the annuity is temporary relative to the period of annuity payment. Note that this makes the annuity paid out between m and n dependent on the duration of the disability. Formalize the payment process in terms of the duration state process. Characterize the statewise reserves by their corresponding PDEs. How does the system change if you further assume that the death intensity among disabled depends on the duration of disability?

1.2 Assume that, instead of dependence on duration since entrance into the current state, we need to model dependence on duration since entrance into a specific state, say, φ. Specify the dynamics of the state process U measuring duration since entrance into state φ. Modify the differential equation (1.2) so that the solution characterizes the value with dependence on duration since entrance into a state φ.

2 Reserve-Dependent Payments and Intensities

In this section we study two other extensions to the general Markov framework studied in Chap. V. As in the previous section, the mathematical generalizations arising from the two extensions are essentially the same, but practically they arise from quite different ideas. On purpose, we present them in the opposite order compared to the previous section, in order to stress the chronological development of the ideas from a practical point of view. One extension is an extension to reserve-dependent payments. This generalizes the contract specification and has nothing to do with generalization of the probability model. Another extension is an extension to reserve-dependent intensities. This generalizes the probability model and has nothing to do with generalization of the contract specification. The reason why the mathematics of the two extensions coincide is that they are both handled by understanding the backward SDE for V in a broader sense than it has been understood so far.

The theory for backward SDEs is beyond the scope of this presentation, but the reserve process $V^{Z(t)}(t)$ studied in Chap. V fulfills a backward SDE with dynamics specified in, e.g., V.(3.9) and with terminal condition $V^{Z(n)}(n) = 0$. Actually, the backward SDE is said to be linear since both in the dt and the d$N(t)$ term of d$V^{Z(t)}(t)$ there appear linear functions of the vector of reserves $(V^0, \ldots, V^j)$ only. It is of technical interest to study when such backward SDEs have solutions if the coefficients are non-linear. Here, we just discuss various linear and non-linear functions of practical interest that may show up in the coefficients of V, without really paying much attention to whether we specify mathematically sensible quantities or not.

The generalization of the contract specification is formalized by allowing for payment coefficients to depend on the reserve itself. We could allow for dependence on the reserves corresponding to all states, both the state in which the policy actually is and the states in which it is not. However, we stick to the slightly simpler version where we allow for dependence on the reserve corresponding to the state in which the policy actually is. However, we allow the lump sum payment upon jump from $Z(t-)$ to k to depend on both the reserve in state $Z(t-)$ and the reserve in state k, respectively. Then the payment process becomes

$$
\mathrm{d}B(t) = \mathrm{d}B^{Z(t)}\big(t, V^{Z(t)}(t)\big)
$$
$$
+ \sum_{k:\, k \neq Z(t-)} b^{Z(t-)k}\big(t, V^{Z(t-)}(t-), V^k(t)\big)\,\mathrm{d}N^k(t),
$$

where

$$
\mathrm{d}B^{Z(t)}\big(t, V^{Z(t)}(t)\big) = b^{Z(t)}\big(t, V^{Z(t)}(t)\big)\,\mathrm{d}t + \Delta B^{Z(t)}\big(t, V^{Z(t)}(t-)\big).
$$

The functions $b^{jk}(t, v, v^k)$, $b^j(t, v)$, and $\Delta B^j(t, v)$ are deterministic and sufficiently regular and specify the contractual payments. The technical issues about existence of a reserve process defined as

$$V^{Z(t)}(t) = \mathbb{E}\left[\int_t^n e^{-\int_t^s r} \, dB(s) \,\Big|\, Z(t) \right]$$

are now hidden in the 'sufficient regularity' of the payment coefficients. We emphasize that these regularity issues are far from trivial.

For reserve-dependence in the continuous payment coefficient, we can think of a situation where the insurance company demands a continuous payment to the owners of the insurance company for providing the equity capital it takes to keep the business going. We assume that they demand a stream of continuous payments linear in the reserve with constant coefficient δ. A payment which is increasing in the reserve could be argued for, unsophisticatedly, by the fact that a large reserve indicates large volume of business and therefore large demand for equity capital and, thus, a large payment to equity providers. Similarly unsophisticatedly, we take the increasing function to be even linear. We assume that the payment to the owners is included in the benefit process although the policy holder does not receive this part. Then

$$b^j(t, v) = b^j(t) + \delta v,$$

where b^j is the contracted benefit paid to the policy holder. It is easily seen from Thiele's differential equation that the term δv in the payment coefficient can be absorbed into the interest rate term by replacing the true interest rate r by a calculation interest rate of $r - \delta$. We now have essentially two reserves. The reserve calculated with the true interest rate r and the payment to the policy holders, $V_r(t)$, where the subscript r denotes calculation with the interest rate r, and the larger reserve $V_{r-\delta}(t)$ calculated with the smaller interest rate $r - \delta$. The difference $V_{r-\delta} - V_r$ between the two reserves reflects the expected present value of payments to the owners. It is important to note that this actually formalizes the situation where the payments to owners are linear in the reserve $V_{r-\delta}$, including the present value of payments to the owners, and not in the reserve V_r, excluding the present value of payments to owners.

Calculating the present value of payments to owners in case these are specified in terms of V_r leads to a different calculation which we perform here, en passant, for the sake of completeness. The value in question is

$$\mathbb{E}\left[\int_t^n e^{-\int_t^s r} \delta V_r^{Z(s)}(s) \, ds \,\Big|\, \mathcal{F}(t) \right] = \delta \mathbb{E}\left[\int_t^n e^{-\int_t^s r} \int_s^n e^{-\int_s^u r} \, dB(u) \, ds \,\Big|\, \mathcal{F}(t) \right]$$

$$= \delta \mathbb{E}\left[\int_t^n \int_s^n e^{-\int_t^u r} \, dB(u) \, ds \,\Big|\, \mathcal{F}(t) \right] = \delta \mathbb{E}\left[\int_t^n \int_t^u ds \, e^{-\int_t^u r} \, dB(u) \,\Big|\, \mathcal{F}(t) \right]$$

$$= \delta \mathbb{E}\left[\int_t^n (u - t) e^{-\int_t^u r} \, dB(u) \,\Big|\, \mathcal{F}(t) \right] = \delta V_r^{Z(t)}(t) D(t),$$

where, in the second equality, we apply the tower property, and where we define

$$D(t) = \frac{\mathbb{E}\left[\int_t^n (u - t) e^{-\int_t^u r} \, dB(u) \,\middle|\, \mathcal{F}(t)\right]}{\mathbb{E}\left[\int_t^n e^{-\int_t^u r} \, dB(u) \,\middle|\, \mathcal{F}(t)\right]}.$$

The process D contains a measure of so-called duration of the future payment stream, not to be confused with the notion of duration discussed in the previous section.

For an example of a transition payment dependent on the reserve, we turn to the survival model. Assume that the contract specifies that upon death, the reserve itself is, simply, paid out as lump sum. Then we have that

$$b^{01}(t, v, v^1) = v,$$

and, recalling that $V^1 \equiv 0$ in the standard survival model, we get the sum at risk,

$$R^{01}(t) = b^{01}(t, V^0(t), V^1(t)) + V^1(t) - V^0(t) = V^0(t) - V^0(t) = 0,$$

such that the risk premium is zero. Thus, the seemingly complicated product design where the reserve is paid out upon death was not so complicated after all. By studying Thiele's differential equation, we realize that mortality risk can be disregarded when calculating the expected present value of future payments, since the solution to Thiele's differential equation with zero risk premium is simply

$$V^0(t) = \int_t^n e^{-\int_t^s r} \, dB^0(t).$$

This contract reflects in a way how pension saving works in a bank that does not have the permission to take on insurance risk. Here, the reserve is simply paid out upon death. The disadvantage is that in the younger years of the policy holder where he may have children that are economically dependent on him, the reserve, and therefore also the death sum, is low and may underserve the demand whereas in the older years of the policy holder where the children can take care of themselves, the reserve, and therefore also the death sum, is high and may overserve the demand. It should be stressed that only for the expected valuation of payments does the risk disappear. If we wish to calculate other characteristics of the payment stream, like the variance or the conditional expected cash flow during a specific time interval in the future, the mortality risk has to be taken correctly into account.

The idea of a reserve-dependent death sum can easily be generalized. In the previous paragraph, the bank designed a contract that serves the purpose of a pension savings contract to the best of their ability within the constraint that they are not allowed to carry insurance risk. The opposite situation may apply. The insurance company may need, for a certain categorization of the contract as an insurance contract, to carry a significant amount of insurance risk but may wish, on the other hand, to resemble the bank pension saving contract as closely as possible. Then they may think of applying

$$b^{01}(t, v, v^1) = (1 - \varepsilon)v,$$

for ε large enough to carry significant insurance risk but low enough to, sufficiently, resemble the bank contract, in practice seen to be 0.01. Then

$$R^{01}(t) = b^{01}(t, V^0(t), V^1(t)) + V^1(t) - V^0(t)$$
$$= (1 - \varepsilon)V^0(t) - V^0(t) = -\varepsilon V^0(t).$$

The risk premium to be paid now equals $-\mu_{01}(t)\varepsilon V^0(t)$. As for the example above with payments to the owners, this term can be absorbed into different artificial policy designs without reserve-dependent payments. We can absorb the sum at risk into the interest rate term and calculate the bank saving contract with an interest rate equal to $r + \varepsilon\mu_{01}(t)$. Or we can allow for insurance risk, have no payments upon death but work with the artificial calculation mortality intensity $\varepsilon\mu_{01}(t)$.

All of the examples above preserve the linearity of the differential equation for V. One example that does not is the design where the larger of a prespecified death sum g^{01} and the reserve V^0 is paid out. Then

$$b^{01}(t, v) = \max(g^{01}, v).$$

The relation between g^{01} and the other payment coefficients, e.g. b^0 and $\Delta B^0(n)$, determines at which time points the maximum is obtained by the death sum g^{01} or the reserve V^0. Now, the death sum is not linear in V^0, but the piecewise linearity can be handled by gluing together pieces of the linear Thiele differential equation where the maximum is attained in one of the arguments.

The generalization of the probability model is formalized by intensities μ_{ij} taking the reserve as argument. As for the reserve-dependent payments, we restrict ourselves to the simple version where the intensity μ_{ij} is allowed to depend on the reserves V^i and V^j and not all other reserves corresponding to other states. Then

$$M^k(t) = N^k(t) - \int_0^t \mu_{Z(s)k}(s, V^{Z(s)}(s), V^k(s)) \, ds$$

forms a martingale. Again, the question arises whether this forms a meaningful reserve process and a meaningful probability model which are two sides of the same story. This relies now on the regularity of the intensities with respect to the reserves and is a non-trivial question.

For a practical motivation of reserve-dependence of intensities, we have to add a new feature to the contract. We assume that the policy holder, from any state of the contract, can put the contract to an end. This act is called surrender or lapse, and we represent its outcome by one of the Markov states $\{0, \ldots, J\}$, taken here as J. In that case, the insurance company is relieved from the obligation which has value V, and they are therefore willing to pay out a surrender value b^{jJ} in case of surrender from state j. We can model this contract feature by adding a surrender state to which the policy can jump and trigger the surrender lump sum payment G, replacing all future payments.

One canonical choice of G is actually V itself, setting us back to the case of reserve-dependent payments. Note that this corresponds in a way to the specification of the bank saving contract above, since if $G = V$ and $V^J = 0$, i.e. all future payments fall away upon surrender, the sum at surrender risk, and therefore also the surrender risk premium, becomes zero. Thus, for the purpose of calculating the expected value of future payments, we can disregard surrender risk. As discussed above, however, when calculating other quantities we cannot disregard the surrender risk.

What we have in mind is the case where G is different from V. In that case, it is reasonable to discuss what drives policy holders to surrender and, therefore, should play a role in the modelling of μ_{jJ}. We return to this question when we discuss more thoroughly the concept of policy holder behavior. For the time being, we just assume, quite reasonably, that the financial gain from surrendering $G - V$, should, or at least could, play a role in the policy holder's tendency to surrender. But this introduces a reserve-dependent intensity. For example, the intensity could be of the form

$$\mu_{jJ}(t, v) = k_1 \exp\{k_2(G^{jJ}(t) - v)\},$$

for constants k_1 and k_2. This intensity supports the natural idea of a surrender intensity which is increasing in $G - V$. The larger the financial gain, the larger the tendency to surrender. This clearly leads to non-linearity in Thiele's differential equation and should cause some technical concern. It is tempting to start out with a case of linearity, e.g.

$$\mu_{jJ}(t, v) = k_2(G^{jJ}(t) - v),$$

which is also increasing in $G - V$. This is, however, not really thought through. First, the intensity is multiplied by the sum at surrender risk to form the surrender risk premium. So, linearity of μ in V leads to a second-order polynomial in V in Thiele's differential equation. Second, while the exponential form of the intensity makes sure that the intensity is non-negative, the linear form does not. So, the linear μ is not at all fruitful. We return to ideas of modelling policy holder behavior such as surrender in a succeeding section.

We finalize this section with the statement of the general non-linear system of differential equations characterizing the reserve in case of reserve-dependent payments and intensities. Introducing

$$R^{jk}(t, V^j(t), V^k(t)) = b^{jk}(t, V^j(t), V^k(t)) + V^k(t) - V^j(t),$$

this reads

$$\frac{d}{dt}V^j(t) = rV^j(t) - b^j(t, V^j(t))$$

$$- \sum_{k:k\neq j} \mu_{jk}\big(t, V^j(t), V^k(t)\big) R^{jk}\big(t, V^j(t), V^k(t)\big), \qquad (2.1)$$

$$V^j(t-) = \Delta B^j(t, V^j(t-)) + V^j(t), \qquad V^j(n) = 0.$$

Exercises

2.1 Thiele's differential equation can easily be generalized to the case where the interest rate also depends in the state, i.e.

$$r(t) = \sum_j I^j(t) r^j(t).$$

We simply decorate $r(t)$ in Thiele's differential equation with a state-specification into $r^j(t)$. It is *not* the exercise to verify this result. Consider now the general non-linear system with reserve-dependent payments and intensities, see (2.1). Assume that ΔB^j does not depend on V and that there exist $\mu_1^{jk}(t) \geq 0$ and state coefficients $\big(b_1^j, b_2^j, b_1^{jk}, b_2^{jk}\big)$ for $j \neq k$ such that

$$b^j\big(t, V^j(t)\big) = b_1^j(t) + b_2^j(t)V^j(t),$$

$$\mu^{jk}\big(t, V^k(t), V^j(t)\big) R^{jk}(t) = \mu_1^{jk}(t)\big(b_1^{jk}(t) + V^k(t) - (1 - b_2^{jk}(t))V^j(t)\big).$$

Show that the statewise reserves can be calculated in a standard way without reserve-dependent payments and intensities, with the artificial statewise payments b_1^j, the artificial transition payments b_1^{jk}, the artificial transition intensities μ_1^{jk}, and the artificial interest rate

$$r^j(t) = r - b_2^j(t) - \sum_{k:k\neq j} \mu_1^{jk} b_2^{jk}(t).$$

2.2 Consider surrender modelling in the case where $G < V$, which can be an integral part of the design of the surrender option. It seems reasonable to model an intensity which is increasing in G and decreasing in V (why?) and we propose the model

$$\mu_{jJ}(t, v) = k_j(t)\frac{1}{1 - G^{jJ}(t)/v},$$

which has exactly this feature. Show that this corresponds to working in a model without surrender but where the interest rate r in state j is replaced by $r + k_j$.

3   Bonds and Forward Interest Rates

In this section, we study some notions and ideas from bond market theory. The exposition is designed for applications to life insurance in two ways. First, the concepts introduced and discussed can be, more or less, directly translated to mortality and other transition rates and markets for trading mortality and other transition risks. We make that translation in the next section. Second, the concepts that are discussed in this section and the next one can be directly implemented in reserve calculations. We make this implementation in the last section.

By dealing with forward interest rates in this chapter on special studies in connection with stochastic mortality rather than in the chapter on financial mathematics, we emphasize the close connection between mortality and interest rates. The connection is highlighted and exploited during the first two sections. The connection is mainly mathematical, though, and it is important, from a conceptual point of view, to highlight also the dissimilarities.

We start out by introducing the notion of zero-coupon bonds. A zero-coupon bond is a financial asset that pays the amount 1 at a future time point. We denote by $P(t, T)$ the price at time t of a zero-coupon bond paying 1 at time T. So far, we have worked with a deterministic interest rate in case of which it is clear that the only arbitrage-free price of 1 at time T equals

$$P(t, T) = e^{-\int_t^T r(s)\,ds},\tag{3.1}$$

where r is the return earned on a bank account

$$dS^0(t) = r(t)S^0(t)\,dt.$$

Note that $P(t, T)$ solves the backward differential equation

$$\frac{\partial P(t, T)}{\partial t} = r(t)P(t, T), \quad P(T, T) = 1.\tag{3.2}$$

We now think of a situation where there is still access to investment in the bank account, i.e. S^0 is a traded asset, but where r is a stochastic process. In that case, it is clear that (3.1) must be replaced by something else since the discount factor is not known at time t but the price, if the zero-coupon bond exists, is. We present here two alternative representations of P that play crucial roles below.

The first expression is

$$P(t, T) = \mathbb{E}^{\mathbb{Q}}\left[e^{-\int_t^T r(s)\,ds}\,\middle|\,\mathcal{F}(t)\right].\tag{3.3}$$

This expression relates to the general understanding that a price is an expectation of the discounted cash flow. For a zero-coupon bond, the cash flow is a lump sum payment of 1 at time T, and therefore the price should be expressible by (3.3). In

Chap. VI the price was equally well formulated as the discounted expected cash flow because the discount factor was deterministic. In this section with stochastic interest rates, it is crucial to put the expectation and the discounting in the correct order. Note that if we use S^0 as the discount factor and P has the representation (3.3), then the discounted price process $e^{-\int_0^t r(s)\,ds} P(t, T)$ is a martingale under $\mathbb{Q}$, since

$$
e^{-\int_0^t r(s)\,ds} P(t, T) = e^{-\int_0^t r(s)ds} \mathbb{E}^{\mathbb{Q}}\left[e^{-\int_t^T r(s)\,ds}\,\Big|\,\mathcal{F}(t)\right]
$$
$$
= \mathbb{E}^{\mathbb{Q}}\left[e^{-\int_0^T r(s)\,ds}\,\Big|\,\mathcal{F}(t)\right],
$$

which is, certainly, a martingale under $\mathbb{Q}$.   ●

It should be emphasized that (3.3) is not a constructive representation for calculating $P(t, T)$, even if we knew the dynamics of r under the objective measure. This is so, because we do not know, up-front, which measure $\mathbb{Q}$ to use. Instead the equation (3.3) should be viewed the other way around. We think of observing the bond prices for all maturities and from these prices we deduce the measure $\mathbb{Q}$ such that (3.3) is fulfilled. Thus, the unknown in equation (3.3) is $\mathbb{Q}$ and not $P(t, T)$, which has to be given to us from outside (from the market prices). Having deduced $\mathbb{Q}$, we can now start to price cash flows other than 1, even interest rate dependent cash flows.

The second expression is

$$
P(t, T) = e^{-\int_t^T f(t,s)\,ds}, \tag{3.4}
$$

where $f(t, s)$ is a function of s known at time t since all $P(t, s)$ are so. It is obvious that the idea of representing P by (3.4) comes from the expression (3.1). The rates f are designed such that we can use the formula (3.1) with $r(s)$ replaced by $f(t, s)$. The rates f are called the *forward rates* and we have

$$
\mathbb{E}^{\mathbb{Q}}\left[e^{-\int_t^T r(s)\,ds}\,\Big|\,\mathcal{F}(t)\right] = e^{-\int_t^T f(t,s)\,ds}. \tag{3.5}
$$

Again, it is crucial to read (3.4) in the right way. Equation (3.4) is not a constructive formula for calculating $P(t, T)$. We think of having observed $P(t, T)$ in the market for all maturities. Then we can calculate the forward interest rate from (3.4). They can then be used for calculating values of general (deterministic) payment streams. The value of a deterministic payment stream B simply equals

$$
V(t) = \int_t^n e^{-\int_t^s f(t,\tau)\,d\tau}\,dB(s). \tag{3.6}
$$

A different way of expressing the relation (3.4) for all T is by differentiating both sides w.r.t. the maturity time (now changed to s),

$$\frac{\partial P(t, s)}{\partial s} = -f(t, s)P(t, s), \qquad P(t, t) = 1. \tag{3.7}$$

This is not viewed as an ODE where we, for a known function f, want to calculate the solution P. Rather it is a continuum of linear equations, one for every $s > t$, that has to be solved with respect to f. The bond prices and its derivatives are observed from market prices. Equivalent to (3.7), we can also write, and this is most often presented as the very definition of a forward rate,

$$f(t, s) = -\frac{\partial \log P(t, s)}{\partial s}.$$

Replacing the interest rates by the forward rates obviously works well when the cash flow is independent of the interest rate as was done in (3.6). We are now going to see that this is true even for cash flows that are linear in the interest rate. First, one may wonder why anyone would be interested in that. A simple motivation comes from the lending market. If you borrow an amount K in a floating interest rate loan with no repayments, then the interest payment for the loan over $(s, s + \mathrm{d}s)$ equals $Kr(s)\,\mathrm{d}s$ and the value of that interest payment is exactly

$$\mathbb{E}^{\mathbb{Q}}\!\left[e^{-\int_t^s r(\tau)\,\mathrm{d}\tau} Kr(s)\,\mathrm{d}s \,\Big|\, \mathcal{F}(t)\right] = K\,\mathbb{E}^{\mathbb{Q}}\!\left[e^{-\int_t^s r(\tau)\,\mathrm{d}\tau} r(s) \,\Big|\, \mathcal{F}(t)\right]\mathrm{d}s\,.$$

Thus, the cash flow $r(s)$ is certainly interesting to evaluate. However, note that

$$\mathbb{E}^{\mathbb{Q}}\!\left[e^{-\int_t^s r(\tau)\,\mathrm{d}\tau} r(s) \,\Big|\, \mathcal{F}(t)\right] = \mathbb{E}^{\mathbb{Q}}\!\left[-\frac{\partial}{\partial s}e^{-\int_t^s r(\tau)\,\mathrm{d}\tau} \,\Big|\, \mathcal{F}(t)\right] \tag{3.8}$$

$$= -\frac{\partial}{\partial s}\mathbb{E}^{\mathbb{Q}}\!\left[e^{-\int_t^s r(\tau)\,\mathrm{d}\tau} \,\Big|\, \mathcal{F}(t)\right] = -\frac{\partial}{\partial s}P(t, s) = e^{-\int_t^s f(t,\tau)\,\mathrm{d}\tau} f(t, s).$$

Thus, even for linear claims, we price by simply replacing the interest rate by the forward rate. This is convenient, for example, when we calculate the swap rate, i.e. the flat constant rate ρ one can choose to pay instead on the loan (without repayments) instead of the floating rate r, at the initial price of zero,

$$0 = \mathbb{E}^{\mathbb{Q}}\!\left[\int_0^T e^{-\int_0^t r(s)\,\mathrm{d}s} r(t)\,\mathrm{d}t\right] - \mathbb{E}^{\mathbb{Q}}\!\left[\int_0^T e^{-\int_0^t r(s)\,\mathrm{d}s} \rho\,\mathrm{d}t\right],$$

which combined with (3.8) and (3.5) for $t = 0$ gives

$$\rho = \frac{\mathbb{E}^{\mathbb{Q}}\!\left[\int_0^T e^{-\int_0^t r(s)\,\mathrm{d}s} r(t)\,\mathrm{d}t\right]}{\mathbb{E}^{\mathbb{Q}}\!\left[\int_0^T e^{-\int_0^t r(s)\,\mathrm{d}s}\,\mathrm{d}t\right]} = \frac{\int_0^T e^{-\int_0^t f(0,s)\,\mathrm{d}s} f(0, t)\,\mathrm{d}t}{\int_0^T e^{-\int_0^t f(0,s)\,\mathrm{d}s}\,\mathrm{d}t}.$$

Thus, the swap rate is simply a weighted average of future forward rates with the zero-coupon bonds as weights. Since plain integration gives

$$\int_0^T e^{-\int_0^t f(0,s)\,ds} f(0,t)\,dt = 1 - P(0,T),$$

we have the even simpler expression for the swap rate

$$\rho = \frac{1 - P(0,T)}{\int_0^T e^{-\int_0^t f(0,s)\,ds}\,dt}.$$

To prepare for later applications, we also introduce an auxiliary quantity

$$\widetilde{p}(t_0, t, T) = e^{-\int_t^T f(t_0,s)\,ds}. \tag{3.9}$$

For $t_0 < t < T$, this does not express the value at time t of a payment of 1 at time T. Only for $t = T$ is the value correctly equal to 1 and for $t = t_0$, correctly equal to $P(t_0, T)$, i.e.

$$P(t_0, T) = \widetilde{p}(t_0, t_0, T).$$

The quantity $\widetilde{p}(t_0, t, T)$ elegantly solves a backward differential equation corresponding to (3.2) with r replaced by the forward rate,

$$\frac{\partial \widetilde{p}(t_0, t, T)}{\partial t} = f(t_0, t)\widetilde{p}(t_0, t, T), \quad P(t_0, T, T) = 1.$$

Such a simple equation does not exist for $P(t, T)$ since the forward rates are generally irregular functions of t. The idea of introducing $\widetilde{p}(t_0, t, T)$ plays a role in rediscovering Thiele-type differential equations when revisiting reserves in life insurance in Sect. 7 below.

4   Survival Probabilities and Forward Mortality Rates

In this section, we discuss many of the concepts and ideas of the previous section but for a different object of study, namely stochastic mortality. This may occur, for example, for an insured particularly susceptible to epidemics or as result of future progress in treatment of serious diseases. Also general longevity may play a role.

The starting point is the conditional survival probability which we now, to stress the relation to the previous section, denote by $p(t, T)$, and which, in the case of a deterministic mortality rate μ, is given by

$$p(t, T) = e^{-\int_t^T \mu(s)\,ds}. \tag{4.1}$$

Further, it solves the backward differential equation

$$\frac{\partial p(t, T)}{\partial t} = \mu(t)p(t, T), \quad p(T, T) = 1. \tag{4.2}$$

The mathematical structure of (4.1) and (4.2) is the same as the mathematical structure of (3.1) and (3.2), although the two quantities express completely different objects. If we now introduce the idea of a stochastic intensity, i.e. the process μ is itself a stochastic process, then it is natural to generalize (4.1) to the survival probability

$$p(t, T) = \mathbb{E}\left[e^{-\int_t^T \mu(s)\,ds} \,\middle|\, \mathcal{F}(t)\right], \tag{4.3}$$

where $\mathcal{F}$ is the filtration generated by μ.

We can note the similarity with (3.3) but also the difference that we have omitted the measure specification. The quantity (4.3) is the true survival probability but for pricing issues one may wish to use a different survival probability. Mortality risk is no longer diversifiable since the lifetime of different members of a population are now dependent through their common dependence on μ and the law of large numbers does not apply. Only conditional on the whole history of μ, independence among members of the population allows for use of the law of large numbers. But when standing at time 0 or at time $t < T$, calculating a measurable financial value, this approach does not apply.

The question is whether there is any market that can inform us in any way about which measure to use in (4.3) for survival probabilities in the context of valuation. Such a financial market for mortality risk has been idealized over the last few decades, but it has been long in coming. For such a market to be informative about which measure to use, it has to be large, liquid and transparent, and neither of these properties have been in place. This does not prevent us from developing the related concepts, though, by assuming that *someone*, be that the financial market, the supervisory authorities, or some other organization, tells us which measure to use. Thus, for valuation purposes, we replace the expression in (4.3) by

$$q(t, T) = \mathbb{E}^{\mathbb{Q}}\left[e^{-\int_t^T \mu(s)\,ds} \,\middle|\, \mathcal{F}(t)\right].$$

If $\mathbb{Q} \neq \mathbb{P}$, it is not the objective survival probability, but we call it a survival probability, nevertheless.

Knowing $\mathbb{Q}$ is essentially the same as knowing all survival probabilities $q(t, T)$ for $T > t$, and this is our idealized standing point in what follows. Based on these survival probabilities, one is inspired from (3.7) to introduce the forward mortality rate as the function solving

$$\frac{\partial q(t, s)}{\partial s} = -m(t, s)q(t, s), \quad q(t, t) = 1, \tag{4.4}$$

so that we have the usual expression for the survival probability in terms of the forward mortality rates,

$$q(t, T) = e^{-\int_t^T m(t,s)\,ds}. \tag{4.5}$$

If we now follow the structure of the previous section, we come to the question about the relevance and the representation of

$$\mathbb{E}^{\mathbb{Q}}\left[e^{-\int_t^T \mu(s)\,ds} \,\middle|\, \mathcal{F}(t)\right].$$

In the interest rate section, we motivated the study with r instead of μ by referring to floating rate loans and swap rates. Here with μ instead of r, the motivation is straightforward as we face the conditional density of the time of death. Calculations identical to the ones in (3.8) lead to the satisfying

$$\mathbb{E}^{\mathbb{Q}}\left[e^{-\int_t^s \mu(\tau)\,d\tau} \mu(s) \,\middle|\, \mathcal{F}(t)\right] = e^{-\int_t^s m(t,\tau)\,d\tau} m(t, s), \tag{4.6}$$

stating that we simply calculate the density of the lifetime by replacing the mortality rate by the mortality forward rate in the usual formula for the density.

Further, with inspiration from (3.9), we can introduce the auxiliary survival function

$$\widetilde{q}(t_0, t, T) = e^{-\int_t^T m(t_0,s)\,ds}.$$

This allows for an ordinary backward differential equation corresponding to (4.2),

$$\frac{\partial \widetilde{q}(t_0, t, T)}{\partial t} = m(t_0, t)\widetilde{q}(t_0, t, T), \quad \widetilde{q}(t_0, T, T) = 1,$$

such that

$$q(t_0, T) = \widetilde{q}(t_0, t_0, T).$$

So far, we have followed the steps and ideas of the forward interest rate section closely. It should be stressed, however, that the practical market conditions for interest rates and mortality rates are very different. While there does exist a large, liquid, and transparent market for interest rate risk from where we can deduce the $\mathbb{Q}$-measure of interest rate risk and the forward interest rates, the bond market, there does not exist a similar market for survival probabilities from where we can deduce the $\mathbb{Q}$-measure of mortality risk and the forward mortality rates. In that respect, the current section is merely a convenient pattern of thinking, an academic mathematical exercise that we now take much further.

We have discussed the survival probability for valuation purposes. However, when pricing, the survival probability should not stand alone but be accompanied by the discount factor. Letting now $\mathcal{F}$ be the filtration generated by r, μ, the true quantities of interest are

$$\mathbb{E}^{\mathbb{Q}}\left[e^{-\int_t^s r(\tau)\,d\tau}\,\Big|\,\mathcal{F}(t)\right], \tag{4.7}$$

$$\mathbb{E}^{\mathbb{Q}}\left[e^{-\int_t^s (r(\tau)+\mu(\tau))\,d\tau}\,\Big|\,\mathcal{F}(t)\right], \tag{4.8}$$

$$\mathbb{E}^{\mathbb{Q}}\left[e^{-\int_t^s (r(\tau)+\mu(\tau))\,d\tau}\mu(s)\,\Big|\,\mathcal{F}(t)\right], \tag{4.9}$$

since these are the ones appearing in our financial valuation of 1 and the actuarial valuation of a pure endowment (the lump sum 1 paid upon survival until time s) and a term insurance (the lump sum 1 paid upon death before time s). The latter expression is just an ingredient in the value of the term insurance contract

$$\mathbb{E}^{\mathbb{Q}}\left[\int_t^n e^{-\int_t^s (r(\tau)+\mu(\tau))\,d\tau}\mu(s)\,ds\,\Big|\,\mathcal{F}(t)\right].$$

However, due to the linearity of the expectation we can turn our interest directly to (4.9). An extended list of objects of interest also includes the value of $r(s)$ and the value of $r(s)$ conditional upon survival, i.e.

$$\mathbb{E}^{\mathbb{Q}}\left[e^{-\int_t^s r(\tau)\,d\tau}r(s)\,\Big|\,\mathcal{F}(t)\right], \tag{4.10}$$

$$\mathbb{E}^{\mathbb{Q}}\left[e^{-\int_t^s (r(\tau)+\mu(\tau))\,d\tau}r(s)\,\Big|\,\mathcal{F}(t)\right]. \tag{4.11}$$

The key question is whether there exist interest and forward mortality rates that match market prices if these are observed for all these contracts. If r and μ are independent, the case is clear. From (3.4), (4.5), and (4.6), we get from the independence that

$$\mathbb{E}^{\mathbb{Q}}\left[e^{-\int_t^s r(\tau)\,d\tau}\,\Big|\,\mathcal{F}(t)\right] = e^{-\int_t^s f(t,\tau)\,d\tau}, \tag{4.12}$$

$$\mathbb{E}^{\mathbb{Q}}\left[e^{-\int_t^s (r(\tau)+\mu(\tau))\,d\tau}\,\Big|\,\mathcal{F}(t)\right] \tag{4.13}$$
$$= \mathbb{E}^{\mathbb{Q}}\left[e^{-\int_t^s r(\tau)\,d\tau}\,\Big|\,\mathcal{F}(t)\right]\mathbb{E}^{\mathbb{Q}}\left[e^{-\int_t^s \mu(\tau)\,d\tau}\,\Big|\,\mathcal{F}(t)\right]$$
$$= e^{-\int_t^s (f(t,\tau)+m(t,\tau))\,d\tau},$$

$$\mathbb{E}^{\mathbb{Q}}\left[e^{-\int_t^s (r(\tau)+\mu(\tau))\,d\tau}\mu(s)\,\Big|\,\mathcal{F}(t)\right] \tag{4.14}$$
$$= \mathbb{E}^{\mathbb{Q}}\left[e^{-\int_t^s r(\tau)}\,\Big|\,\mathcal{F}(t)\right]\mathbb{E}^{\mathbb{Q}}\left[e^{-\int_t^s \mu(\tau)\,d\tau}\mu(s)\,\Big|\,\mathcal{F}(t)\right]$$
$$= e^{-\int_t^s (f(t,\tau)+m(t,\tau))\,d\tau}m(t,s),$$

so that we can use the usual expression for evaluating simple contracts as before with both r and μ replaced by f and m. We also get that

$$\mathbb{E}^{\mathbb{Q}}\left[e^{-\int_t^s r(\tau)\,d\tau} r(s)\,\Big|\,\mathcal{F}(t)\right] = e^{-\int_t^s f(t,\tau)\,d\tau} f(t,s),$$

$$\mathbb{E}^{\mathbb{Q}}\left[e^{-\int_t^s (r(\tau)+\mu(\tau))\,d\tau} r(s)\,\Big|\,\mathcal{F}(t)\right] = e^{-\int_t^s (f(t,\tau)+\mu(t,\tau))\,d\tau} f(t,s).$$

Substantial consistency difficulties arise if the interest rate r and the mortality rates μ are dependent. We discuss these in Sect. 5 below, but assume independence in the rest of this section. We turn to a different challenge with actuarial use of forward mortality rates. Thinking of our Markov models, what we really need are general forward transition rates in a setting where the transition intensities $\mu_{jk}(t)$ are stochastic. The question is whether there exists a meaningful generalization of the forward mortality rate m to the multistate case. For this we let the transition probability be given by

$$q_{jk}(t,T) = \mathbb{E}^{\mathbb{Q}}\left[\pi_{jk}(t,T)\,\big|\,\mathcal{F}(t)\right],$$

where $\mathcal{F}$ is the filtration generated by μ and

$$\pi_{jk}(t,T) = \mathbb{P}\big(Z(T) = k\,\big|\,Z(t) = j,\ \mathcal{F}(T)\big)$$

is the transition probability given the transition intensities. This generalizes the probability

$$q(t,T) = \mathbb{E}^{\mathbb{Q}}\left[\pi(t,T)\,\big|\,\mathcal{F}(t)\right], \quad \text{where } \pi(t,T) = e^{-\int_t^T \mu(s)\,ds}.$$

We now consider a generalization of (4.4) via Kolmogorov's forward differential equations for the transition probabilities:

Definition 4.1 The forward transition intensities $m_{jk}(t,s)$ are defined by

$$\frac{\partial}{\partial s} q_{jk}(t,s) = \sum_{i:i\neq k} q_{ji}(t,s)m_{ik}(t,s) - q_{jk}(t,s)m_k(t,s), \tag{4.15}$$

$$q_{jk}(t,t) = \mathbb{1}_{j=k}, \quad \text{where } m_k(t,s) = \sum_{j\neq k} m_{kj}(t,s).$$

Recall that we think of the probabilities $\mathbb{Q}$ as given from an outside mortality market or insurance market. The unknowns in the Kolmogorov forward equation above are the forward transition rates m_{jk}. This forms a system of linear equations with, in general, $J(J-1)$ equations, namely q_{jk} for $j \neq k$ (q_{jj} is determined as $1 - q_j$), with $J(J-1)$ unknowns, namely the m_{jk} for $j \neq k$. Although this definition of the multistate forward rates has its merits, e.g. that these forward rates are independent of j, it also has serious drawbacks. Above all, it is *not* generally true that

$$E^Q\left[\pi_{jk}(t,s)\,\mu_{kl}(s)\big|\,\mathcal{F}(t)\right] = q_{jk}(t,s)\,m_{kl}(t,s).\qquad(4.16)$$

Other multistate forward rates have been proposed in the literature. For example, one could take (4.16) as the very definition and set the forward rates in accordance with

$$m_{kl}^{j}(t,s) = \frac{E^Q\left[\pi_{jk}(t,s)\,\mu_{kl}(s)\big|\,\mathcal{F}(t)\right]}{q_{jk}(t,s)}.\qquad(4.17)$$

The drawback of this definition is already made visible, namely dependence on the state j. Actually, one can plug these forward rates into Kolmogorov's forward equation and then obtain the transition probabilities $q_{jk}(t,s)$. But they work only if the initial state is j. If the initial state is different from j, we get a different set of forward transition rates that match the transition probabilities from that state.

Yet, a third definition proposed in the literature is the forward rates falling out of the following equation,

$$e^{-\int_t^s m_{kl}(t,\tau)\,d\tau} = \mathbb{E}\left[e^{-\int_t^s \mu_{kl}(t,\tau)\,d\tau}\,\bigg|\,\mathcal{F}(t)\right].\qquad(4.18)$$

In contrast to the two other definitions proposed in Definition 4.1 and (4.17), the defining equation (4.18) is, exclusively, related to the distribution of μ and only indirectly related to the distribution of Z. All the definitions have their own merits and their own drawbacks. It should be mentioned that all definitions coincide in the case of a so-called *multiple causes of death model* where several absorbing death states can be reached after sojourn in the state alive since birth. The probabilistic explanation of why this model forms a corner case, is that this model is Markov (with respect to the filtration generated by Z) even if the mortality intensities are stochastic. This is easily seen since, in that state model, either the past or the future is trivial.

In the rest of the section, we continue working on the forward rates introduced in Definition 4.1, keeping in mind that (4.16) is not true.

We can even generalize the auxiliary functions $\widetilde{q}$ as the solution to Kolmogorov's backward differential equations

$$\frac{\partial}{\partial t}\widetilde{q}_{jk}(t_0,t,T) = -\sum_{i:i\neq j} m_{ji}(t_0,T)\big(\widetilde{q}_{ik}(t_0,t,T) - \widetilde{q}_{jk}(t_0,t,T)\big),$$

$$\widetilde{q}_{jk}(t_0,T,T) = \mathbb{1}_{j=k},$$

and note that, although it is not itself in general a true probability, indeed we have that

$$q_{jk}(t_0, T) = \tilde{q}_{jk}(t_0, t_0, T).$$

Example 4.2 Let us reconsider the disability model from Sect. V.2.3. We can now write the Kolmogorov forward equation as

$$\frac{\partial}{\partial s} q_{00}(t, s) = q_{01}(t, s) m_{10}(t, s) - q_{00}(t, s)\big(m_{01}(t, s) + m_{02}(t, s)\big),$$

$$\frac{\partial}{\partial s} q_{01}(t, s) = q_{00}(t, s) m_{01}(t, s) - q_{01}(t, s)\big(m_{10}(t, s) + m_{12}(t, s)\big),$$

$$\frac{\partial}{\partial s} q_{10}(t, s) = q_{11}(t, s) m_{10}(t, s) - q_{10}(t, s)\big(m_{01}(t, s) + m_{02}(t, s)\big),$$

$$\frac{\partial}{\partial s} q_{11}(t, s) = q_{10}(t, s) m_{01}(t, s) - q_{11}(t, s)\big(m_{10}(t, s) + m_{12}(t, s)\big),$$

with boundary conditions $q_{00}(t, t) = q_{11}(t, t) = 1$, $q_{01}(t, t) = q_{10}(t, t) = 0$. This forms a linear system of four equation with four unknowns, $m_{01}(t, s)$, $m_{02}(t, s)$, $m_{10}(t, s)$, and $m_{12}(t, s)$. The dimension is smaller than the general $J(J - 1) = 6$ for a general three-state model. This is because of the absorbing death state. After observing that $\frac{\partial}{\partial s} q_{20}(t, s) = \frac{\partial}{\partial s} q_{21}(t, s) = 0$ and since $q_{20}(t, t) = q_{21}(t, t) = 0$ we immediately have that $m_{20}(t, s) = m_{21}(t, s) = 0$. This excludes the death state from the system, which becomes $(J - 1)(J - 1) = 4$-dimensional. Note again here that it is *not* generally true that the conditional expected densities can be written as $q_{jk}(t, s) m_{kl}(t, s)$.   ◊

5   Dependent Interest and Mortality Rates

The analysis of the interplay between stochastic r and μ in Sect. 4 assumed independence. However, the real troubles arise if they are dependent. Whereas this may be unrealistic if μ is actually modelling death, we ask the reader to think of the surrender case with an interest rate dependent surrender intensity μ. Put briefly, the trouble is that we have (for every T) three prices to match (4.7), (4.8), and (4.9) but only two rates, f and m. This is (for every s) three equations with two unknowns. Three different opinions about this have been expressed in the literature in the following chronological order:

- (The financial point of view, 2005): From all the equations and unknowns, the zero-coupon bond price and the forward interest rate are the most important ones and we determine f by

$$\mathbb{E}^{\mathbb{Q}}\left[e^{-\int_t^s r(\tau)\, d\tau} \,\Big|\, \mathcal{F}(t)\right] = e^{-\int_t^s f(t,\tau)\, d\tau}.$$

Thereafter we can determine residually the forward mortality rates of the two insurance products. In case of dependence between r and μ, the two forward mortality rates of the two insurance products differ. Further, none of the product-specific forward mortality rates coincide with the purely probabilistic forward rate m. From a financial point of view, we live with that inconsistency within insurance pricing and towards probabilities. We determine the *pure endowment forward mortality rate* $m^{pe}(t, s)$ by

$$\mathbb{E}^Q\left[e^{-\int_t^s (r(\tau)+\mu(\tau))\, d\tau} \,\Big|\, \mathcal{F}(t) \right] = e^{-\int_t^s \left(f(t,\tau)+m^{pe}(t,\tau) \right) d\tau},$$

and the *term insurance forward mortality rate* m^{ti} by

$$\mathbb{E}^Q\left[e^{-\int_t^s (r(\tau)+\mu(\tau))\, d\tau} \,\Big|\, \mathcal{F}(t) \right] = e^{-\int_t^s \left(f(t,\tau)+m^{ti}(t,\tau) \right) d\tau} m^{ti}(t, s).$$

Simple calculations show how m^{pe} can be expressed in terms of $f(t, s), m(t, s)$ as defined in (4.6), and

$$\mathbb{C}\text{ov}^Q\left[\left(e^{-\int_t^s r(\tau)\, d\tau}, e^{-\int_t^s \mu(\tau)\, d\tau} \right) \Big| \mathcal{F}(t) \right].$$

Similar calculations show how m^{ti} can be expressed in terms of $f(t, s), m(t, s)$, and

$$\mathbb{C}\text{ov}^Q\left[\left(e^{-\int_t^s r(\tau)\, d\tau}, e^{-\int_t^s \mu(\tau)\, d\tau}\mu(s) \right) \Big| \mathcal{F}(t) \right].$$

For the extended list of objects (4.10) and (4.11), we get that (4.10) can be expressed consistently with (4.12) as the, by now, usual

$$\mathbb{E}^Q\left[e^{-\int_t^s r(\tau)\, d\tau} r(s) \,\Big|\, \mathcal{F}(t) \right] = e^{-\int_t^s f(t,\tau)\, d\tau} f(t, s).$$

The object (4.11) requires, in contrast, yet a third forward mortality rate based on $f(t, s), m(t, s)$ and

$$\mathbb{C}\text{ov}^Q\left[\left(e^{-\int_t^s r(\tau)\, d\tau} r(s), e^{-\int_t^s \mu(\tau)\, d\tau} \right) \Big| \mathcal{F}(t) \right].$$

- (The pessimistic point of view, 2010): The fact that we cannot represent the price of a 1 or the price of a pure endowment and a term insurance consistently by means of forward interest and mortality rates makes the whole concept of a forward mortality rate shaky and maybe we should give it all up.

- (The actuarial point of view, 2014): From all the equations and unknowns, the consistency within insurance pricing is the most important one and we determine f^{ins} and m^{ins} (superscript ins for *insurance*) simultaneously in accordance with

$$\mathbb{E}^{\mathbb{Q}}\left[e^{-\int_t^s (r(\tau)+\mu(\tau))\,d\tau}\,\middle|\,\mathcal{F}(t)\right] = e^{-\int_t^s [f^{ins}(t,\tau)+m^{ins}(t,\tau)]\,d\tau}, \tag{5.1}$$

$$\mathbb{E}^{\mathbb{Q}}\left[e^{-\int_t^s r(\tau)+\mu(\tau)\,d\tau}\mu(s)\,\middle|\,\mathcal{F}(t)\right] = e^{-\int_t^s [f^{ins}(t,\tau)+m^{ins}(t,\tau)]\,d\tau}m^{ins}(t,s). \tag{5.2}$$

Since we give up consistency with (4.12), we have (for every s) only two equations for our two unknowns f^{ins} and m^{ins}. It is less important that the *insurance forward interest rate* f^{ins} is not identical to the financial forward interest rate f determined by (4.12) and the *insurance mortality forward rate* is not identical to the purely probabilistic mortality forward rate m. With the actuarial point of view, we live with that inconsistency towards financial pricing and probabilities.

For the extended list of objects (4.10) and (4.11), we get that (4.10) can be expressed consistently with (4.12), whereas (4.11) can actually be expressed as

$$\mathbb{E}^{\mathbb{Q}}\left[e^{-\int_t^T (r(s)+\mu(s))\,ds}r(t)\,\middle|\,\mathcal{F}(t)\right] = e^{-\int_t^T [f^{ins}(t,s)+m^{ins}(t,s)]\,ds}f^{ins}(t,T).$$

This follows from (5.1) and (5.2) by applying the usual differential argument,

$$\mathbb{E}^{\mathbb{Q}}\left[e^{-\int_t^T (r(s)+\mu(s))\,ds}r(T)\,\middle|\,\mathcal{F}(t)\right]$$

$$= \mathbb{E}^{\mathbb{Q}}\left[e^{-\int_t^T (r(s)+\mu(s))\,ds}(r(T)+\mu(T))\,\middle|\,\mathcal{F}(t)\right]$$

$$\quad - \mathbb{E}^{\mathbb{Q}}\left[e^{-\int_t^T (r(s)+\mu(s))\,ds}\mu(T)\,\middle|\,\mathcal{F}(t)\right]$$

$$= \mathbb{E}^{\mathbb{Q}}\left[-\frac{\partial}{\partial T}e^{-\int_t^T (r(s)+\mu(s))\,ds}\,\middle|\,\mathcal{F}(t)\right]$$

$$\quad - e^{-\int_t^T [f^{ins}(t,s)+m^{ins}(t,s)]\,ds}m^{ins}(t,T)$$

$$= \frac{\partial}{\partial T}\mathbb{E}^{\mathbb{Q}}\left[e^{-\int_t^T (r(s)+\mu(s))\,ds}\,\middle|\,\mathcal{F}(t)\right]$$

$$\quad - e^{-\int_t^T [f^{ins}(t,s)+m^{ins}(t,s)]\,ds}m^{ins}(t,T)$$

$$= -\frac{\partial}{\partial T}e^{-\int_t^T [f^{ins}(t,s)+m^{ins}(t,s)]ds}$$

$$\quad - e^{-\int_t^T [f^{ins}(t,s)+m^{ins}(t,s)]\,ds}m^{ins}(t,T)$$

$$= e^{-\int_t^T [f^{ins}(t,s)+m^{ins}(t,s)]\,ds} \left(f^{ins}(t,T) + m^{ins}(t,T) \right)$$

$$- e^{-\int_t^T [f^{ins}(t,s)+m^{ins}(t,s)]\,ds} m^{ins}(t,T)$$

$$= e^{-\int_t^T [f^{ins}(t,s)+m^{ins}(t,s)]ds} f^{ins}(t,T).$$

6 Stochastic Interest and Mortality Rate Models

The many different conditional expectations above, including (4.7), (4.8), (4.9), (4.10), and (4.11), serve as representations. We never really got to calculate these quantities for specific stochastic models for r and μ. This may be needed for various purposes. If we disregard the complications arising from dependence between r and μ, we could start by concentrating on the objects of the type

$$\mathbb{E}^{\mathbb{Q}}\left[e^{-\int_t^T r(\tau)\,d\tau} \,\Big|\, \mathcal{F}(t) \right], \tag{6.1}$$

$$\mathbb{E}^{\mathbb{Q}}\left[e^{-\int_t^T \mu(s)\,ds} \,\Big|\, \mathcal{F}(t) \right], \tag{6.2}$$

bearing in mind that the notion of forward rates immediately brings us to the remaining objects of study. So, let us have a closer look at (6.1) and (6.2).

The textbook starting point is to describe r by a diffusive SDE,

$$dr(t) = \alpha(t,r(t))\,dt + \sigma(t,r(t))\,dW(t), \quad r(0)=r_0.$$

Since we are interested in calculating expectations under the measure $\mathbb{Q}$, we take W to be Brownian motion under the measure $\mathbb{Q}$. Thus, α is the drift of r under the measure $\mathbb{Q}$. We can now look for convenient functions α and σ that make quantities like (6.1) more or less easy to calculate. The SDE model for r is a special case of a so-called m-factor model where $r(t) = r(t, X(t))$ for an m-dimensional SDE,

$$dX(t) = \alpha(t, X(t))\,dt + \sigma(t, X(t))\,dW(t), \quad X(0)=x_0. \tag{6.3}$$

The special case appears if $m = 1$ and we take $r(t) = X(t)$.

One class of models that has received special attention is the class of affine models, where

$$dr(t) = (a(t)+br(t))\,dt + \sqrt{c(t)+\delta r(t)}\,dW(t). \tag{6.4}$$

The affine version of the multi-factor model is simply constructed by letting X follow an m-dimensional version of (6.4) and letting r be an affine function of X. In order to see why this class of models is particularly attractive we establish the PDE characterizing the value (6.1). Since r is a Markov process under $\mathbb{Q}$, we know that there exists a function $p(t, r)$, skipping the specification of the terminal time point T in the notation for the function p, such that

$$P(t, T) \;=\; p(t, r(t)) \;=\; \mathbb{E}^Q\!\left[e^{-\int_t^T r(\tau)\,d\tau}\,\middle|\, r(t)\right].$$

Calculations similar to, but simpler than, the ones leading to Proposition VI.3.3 gives the following PDE characterizing the function p,

$$p_t(t, r) \;=\; -rp(t, r) - \alpha(t, r)p_r(t, r) - \frac{1}{2}\sigma^2(t, r)p_{rr}(t, r), \tag{6.5}$$

$$p(T, r) \;=\; 1. \tag{6.6}$$

What is so special about the affine structure (6.4), i.e.

$$\alpha(t, r) \;=\; a(t) + b(t)r, \tag{6.7}$$

$$\sigma(t, r) \;=\; \sqrt{c(t) + \delta(t)r}, \tag{6.8}$$

is that it allows for a certain separation of the dependence of p on the two variables t and r. We guess a solution to the differential equation in the form

$$p(t, r) \;=\; e^{f(t)r + g(t)}, \tag{6.9}$$

with derivatives

$$p_t(t, r) \;=\; p(t, r)\big(f'(t)r + g'(t)\big),$$

$$p_r(t, r) \;=\; p(t, r)f(t), \qquad p_{rr}(t, r) \;=\; p(t, r)f^2(t).$$

Plugging these derivatives into (6.5) with α and σ given by (6.7) and (6.8), we get after dividing by $p(t, r)$,

$$f'(t)r + g'(t) \;=\; -r - \big(a(t) + b(t)r\big)f(t) - \frac{1}{2}\big(c(t) + \delta(t)r\big)f^2(t).$$

Here we can collect terms with and without r to form a two-dimensional system of ODEs,

$$f'(t) \;=\; -1 - b(t)f(t) - \frac{1}{2}\delta(t)f^2(t), \tag{6.10}$$

$$g'(t) \;=\; -a(t)f(t) - \frac{1}{2}c(t)f^2(t). \tag{6.11}$$

These have to be solved with terminal conditions following from (6.6)

$$f(T) = g(T) = 0.$$

Depending on a, b, c, δ, and specifically on whether any of them are constant or zero, this system is more or less tractable. But that is not really the point. The point is that we reduced the representation in terms of a PDE to a representation in terms of a system of ODEs. This is in general a substantial mathematical simplification.

One standard exercise in bond market theory is now to calibrate the parameters in the interest rate model to the observed market prices P of bonds. If we now decorate f and g with a T to stress that they depend on T, calibration means to solve the equation, with P being the observed price,

$$P(t, T) = e^{f^T(t)r + g^T(t)},$$

w.r.t. the functions a, b, c, δ. If these functions are constant and we only want to match the price for one maturity T, the model is clearly overparametrized as we then have one equation with four unknowns. These then have to be specified from other sources. For example, historical volatility may be used to estimate c and δ, as it is unchanged by the change of measure and can therefore be estimated from realized volatility. If we want to match the price for many maturities, perhaps even a continuum of maturities, the model is clearly underparametrized as we have then a continuum of equations with four unknowns. Then, an obvious idea is to take one of the parameters in the drift, a or b, to be a function of time so that we also have a continuum of unknowns in order to match the observed price with the theoretical model-based value for every maturity. Note, however, that since the l.h.s. of the calibration equation is the truth and the r.h.s. is just a model value, then the day (or an instant) after the calibration exercise is made, a new calibration leads to a different set of parameters or functions. The change in calibrated parameters as the calibration exercise is performed through time expresses in a sense how badly the model models reality.

Let us now turn to (6.2). Although it looks the same, it represents something completely different. Thinking more carefully about what it means that 'time passes by' under the integral in (6.2) makes it a much more involved object. The point is that 'time passing by' means both that calendar time passes by and that the individual of whom we are measuring the survival probability gets older. This is not in itself a problem. It just means that the integral is taken over mortality intensities of the same person at different ages that he lives through at different calendar time points. It only places high demand on the process μ to capture these effects as we want them to be captured. In particular, we need to consider what kind of consistency constraints across ages and/or calendar times we wish to impose on μ. What we mean by consistency depends to some extent on whether we think in terms of mathematics, mathematical consistency, mathematical convenience, calculation, etc. or whether we think in terms of biological aging, biological consistency, or demographic consistency in the sense of being meaningful, biologically, historically, and demographically. The studies of mortality rates are divided into these two directions and we continue along the first route which is also said to be pure data-driven. This means that the only input we have is the mortality data we observe. No further expertise from other areas of science is introduced to deem whether a model is good or bad.

Even among the data-driven models formed mainly with regard to mathematics and statistics, there seems to be a division of the subject area depending on whether the first source of inspiration is mathematical finance or statistics. Some models are much inspired from the decades of studying interest rate models, interest rate theory, and bond market theory in mathematical finance. As in the example of the affine interest rate model above, this approach is much driven by mathematical convenience and less by regard to trends and uncertainty of observations. Other models take the statistical modelling with mortality data as a starting point and build up statistical models that describe historical developments of mortality rates over years and ages well. The ultimate purpose in both cases is to forecast the development of mortality rates in the future and calculate probabilities based on them. In any case, this is a dubious ambition since the future is in many senses not going to look like the past at all. So, why should this be true for mortality trends and uncertainty. This is why it is said that stochastic mortality modelling and forecasting is a dirty job, but somebody has to do it. Below, we start out with the statistical approach, go towards the models inspired from interest rate theory, and end with a way of unifying the two directions.

Among the statistical models, the theoretically and practically dominant example is the Lee–Carter model. Our version of the Lee–Carter model and generalizations look a bit different than the original version, but this is just because we wish to stick to continuous-time modelling. In most presentations and applications, the discrete-time version appears. The main reason is that data typically consists of occurrences and exposures of deaths on a year-to-year basis, in both the calendar time and the age dimension. This gives time series of mortality estimators for every integer age in every integer year. We are not going to do the statistical data work here but just want to understand the structures in the proposed models and similarities with continuous-time models inspired from finance. Therefore, we present their continuous-time version. The Lee–Carter (age-period) model specifies

$$\mu(x, t) = e^{\beta_1(x) + \beta_2(x)\kappa(t)}$$

as the mortality intensity of a person at age x at calendar time t. If there were no calendar time effects we would just have $\mu(x) = e^{\beta_1(x)}$ of which the Gompertz–Makeham mortality model is a special case with a linear function β_1. The term $e^{\beta_2(x)\kappa(t)}$ captures the calendar time effects. The process $\kappa(t)$ contains calendar time improvements whereas the function $\beta_2(x)$ distributes these calendar time improvements over the different ages. In some simple versions, κ is a deterministic function but to capture uncertainty in improvements, it has to be modeled as a stochastic process. In the discrete-time version, it could be a random walk with drift or, in continuous-time, Brownian motion with drift. The functions β^1 and β^2 were originally introduced as non-parametric functions but could of course take a parametric form. Note that non-parametric in the original discrete version is actually parametric in the sense of one parameter for every age. The model appears quite general. But it is important to understand its limitations that have been the starting point for many generalizations in the academic literature. Above all, the calendar time effect is distributed over years of age via β_2 in the same way at every calendar time point. This rules out that certain improvements at some historic time points affect some ages whereas other improvements at different historic time points affect other ages. Another drawback is that the function β_2 distributes the level of improvement and the uncertainty of that improvement over ages in the same way. Thus, the level and the uncertainty of the calendar time effect at a given age are fully entangled.

As mentioned, the Lee–Carter model has been generalized in numerous direc-tions. One important generalization is the introduction of cohort effects in the Renshaw–Haberman (age-period-cohort) model,

$$\mu(x, t) = \exp\{\beta_1(x) + \beta_2(x)\kappa(t) + \beta_3(x)\gamma(t - x)\},$$

for a stochastic cohort process γ. The cohort effect allows for a separate dependence on year of birth. Depending on the parametrization of the involved functions and processes, the number of parameters may be relatively high, and in order to maintain robustness, it has been suggested to give up the age-heterogeneity of calendar time and cohort effects now that both effects are present, and simply put $\beta_2 = \beta_3 = 1$.

We now turn to the models inspired by interest rate and bond market theory. They are of the form

$$d\mu(t, x) = \alpha(t, x, \mu(t, x)) dt + \sigma(t, x, \mu(t, x)) dW(t), \qquad (6.12)$$

with initial condition $\mu(0, x) = \mu_0(x)$, which is actually the initial mortality curve as a function of ages. The function σ and Brownian motion W can be of higher dimension. We stress that as time passes by, not only t but also x, which is the age at time t, is moving. A different way of writing the differential equation for a specific individual with age x_0 at time 0 is therefore

$$d\mu(t, x_0 + t)$$
$$= \alpha(t, x_0 + t, \mu(t, x_0 + t)) dt + \sigma(t, x_0 + t, \mu(t, x_0 + t)) dW(t), \qquad (6.13)$$

$\mu(0, x_0) = \mu_0$, with μ_0 being that person's mortality rate at time 0. It is important to understand that for a specific individual, the second argument, the age, of μ, α, and σ moves linearly in time, although this is somewhat hidden when written in the formulation (6.12). The differential equation for μ describes the development of the mortality intensity of an individual as both calendar time and age goes by.

It is clear that one has to be careful about the specification of α and σ in order to get reasonable developments of the relation between the mortality intensities of individuals at different ages at the same calendar time or, equivalently, at the same age at different calendar times. For example, what is the proper relation between the initial mortality curve $\mu(0, x)$ and the drift and volatility of the improvements from there? By choosing the dimension of W, one can create different correlation structures between mortality improvements of individuals at different ages at the same time point. Some of these correlation structures may be meaningful whereas others may not. Thus, the discussion about consistency in mortality improvements from calendar time across ages shows up again. An individual who is x_0 years old at time 0 has a mortality intensity of $\mu(s, x_0 + s)$ at time s such that the survival probability becomes

$$p(t, m) = \mathbb{E}^Q\left[e^{-\int_t^T \mu(s, x_0+s)\, ds} \,\middle|\, \mu(t, x_0 + t) = m\right].$$

One can impose conditions on α and σ similar to (6.7) and (6.8),

$$\alpha(t, x) = a(t, x)\mu(t, x) + b(t, x),$$

$$\sigma(t, x) = \sqrt{c(t, x)\mu(t, x) + \delta(t, x)}.$$

These allow a representation of the survival probability similar to (6.9),

$$p(t, x, m) = e^{f(t,x)m + g(t,x)}, \tag{6.14}$$

where f and g follow ODEs similar to (6.10) and (6.11),

$$\frac{\partial}{\partial t} f(t, x) = -1 - b(t, x)f(t, x) - \frac{1}{2}\delta(t, x)f^2(t, x), \quad f(T, x) = 0,$$

$$\frac{\partial}{\partial t} g(t, x) = -a(t, x)f(t, x) - \frac{1}{2}c(t, x)f^2(t, x), \quad g(T, x) = 0.$$

Note that here we again suppress the dependence of p, f, and g on T.

Such classes of mortality models are called affine models, similarly to the class of affine interest rate models. They have been popular for the same reason, namely that the dependence on time and current mortality are separated and survival probabilities are calculated by solving the system of ODEs for f and g rather than the more involved PDE

$$p_t(t, x, m) = -mp(t, x, m) - \alpha(t, x, m)p_m(t, x, m) - \frac{1}{2}\sigma^2(t, x, m)p_{mm}(t, x, m),$$

$$p(m, x, m) = 1.$$

A special one-factor model has been proposed to combine the best of Lee–Carter, namely the explicit modelling of a mortality improvement process, with the best of affine models, namely access to separability of the variables in the survival probability. Consider a mortality curve at time 0 given by $\mu(0, x)$ and now build the mortality rate $\mu(t, x)$ as

$$\mu(t, x, \xi(t, x)) = \mu(0, x)\xi(t, x),$$

where ξ is an improvement process that models both improvements and their distribution over ages. We now let ξ follow a process similar to (6.4), namely,

$$
\begin{aligned}
&\mathrm{d}\xi(t, x) \\
&\quad = \big(a(t, x) + b(t, x)\xi(t, x)\big)\,\mathrm{d}t + \sqrt{c(t, x) + \delta(t, x)\xi(t, x)}\,\mathrm{d}W(t)
\end{aligned}
\tag{6.15}
$$

with $\xi(0, x) = 1$. As when writing (6.13) instead of (6.12), we can also write (6.15) differently by replacing x by $x_0 + t$ in the differential equation and x by x_0 in the initial condition. One can now find functions f and g, such that (6.14) is fulfilled in the sense of

$$
\begin{aligned}
p(t, x_0 + t, y) &= \mathbb{E}^Q\!\left[e^{-\int_t^T \mu(s, x_0 + s)\,\mathrm{d}s} \,\middle|\, \xi(t, x_0 + t) = y\right] \\
&= \exp\{f(t, x_0 + t)\mu(0, x_0 + t)y + g(t, x_0 + t)\}.
\end{aligned}
$$

Because of the extra dimension compared to the interest rate model, the notation becomes somewhat heavier, but the idea here is really to build a one-factor model corresponding to (6.3) with affine coefficients. Since the letter x in mortality models is used for the age, the factor is now called ξ. Instead of the immediate one-factor construction, corresponding to $r = X$ in the interest rate model, we take $\mu = \mu_0(x)\xi$ with μ_0 as the initial mortality curve. If ξ is affine, then also μ becomes affine, which can be seen by the calculation

$$
\begin{aligned}
\mathrm{d}\mu(t, x_0 + t, \xi(t, x_0 + t)) &= \mathrm{d}\big(\mu_0(x_0 + t)\xi(t, x_0 + t)\big) \\
&= \xi(t, x_0 + t)\frac{\mathrm{d}}{\mathrm{d}t}\mu_0(x_0 + t)\,\mathrm{d}t + \mu_0(x_0 + t)\,\mathrm{d}\xi(t, x_0 + t) \\
&= \left(\mu_0(x_0 + t)a + \left[\frac{\frac{\mathrm{d}}{\mathrm{d}t}\mu_0(x_0 + t)}{\mu_0(x_0 + t)} + b\right]\mu(t, x_0 + t, \xi)\right)\mathrm{d}t \\
&\quad + \sqrt{\mu_0^2(x_0 + t)c + \mu_0(x_0 + t)\delta\mu(t, x_0 + t, \xi)}\,\mathrm{d}W(t),
\end{aligned}
$$

where in the two last lines ξ, a, b, c, δ are all evaluated at $(t, x_0 + t)$. To be sure that ξ and μ stay positive, we need $c(t, x) = 0$ and

$$2a(t, x_0 + t) \geq |\delta(t, x_0 + t)|.$$

This is a condition for giving μ the interpretation as an intensity but not for the calculations above to hold true. If the condition is fulfilled, ξ shows mean-reversion to a positive level, which means that sufficiently many mortality improvements in the future are followed by an upward trend in mortality. This is a natural place to stop our short tour of mortality modelling. This exemplifies rather concretely how properties of μ that may be beneficial from one (mathematical) point of view may be considered as inconsistent from another (biological). The appropriate mortality model to use probably depends on the specific situation and what we want to use the model for.

We finish by mentioning two generalizations that have received considerable attention over the last decade. Looking back at our Markov process model, one may clearly be interested in a stochastic model not just for the mortality rate but also for the vector of transition rates. This adds dimensions to the issues of consistency which should now not be across only calendar time and age but also transitions. For example, in a disability model, how do the mortality intensities as active and disabled, respectively, co-move as stochastic processes? Some co-movement could be expected since it is in a sense the *same risk event*, death, that drives the intensity, but from two different states of health. Are they correlated or, perhaps, co-integrated? Or are they both functions of the same underlying single stochastic factor? This concerns the dependence structure between μ_{ik} and μ_{jk} for $i \neq j$. Also, in a disability model, how do the disability and mortality intensity, respectively, as active co-move as stochastic processes? Again, some co-movement could be expected since it is in a sense the *same risk exposure*, as active, that drives the intensity, but to two different states. Is it a correlated, co-integrated, one-factor model or something completely different? This concerns the dependence structure between μ_{jk} and $\mu_{j\ell}$ for $k \neq \ell$. And what about the dependence structure between μ_{ij} and $\mu_{k\ell}$ for $i \neq k$ and $j \neq \ell$? What kind of structure do we find in the data? What kind of structure would we prefer to work with from a (financial) mathematical point of view? And what structure do we believe in from a biological/demographic point of view? There exist generalizations of affine models to multistate models. They aim at dimension reduction of the differential equations characterizing transition probabilities instead of the survival probability. One can find conditions for matrix coefficients of a stochastic differential equation describing μ, such that the transition probability is characterized by functions that are determined by systems of ODEs rather than PDEs.

In the previous paragraph, we discussed co-movement between intensities within the same state process, typically describing the life history of a single individual. A different type of co-movement appears if we consider the intensities of two different state processes describing two different individuals. When they are not just governed by the same intensity, it is because certain risk characteristics among them could be different. The purpose is to model co-movement of mortality, e.g., among males and females within one population. Are these correlated? co-integrated? or something else?

More often, the idea is applied to modelling co-movement of mortality in different countries. Are they correlated, co-integrated, or something else? Some co-movement could be expected since the different populations, whether that be males and females within one population or citizens in different countries, in some ways are influenced by the same demographic-medical-economic conditions but in other ways they are not, precisely because they belong to different populations.

All these questions attract much attention and there is still a lot of statistical modelling to be done and understanding to be obtained.

7 Reserves Revisited

For a general payment stream, we introduce the market reserve

$$
V(t) = \mathbb{E}^{\mathbb{Q}}\Big[\int_t^n e^{-\int_t^s r} \, dB(s) \,\Big|\, \mathcal{F}(t) \Big],
$$

where, as in Sect. VI.1, the measure $\mathbb{Q}$ is deduced from all relevant information like market prices of bonds, mortality markets, and, possibly, assuming diversifiability of insurance risk conditional on μ. If we now consider payments driven by an underlying Markov process as in Chap. V, the payment stream is independent of the interest rate and we get

$$
\begin{aligned}
V(t) &= \int_t^n \mathbb{E}^{\mathbb{Q}}\Big[e^{-\int_t^s r} \, dB(s) \,\Big|\, \mathcal{F}(t) \Big] \\
&= \int_t^n \mathbb{E}^{\mathbb{Q}}\Big[e^{-\int_t^s r} \,\Big|\, \mathcal{F}(t) \Big] \mathbb{E}^{\mathbb{Q}}\big[dB(s) \,\big|\, \mathcal{F}(t) \big].
\end{aligned} \tag{7.1}
$$

In order to avoid the drawbacks from not having access to the relation (4.11), we disregard transition payments and allow for statewise payments only. Then, using the results from the previous sections and performing calculations similar to those of Proposition V.3.1, we can write this as

$$
\int_t^n e^{-\int_t^s f(t,\tau)d d\tau} \sum_k q_{jk}(t,s) \, dB^k(s),
$$

which is just the usual definition of a reserve, in case of statewise payments only, with the interest and transition rates (r, μ) replaced by the forward interest and transition rates (f, m).

The calculation shows the importance of the (conditional expected) cash flows

$$\mathbb{E}^{\mathbb{Q}}\big[dB(s) \,\big|\, Z(t) = j \big].$$

Knowing these, one can easily calculate the reserve for different interest rate assumptions. Also, the idea of a hedging strategy in the bond market is now to match the coupon payments from the bond positions with the conditional expected cash flows from the insurance contract by investing in $\mathbb{E}^{\mathbb{Q}}\big[dB(s) \,\big|\, Z(t) = j \big]$ zero-coupon bonds with maturity s. It is clear that this is not a true hedge in the sense that the insurance company is risk-free. However, the hedge works in the sense that the loss left to the owners of the company to cover is expressed by the diversifiable jump martingales $M^k(t)$. The forward rates appear in the coefficient of this jump martingale but possible martingale terms from the bond market, through dP, dr or df are eliminated by the suggested hedging strategy. It is beyond the scope of this exposition to carry out the details here.

We can even adopt the idea of the auxiliary function $\tilde{p}(t_0, t, T)$ and $\tilde{q}(t_0, t, T)$ and introduce

$$\widetilde{V}^j(t_0, t) \;=\; \int_t^n e^{-\int_t^s f(t_0, \tau)\, d\tau} \sum_k \tilde{q}_{jk}(t_0, t, s)\, dB^k(s).$$

This is the expected present value of payments from t and onwards but evaluated with the forward rates drawn at time t_0 and therefore not a real economic value. However, it solves the usual Thiele differential equation in t with $r(t)$ replaced by the forward rate $f(t_0, t)$, and at time t_0 it is actually the true real economic market value itself

$$\widetilde{V}^j(t_0, t_0) \;=\; V^j(t_0).$$

Calculations similar to those proving Proposition V.3.2 gives us the generalized Thiele differential equation

$$\frac{d}{dt} \widetilde{V}^j(t_0, t)$$

$$= \; f(t_0, t)\widetilde{V}^j(t_0, t) - b^j(t) \;-\; \sum_{k:k \neq j} m_{jk}(t_0, t)\big(\widetilde{V}^k(t_0, t) - \widetilde{V}^j(t_0, t)\big),$$

$\widetilde{V}^j(t_0, n) = 0$, and the gluing condition

$$\widetilde{V}^j(t_0, t-) \;=\; \widetilde{V}^j(t_0, t) + \Delta B^j(t).$$

This can be used to implement discounting by forward rates in calculations based on ordinary Thiele differential equations.

Note, however, that if payments in B depend on the financial risk, e.g. if they are interest rate dependent, then the whole idea breaks down since (7.1) is not true anymore. Only if the payments are linear in the interest rate can we perform calculations similar to (3.8) to obtain a full replacement of interest rates r by forward rates similar to our replacement of μ by m.

8   Incidental Policy Holder Behavior

The interest in modelling policyholder behavior has grown rapidly during the last decades. There are various approaches to modelling policy holder behavior, depending on assumptions about what drives the policy holders to behave in a particular way. Behavior is relevant because most insurance contracts allow the policy holder to intervene prematurely and change the conditions of the contract. One option which is almost always there is the option to stop paying the premiums but keep the contract on new terms with recalculated (lower) benefits. This is a natural option, since the policy holder may not be able to afford the premiums and, so, the option just says that you are not forced to pay something that you are not able to. The recalculation of benefits has to be stipulated in the contract or the law such that both parties know, in advance, the consequence of stopping the premium payments. Although the option to stop paying premiums (also called the *free policy option* or the *paid-up policy option*) is essential, another option is the standard one to think of when talking about policy holder behavior. This is the so-called *surrender* or *lapse option* which allows the policy holder to fully cancel his contract. During the term of the contract, the insurance company is in debt to the policy holder and is therefore released from an obligation in case of surrender. Therefore, the company is willing to pay an amount to the policy holder if he surrenders. This is called the surrender value.

We already touched upon the surrender option in Sect. 2. As a motivation for reserve-dependent intensities, we introduced a state representing that the policy has been surrendered and discussed reserve-dependence of the intensity into that state. This intuitively formalizes the idea that the financial gain of surrender, $G - V$, the difference between the surrender value paid out upon surrender G and the value of the contract V should, or at least could, play a role in the policy holder's tendency to surrender. Actually, this idea of a reserve-dependent surrender intensity is a rather delicate interpolation of two extreme ideas that are explored more intensively in the academic literature and frequently implemented in practice in different settings. The two extremes can be thought of as completely incidental behavior and completely rational behavior. In this section, we formalize the idea of incidental behavior and work out the details for the case with two behavior options, namely the option to surrender and the option to stop paying the premiums. In the next section, we formalize the idea of rational behavior. At the end of that section, we return to the reserve-dependent intensities and discuss in what sense this can be viewed as a delicate interpolation of these two extremes.

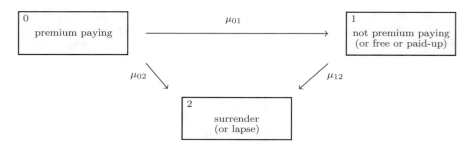

Fig. 1 Behavioral model

The idea of the incidental behavior is to extend the Markov model of the insurance risk discussed in Chap. V to also encompass behavioral risk. By behavioral risk, we mean the risk of policy holder intervention in the contract in a specific way that influences the payment process of the contract. A pure behavioral model of the free policy and surrender options could be illustrated as in Fig. 1. The idea of incidental behavior is that there exists deterministic intensities such that the behavior of the policy holder is reasonably well described by such a model. This means that the policy holder behaves in a particular way with a randomness that is similar to the insurance risk.

The option to surrender or transcribe into a free policy is only present when the policy holder is in certain states of the risk process described in Chap. V. This links the two chains arising from risk and behavior to each other. We form a model with a mix of the usual risk events and behavior events and study the special case where the usual risk model is the disability model. Then, we get the mixed model illustrated in Fig. 2. Here, we note that the free policy option is held only as long as the policy holder is active and premium paying, i.e. the only transition from the upper, premium payment states (blue) to the lower free policy states (red) goes via the green transition from state $(0, p)$ to state $(0, f)$.

We are now going to specify the payment stream such that payments depend on the state and transitions of the joint multi-state risk and behavior model. We assume that we have deterministic statewise payment processes B_p^j and transition payments b_p^{jk} for the upper premium payment states $j, k \in \{(0, p), (1, p), (2, p), (3, p)\}$. For the states $j, k \in \{(0, p), (1, p), (2, p)\}$, these are the usual payments thought of in Chap. V. The transition into $(3, p)$ corresponds to surrendering as premium paying. This happens with an intensity μ_p^{03} and triggers a surrender value b_p^{03}. After a possible transcription into free policy, we have a different set of statewise payment processes B_f^j and transition payments b_f^{jk} for the lower free policy states $j, k \in \{(0, f), (1, f), (2, f), (3, f)\}$. A special feature of the free policy option is the link between the payment coefficients before and after free policy transcription. Upon transition from $(0, p)$ to $(0, f)$, happening with intensity μ_f^{00}, the original premiums fall away and the original benefits are reduced by a so-called free policy factor f. The free policy factor is a function of time and is drawn upon transition, which gives

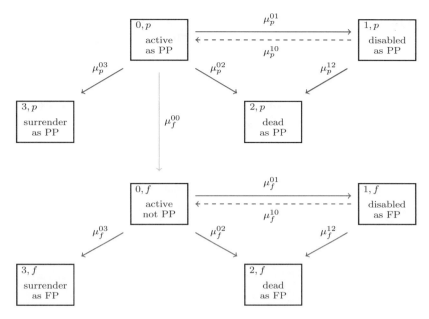

Fig. 2 Disability model with free policy and surrender options. PP = premium paying, FP = free policy

the payment coefficients after transcription two time arguments, current time t and the time of transcription τ, i.e., we have the following link

$$\mathrm{d}B_f^j(t, \tau) \;=\; f(\tau)\,\mathrm{d}B_p^{j+}(t), \qquad b_f^{jk}(t, \tau) \;=\; f(\tau)b_p^{jk+}(t).$$

Instead of working with the transcription time τ as time input, we can work with the duration since transcription. If we denote this by u, we have that $t = \tau + u$. Note that this is a slightly different measure of duration than the one we had in Sect. 1, since that duration measured duration of sojourn in the current state. Here, if the policy is in state $(1, f)$, u does not measure duration of sojourn in $(1, f)$ but duration since transition to $(0, f)$. If we specify payments in terms of duration instead of time of free policy transition, we have

$$\mathrm{d}B_f^{j+}(t, u) \;=\; f(t - u)\,\mathrm{d}B_p^{j+}(t), \tag{8.1}$$

$$b_f^{jk+}(u, t) \;=\; f(t - u)b_p^{jk+}(t). \tag{8.2}$$

We can now plug these payment coefficients more or less directly into the case of duration-dependent payments. It turns out, though, that this special case of duration dependence allows for considerable simplifications that we consider below.

Typically, the surrender value b_p^{03} and the free policy factor are themselves functions of the statewise payment processes B_p^i and transition payments b_p^{ij} for $i, j \in \{(0, p), (1, p), (2, p)\}$ in the following way. Assume a set of technical assumptions on the interest and transition rate in the disability model consisting of the states 0, 1, 2. This allows for calculation of a technical reserve in the state active, V^{0*}. The surrender value is typically specified in terms of the technical reserve, i.e. b_p^{03} is a function of V^{0*}. The free policy factor is typically formed as a ratio of first-order reserves

$$f(t) = \frac{V^{0*}(t)}{V^{0*+}(t)},$$

where V^{0*+} is the technical value of future benefits only. One can see that simply taking $b_p^{03}(t) = V^{0*}(t)$ and $f(t) = V^{0*}(t)/V^{0*+}(t)$ makes inclusion of behavioral options in the technical model redundant. It makes no difference to the technical value in the state $(0, 1, 2)$ if the behavior options are there or not. However, it does make a difference when evaluating under the market basis.

There are special issues concerning the payment coefficients in the absorbing states, $(2, p)$, $(2, f)$, $(3, p)$, $(3, f)$. Usually, there are no payments in the states $(2, p)$ and $(2, f)$. In that case, we can collapse these states into one state and still allow for different transition payments upon entering that state, $b_p^{02} \neq b_f^{02}$ and $b_p^{12} \neq b_f^{12}$. However, there do exist insurances where a payment rate is in force after entering the death state and in that case, we need the distinction between $(2, p)$ and $(2, f)$. In contrast, there are never payments in the state $(3, p)$ and $(3, f)$ since surrender literally means that the whole contract is cancelled. Therefore, we could freely collapse $(3, p)$ and $(3, f)$ into one state 3 and still allow for different transition payments upon entering that state, $b_p^{03} \neq b_f^{03}$. We just chose to distinguish $(3, p)$ and $(3, f)$ for reasons of symmetry. If we maintain two different death states for a premium paying contract and a free policy contract, respectively, but collapse into one state of surrender, we obtain the 7-state model that has been the starting point in numerous publications.

There are special issues about the transition intensities. One can imagine that the transitions between active, disabled, and dead do not depend on whether a premium is paid or not. This corresponds to setting $\mu_p^{ij} = \mu_f^{ij}$ for $i, j \in \{0, 1, 2\}$. We have just allowed these to be different for the sake of generality. In sharp contrast, one may indeed imagine that the transition rates into the surrender state heavily depend on whether a premium is paid or not. It is natural to believe that $\mu_f^{03} > \mu_p^{03}$ since policy holders that already stopped paying premiums are more likely to cancel the contract than policy holders that pay premiums. Actually, the idea that the surrender intensity for free policies is very high has been used as an argument for not considering the free policy state at all. Behavior options are appropriately taken into account via surrender and intermediary sojourn in the free policy state is just a negligible detail.

Let us now turn to the valuation after and before transition into the free policy state, respectively. We consider the situation after transition first. Then τ is known and constant and the future payments can be directly evaluated as

$$V_f^j(t, u) = \mathbb{E}\left[\int_t^n e^{-\int_t^s r}\, dB(s, u)\,\middle|\, Z(t-) = (j, f)\right]$$

$$= \mathbb{E}\left[\int_t^n e^{-\int_t^s r} f(t - u)\, dB^+(s)\,\middle|\, Z(t) = (j, f)\right]$$

$$= f(t - u)V^{j+}(t) = f(\tau)V^{j+}(t).$$

For the sake of completeness, and as an example of duration dependence, we next derive the same expression from the PDE in Sect. 1. In this case where the duration process counts time from entering into state $(0, f)$ but is not set to zero upon transitions out of that state, one can see that the PDE characterizing the reserve reads as

$$\frac{\partial}{\partial t} V_f^j(t, u) + \frac{\partial}{\partial u} V_f^j(t, u)$$

$$= rV_f^j(t, u) - b_f^j(t, u) - \sum_{k: k \neq j} \mu_f^{jk}(t)\left(b_f^{jk}(t, u) + V_f^k(t, u) - V_f^j(t, u)\right),$$

where $b^j(t, u)$ and $b^{jk}(t, u)$ are given by (8.1) and (8.2),

$$b_f^j(t, u) = f(t - u)b_p^{j+}(t), \qquad b_f^{jk}(t, u) = f(t - u)b_p^{jk+}(t).$$

We now guess a solution to the PDE in the form

$$V_f^j(t, u) = f(t - u)g^j(t).$$

The special thing about this guess is, of course, the multiplicative separation of the variables u and j. We plug this solution into the PDE to obtain

$$f'(t - u)g^j(t) + f(t - u)\frac{d}{dt}g^j(t) - f'(t - u)g^j(t)$$

$$= rf(t - u)g^j(t) - b_f^j(t, u)$$

$$- \sum_{k: k \neq j} \mu_f^{jk}(t)\left(b_f^{jk}(t, u) + f(t - u)g^k(t) - f(t - u)g^j(t)\right).$$

Rearranging, plugging in $b_f^j(t, u)$ and $b_f^{ij}(t, u)$, and dividing by $f(t - u)$ we get

$$\frac{d}{dt} g^j(t) = rg^j(t) - b_p^{j+}(t) - \sum_{k:k\neq j} \mu_f^{jk}(t)\left(b_p^{jk+}(t) + g^k(t) - g^j(t)\right).$$

But this differential equation characterizes $V^{j+}(t)$ so that we, again, arrive at

$$V_f^j(t, u) = f(t - u)V^{j+}(t).$$

Let us now consider valuation before transition. Then duration does not matter and we get the simple equation for valuation

$$\frac{d}{dt} V_p^j(t) = rV_p^j(t) - b_p^j(t) - \sum_{k:k\neq j} \mu_p^{jk}(t)\left(b_p^{jk}(t) + V_p^k(t) - V_p^j(t)\right)$$

$$- 1_{j=0}\, \mu_f^{00}(t)\left(V_f^0(t, 0) - V_p^0(t)\right).$$

Note that since the free policy states can be reached only via the transition from $(0, p)$ to $(0, f)$, the free policy risk premium $\mu_f^{00}(t)\left(V_f^0(t, 0) - V_p^0(t)\right)$ is paid only if $j = 0$. The quantity $V_f^0(t, 0)$ is the reserve upon transition to free policy and $V_f^0(t, 0) - V_p^0(t)$ is the sum at free policy risk. The time arguments of $V_f^0(t, 0)$ reflect time and duration in the set of free policy states such that

$$V_f^0(t, 0) = f(t - 0)V^{0+}(t) = f(t)V^{0+}(t).$$

Since $f(t)$ and $V^{0+}(t)$ can be calculated first and plugged in, the differential equation for $V_p^j(t)$ is solvable.

We now present a trick that allows for further simplification. Note that the free policy risk premium can be rewritten as

$$-\mu_f^{00}(t)f(t)\left(V^{0+}(t) - V_p^0(t)\right) - \mu_f^{00}(t)(1 - f(t))\left(- V_p^0(t)\right).$$

This decomposition has the following interpretation. One can think of a state reached by the intensity $\mu_f^{00}(t)f(t)$ and in which the policy holder stops paying the premiums but receives the full benefits valuated in $V^{0+}(t)$. This corresponds to a risk premium $\mu_f^{00}(t)f(t)\left(V^{0+}(t) - V_p^0(t)\right)$. Another state can be reached by the intensity $\mu_f^{00}(t)(1 - f(t))$ and in that state all payments stop and everything is lost for the policy holder (no surrender value). The risk premium for this transition is $\mu_f^{00}(t)(1 - f(t))\left(- V_p^0(t)\right)$. In this way, we have transferred the appearance of the free policy ratio f from the payment structure to the structure of risk and states. This may be computationally beneficial.

The idea turns out to be powerful enough to even handle the calculation of cash flows. As long as the policy holder is in one of the premium paying states $(0, p)$, $(1, p)$, $(2, p)$, $(3, p)$, cash flows in these states are easy to calculate from the transition probabilities. Also, given that the policy holder is in one of the free policy states, future cash flows are easy to calculate from transition probabilities. The difficulty arises when the policy holder is in one of the premium paying states and we want to calculate the future cash flows in the free policy states. Given that we are in state (j, p) and we are studying the cash flow in state (k, f), this involves the calculation of

$$\int_t^s p_p^{j0}(t, \tau) \mu_f^{00}(\tau) f(\tau) p_f^{0k}(\tau, s) \, d\tau \,. \tag{8.3}$$

The integral is not a usual transition probability due to the appearance of f. However, it turns out that the trick above can be used again so that a standard transition probability calculation in an auxiliary model can be used. The integral is equal to the transition probability $\tilde{p}^{jk}(t, s)$ where j is one of the premium payment states and k is one of the free policy states in an auxiliary model where the intensity from $(0, p)$ to $(0, f)$ equals $\mu_f^{00}(t) f(t)$ and where an additional artificial absorbing state $(0', f)$ can be reached by the intensity $\mu_f^{00}(t)(1 - f(t))$. If we now assume that the full unreduced benefits are paid in the free policy states, we actually get to the correct cash flow calculation, i.e.

$$\int_t^s p_p^{j0}(t, \tau) \mu_f^{00}(\tau) f(\tau) p_f^{0k}(\tau, s) \, d\tau \;=\; \tilde{p}^{jk}(t, s) \,.$$

This is most easily seen by considering the differential equations for the l.h.s. and the r.h.s. separately. Noting that the transition probability from the free policy state $(0, f)$ to the premium payment state $(0, p)$ is zero, we get from Kolmogorov's backward differential equation that the derivative of the l.h.s. equals

$$- \mathbb{1}_{j=0} \, \mu_f^{00}(t) f(t) p_f^{0k}(t, s)$$

$$+ \mathbb{1}_{j=0} \, \mu_f^{00}(t) \int_t^s p_p^{00}(t, \tau) \mu_f^{00}(\tau) f(\tau) p_f^{0k}(\tau, s) \, d\tau$$

$$- \sum_{\ell: \ell \neq j} \mu_p^{j\ell}(t) \int_t^s \left(p_p^{\ell0}(t, \tau) - p_p^{j0}(t, \tau) \right) \mu_f^{00}(\tau) f(\tau) p_f^{0k}(\tau, s) \, d\tau \,.$$

For calculating the differential equation for the r.h.s., we note in addition that no free policy states can be reached from the additional artificial absorbing state $(0', f)$, and get

$$
\frac{d}{dt} \widetilde{p}^{jk}(t, s) = - \sum_{\ell: \ell \neq j} \mu_{jl}(t) \left(\widetilde{p}^{\ell k}(t, s) - \widetilde{p}^{jk}(t, s) \right)
$$

$$
= - \sum_{\ell: \ell \neq j} \mu_{jl}^{p}(t) \left(\widetilde{p}^{\ell k}(t, s) - \widetilde{p}^{jk}(t, s) \right)
$$

$$
- \mathbb{1}_{j=0} \, \mu_{00}^{f}(t) f(t) \left(p_{f}^{0k}(t, s) - \widetilde{p}^{0k}(t, s) \right)
$$

$$
+ \mathbb{1}_{j=0} \, \mu_{00}^{f}(t)(1 - f(t)) \widetilde{p}^{0k}(t, s)
$$

$$
= - \sum_{\ell: \ell \neq j} \mu_{j\ell}^{p}(t) \left(\widetilde{p}^{\ell k}(t, s) - \widetilde{p}^{jk}(t, s) \right)
$$

$$
- \mathbb{1}_{j=0} \, \mu_{00}^{f}(t) f(t) p_{f}^{0k}(t, s) + \mathbb{1}_{j=0} \, \mu_{00}^{f}(t) \widetilde{p}^{0k}(t, s).
$$

Since the two differential equations are the same and the boundary condition at $t = 0$ is zero, the two quantities are the same. Calculation of the cash flows via $\widetilde{p}^{jk}(t, s)$ essentially means that we have moved the free policy scaling of payments to a scaling of intensities of transition into the free policy states. This is only possible since the free policy states can be reached by a transition from $(0, p)$ to $(0, f)$ only, and this transition affects payments uniformly in the free policy states. As mentioned, it may be a computationally convenient translation for calculating cash flows since (8.3) can be calculated as a usual transition probability.

Taking policy holder behavior into account for calculating cash flows is particularly relevant. If the market basis and the first-order bases are not too far apart, taking policy holder options into account has only marginal numerical influence on the reserves. However, the cash flows may still be very different. Since these may be used for internal risk management decisions, policy holder behavior plays an important role there.

Exercises

8.1 Write down an integral version of the statewise reserve $V_p^j(t)$, based on the probabilities $\widetilde{p}_{jk}(t, s)$

8.2 Consider a stochastic retirement model, i.e. a disability model without recovery, where the disability state is replaced by a retired state. In state a, i.e. during activity at the labor market, the policy holder pays premiums at rate π. In state i, i.e. during retirement, the policy holder receives a life annuity at rate $f(t - u)b$, where u measures duration of retirement, i.e. the ratio f is settled upon retirement, similar to the free policy ratio. Draw a sketch of the model and write down the payment process as a function of time and duration. Characterize the reserve in state a via an ordinary differential equation. Explain how you can characterize the reserve via a model with two different retirement states, one in which the policy holder receives

b and one in which the policy holder receives 0. What are the intensities in the artificial model?

9 Rational Policy Holder Behavior

In Sect. 8, we studied the case where policy holder behavior is assumed to be arbitrary. In this section, we study the case where the policy holder is assumed to be rational. By rational, we mean that policy holders behave in a particular way in order to maximize their financial position. This raises some interesting questions. On the basis of what kind of information does the policy holder make his decisions? What is his financial position in relation to the specific contract that we are considering?

The standard approach to these questions is a marginal approach in the following sense. One can think of a contract without a specific behavioral option. When evaluating the contract including the behavioral option, we assume that the decision to exercise the option is based on the same information that we used to evaluate the contract without the option. Said in a different way, the fact that we add decisions to the contract does not in itself mean that we add information on which the decision is to be based. The only information one can say is added, is whether the option is exercised or not. Also, the approach to the financial position is marginal in the sense that we simply seek to maximize the value of the contract that we are considering. How conditions in the contract in focus interact with conditions in other contracts that the policy holder may have is kept out of the equation. Of course, one can generalize to a more holistic view on information and financial positions, but the fundamental ideas and structures are well described in the marginal approach.

Later in the section, we return to insurance contracts with behavioral options. However, the fundamental idea and structure is originally studied for purely financial contracts. Since certain aspects are simpler there, this is where we choose to start. The context is so-called American option pricing and its relation to the theory of optimal stopping. We reconsider the Black–Scholes market

$$dS^0(t) = r S^0(t) \, dt,$$

$$dS(t) = \alpha S(t) \, dt + \sigma S(t) \, dW(t) = r S(t) \, dt + \sigma S(t) \, dW^Q(t),$$

with the payment stream

$$dB(t) = b(t, S(t)) \, dt + \Delta B(t, S(t)),$$

with $\mathcal{T} = \{0, n\}$. Thus, we allow in the first place for lump sum payments at time 0 and n, respectively. In VI.(2.2) we considered the valuation formula

$$V(t, S(t)) = \mathbb{E}^Q \left[\int_t^n e^{-\int_t^s r} \, dB(s) \,\middle|\, S(t) \right].$$

Let us now say that the policy holder is allowed to cancel the contract prematurely at time τ and, in that case, he receives at time τ the payoff $\Phi(\tau) = \Phi(\tau, S(\tau))$ [with a slight misuse of notation since we use the same letter Φ for the stochastic process determining the payoff and for the function of $(\tau, S(\tau))$ from which its value can be calculated]. The payment of $\Phi(\tau, S(\tau))$ replaces then, upon exercise of the option, all future payments otherwise agreed upon in B. If we now take the marginal approach explained in the first paragraph, it is now possible to argue, by arbitrage arguments, that the unique contract value at time t can be written as

$$V(t, S(t)) = \sup_{\tau \in [t,n]} \mathbb{E}^{\mathbb{Q}} \left[\int_t^\tau e^{-\int_t^s r}\, dB(s) + e^{-\int_t^\tau r} \Phi(\tau) \,\middle|\, S(t) \right], \tag{9.1}$$

where the supremum is taken over all stopping times. The marginal approach to information corresponds to the idea that the stopping time is actually, since we did not state anything else, a stopping time w.r.t. the information generated by S. The marginal approach to financial positioning corresponds to the idea that whether the option has been exercised or not does not influence any other payment streams relevant for pricing of the specific payment stream.

The decision not to exercise prematurely is included in the supremum in (9.1) by specifying

$$\Phi(n) = 0, \tag{9.2}$$

and letting the decision be given by $\tau = n$. Then, in case of no premature exercise, the policy holder receives the contracted $\Delta B(n, S(n))$ as a lump sum at time n.

One cannot see immediately from (9.1) how the supremum over stopping times affects the price representation in terms of a solution to a differential equation. We recall the structure of the representation in the case of no exercise option from Proposition VI.2.2. For a fixed τ we are back to that representation with τ replacing n as maturity date. But what happens to the representation when we take the supremum over τ? Actually, the conditional expectation with a supremum can be represented by a so-called variational inequality.

Proposition 9.1 *Assume that there exists a function V solving*

$$V_t(t, s) \leq rV(t, s) - b(t, s) - V_s(t, s)rs - \frac{1}{2}V_{ss}(t, s)\sigma^2 s^2, \tag{9.3a}$$

$$V(t, s) \geq \Phi(t, s), \tag{9.3b}$$

$$0 = \left[V_t(t, s) - rV(t, s) + b(t, s) + V_s(t, s)rs + \frac{1}{2}V_{ss}(t, s)\sigma^2 s^2 \right]$$
$$\times \left(V(t, s) - \Phi(t, s) \right), \tag{9.3c}$$

$$V(n-, s) = \Delta B(n, s) . \tag{9.3d}$$

Then this function is indeed equal to the reserve defined in (9.1), and the optimal stopping time at time t is given by

$$\tau_t^* = \inf\{s \in [t, n] : V(s, S(s)) = \Phi(s, S(s))\}. \tag{9.4}$$

The proof of Proposition 9.1 generalizes the proof of Proposition VI.2.2. Before we prove Proposition 9.1, we make a few remarks. This variational inequality (9.3) should be compared with Proposition VI.2.2. First, VI.(2.4) is replaced by (9.3a)–(9.3c). The equation VI.(2.4) turns into an inequality in (9.3a). An additional inequality (9.3b) states that the contract value always exceeds the exercise payoff. This is reasonable, since one of the possible exercise strategies is to exercise immediately and this would give an immediate exercise payoff. The equality (9.3c) is the mathematical version of the following statement: at any point in the state space (t, s) at least one of the inequalities in (9.3a) and (9.3b) must be an equality. Second, VI.(2.5) is replaced by (9.3d) since we disregard lump sum payment outside time τ and time n. If there are no premature payments b, this goes out of the equation and we have a classical variational inequality characterizing the value of an American option.

By the variational inequality (9.3), one can divide the state space into two regions, possibly intersecting. In the first region, (9.3a) is an equality. This region consists of the states (t, s) where the optimal stopping strategy for the contract holder is not to stop. In this region, the contract value follows the differential equation VI.(2.4) as if there were no exercise option. In the second region, (9.3b) is an equality. This region consists of the states (t, s) where the optimal stopping strategy for the contract holder is to stop. Thus, in this region the value of the contract equals the exercise payoff.

Proof Given a function $V(t, s)$ we can, by replacing n by τ in VI.(2.6), write

$$e^{-rt} V(t, S(t)) = -I_1(t, \tau) - I_2(t, \tau) + e^{-r\tau} V(\tau, S(\tau)),$$

where

$$I_1(t, \tau) = \int_t^\tau e^{-rs} \Big[-rV(s, S(s)) + V_s(s, S(s)) $$
$$+ V_s(s, S(s)) rS(s) + \frac{1}{2} V_{ss}(s, S(s)) \sigma^2 S^2(s) \Big] ds,$$

$$I_2(t, \tau) = \int_t^\tau e^{-rs} V_s(s, S(s)) \sigma S(s) \, dW^Q(s).$$

Now, consider an arbitrary stopping time τ. Using (9.3a) with s replaced by $S(t)$ to bound the integrand in I_1 and a similar bound for $e^{-r\tau} V(\tau, S(\tau))$ based on (9.3b) gives

$$e^{-rt} V(t, S(t)) \geq \int_t^\tau e^{-rs} dB(s) - I_2(t, \tau) + e^{-r\tau} \Phi(\tau, S(\tau)).$$

Taking conditional $\mathbb{Q}$-expectation on both sides, multiplying by e^{rt} and, finally, taking the supremum on both sides over τ, we get

$$V(t, S(t)) \geq \sup_{\tau \in [t,n]} \mathbb{E}^{\mathbb{Q}}\left[\int_t^\tau e^{-\int_t^s r}\, dB(s) + e^{-\int_t^\tau r}\Phi(\tau) \,\Big|\, S(t) \right]. \qquad (9.5)$$

Now, consider instead the stopping time τ_t^* defined in (9.4). This stopping time is indeed well defined since, from (9.2) and (9.3d), $V(n, S(n)) = \Phi(n, S(n)) = 0$. We now know from (9.3c) that

$$V_u(u, s) = rV(u, s) - b(u, s) - V_s(u, s)rs - \frac{1}{2}V_{ss}(u, s)\sigma^2 s^2, \quad u \in [t, \tau^*],$$

so that

$$e^{-rt}V(t, S(t))$$

$$= \int_t^{\tau^*} e^{-rs}\left(dB(s) - V_s(s, S(s))\sigma(s, S(s))S(s)\, dW^{\mathbb{Q}}(s) \right)$$

$$+ e^{-r\tau^*}\Phi(\tau^*, S(\tau^*)).$$

Taking, first, conditional expectation on both sides and then comparing with an arbitrary strategy gives that

$$V(t, S(t)) \leq \sup_{\tau \in [t,n]} \mathbb{E}^{\mathbb{Q}}\left[\int_t^\tau e^{-\int_t^s r}\, dB(s) + e^{-\int_t^\tau r}\Phi(\tau) \,\Big|\, S(t) \right]. \qquad (9.6)$$

Involving (9.5) shows that in fact equality holds in (9.6). Also, the supremum is obtained by the stopping strategy (9.4). $\square$

Note that, in order to know when to exercise, one must be able to calculate the value. Only rarely, the variational inequality (9.3) has an explicit solution. However, there are several numerical procedures developed for this purpose. One may, for example, use targeted Monte Carlo techniques, general PDE approximations, or certain specific approximations developed for specific functions Φ. Some of these methods approximate, directly or indirectly, the division of the state space into regions where it is optimal to stop and regions where it is not.

Based on the ideas and insight obtained above, we now return to the surrender problem of the insurance policy holder. The surrender option is similar to the American option in financial contracts since it also cancels all future payments in the contract against a lump sum paid immediately. In Chaps. V and VI, we considered various contract designs and we must decide on a specific contract design to which we add the surrender option. We consider the unit-link insurance contract studied in Sect. VI.3, with the set-up of payment streams and state processes as there. Assume now that the contract holder can terminate his policy at any point in time. Given that he does so at time t when the policy is in state $Z(t)$ and the value of the state process X is $X(t)$, he receives the *surrender value*

$$\Phi(t) = \Phi^{Z(t)}(t, X(t))$$

for a sufficiently regular function $\Phi^j(t, x)$. Recall, that X can be thought of as the retrospective savings account of the policy, see also Remark VI.3.1. One possibility could be, simply, to have $\Phi^j(t, x) = x$.

We are now interested in calculating the value of future payments specified in the policy. We consider the reserve defined by

$$V^{Z(t)}(t, X(t)) = \sup_{\tau \in [t,n]} \mathbb{E}^{\mathbb{P} \otimes \mathbb{Q}} \left[\int_t^\tau e^{-\int_t^s r} \, dB(s) + e^{-\int_t^\tau r} \Phi(\tau) \,\Big|\, Z(t), X(t) \right].$$
$$(9.7)$$

We know from Proposition VI.3.3 how to characterize this value in the case without a supremum. The question is how the supremum over stopping times generalizes that representation. The ideas and results for the American option above indicate how the differential equation in Proposition VI.3.3 is replaced by a variational inequality. We define the differential operator $\mathcal{D}$ and the sum at risk R like in VI.(3.6).

Proposition 9.2 *Assume that there exists a function $V^j(t, x)$ satisfying $V^j(n, x) = 0$ and solving*

$$V_t^j(t, x) \le r V^j(t, x) - b^j(t, x) - \sum_{k: k \ne j} \mu_{jk}(t) R^{jk}(t, x) - \mathcal{D}_x V^j(t, x),$$

$$V^j(t, x) \ge \Phi^j(t, x),$$

$$0 = \left[V_t^j(t, x) - r V^j(t, x) + b^j(t, x) + \sum_{k: k \ne j} \mu_{jk}(t) R^{jk}(t, x) + \mathcal{D}_x V^j(t, x) \right]$$
$$\times \left[V^j(t, x) - \Phi^j(t, x) \right]$$

for $t \notin \mathcal{T}$, whereas

$$V^j(t-, x) \ge \Delta B^j(t, x) + V^j\big(t, x - \Delta A^j(t, x)\big),$$

$$V^j(t-, x) \ge \Delta B^j(t, x) + \Phi(t, x),$$

$$0 = \left[V^j(t-, x) - \Delta B^j(t, x) - V^j\big(t, x - \Delta A^j(t, x)\big) \right]$$
$$\times \left[V^j(t-, x) - \Delta B^j(t, x) - \Phi(t, x) \right]$$

for $t \in \mathcal{T}$. *Then this function is indeed equal to the reserve defined in* (9.7), *and the optimal stopping time at time t is given by*

$$\tau_t^* = \inf\{s \in [t, n] : V^{Z(s)}(s, X(s)) = \Phi^{Z(s)}(s, X(s))\}.$$

This variational inequality can be compared with Proposition VI.3.3 in the same way as (9.3) was compared with Proposition VI.2.2. Its verification goes in the same way as the verification of (9.3), although it becomes somewhat more involved. We do not go through this here. As in the previous section, one can now divide the state space into two regions, possibly intersecting. In the first region, the reserve follows a differential equation as if surrender were not possible. This region consists of states from where immediate surrender is suboptimal and the differential system is identical with VI.(3.7). In the second region, the reserve equals the surrender value. This region consists of the states where immediate surrender is optimal.

We have handled the unit-link contract. Extensions related to with-profit contracts are straightforward. In the case of cash dividends formalized in Sects. VI.5–6, the surrender value is in practice typically given by the first-order reserve defined in VI.(4.1), in the sense that

$$\Phi^j(t, x) = V^{j*}(t).$$

In the case of bonus payments formalized in Sects. VI.7–8, the surrender value is in practice typically given by the technical value of future payments, including the ones that have been added through dividend payouts until time t, i.e.

$$\Phi^j(t, x) = x.$$

Thus, in both cases, one typically has to give up the participation in the surplus Y when surrendering. For that particular reason, in certain cases it is never optimal for the policy holder to surrender. In that case, valuation of the surrender option is trivially equal to zero.

So far in this section we have only worked with the surrender option. In Sect. 8, a different option was the main challenge, namely the free policy option. Also, we stressed that these options are typically available in only certain states of the underlying risk process. Thus, the whole process of making decisions and exercising options is clearly more complicated than indicated in our discussion about rational behavior so far. But the structure of the variational inequalities inspires a generalization to these more general decision processes. We present here the solution in the specific model illustrated in Fig. 2. As in Sect. 8, we disregard participation in the financial performance by considering a policy where nominal payments are simply linked to the risk process Z exemplified by the disability model. Thus, compared to Proposition 9.2, we disregard the X-dependence of payments and reserves, and specify the Z process to correspond to the disability model, but, on the other hand, generalize the behavior process to be of state type corresponding to the one illustrated in Fig. 1. Most importantly, however, there do not exist intensities between these behavioral states. Transitions are decided at the discretion of the policy holder. However, the surrender state can only be reached as long as the policy holder is still active, no matter whether he is premium paying or in the free policy states. Also, the free policy states can only be reached as long as the policy holder is active. Comparing with Fig. 2, the transitions $(0, p) \rightarrow (3, p)$, $(0, p) \rightarrow (0, f)$, and $(0, f) \rightarrow (3, f)$, are behavioral transitions whereas the rest are usual risk transitions with transition intensities.

We can now see that the value of the contract based on rational behavior, as long as the policy holder is in the free policy states and has been so in a period for a time period of length u, is

$$V_f^j(t, u) = \sup_{\tau_{fs} \in [t,n]} \mathbb{E}\left[\int_t^{\tau_{fs}} e^{-\int_t^s r} \, dB(s, u) + e^{-\int_t^{\tau_{fs}} r} \Phi^f(\tau_{fs}, u) \,\Big|\, Z(t) = (j, f) \right],$$

where τ_{fs} is the stopping time indicating transition from free policy to surrender, the argument u is the duration since entrance into the free policy state, and $\Phi^f(\tau_{fs}, u)$ is the surrender value from the free policy state at time τ_{fs} if the policy has then been in the free policy state for a duration of u. If all payments in the free policy states are scaled, upon transition to free policy with the free policy factor f, we can, as in Sect. 8, get to

$$V_f^j(t, \tau) = f(\tau) V^{j+}(t),$$

where now

$$V^{j+}(t) = \sup_{\tau_{fs} \in [t,n]} \mathbb{E}\left[\int_t^{\tau_{fs}} e^{-\int_t^s r} \, dB^+(s) + e^{-\int_t^{\tau_{fs}} r} \Phi^f(\tau_{fs}) \,\Big|\, Z(t) = j \right],$$

where $\Phi(\tau_{fs})$ is the surrender value from the premium payment state, but evaluated at the time of surrender from the free policy state, τ_{fs}. Since the surrender option in the free policy states can be exercised only if the policy holder is active, the policy is not stopped as long as the policy holder is disabled or dead. Thus, in those states $V^{j+}(t)$ just follows the standard Thiele differential equation. Only in the state $(0, f)$ do we need to characterize the value via a variational inequality. We have that

$$b_f^{03}(t, \tau) = f(\tau) b_p^{03}(t) = f(\tau) \Phi(t).$$

Then V^{0+} is characterized at points of differentiability by

$$\frac{d}{dt} V^{0+}(t) \leq r V^{0+}(t) - b^{0+}(t) - \sum_{k:k\neq 0} \mu_{0k}(t) R^{0k+}(t),$$

$$V^{0+}(t) \geq \Phi(t),$$

$$0 = \left[\frac{d}{dt} V^0(t) - r V^{0+}(t) + b^{0+}(t) + \sum_{k: k \neq 0} \mu_{0k}^P(t) R^{0k+}(t) \right]$$

$$\times \left[V^{0+}(t) - \Phi(t) \right],$$

$$V^{0+}(n) = 0.$$

At deterministic points with lump sum payments and thus no differentiability, the system is extended corresponding to the second part of Proposition 9.2.

Now, we consider valuation before free policy transcription. Before the free policy states have been reached, we again distinguish between states in which no behavioral options are present, i.e. disabled $(1, p)$ and dead $(2, p)$, and states with behavioral options. In $(1, p)$ and $(2, p)$, neither the premium payments nor the policy as such can be stopped. Therefore, in those states, a usual Thiele differential equation characterizes the solution. The difficult part is the value in state $(0, p)$ which, of course, appears in the differential equations for the states $(1, p)$ and $(2, p)$. In that state, we have two competing stopping times, namely one for surrendering τ_{ps} and one for transcription to free policy τ_{pf}. One can see that the value based on rational behavior is given by

$$V_p^0(t) = \sup_{\tau_{ps}, \tau_{pf} \in [t,n]} \mathbb{E}\left[T_1 + T_2 + T_3 \mid Z(t) = (0, p) \right],$$

where

$$T_1 = \int_t^{\min(\tau_{ps}, \tau_{pf})} e^{-\int_t^s r} \, dB(s),$$

$$T_2 = e^{-\int_t^{\tau_{ps}} r} \Phi(\tau_{ps}) \mathbb{1}_{\tau_{ps} < \tau_{pf}},$$

$$T_3 = e^{-\int_t^{\tau_{pf}} r} V_f^0(\tau_{pf}, 0) \mathbb{1}_{\tau_{ps} > \tau_{pf}}.$$

The question is how the variational inequality in the case of one stopping time generalizes to the case with two stopping times. One can characterize the value $V_p^0(t)$ at time points of differentiability by

$$\frac{d}{dt} V_p^0(t) \leq r V_p^0(t) - b^0(t) - \sum_{k: k \notin \{0,3\}} \mu_p^{0k}(t) R^{0k}(t), \tag{9.9a}$$

$$V_p^0(t) \geq \Phi(t), \tag{9.9b}$$

$$V_p^0(t) \geq V_f^0(t, 0), \tag{9.9c}$$

$$0 = \left[\frac{d}{dt} V_p^0(t) - r V_p^0(t) + b^0(t) + \sum_{k:\, k \notin \{0,3\}} \mu_p^{0k}(t) R^{0k}(t)\right]$$

$$\times \left[V_p^0(t) - \Phi(t)\right] \times \left[V_p^0(t) - V_f^0(t, 0)\right], \tag{9.9d}$$

$$V_p^0(n) = 0.$$

At deterministic points with lump sum payments and, thus, no differentiability, the system is extended corresponding to the second part of Proposition 9.2.

The variational inequality (9.9) contains an extra layer compared to the variational inequalities above. The inequality in (9.9a) is the inequality version of the usual differential equation. The inequality (9.9b) reflects that the value at any point in time is larger than the surrender value. This is clear because one of the decisions you can make at that point in time is to surrender. The inequality (9.9c) reflects that the value at any point in time is larger than the value given that the policy is transcribed into a free policy. This is clear because another decision you can make at that time is to transcribe into free policy. The condition (9.9d) is the mathematical formulation of the constraint that at any point in time, at least one of these three inequalities must be an equality. The state space, which is now just in the time dimension but which could be generalized to more dimensions, can now be split into three instead of two regions, possibly intersecting. In the continuation region, (9.9a) is an equality, the value develops as a usual differential equation, and it is optimal neither to surrender nor to transcribe into free policy. In the surrender region, (9.9b) is an equality, the value is equal to the surrender value, and it is optimal to surrender. In the free policy region, (9.9c) is an equality, the value is equal to the value given transcription to free policy, and it is optimal to transcribe to free policy. When the regions intersect, this simply reflects that more decisions are optimal and the value is independent of which one of these decisions are made. At some points, two regions may intersect whereas at other points, all three regions may intersect.

We finish the two sections on policy holder behavior by returning to the idea of reserve-dependent intensities and the idea that this formalizes a delicate interpolation between the two extremes. Reserve-dependent intensities were introduced as a way to reflect that the financial gain upon a transition influences the policy holder's tendency to make this transition. In case of surrender with the surrender value G, we proposed in the survival model to work with a surrender intensity as a function of $G - V$, where V is the reserve of interest. It is clear that incidental behavior is a special case, since this is also based on intensities that just happen not to depend on the value of the contract itself. The connection with optimal stopping is less clear. However, one can show that the value of the optimal stopping problem or, equivalently, the solution to a variational inequality, can be approximated by solutions to differential equations, where the intensity of surrendering becomes large whenever $G > V$ and becomes small whenever $G < V$. An example of such an intensity is

$$\mu(t, V(t)) = \exp\{\beta(G(t) - V(t))\},$$

which behaves exactly like this, when we let $\beta \to \infty$. In this sense, the reserve-dependent intensities can reach both the incidental behavior and, in the limit, rational behavior. As such, this is one of the numerical approximation procedures available for approximating the solution to an optimal stopping problem, whether that appears in an American option in a purely financial setting or in a surrender problem in life insurance. But more importantly from a practical point of view, for a given β, the reserve-dependent intensity can be considered as a well-balanced compromise between the two extremes, incidental and rational behavior, that may be useful for a practical numerical calculation.

Let us implement this idea in the two main examples in this section. The American option price characterized in (9.3) can be approximated by the solution to

$$V_t(t, s) = rV(t, s) - b(t, s) - V_s(t, s)rs - \frac{1}{2}V_{ss}(t, s)\sigma^2 s^2$$
$$- \mu(t, s, V(t, s))(\Phi(t, s) - V(t, s)),$$

$V(n-, s) = \Delta B(n, s)$, where

$$\mu(t, s, v) \to \begin{cases} \infty & \text{for } \Phi(t, s) > v, \\ 0 & \text{for } \Phi(t, s) < v. \end{cases}$$

For a parametrized function μ like the exponential one suggested above, we have an approximation procedure for calculation of the American option price. But more importantly from a practical point of view, the idea can be used as a well-balanced, realistic compromise between incidental and rational behavior.

For the policy holder options, we go directly to the case with both free policy and surrender option in order to illustrate the most general case. We can approximate the solution to (9.9) by

$$\frac{d}{dt}V_p^0(t) = rV_p^0(t) - b^0(t) - \sum_{k:\, k \notin \{0,3\}} \mu_p^{0k}(t)R^{0k}(t)$$
$$- \mu_p^{03}(t, V_p^0(t))(\Phi(t) - V_p^0(t)) - \mu_f^{00}(t, V_p^0(t))(V_f^0(t, t) - V_p^0(t)),$$

where

$$\mu_p^{03}(t, v) \to \begin{cases} \infty & \text{for } \Phi(t) > v, \\ 0 & \text{for } \Phi(t) < v, \end{cases} \qquad \mu_f^{00}(t, v) \to \begin{cases} \infty & \text{for } V_f^0(t, 0) > v, \\ 0 & \text{for } V_f^0(t, 0) < v. \end{cases}$$

For parametrized functions for μ like the exponential one suggested above, we have an approximation procedure for calculation of the policy value based on rational behavior. But more importantly from a practical point of view, the idea can be used as a well-balanced, realistic compromise between incidental and rational behavior.

Exercises
9.1 In Proposition 9.2, argue that for $t \in \mathcal{T}$, we can alternatively write the gluing condition as

$$V^j(t-, x) = \Delta B^j(t, x) + \max\left(V^j(t, x - \Delta A^j(t, x)), \Phi(t, x)\right). \qquad (9.10)$$

This is true since the policy holder has the option to surrender specifically at time $t \in \mathcal{T}$. This is also the right gluing condition even if the surrender option can be exercised at time points $t \in \mathcal{T}$, only. In that case, the variational inequality for $t \notin \mathcal{T}$ collapses into the usual differential equation but with the gluing condition (9.10). In the Black–Scholes model, characterize the value of a so-called Bermudan option. The Bermudan option (named oddly after Bermuda's geographical position between Europe and the US, but closest to the US) allows the holder to exercise prematurely but only at predetermined deterministic dates, say, when $t \in \mathcal{T}$.

9.2 Consider a stochastic retirement model where, instead of the situation described in Exercise 8.2, the retirement state is reached upon request by the policy holder who seeks to maximize the present value of consumption. Thus, there exists no retirement intensity. In state a, i.e. during activity at the labor market, the policy holder pays at rate π and consumes therefore at rate $a - \pi$, where a is the salary rate. In state i, i.e. during retirement, the policy holder receives and consumes a life annuity at rate $f(\tau)b$, where τ is the stopping time where the policy holder retires. A proportional appreciation, e.g. if the individual likes leisure time, of consumption as retired compared to consumption as active is here just an integral part of f. Define the present value of future consumption in terms of an optimal stopping problem where the policy holder retires to maximize the present value of future consumption. Characterize the present value of future consumption in terms of a variational inequality.

9.3 Disregard X dependence in the payment stream of an insurance contract based on the model illustrated in Fig. 2. Characterize the value of the payment stream in the mixed case where the free policy option is modeled via incidental behavior and the surrender option is modeled via rational behavior.

10   Higher Order Moments. Hattendorf's Theorem

So far, we have, in the context of life insurance, concentrated exclusively on the first moment when characterizing the present value of future payments. In classical insurance the LLN makes the plain expectation the most important characteristic. In finance, the theory of risk-neutral pricing makes the plain expectation the most important characteristic. However, in particular outside the theory of risk-neutral pricing, other aspects of the distribution of the present value of future payments may also be of relevance. Certain risks cannot be diversified away by building large portfolios and even large portfolios are not infinitesimally large. In such cases, one might be interested in, for instance, higher order moments of present values of payment streams.

If we consider the classical payment stream in formula V.(1.1), we might be interested in calculating the conditional qth moment,

$$V^{j(q)}(t) = \mathbb{E}\left[\left(\int_t^n e^{-\int_t^s r} dB(s)\right)^q \bigg| Z(t) = j\right]. \tag{10.1}$$

We state here, without proof, the generalization of the Thiele differential equation characterizing the qth moment.

Proposition 10.1 *Assume that there exists a function $V^{j(q)}$ such that*

$$\frac{d}{dt}V^{j(q)}(t) = (qr + \mu_{j\cdot}(t))V^{j(q)}(t) - qb^j(t)V^{j((q-1))}(t) - \sum_{k:k\neq j}\mu_{jk}(t)R^{jk(q)}(t)$$

for $t \notin \mathcal{T}$, and

$$V^{j(q)}(t-) = \sum_{p=0}^q \binom{q}{p}(\Delta B^j(t))^p V^{j(q-p)}(t)$$

for $t \in \mathcal{T}$, where

$$R^{jk(q)}(t) = \sum_{p=0}^q \binom{q}{p}(b^{jk}(t))^p V^{k(q-p)}(t).$$

Then $V^{j(q)}$ is indeed the statewise qth moment defined in (10.1).

First, note that the differential equation stated in Proposition 10.1 generalizes the Thiele differential equation. Namely, for $q = 1$, we realize that for $t \notin \mathcal{T}$,

$$R^{jk(1)}(t) = \sum_{p=0}^1 \binom{1}{p}(b^{jk}(t))^p V^{k(1-p)}(t) = V^k(t) + b^{jk}(t) = R^{jk}(t),$$

and for $t \in \mathcal{T}$

$$V^{j(1)}(t-) = \sum_{p=0}^1 \binom{1}{p}(\Delta B^j(t))^p V^{j(1-p)}(t) = V^j(t) + \Delta B^j(t).$$

Since the 0th moment is always 1, the classical Thiele differential equation falls out.

Second, note that the system of differential equations is actually a system of $(J+1) \times q$ equations, namely one equation for each state and moment up to the qth moment. However, since only moments lower than, say, the pth moment are needed to find the pth moment, the $(J + 1) \times q$-dimensional system can be reduced to q $(J + 1)$-dimensional systems, which is much easier to solve. We start with the 0th, which by definition is 1. Thereafter, we solve the 1st moment, which is just the classical Thiele equation of dimension $J + 1$. Then we can draw on the 0th and the 1st order moments in the $(J + 1)$-dimensional equation for the second-order moment, and so on. If the underlying Markov chain is of feed-forward type, then for each moment, we do not even have to solve one $(J + 1)$-dimensional system but simply solve $J + 1$ one-dimensional equations. Thus, in that case, to find the qth moment, the q $(J + 1)$-dimensional systems are reduced even further to $q(J + 1)$ one-dimensional equations.

Having established the differential equation for higher order moments, we can also derive differential equations for higher order central moments. Of particular interest is of course the conditional variance

$$\mathbb{V}\mathrm{ar}^j(t) = \mathbb{E}\left[\left(\int_t^n e^{-\int_t^s r}\, dB(s) - V^j(t)\right)^2 \Big| Z(t) = j\right]. \tag{10.2}$$

Proposition 10.2 *Assume that $\Sigma^j(t)$ is a function such that*

$$\frac{d}{dt}\Sigma^j(t) = 2r\Sigma^j(t) - \sum_{k:k\neq j} \mu^{jk}(t)\big(R^{jk}(t)^2 + \Sigma^k(t) - \Sigma^j(t)\big)$$

and $\Sigma(n) = 0$. Then indeed $\Sigma^j(t) = \mathbb{V}\mathrm{ar}^j(t)$, the statewise reserve variance defined in (10.1).

Proof We differentiate plainly the conditional variance

$$\mathbb{V}\mathrm{ar}^j(t) = V^{j(2)}(t) - V^j(t)^2$$

and get, by reference to the differential equations for the 1st and 2nd moments,

$$\frac{d}{dt}\mathbb{V}\mathrm{ar}^j(t) = \frac{d}{dt}V^{j(2)}(t) - \frac{d}{dt}V^j(t)^2$$

$$= \big(2r + \mu_{j\bullet}(t)\big)V^{j(2)}(t) - 2b^j(t)V^j(t)$$

$$- \sum_{k:k\neq j}\mu_{jk}(t)\sum_{p=0}^{2}\big(V^{k(2)}(t) + 2b^{jk}(t)V^k(t) + b^{jk}(t)^2\big) - 2V^j(t)\frac{d}{dt}V^j(t)$$

$$= 2r V^{j(2)}(t) - 2b^j(t)V^j(t)$$

$$- \sum_{k:k\neq j}\mu_{jk}(t)\big(V^{k(2)}(t) + 2b^{jk}(t)V^k(t) + b^{jk}(t)^2 - V^{j(2)}(t)\big)$$

$$- 2V^j(t)\Big(rV^j(t) - b^j(t) - \sum_{k:k\neq j}\mu_{jk}(t)\big(b^{jk}(t) + V^k(t) - V^j(t)\big)\Big)$$

$$= 2r \operatorname{Var}^j(t) - \sum_{k: k \neq j} \mu_{jk}(t) \Big(V^{k(2)}(t) + 2b^{jk}(t) V^k(t) + b^{jk}(t)^2$$

$$- V^{j(2)}(t) - 2V^j(t) \big(b^{jk}(t) + V^k(t) - V^j(t) \big) \Big),$$

where $(\cdot)$ in the last sum equals

$$b^{jk}(t)^2 + V^k(t)^2 + \big(V^j(t) \big)^2 + 2b^{jk}(t) V^k(t) - 2V^j(t) b^{jk}(t) - 2V^j(t) V^k(t)$$

$$+ V^{k(2)}(t) - \big(V^k(t) \big)^2 - \big[V^{j(2)}(t) - \big(V^j(t) \big)^2 \big]$$

$$= R^{jk}(t)^2 + \operatorname{Var}^k(t) - \operatorname{Var}^j(t).$$

For $t \in T$, we have that

$$\operatorname{Var}^j(t-) = V^{j(2)}(t-) - \big(V^j(t-) \big)^2$$

$$= \sum_{p=0}^{2} \binom{2}{p} \big(\Delta B^j(t) \big)^p V^{j(2-p)}(t) - \big(\Delta B^j(t) + V^j(t) \big)^2$$

$$= V^{j(2)}(t) + 2\Delta B^j(t) V^j(t) + \big(\Delta B^j(t) \big)^2 - \big(\Delta B^j(t) + V^j(t) \big)^2$$

$$= \operatorname{Var}^j(t),$$

so that the gluing condition just stresses continuity at these specific time points. The intuition is that, of course, across *deterministic* time points with statewise lump sum payments, there is no accumulation of risk, so therefore the variance is continuous across these time points.   □

Considering the differential equation for $\operatorname{Var}^j(t)$, we immediately recognize this as a regular Thiele equation for the case where the interest rate is $2r$ and corresponding to the special payment stream

$$d\widetilde{B}(t) = \sum_\ell R^{Z(s-)\ell}(t)^2 \, dN^\ell(t).$$

Thus, we can represent the conditional variance as

$$\operatorname{Var}^j(t) = \mathbb{E}\left[\int_t^n e^{-2r(s-t)} \sum_\ell R^{Z(s-)\ell}(s)^2 \, dN^\ell(s) \,\Big|\, Z(t) = j \right] \quad (10.3)$$

$$= \int_t^n e^{-2r(s-t)} \sum_{k \neq \ell} p^{jk}(t, s) \mu_{k\ell}(s) R^{k\ell}(s)^2 \, ds.$$

Results like the representation (10.3) are spoken of as the Hattendorff Theorem after the Danish actuary Hattendorff, who discovered it in the case of the survival model.

It is obvious that we can now take this idea further and derive a differential equation for even higher order conditional central moments,

$$\mathbb{E}\left[\left(\int_t^n e^{-\int_t^s r}\, dB(s) - V^j(t)\right)^q \,\middle|\, Z(t) = j\right]$$

$$= \sum_{p=0}^q \binom{q}{p}(-1)^{q-p} V^{j(p)}(t) V^j(t)^{q-p},$$

and from there go on to differential equations for the conditional skewness

$$\left(\mathbb{V}\mathrm{ar}^j(t)\right)^{-3/2} \sum_{p=0}^3 \binom{3}{p}(-1)^{3-p} V^{j(p)}(t) V^j(t)^{3-p},$$

or the conditional kurtosis. The characterizing differential equations obtained thereby may be of little relevance in actuarial practice. But note that the original payment stream B generalizes many different objects of interest when studying different probabilities based on an underlying Markov chain. So, the application of formulas obtained within the area of life insurance go far beyond that domain.

We finish the section with yet another generalization to higher order conditional multivariate moments but stick to the second-order moment. The situation is the following. We have two payment streams B_1 and B_2 driven by the same underlying Markov chain. The simplest case is to think of a term life insurance and a temporary life annuity with respect to the same uncertain lifetime. Generalizations to multi-state examples follow. Now, how do we characterize the conditional covariance $\mathbb{C}\mathrm{ov}^j(t)$ between the present value of B_1 and the present value of B_2, given as

$$\mathbb{E}\left[\left(\int_t^n e^{-\int_t^s r}\, dB_1(s) - V_1^j(t)\right)\left(\int_t^n e^{-\int_t^s r}\, dB_2(s) - V_2^j(t)\right)\,\middle|\, Z(t) = j\right] \quad (10.4)$$

where, obviously, V_1^j and V_2^j are the statewise reserves for the two payment processes, respectively? Again, we state a final result without proof.

Proposition 10.3 *Assume that $\Upsilon^j(t)$ is a function such that*

$$\frac{d}{dt}\Upsilon^j(t) = 2r\,\Upsilon^j(t) - \sum_{k:k\neq j} \mu^{jk}(t)\left(R_1^{jk}(t) R_2^{jk}(t) + \Upsilon^k(t) - \Upsilon^j(t)\right)$$

and $\Upsilon^j(n) = 0$. Then indeed $\Upsilon^j(t) = \mathbb{C}\mathrm{ov}^j(t)$, the statewise covariance given by (10.4).

(Here obviously R_1^{jk} and R_2^{jk} are the statewise sums at risk for the two payment processes, respectively.)

From there, we can even derive differential equations for the statewise correlation coefficient

$$\rho_{12}^{j}(t) = \frac{\mathbb{C}\mathrm{ov}^{j}(t)}{\sqrt{\mathbb{V}\mathrm{ar}_1^{j}(t)}\sqrt{\mathbb{V}\mathrm{ar}_2^{j}(t)}},$$

where $\mathbb{V}\mathrm{ar}_i^{j}$ is the conditional variance for the present value of payment stream i.

Exercises

10.1 Characterize by differential equations the second-order moment and the variance, conditional on survival, of an elementary life annuity benefit in a survival model.

10.2 Characterize by differential equations the statewise second-order moments and the statewise variance of an elementary disability annuity with an insurance sum S paid out upon death, independent of whether death occurs as active or disabled. All coverages terminate at time n.

10.3 Earlier we worked with reserve-dependent payments and intensities. Discuss the idea of conditional variance-dependent payments and intensities. Can we handle this from a mathematical point of view? What makes this situation particularly easy or complicated to work with from a mathematical point of view? Can you imagine meaningful applications of this idea from an economic point of view? What makes this situation particularly easy or complicated to work with from an economic point of view?

Notes

In this chapter, we have studied a series of special topics that have all attracted attention in the recent literature. They are mainly driven by (a) the ongoing integration of the theory of finance into the theory of insurance, (b) an increased need for demographic modelling in actuarial science, and (c) focus on behavioral options, not only but also from regulators.

Models with duration-dependence applied to insurance have attracted increasing attention over the last decade. Duration-dependence in the intensities, also known as the semi-Markov model, has been well studied in applied probability for a long time. Applications to insurance were introduced in actuarial science by Hoem [89]. An overview of applications to insurance and finance is found in the textbook by Janssen and Manca [94] and applications to insurance are studied by Christiansen [49] and by Buchardt et al. [39] in modern insurance mathematical frameworks.

Note the distinction between duration-dependent intensities, which is a general class of models applied in various areas, and duration-dependent payments, which is a topic quite specific to life insurance. The distinction between reserve-dependent intensities and reserve-dependent payments can be viewed similarly. Reserve-dependent intensities introduce classes of probability models that can be applied generally to problems related to (sub-)optimal stopping. The presentation in Gad et al. [78] is specific to life insurance but the idea can be, and has been, applied outside. The idea of reserve-dependent payments is, in contrast, quite specific to life insurance, see e.g. Christiansen et al. [50].

Forward rates is a classical object of study in financial mathematics. The structure was recognized in life insurance and the forward mortality rate was introduced by Milevsky and

Promislow [123]. Some issues of concern were raised by Miltersen and Persson [120] (the financial point of view). The drawbacks of the concept were discussed in Norberg [134] (the pessimistic point of view). The issues with dependence between transition rates and interest rates have been studied by Buchardt [35] (the actuarial point of view). Several generalizations to multistate models have been proposed in the literature. A literature overview, a new definition, and a comparison with the different proposals by Christiansen and Niemeyer [51] and by Buchardt [36] are given in Buchardt, Furrer and Steffensen [37].

Stochastic mortality rate models is a true neoclassic in life insurance mathematics. Journals and conferences are well-equipped with contributions on the subject, sometimes with radical ideas, sometimes with marginal variations, and often with a good amount of statistical inferences. Databases with population mortality data are relatively accessible whereas real data for multistate models is difficult to get access to. A few classical contributions, also in relation to the idea of affine rates discussed here, are Dahl [52] and Biffis [24]. More generally, we refer to the survey papers by Cairns et al. [45] and Barrieu et al. [19] but the list of publications within the area, also after 2012, is long, and the area is still under development.

Policyholder behavior is another neoclassic in life insurance mathematics although much less studied than stochastic mortality rates. The rational policy holder was studied in Steffensen [161], incidental behavior in Buchardt and Møller [38] and Henriksen et al. [86]. One approach to spanning models in between these extremes in simple parametric models is the work on surrender rates by Gad, Juhl and Steffensen [78].

The vast majority of the literature concerns the characterization of the conditional expectation of the present value of future payments. But both higher order moments and the probability distribution as such have been characterized in a few works. Ramlau-Hansen [146] relates to our multistate version of Hattendorff's theorem. Norberg [132] characterizes all existing moments by differential equations and recently, a general theory based upon the product integral has been developed in Bladt, Asmussen and Steffensen [28].

Chapter VIII: Orderings and Comparisons

1 Stochastic Ordering of Risks

We consider orderings between one-dimensional r.v.s X, Y (risks). An obvious example is a.s. ordering, $X \leq_{\text{a.s.}} Y$. We shall, however, mainly be concerned with orderings which only involve the distributions, i.e., which are law invariant. The orderings, written $X \preceq Y$ with a subscript on $\preceq$, will be given various meanings. The definitions typically come in three equivalent forms:

I) in terms of expectations, $\mathbb{E} f(X) \leq \mathbb{E} f(Y)$ for all $f \in \mathcal{C}$, where $\mathcal{C}$ is a class of functions chosen according to the desired properties of the specific order;

II) in terms of the c.d.f.s F_X, F_Y;

III) in terms of a *coupling* with a prespecified property. By a coupling we understand r.v.s X', Y' defined on a common probability space $(\Omega', \mathcal{F}', \mathbb{P}')$, such that the $\mathbb{P}'$-distribution of X' is the same as the given distribution of X, and the $\mathbb{P}'$-distribution of Y' is the same as the given distribution of Y.[1] To facilitate notation, we will often omit the primes in the definition of coupling, and when we talk about a coupling of X, Y, we will understand a redefinition of X, Y on a common probability space that does not change the marginal distributions.

The implication II) $\Rightarrow$ III) and/or I) $\Rightarrow$ III) is often referred to as *Strassen's theorem* after his 1966 paper [165], but it should be noted that the result there is much more general and abstract than the cases we will encounter here. In fact, the older proofs of the result of [165] require the Hahn–Banach theorem and hence

---

[1] To clarify the relevance of this construction, note that the first two ways I), II) to define an ordering only involve the distributions of X, Y, and that X, Y could in principle be defined on different probability spaces $(\Omega_X, \mathcal{F}_X, \mathbb{P}_X)$, $(\Omega_Y, \mathcal{F}_Y, \mathbb{P}_Y)$. Even if X, Y were defined on a common probability space $(\Omega, \mathcal{F}, \mathbb{P})$, the desired coupling property would not necessarily hold, cf. Exercise 1.1.

© Springer Nature Switzerland AG 2020
S. Asmussen, M. Steffensen, *Risk and Insurance*, Probability Theory and Stochastic Modelling 96, https://doi.org/10.1007/978-3-030-35176-2_8

the axiom of choice, whereas the proofs we give here are fairly elementary. For a discussion of these matters, see Müller and Stoyan [126, p. 23].

Our first and simplest example is *stochastic ordering*. We say that X is smaller than Y in stochastic ordering, or just that X is stochastically smaller than Y, written $X \preceq_{st} Y$, if

I) $\mathbb{E}f(X) \leq \mathbb{E}f(Y)$ for all $f \in \mathcal{C}_{incr}$ where $\mathcal{C}_{incr}$ is the class of all non-decreasing functions.[2]

The intuitive content is simply that Y is in some sense larger than X, cf. III) of the following result:

Theorem 1.1 *Properties* I) *and properties* II), III) *are equivalent, where:*
II) $F_X(x) \geq F_Y(x)$ *for all x. Equivalently, formulated in terms of the tails,* $\overline{F}_X(x) \leq \overline{F}_Y(x)$ *for all x;*
III) *There exists a coupling of X, Y such that $X \leq_{a.s.} Y$.*

Proof I) $\Rightarrow$ II): note that $f(y) = \mathbb{1}_{y > x}$ is non-decreasing and

$$\mathbb{E}f(X) = \mathbb{P}(X > x) = \overline{F}_X(x), \quad \mathbb{E}f(Y) = \mathbb{P}(Y > x) = \overline{F}_Y(x).$$

II) $\Rightarrow$ III): Let U be uniform$(0,1)$. Since $F_X^{\leftarrow} \leq F_Y^{\leftarrow}$ (generalized inverse c.d.f., cf. Sect. A.8 of the Appendix), we have $X' \leq_{a.s.} Y'$ where $X' = F_X^{\leftarrow}(U)$, $Y' = F_Y^{\leftarrow}(U)$. That $X' \overset{\mathcal{D}}{=} X$, $Y' \overset{\mathcal{D}}{=} Y$ follows from Proposition A.3.

III) $\Rightarrow$ I): Obvious, since for a given coupling X', Y' we have $f(X') \leq_{a.s.} f(Y')$ for $f \in \mathcal{C}_{incr}$ and hence

$$\mathbb{E}f(X) = \mathbb{E}f(X') \leq \mathbb{E}f(Y') = \mathbb{E}f(Y). \qquad \square$$

Example 1.2 Let X be Bernoulli(p_X) (i.e. $\mathbb{P}(X = 0) = 1 - p_X$, $\mathbb{P}(X = 1) = p_X$) and Y Bernoulli(p_Y), where $p_X \leq p_Y$. Then $F_X(x)$ and $F_Y(x)$ are both 0 when $x < 0$ and both 1 when $x \geq 1$, whereas

$$F_X(x) = 1 - p_X \geq 1 - p_Y = F_Y(x)$$

when $0 \leq x < 1$. Thus II) holds, so $X \preceq_{st} Y$. $\qquad\qquad \diamond$

Example 1.3 If X_λ is Poisson(λ) then $X_\lambda \preceq_{st} X_\mu$ when $\lambda \leq \mu$. Indeed, assume that X'_λ, $X'_{\mu - \lambda}$ are defined on the same probability space and independent. Then $X'_\mu = X'_\lambda + X'_{\mu - \lambda}$ is Poisson(μ) so X'_λ, X'_μ is the desired coupling of X_λ, X_μ. $\qquad \diamond$

In economics, one sometimes uses the term *first-order stochastic dominance* instead of stochastic ordering.

---

[2]In the definition of $\mathcal{C}_{incr}$ and similar classes in the following, it is tacitly understood that the domain is some subset of $\mathbb{R}$ containing both the support of X and the support of Y. Also, it is implicit that $\mathbb{E}f(X) \leq \mathbb{E}f(Y)$ is only required for functions $f \in \mathcal{C}_{incr}$ such that the expectations are well defined.

Exercises

1.1 Give one or more examples of r.v.s X, Y defined on the same probability space $(\Omega, \mathscr{F}, \mathbb{P})$ such that $X \preceq_{st} Y$ but not $X \leq_{a.s.} Y$.

1.2 Assume X, Y have densities $h_X(x), h_Y(x)$ and there exists an x_0 such that $h_X(x) \geq h_Y(x)$ for $x < x_0$ and $h_X(x) \leq h_Y(x)$ for $x > x_0$. Show that then $X \preceq_{st} Y$.

1.3 Assume $\mathbb{E}X = \mathbb{E}Y$ and that X, Y do not have the same distribution. Show that one cannot have $X \preceq_{st} Y$.

1.4 Let X, Y be gamma(α_1, λ_1), gamma(α_2, λ_2). Show by coupling that if $\alpha_1 = \alpha_2$, $\lambda_1 > \lambda_2$, then $X \preceq_{st} Y$, and that the same conclusion holds if instead $\alpha_1 < \alpha_2$, $\lambda_1 = \lambda_2$.

1.5 Show that the binomial(n, p) distribution is increasing w.r.t. p in stochastic order, e.g. by constructing an explicit coupling.

Notes
Classical texts on orderings of r.v.s are Müller and Stoyan [126] and Shaked and Shantiku-mar [157]. The orderings treated in this text are the most important for insurance and finance, but there is an abundance of different ones. Many of these have been developed for *ad hoc* purposes in different areas of application such as reliability theory.

2   Convex and Increasing Convex Ordering

2.1   Preliminaries on Stop-Loss Transforms and Convexity

Before proceeding to further examples of order relations between r.v.s, we need some background.

A restatement of the definition of a function $f(x)$ defined on a real interval I to be convex is that to each $x_0 \in I$ there is a *supporting line*, i.e., a line $\ell_{x_0}(x) = a + bx$ with $\ell_{x_0}(x_0) = f(x_0)$ and $\ell_{x_0}(x) \leq f(x)$ for all $x \in I$. A convex function is continuous, the right and left derivatives $f_r(x)$ and $f_l(x)$ exist for all $x \in I$ and are non-decreasing, and $f_r(x) = f_l(x)$ (so that f is differentiable at x) except at an at most countable set of points. The supporting line $\ell_{x_0}(x)$ is unique at a point x_0 where f is differentiable and then equal to the tangent $f(x_0) + f'(x_0)(x - x_0)$ (as for x_1 on Fig. 1). If $f_r(x_0) \neq f_l(x_0)$ (as for x_2), one has necessarily $f_r(x_0) > f_l(x_0)$, and there exists a bundle of supporting lines $\ell_{x_0}(x)$. More precisely, one may take $\ell_{x_0}(x) = f(x_0) + b(x - x_0)$ for any b satisfying $f_l(x_0) \leq b \leq f_r(x_0)$.

We shall need the following property, which is obvious from the definition of a convex function f and its continuity:

Proposition 2.1 *For a convex function f on I, it holds for any countable dense set* $\{x_n\}_{n=1,2,\dots} \subset I$ *that* $f(x) = \sup_{n=1,2,\dots} \ell_{x_n}(x)$ *for all* $x \in I$.

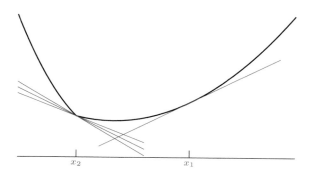

Fig. 1 Supporting lines

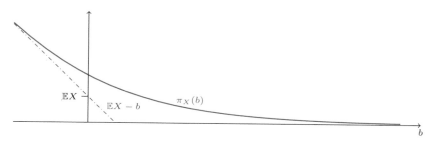

Fig. 2 A stop-loss transform

For an r.v. X with $\mathbb{E}|X| < \infty$, we define the *stop-loss transform* π_X of X by

$$\pi_X(b) \;=\; \mathbb{E}[X - b]^+ \;=\; \int_b^\infty \overline{F}_X(y)\,\mathrm{d}y\,.$$

Some of the main properties of the stop-loss transform are given in Proposition 2.2 and graphically illustrated in Fig. 2.

Proposition 2.2

(a) $\pi_X(b)$ *is continuous, non-increasing and convex in* b *with right derivative* $-\overline{F}_X(b)$;
(b) $\lim_{b\to\infty}\pi_X(b) = 0$, $\lim_{b\to-\infty}(\pi_X(b) + b) = \mathbb{E}X$;
(c) *A non-increasing and convex function* $\pi(b)$ *such that* $\lim_{b\to\infty}\pi(b) = 0$ *and that* $\lim_{b\to-\infty}(\pi_X(b) + b)$ *exists is the stop-loss transform of an r.v.* X *with* $F_X(b) = 1 + \pi'(b)$, *with* $\pi'(b)$ *meaning right derivative*;
(d) *If* $\pi_{X_n}(b) \to \pi_X(b)$ *for all* b, *then* $X_n \xrightarrow{\mathcal{D}} X$.

Proof In (a), it is obvious from right continuity of $\overline{F}_X(b)$ that the right derivative of $\pi_X(b)$ is $-\overline{F}_X(b)$. From this, the non-increasing property is clear, as is convexity because $-\overline{F}_X(b)$ is non-decreasing. In (b), the first assertion follows by monotone convergence since $(X - b)^+ \downarrow 0$ as $b \uparrow \infty$. For the second, note that $(X - b)^+ + b = \max(X, b) \uparrow X$ as $b \downarrow -\infty$, and use again monotone convergence.

In (c), define $F(b) = 1 + \pi'(b)$. Then the given properties ensure that F is non-decreasing and right continuous with limits 0 and 1 at $-\infty$, resp. ∞. From this the claim follows.

Finally for (d), we note that for any b and $\epsilon > 0$ we have

$$\pi_{X_n}(b) - \pi_{X_n}(b + \epsilon) \rightarrow \pi_X(b) - \pi_X(b + \epsilon),$$

i.e.

$$\int_b^{b+\epsilon} \overline{F}_{X_n}(y) \, dy \rightarrow \int_b^{b+\epsilon} \overline{F}_X(y) \, dy.$$

The bounds $\overline{F}_X(b + \epsilon) \leq \overline{F}_X(y) \leq \overline{F}_X(b)$ for $b < y \leq b + \epsilon$ (and similarly for X_n) therefore yield $\liminf \epsilon \overline{F}_{X_n}(b) \geq \epsilon \overline{F}_X(b + \epsilon)$, and therefore by right continuity $\liminf \overline{F}_{X_n}(b) \geq \overline{F}_X(b)$. Replacing b by $b - \epsilon$, we get in a similar manner that $\limsup \epsilon \overline{F}_{X_n}(b) \leq \epsilon \overline{F}_X(b - \epsilon)$. Thus, if b is a continuity point of F, we have $\limsup \overline{F}_{X_n}(b) \leq \overline{F}_X(b)$, and therefore $\overline{F}_{X_n}(b) \rightarrow \overline{F}_X(b)$, which implies $X_n \xrightarrow{\mathcal{D}} X$. □

Exercises

2.1 Show that $X \preceq_{\mathrm{st}} Y$ if and only if $\pi_Y(b) - \pi_X(b)$ is non-increasing [hint: $\pi_Y(b) - \pi_X(b)$ is differentiable].

2.2 Show the following converse of Proposition 2.2(d): if $X_n \xrightarrow{\mathcal{D}} X$ and the X_n are uniformly integrable, then $\pi_{X_n}(b) \rightarrow \pi_X(b)$ for all b.

2.3 Show that

$$\pi_X(b) = \int_{-\infty}^b F_X(y) \, dy - b + \mathbb{E}X.$$

2.2 Convex Ordering

We say that X is smaller than Y in *convex ordering*, written $X \preceq_{\mathrm{cx}} Y$, if
I) $\mathbb{E}f(X) \leq \mathbb{E}f(Y)$ for all $f \in \mathcal{C}_{\mathrm{cx}}$ where $\mathcal{C}_{\mathrm{cx}}$ is the class of all convex functions. The intuitive content is that $\preceq_{\mathrm{cx}}$ is a dispersion order, so that $X \preceq_{\mathrm{cx}} Y$ means that Y is more variable around the mean than X. That is, if X, Y are risks, we would consider Y as more dangerous than X, and in utility terms, $u(x + X)$ has larger expected utility than $u(x + Y)$. For example:

Proposition 2.3 $X \preceq_{\mathrm{cx}} Y \Rightarrow \mathbb{E}X = \mathbb{E}Y$, $\mathrm{Var}\, X \leq \mathrm{Var}\, Y$.

Proof Since $f(x) = x$ is convex, we have $\mathbb{E}X \leq \mathbb{E}Y$, and since $f(x) = -x$ is convex, we have $-\mathbb{E}X \leq -\mathbb{E}Y$. Hence $\mathbb{E}X = \mathbb{E}Y$. The convexity of x^2 then yields

$$\mathbb{V}\mathrm{ar}\, X \;=\; \mathbb{E}X^2 - (\mathbb{E}X)^2 \;\leq\; \mathbb{E}Y^2 - (\mathbb{E}X)^2 \;=\; \mathbb{E}Y^2 - (\mathbb{E}Y)^2 \;=\; \mathbb{V}\mathrm{ar}\, Y\,. \qquad \square$$

The analogue of Theorem 1.1 is:

Theorem 2.4 *Properties* I) *and properties* II), II'), III) *are equivalent, where:*
II) $\pi_X(b) = \mathbb{E}[X - b]^+ \leq \pi_Y(b) = \mathbb{E}[Y - b]^+$ *for all* b *and* $\mathbb{E}X = \mathbb{E}Y$;
II') $\pi_X(b) = \mathbb{E}[X - b]^+ \leq \pi_Y(b) = \mathbb{E}[Y - b]^+$ *and*[3] $\mathbb{E}[b - X]^+ \leq \mathbb{E}[b - Y]^+$ *for all* b;
III) *There exists a coupling of* X, Y *such that* $X = \mathbb{E}[Y \mid X]$. *Equivalently, the ordered pair* (X, Y) *is a martingale.*

Proof III) $\Rightarrow$ I): easy, since by Jensen's inequality for conditional expectations,

$$\mathbb{E}f(X) \;=\; \mathbb{E}f\big(\mathbb{E}[Y \mid X]\big) \;\leq\; \mathbb{E}\big[\mathbb{E}[f(Y) \mid X]\big] \;=\; \mathbb{E}f(Y)$$

for $f \in \mathcal{C}_{\mathrm{cx}}$.

I) $\Rightarrow$ II'): trivial, since the functions $(x - b)^+$ and $(b - x)^+$ are convex.
I) $\Rightarrow$ II): that $\pi_X(b) \leq \pi_Y(b)$ follows just as in the proof of I) $\Rightarrow$ II'), and $\mathbb{E}X = \mathbb{E}Y$ was shown above.
II') $\Rightarrow$ II): recalling the asymptotics of stop-loss transforms as $b \to \infty$ (Proposition 2.2(b)), $\pi_X(b) \leq \pi_Y(b)$ implies $\mathbb{E}X \leq \mathbb{E}Y$. Similarly, $\pi_{-X}(b) \leq \pi_{-Y}(b)$ gives $-\mathbb{E}X \leq -\mathbb{E}Y$.
II') $\Rightarrow$ III): This is the difficult part of the proof and is performed by creating a martingale sequence $\{X_n\}$ such that $X_0 = X$, $X_n \overset{\mathcal{D}}{\to} Y$ and $\pi_n \to \pi_Y$ where $\pi_n = \pi_{X_n}$. Choose a dense countable set $x_1, x_2, \ldots \in \mathbb{R}$ and let ℓ_{x_n} be a supporting line of π_Y at x_n. Then

$$\pi_Y(x) \;=\; \sup_{n=1,2,\ldots} \ell_{x_n}(x) \quad \text{for all } x \in \mathbb{R}\,. \tag{2.1}$$

On a suitable probability space equipped with enough uniform$(0, 1)$ r.v.s to perform the randomizations described below, we now define a sequence $X_0, X_1, \ldots$ of r.v.s with stop-loss transforms $\pi_0, \pi_1, \ldots$ satisfying $\pi_n \leq \pi_Y$. We first take $X_0 \overset{\mathcal{D}}{=} X$; $\pi_0 \leq \pi_Y$ then follows by assumption. Recursively, we then define X_n from X_{n-1} as follows. We first note graphically that the set $\big\{b : \pi_{n-1}(b) < \ell_{x_n}(b)\big\}$ is either empty or an open interval, say (a_n, b_n) (cf. Fig 3). In the first case, we take simply $X_n = X_{n-1}$. Otherwise, let $X_n = X_{n-1}$ if $X_{n-1} < a_n$ or $X_{n-1} > b_n$. If $a_n \leq X_{n-1} = x \leq b_n$, we let $X_n = a_n$ w.p. $(b_n - x)/(b_n - a_n)$ and $X_n = b_n$ w.p.

---

[3] The second condition can be restated that the stop-loss-transform of $-X$ is dominated by the stop-loss-transform of $-Y$ (replace b by $-b$).

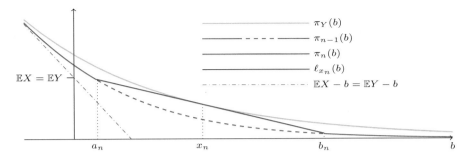

Fig. 3 Updating from π_{n-1} to π_n

$(x - a_n)/(b_n - a_n)$. Then for $a_n \le x \le b_n$,

$$
\begin{aligned}
\mathbb{E}\big[X_n \mid X_{n-1} = x\big] &= a_n \frac{b_n - x}{b_n - a_n} + b_n \frac{x - a_n}{b_n - a_n} \\
&= \frac{a_n b_n - x a_n + x b_n - a_n b_n}{b_n - a_n} = x .
\end{aligned}
$$

Since the same is obviously true for $x < a_n$ or $x > b_n$, the sequence $\{X_n\}$ is therefore a martingale.

For $b > b_n$, $X_n - b$ can only be positive if $X_n = X_{n-1}$, and so $\pi_n(b) = \pi_{n-1}(b)$. Similarly, for $b < a_n$, $b - X_n$ can only be positive if $X_n = X_{n-1}$, and the identity $x^+ = x + x-$ then gives

$$
\begin{aligned}
\pi_n(b) &= \mathbb{E}\big[X_n - b + (X_n - b)^-\big] \\
&= \mathbb{E}\big[X_{n-1} - b + (X_{n-1} - b)^-\big] = \pi_{n-1}(b) ,
\end{aligned}
$$

where we used that $\mathbb{E}X_n = \mathbb{E}X_{n-1}$ by the martingale property. The same is also true by continuity for $b = a_n$ or $x = b_n$ so that $\pi_n(a_n) = \pi_{n-1}(a_n)$ and $\pi_n(b_n) = \pi_{n-1}(b_n)$. Since X_n has no mass on (a_n, b_n), $\pi_n(b)$ on this interval must be the straight line connecting the points

$$
\begin{aligned}
\big(a_n, \pi_n(a_n)\big) &= \big(a_n, \pi_{n-1}(a_n)\big) = \big(a_n, \ell_{x_n}(a_n)\big) , \\
\big(b_n, \pi_n(b_n)\big) &= \big(b_n, \pi_{n-1}(b_n)\big) = \big(b_n, \ell_{x_n}(b_n)\big) .
\end{aligned}
$$

Combining these facts gives $\pi_n(b) = \max\big(\pi_{n-1}(b), \ell_{x_n}(b)\big)$. Then (2.1) implies $\pi_n(b) \uparrow \pi_Y(b)$ so that $X_n \xrightarrow{\mathcal{D}} Y$ by Proposition 2.2(d).

Now

$$
\begin{aligned}
\mathbb{E}|X_n| &= 2\mathbb{E}X_n^+ - \mathbb{E}X_n = 2\pi_n(0) - \mathbb{E}X \\
&\to 2\pi_Y(0) - \mathbb{E}X = 2\mathbb{E}Y^+ - \mathbb{E}Y = \mathbb{E}|Y| .
\end{aligned}
$$

Therefore the sequence $\{X_n\}$ is L^1-bounded, so by the martingale convergence theorem there exists an r.v. Y' such that $X_n \overset{\text{a.s.}}{\to} Y'$. It follows that $Y' \overset{\mathscr{D}}{=} Y$ so that $\mathbb{E}X_n = \mathbb{E}X = \mathbb{E}Y'$. This implies uniform integrability so that $\mathbb{E}[Y'|X_0] = X_0$ (again by martingale theory) and that the desired coupling is (X_0, Y').   □

Corollary 2.5 *Assume $X \preceq_{\text{cx}} Y$. Then there exists a coupling of (X, Y) such that $\mathbb{E}[Y|X] = X$ and the conditional distribution of Y given $X = x$ is increasing in stochastic order in x.*

Proof In the proof of Theorem 2.4, let $\{X_n(x)\}$ denote the X_n-sequence started from $X_0 = x$ and let $x_1 \le x_2$. We may construct the $X_n(x_1), X_n(x_2)$ from the same sequence of uniform$(0, 1)$ r.v.s $U_1, U_2, \ldots$ by taking

$$X_n(x_i) = \begin{cases} a_n & \text{if } U_n \le (b_n - X_{n-1}(x_i))/(b_n - a_n) \\ b_n & \text{if } U_n > (X_{n-1}(x_i) - a_n)/(b_n - a_n) \end{cases}$$

for $a_n \le X_{n-1}(x_i) \le b_n$. It is then readily checked by induction that $X_n(x_1) \le X_n(x_2)$ a.s. Hence the limits satisfy $Y'(x_1) \le Y'(x_2)$ a.s. But these limits have the desired conditional distributions, so that we have constructed a coupling, implying stochastic order.   □

An often easily verified sufficient condition for convex ordering is the following:

Proposition 2.6 *Assume that $\mathbb{E}X = \mathbb{E}Y$, and there exists an x_0 such that $F_X(x) \le F_Y(x)$ for $x < x_0$ and $F_X(x) \ge F_Y(x)$ for $x \ge x_0$. Then $X \preceq_{\text{cx}} Y$.*

The result is known in the actuarial sciences as *Ohlin's lemma*. The intuition is that the condition on F_X, F_Y says that the left tail of X is lighter than the left tail of Y, and that the right tails are so, both when viewed from x_0, thereby expressing the greater variability of Y. For a graphical illustration, see Fig. 4a.

Proof Let $f \in \mathcal{C}_{\text{cx}}$ and let $\ell_{x_0}(x) = f(x_0) + k(x - x_0)$ be a supporting line, i.e., $\ell_{x_0}(x) \le f(x)$ for all x and $\ell_{x_0}(x_0) = f(x_0)$. Define $X' = X \wedge x_0$, $X'' = X \vee x_0$

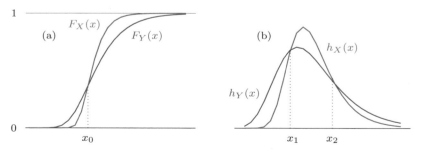

Fig. 4 Convex ordering via (**a**) the c.d.f., (**b**) the density

and similarly for Y', Y''. Since $X' \le x_0$, $Y' \le x_0$, we have $F_{X'}(x) = F_{Y'}(x) = 1$ for $x \ge x_0$, whereas

$$F_{X'}(x) = F_X(x) \le F_Y(x) = F_{Y'}(x)$$

for $x < x_0$. Hence $Y' \preceq_{\mathrm{st}} X'$. Similarly, $X'' \preceq_{\mathrm{st}} Y''$.

Now consider $g(x) = f(x) - \ell_{x_0}(x)$. The right derivative is $f'_{\mathrm{r}}(x) - k$, which is non-decreasing. Hence (since $g(x_0) = 0$ and $g \ge 0$) $g(x)$ is non-increasing on $(-\infty, x_0]$, so that $\mathbb{E}g(X') \le \mathbb{E}g(Y')$ because $X', Y' \in (-\infty, x_0]$ and $Y' \preceq_{\mathrm{st}} X'$. Similarly, $g(x)$ is non-decreasing on (x_0, ∞) and hence $\mathbb{E}g(X'') \le \mathbb{E}g(Y'')$. Since $g(X') + g(X'') = g(X) + g(x_0) = g(X)$, it follows that $\mathbb{E}g(X) \le \mathbb{E}g(Y)$. In view of the definition of g and $\mathbb{E}X = \mathbb{E}Y$, this entails $\mathbb{E}f(X) \le \mathbb{E}f(Y)$. □

A density version of Ohlin's lemma follows and is illustrated in Fig. 4b (it should be compared with Exercise 1.2 on stochastic ordering).

Corollary 2.7 *Assume X, Y have densities $h_X(x), h_Y(x)$ such that there exists $x_1 < x_2$ with*

$$h_X(x) \le h_Y(x), \quad x < x_1,$$
$$h_X(x) \ge h_Y(x), \quad x_1 < x < x_2,$$
$$h_X(x) \le h_Y(x), \quad x > x_2,$$

and that $\mathbb{E}X = \mathbb{E}Y$. Then $X \preceq_{\mathrm{cx}} Y$.

Proof Define

$$\psi(x) = F_Y(x) - F_X(x) = \int_{-\infty}^{x} \big[h_Y(y) - h_X(y) \big] \, dy.$$

Then $\psi(x)$ is non-negative and non-decreasing for $x \le x_1$ and $x \ge x_2$, and non-increasing for $x_1 \le x \le x_2$. Noting that $\psi(x) \to 0$ as either $x \to -\infty$ or $x \to \infty$, graphical inspection shows that this implies the existence of a unique zero $x_0 \in [x_1, x_2]$, which has the property $\psi(x) \ge 0$ for $x \le x_0$ and $\psi(x) \le 0$ for $x \ge x_0$. Thus the conditions of Ohlin's lemma hold. □

Corollary 2.8 *Let $f_1(x), f_2(x)$ be non-decreasing continuous functions such that there exists an x_0 with $f_1(x) \ge f_2(x)$ for $x < x_0$ and $f_1(x) \le f_2(x)$ for $x \ge x_0$. Assume that for a risk X, we have $\mathbb{E}f_1(X) = \mathbb{E}f_2(X)$. Then $f_1(X) \preceq_{\mathrm{cx}} f_2(X)$.*

Proof Let $y_0 = f_2(x_0)$ and $Y_1 = f_1(X)$, $Y_2 = f_2(X)$. If $y < y_0$ and $Y_2 \le y$, then $f_2(X) < f_2(x_0)$, which implies $X < x_0$ and $Y_1 = f_1(X) \ge f_2(X) = Y_2$, $F_{Y_1}(y) \le F_{Y_2}(y)$. Similarly, if $y \ge y_0$ and $Y_2 > y$, then $X > x_0$ and hence $Y_1 \le Y_2$, so that $\overline{F}_{Y_1}(y) \le \overline{F}_{Y_2}(y)$ implying $F_{Y_1}(y) \ge F_{Y_2}(y)$. By right-continuity, this also holds for $y = y_0$. Hence the assumptions of Ohlin's lemma holds for Y_1, Y_2. □

Example 2.9 Suppose one can invest a total amount 1 in n possible stocks (or funds). These stocks produce i.i.d. yields X_i per share. How should one choose the fraction p_i to invest in stock i for the purpose of maximizing the expected utility? That is, what is the maximizer $p_1^*, \ldots, p_n^*$ of $\mathbb{E}u(p_1 X_1 + \cdots + p_n X_n)$ subject to $p_1^* + \cdots + p_n^* = 1$?

We will prove that (not unexpectedly!) $p_i^* = 1/n$. The corresponding yield is $A^* = X_1/n + \cdots + X_n/n$, and the proof of optimality proceeds by showing that $A^* \preceq_{\mathrm{cx}} A = p_1 X_1 + \cdots + p_n X_n$ for any $p_1, \ldots, p_n$; the concavity of u then yields $\mathbb{E}u(A^*) \geq \mathbb{E}u(A)$.

To this end, observe first that $\mathbb{E}[X_i | A^*]$ must be independent of i by symmetry. Since the sum over i is nA^*, we have $\mathbb{E}[X_i | A^*] = A^*$, and therefore

$$\mathbb{E}[A \mid A^*] = \sum_{i=1}^{n} p_i \mathbb{E}[X_i | A^*] = \sum_{i=1}^{n} p_i A^* = A^*.$$

That $A^* \preceq_{\mathrm{cx}} A$ therefore follows from characterization III) of $\preceq_{\mathrm{cx}}$.   ◇

Example 2.10 As a generalization of Example 2.9, let again $X_1, \ldots, X_n$ be i.i.d., but consider now a more general combination $A = \sum_1^n \rho_i(X_i)$ where $\rho_1, \ldots, \rho_n$ are non-negative functions. Letting

$$\rho^*(x) = \frac{1}{n} \sum_{i=1}^{n} \rho_i(x), \quad A^* = \sum_{j=1}^{n} \rho^*(X_j),$$

we shall show that $A^* \preceq_{\mathrm{cx}} A$. In particular, this gives $\mathbb{E}u(A^*) \geq \mathbb{E}u(A)$ for any concave utility function.

The argument is similar to the one in Example 2.9. We have

$$\mathbb{E}[A \mid A^*] = \sum_{i=1}^{n} \mathbb{E}[\rho_i(X_i) \mid A^*] = \sum_{i=1}^{n} \mathbb{E}[\rho_i(X_j) \mid A^*] = n\mathbb{E}[\rho^*(X_j) \mid A^*],$$

where the second equality follows since A^* is symmetric in $X_1, \ldots, X_n$. Multiplying by $1/n$ and summing over $j = 1, \ldots, n$, the r.h.s. becomes $\mathbb{E}[A^* \mid A^*] = A^*$ and the l.h.s. stays $\mathbb{E}[A \mid A^*]$. Thus indeed $\mathbb{E}[A \mid A^*] = A^*$.   ◇

Exercises

2.4 Fill in the proof that $X'' \preceq_{\mathrm{st}} Y''$ in the proof of Theorem 2.4.

2.5 Let X, Y be gamma(α_1, λ_1), gamma(α_2, λ_2) with $\alpha_2 > \alpha_1$ and the same mean, i.e. $\alpha_1/\lambda_1 = \alpha_2/\lambda_2$. Show that the ratio of densities $f_2(x)/f_1(x)$ is increasing on $(0, x_0)$ and decreasing on (x_0, ∞), where $x_0 = (\alpha_2 - \alpha_1)/(\lambda_2 - \lambda_1)$, with limits 0 at $x = 0$ and $x = \infty$. Show thereby that $Y \preceq_{\mathrm{cx}} X$.

2.6 Let $X_1, X_2, \ldots$ be i.i.d. and $\mu = \mathbb{E}X_1$, $\overline{X}_n = (X_1 + \cdots + X_n)/n$. Compute $\mathbb{E}[\overline{X}_n \mid \overline{X}_{n+1}]$, and show thereby that

$$\overline{X}_1 \succeq_{cx} \overline{X}_2 \succeq_{cx} \overline{X}_3 \succeq_{cx} \cdots \succeq_{cx} \mu \,.$$

2.7 Inspired by Ohlin's lemma, one could define an ordering $X \preceq_{ht} Y$ (X is more heavy-tailed than Y) by the requirement that there exists an x_0 such that $F_X(x) \leq F_Y(x)$ for $x < x_0$ and $F_X(x) \geq F_Y(x)$ for $x \geq x_0$. Show that this relation is not transitive, i.e. there exist X, Y, Z such that $X \preceq_{ht} Y$, $Y \preceq_{ht} Z$, but not $X \preceq_{ht} Z$.

2.8 Let X, Y be risks with finite second moment and $X \preceq_{cx} Y \preceq_{cx} X$. Compute $\mathbb{E}(X' - Y')^2$ and show thereby that $X \overset{\mathcal{D}}{=} Y$.

2.9 Consider the set $\mathscr{X}$ of all risks X with $a \leq X \leq b$ for some $b > -\infty, a < \infty$ and $\mathbb{E}X = \mu$ for a given $\mu \in [a, b]$, and let X_0 be the risk with $\mathbb{P}(X_0 = a) = p = 1 - \mathbb{P}(X_0 = b)$ with p chosen such that $X_0 \in \mathscr{X}$. Show that $X \preceq_{cx} X_0$ for all $X \in \mathscr{X}$.

Notes
The abstract proof of Property III) of convex ordering in Theorem 2.4 due to Strassen [165] was for many years the only one around. The present more elementary proof is from Müller and Rüschendorf [125]. Ohlin's lemma also goes under the name of the Karlin–Novikov cut criterion.

2.3   Increasing Convex Ordering

We say that X is smaller than Y in *increasing convex ordering*, written $X \preceq_{icx} Y$, if
I) $\mathbb{E}f(X) \leq \mathbb{E}f(Y)$ for all $f \in \mathcal{C}_{icx}$, where $\mathcal{C}_{icx}$ is the class of all non-decreasing convex functions.
    The analogue of Theorems 1.1, 2.4 is:

Theorem 2.11 *Properties* I) *and properties* II), III) *are equivalent where*
II) $\pi_X(b) = \mathbb{E}[X - b]^+ \leq \pi_Y(b) = \mathbb{E}[Y - b]^+$ *for all b.*
III) *There exists a coupling of* X, Y *such that* $X \leq \mathbb{E}[Y \mid X]$. *Equivalently, the ordered pair* (X, Y) *is a submartingale.*

In view of II), the increasing convex order is often referred to as the *stop-loss ordering* in the insurance / finance literature.
    The proof of Theorem 2.11 uses the following result of independent interest:

Proposition 2.12 *If* $X \preceq_{icx} Y$, *then there exists a* Z *such that* $X \preceq_{st} Z$ *and* $Z \preceq_{cx} Y$.

Proof of Proposition 2.12 From $X \preceq_{icx} Y$ it follows that $\pi_X(b) \leq \pi_Y(b)$ and that $\mathbb{E}X \leq \mathbb{E}Y$ (note that $(x - b)^+$ and x are non-decreasing and convex). Define $\pi(b) = \max\big(\pi_X(b), \mathbb{E}Y - b\big)$. Then $\pi(b) + b = \max\big(\pi_X(b) + b, \mathbb{E}Y\big)$ has limit $\max(\mathbb{E}X, \mathbb{E}Y) = \mathbb{E}Y$ at $-\infty$. The remaining conditions of Proposition 2.2(c) are trivial to check, and hence $\pi = \pi_Z$ for some r.v. Z. The asymptotics at $b = -\infty$ that we just derived imply $\mathbb{E}Z = \mathbb{E}Y$. Also, Jensen's inequality yields

$$\mathbb{E}Y - b \le (\mathbb{E}Y - b)^+ \le \mathbb{E}(Y - b)^+ = \pi_Y(b)$$

and therefore $\pi_Z \le \pi_Y$. That $Z \preceq_{\mathrm{cx}} Y$ now follows by criterion II) of Theorem 2.4.

To show $X \preceq_{\mathrm{st}} Z$, it suffices by Exercise 2.1 to show that $\pi_Z - \pi_X$ is non-increasing. This follows immediately from

$$\pi_Z(b) - \pi_X(b) = \big(\mathbb{E}Y - b - \pi_X(b)\big)^+ = \big(\mathbb{E}Y - \mathbb{E}\max(X, b)\big)^+. \qquad \square$$

Proof of Theorem 2.11 That I) $\Rightarrow$ II) follows since $(x - b)^+$ is non-decreasing and convex. That III) $\Rightarrow$ I) is easy since by Jensen's inequality for conditional expectations,

$$\mathbb{E}f(X) \le \mathbb{E}f\big(\mathbb{E}[Y \mid X]\big) \le \mathbb{E}\big[\mathbb{E}[f(Y) \mid X]\big] = \mathbb{E}f(Y)$$

for $f \in \mathcal{C}_{\mathrm{icx}}$.

It remains to prove II) $\Rightarrow$ III). We first note that the proof of Proposition 2.12 remains valid under assumption II) (for $\mathbb{E}X \le \mathbb{E}Y$, use the asymptotics of π_X, π_Y at $-\infty$). We can now proceed as in the proof of II) $\Rightarrow$ III) in Theorem 2.4 to construct a coupling of (Z, Y) such that $\mathbb{E}[Y|Z] = Z$. Let V be a uniform$(0, 1)$ r.v. independent of Y and Z and $U = F_Z(Z-) + V \Delta F_Z(Z)$ the distributional transform of Z, cf. Sect. A.9 of the Appendix. Then U is uniform$(0, 1)$ and $Z = F_Z^\leftarrow(U)$ a.s. Further $X \preceq_{\mathrm{st}} Z$ implies $F_X \ge F_Z$ and therefore $F_X^\leftarrow \le F_Z^\leftarrow$. Taking $X = F_X^\leftarrow(U)$, we therefore have $X \le Z$ a.s. and $\sigma(X, Z) = \sigma(Z)$ so that

$$\mathbb{E}[Y|X] = \mathbb{E}\big[\mathbb{E}[Y \mid X, Z] \,\big|\, X\big] = \mathbb{E}\big[\mathbb{E}[Y|Z] \,\big|\, X\big] = \mathbb{E}[Z|X] \ge X.$$

Thus, we have obtained the desired coupling for the submartingale property. $\square$

A comparison of parts II) of Theorems 2.4, 2.11 yields at once

Proposition 2.13 $X \preceq_{\mathrm{cx}} Y$ *if and only if* $X \preceq_{\mathrm{icx}} Y$ *and* $\mathbb{E}X = \mathbb{E}Y$.

Exercises

2.10 In Remark I.2.3, the concept of second-order stochastic dominance was mentioned. Discuss the relation to increasing convex order.

2.11 ([58] p. 157) Give an alternative proof of Proposition 2.12 by showing that $Z = \max(X, a)$ satisfies $X \preceq_{\mathrm{st}} Z \preceq_{\mathrm{cx}} Y$ when a is chosen with $\mathbb{E}Z = Y$ [hint: when checking $\pi_Z(b) \le \pi_Y(b)$, consider the cases $a \le b$ and $a > b$ separately].

2.12 Show as a generalization of Ohlin's lemma that if $X \prec_{\mathrm{ht}} Y$ in the sense of Exercise 2.7 and $\mathbb{E}X \le \mathbb{E}Y$, then $X \preceq_{\mathrm{icx}} Y$.

3   Closure Properties of Orderings

Proposition 3.1 *The relations $\preceq_{st}$, $\preceq_{cx}$, $\preceq_{icx}$ are closed under convolutions. That is, if $X_1, \ldots, X_n$ are independent, $Y_1, \ldots, Y_n$ are independent, and $X_i \preceq_{st} Y_i$ for all i, then*

$$X_1 + \cdots + X_n \preceq_{st} Y_1 + \cdots + Y_n$$

and similarly for $\preceq_{cx}$, $\preceq_{icx}$.

Proof We give the proof for $\preceq_{cx}$ and leave $\preceq_{st}$, $\preceq_{icx}$ as Exercise 3.2. Assume that X_i, Y_i are coupled as a martingale on $(\Omega_i, \mathscr{F}_i, \mathbb{P}_i)$ and let

$$(\Omega, \mathscr{F}, \mathbb{P}) = \prod_{i=1}^{n} (\Omega_i, \mathscr{F}_i, \mathbb{P}_i) \,.$$

Then for $\omega = (\omega_1, \ldots, \omega_n)$, we can let $X_i'(\omega) = X_i(\omega_i)$. Then $X_i' \overset{\mathcal{D}}{=} X_i$ and the X_i' are independent. Similar properties of the Y_i' hold, and $\mathbb{E}[Y_i' \mid X_i'] = X_i'$. Hence

$$\mathbb{E}\big[Y_1' + \cdots + Y_n' \mid X_1', \ldots, X_n'\big]$$
$$= \mathbb{E}\big[Y_1' \mid X_1', \ldots, X_n'\big] + \cdots + \mathbb{E}\big[Y_n' \mid X_1', \ldots, X_n'\big]$$
$$= \mathbb{E}\big[Y_1' \mid X_1'\big] + \cdots + \mathbb{E}\big[Y_n' \mid X_n'\big] = X_1' + \cdots + X_n' \,,$$

so property III) holds.   □

Proposition 3.2 *The relations $\preceq_{st}$, $\preceq_{cx}$, $\preceq_{icx}$ are closed under mixtures. That is, if $X(\lambda) \preceq_{st} Y(\lambda)$ for all λ and Λ is an r.v. such that the conditional distribution of X given $\Lambda = \lambda$ is the same as the distribution of $X(\lambda)$ and similarly for Y, then $X \preceq_{st} Y$. The same holds for $\preceq_{cx}$, $\preceq_{icx}$.*

Proof Again, we only consider $\preceq_{cx}$. For $f \in \mathcal{C}_{cx}$, we get

$$\mathbb{E}f(X) = \mathbb{E}\,\mathbb{E}\big[f(X) \mid \Lambda\big] = \int \mathbb{E}f\big(X(\lambda)\big)\,\mathbb{P}(\Lambda \in d\lambda)$$
$$\leq \int \mathbb{E}f\big(Y(\lambda)\big)\,\mathbb{P}(\Lambda \in d\lambda) = \mathbb{E}f(Y) \,.$$

□

Corollary 3.3 *Consider compound sums*

$$A_{X,N} = X_1 + \cdots + X_N \,, \qquad A_{Y,N} = Y_1 + \cdots + Y_N$$

with the X_i i.i.d. and independent of N, and the Y_i i.i.d. and independent of N. If $X \preceq_{st} Y$, then $A_{X,N} \preceq_{st} A_{Y,N}$.

Proof Take $\Lambda = N$ in Proposition 3.2 and use that

$$X_1 + \cdots + X_n \preceq_{\text{st}} Y_1 + \cdots + Y_n$$

by Proposition 3.1.   □

The issue of ordering w.r.t. N is more tricky:

Proposition 3.4 *Consider compound sums*

$$A_N = X_1 + \cdots + X_N, \quad A_M = X_1 + \cdots + X_M$$

with the X_i i.i.d., non-negative and independent of N, M. If $N \preceq_{\text{st}} M$, then $A_N \preceq_{\text{st}} A_M$. The same holds for $\preceq_{\text{cx}}, \preceq_{\text{icx}}$.

Proof The case of $\preceq_{\text{st}}$ is easy by coupling and omitted. For $\preceq_{\text{cx}}, \preceq_{\text{icx}}$, define $S_n = X_1 + \cdots + X_n$, $a_n(d) = \mathbb{E}[S_n - d]^+$. We first note that $a_n(d)$ is convex in n, i.e. that

$$a_{n+2}(d) - a_{n+1}(d) \ \geq \ a_{n+1}(d) - a_n(d). \tag{3.1}$$

Indeed, $a_{n+1}(d)$ on the r.h.s. can be written as $\mathbb{E}[S_n + X_{n+2} - d]^+$ (use that $X_{n+1} \overset{\mathscr{D}}{=} X_{n+2}$), so (3.1) follows from

$$(s + x_{n+1} + x_{n+2} - d)^+ - (s + x_{n+1} - d)^+ \ \geq \ (s + x_{n+2} - d)^+ - (s - d)^+ \tag{3.2}$$

which in turn follows from $\varphi'(y)$ being increasing where $\varphi(y) = (s + y)^+$.

If $N \preceq_{\text{icx}} M$, $A_N \preceq_{\text{icx}} M$ now follows since $a_n(d)$ is convex and increasing in n so that

$$\mathbb{E}[A_N - d]^+ \ = \ \mathbb{E}a_N(d) \ \leq \ \mathbb{E}a_M(d) \ = \ \mathbb{E}[A_M - d]^+,$$

so that II) in the definition of $\preceq_{\text{icx}}$ holds. If additionally $N \preceq_{\text{cx}} M$, then $\mathbb{E}N = \mathbb{E}M$ and therefore $\mathbb{E}A_N = \mathbb{E}N \, \mathbb{E}X = \mathbb{E}M \, \mathbb{E}X = \mathbb{E}A_M$ so that $A_N \preceq_{\text{cx}} A_M$ by Proposition 2.13.   □

Example 3.5 Consider the classical Bayesian case (e.g. Sect. II.1) where N has a mixed Poisson distribution, $N = N(\Lambda)$, where $N(\lambda)$ is a Poisson(λ) r.v. independent of Λ. We will show that $A_1 \preceq_{\text{icx}} A_2$ when $\Lambda_1 \preceq_{\text{icx}} \Lambda_2$ and $A_i = X_1 + \cdots + X_{N(\Lambda_i)}$ (similar assertions hold for $\preceq_{\text{st}}$ and $\preceq_{\text{cx}}$, with the proof for $\preceq_{\text{st}}$ being almost trivial and the one for $\preceq_{\text{cx}}$ a minor extension of the one for $\preceq_{\text{icx}}$).

According to Proposition 3.4, it suffices to show $N(\Lambda_1) \preceq_{\text{icx}} N(\Lambda_2)$. This holds if $\mathbb{E}\big[N(\Lambda_1) - d\big]^+ \leq \mathbb{E}\big[N(\Lambda_2) - d\big]^+$ for any given d, i.e. if

$$f_d(\lambda) \ = \ \mathbb{E}[N(\lambda) - d]^+ \ = \ \int_d^\infty \mathbb{P}\big(N(\lambda) > t\big) \, \mathrm{d}t \ = \ \int_d^\infty \sum_{n > t} \mathrm{e}^{-\lambda} \frac{\lambda^n}{n!} \, \mathrm{d}t$$

is convex and increasing, which in turn holds if $f'_d(\lambda)$ is non-negative and increasing. But

$$f'_d(\lambda) = \int_d^\infty \sum_{n>t} \frac{\mathrm{d}}{\mathrm{d}\lambda} \mathrm{e}^{-\lambda} \frac{\lambda^n}{n!}\, \mathrm{d}t \; = \; \int_d^\infty \sum_{n>t} \left(\mathrm{e}^{-\lambda} \frac{\lambda^{n-1}}{(n-1)!} - \mathrm{e}^{-\lambda} \frac{\lambda^n}{n!} \right) \mathrm{d}t$$

$$= f_{d-1}(\lambda) - f_d(\lambda) \; = \; \int_{d-1}^d \mathbb{P}\big(N(\lambda) > t\big)\, \mathrm{d}t \, .$$

Obviously, the last expression is non-negative, and it is also increasing in λ since $\mathbb{P}\big(N(\lambda) > t\big)$ is increasing in λ (the Poisson(λ) distribution is stochastically increasing in λ).   ◇

Exercises

3.1 Verify the conclusion of Example 3.5 for $\preceq_{\mathrm{st}}$ and $\preceq_{\mathrm{cx}}$.

3.2 Show the $\preceq_{\mathrm{st}}$, $\preceq_{\mathrm{icx}}$ parts of Proposition 3.1.

4   Utility, Deductibles and Reinsurance

Throughout this section, u is an increasing concave utility function, cf. Sect. I.2. The following result follows essentially just from observations made there, but is formulated here in terms of the orderings studied earlier in the chapter for the sake of conformity and easy reference:

Proposition 4.1 *For risks X, Y,*

  (i) $X \preceq_{\mathrm{st}} Y$ *implies* $\mathbb{E}u(w + X) \le \mathbb{E}u(w + Y)$, $\mathbb{E}u(w - X) \ge \mathbb{E}u(w - Y)$,
  (ii) $X \preceq_{\mathrm{cx}} Y$ *implies* $\mathbb{E}u(w + X) \ge \mathbb{E}u(w + Y)$, $\mathbb{E}u(w - X) \ge \mathbb{E}u(w - Y)$,
 (iii) $X \preceq_{\mathrm{icx}} Y$ *implies* $\mathbb{E}u(w - X) \ge \mathbb{E}u(w - Y)$,
 (iv) *For any risk X, one has*

$$\mathbb{E}u(w + X) \le \mathbb{E}u(w + \mathbb{E}X)\,, \qquad \mathbb{E}u(w - X) \le \mathbb{E}u(w - \mathbb{E}X)\,.$$

Proof Part (i) is immediate from u being increasing and part (iii) from $-u(w-x)$ being increasing and convex. The second statement of (ii) is immediate, and the first then follows from $-X \preceq_{\mathrm{cx}} -Y$. Part (iv) then follows from $\mathbb{E}Z \preceq_{\mathrm{cx}} Z$ applied to $\pm X$.   □

4.1   Underinsurance and Coinsurance

Example 4.2 Assume the insurable value (the maximal possible value of a claim, say the cost to rebuild your house if it is totally damaged by a fire) is v, but that only

$S = (1 - k)v$ with $0 \leq v < 1$ is written in the insurance policy (underinsurance). In case of this disastrous event, the insurance company will have a closer look at the value of the house and discover that you have only paid the premium corresponding to a rebuilding cost of $(1 - k)v$. Accordingly, only $(1 - k)v$ is reimbursed. The same principle is used in case of partial damage, i.e. the claim X satisfies $X < v$, resulting in the pro-rata cover $(1 - k)X$. Assume further that the premium is proportional to S, that is, a function of k of the form $(1 - k)H$ with $H > \mathbb{E}X$. We will show that taking full coverage ($k = 0$) is not optimal in the sense of utility, but that the insured obtains a better utility by choosing $0 < k < \epsilon$ for some suitable $\epsilon > 0$.

The expected utility of the insured is

$$g(k) = \mathbb{E}u\big(w - (1 - k)H - kX\big).$$

Assume for simplicity that the utility function u is differentiable. Then

$$g'(k) = \mathbb{E}\big[(H - X)u'\big(w - (1 - k)H - kX\big)\big], \quad g'(0) = (H - \mathbb{E}X)u'\big(w - H\big).$$

Thus $u' > 0$ and $H > \mathbb{E}X$ implies that $g'(k) > 0$, i.e. that $g(k) > g(0)$ for all $0 < k < \epsilon$ for some suitable ϵ. $\Diamond$

Example 4.3 (Coinsurance) Consider n insurers charging utility premiums based on the utility functions $(1 - e^{-a_i x})/a_i$, $i = 1, \ldots, n$. They want to jointly insure a risk X by splitting it up as $X = X_1 + \cdots + X_n$ with X_i the part to be covered by insurer i; X may be a new risk, or the total risk of all companies that they want to redistribute. How should $X_1, \ldots, X_n$ be chosen in order to maximize the competitive performance of the pool, i.e. in order to minimize the total premium H? We will show that the optimal arrangement is

$$X_i^* = w_i X \quad \text{where} \quad w_i = \frac{a}{a_i}, \quad \frac{1}{a} = \sum_{i=1}^{n} \frac{1}{a_i} \tag{4.1}$$

(note that $w_1 + \cdots + w_n = 1$).

According to Sect. I.3, we have

$$H = \sum_{i=1}^{n} \log \mathbb{E}e^{a_i X_i}/a_i, \quad H^* = \sum_{i=1}^{n} \log \mathbb{E}e^{a_i w_i X}/a_i = \frac{1}{a} \log \mathbb{E}e^{aX},$$

where H is the premium under any arrangement and H^* the premium under (4.1). Thus the claim $H^* \leq H$ is equivalent to

$$\big[\mathbb{E}e^{aX}\big]^{1/a} \leq \prod_{i=1}^{n} \big[\mathbb{E}e^{a_i X_i}\big]^{1/a_i},$$

i.e. to

$$\mathbb{E}\big[Y_1 \cdots Y_n\big] \le \prod_{i=1}^{n} \big[\mathbb{E}e^{a_i X_i}\big]^{w_i} = \prod_{i=1}^{n} \big[\mathbb{E}Y_i^{p_i}\big]^{1/p_i} = \prod_{i=1}^{n} \|Y_i\|_{p_i},$$

where $Y_i = e^{a X_i}$ (note that $Y_1 \cdots Y_n = e^{a X}$), $p_i = 1/w_i = a_i/a$. But this is just Hölder's inequality.   ◇

4.2   Utility and Reinsurance

In this subsection, we consider a partitioning of a risk X into two parts $r(X), s(X)$ such that $X = r(X) + s(X)$. More precisely, r, s are assumed non-decreasing (unless otherwise stated) with

$$r(x) \ge 0, \ s(x) \ge 0, \quad x = r(x) + s(x), \quad x \ge 0. \tag{4.2}$$

There are two main examples. The first is reinsurance in the setting of Sect. I.4, where $r(X)$ is paid by the reinsurer and $s(X)$ by the cedent (one may memorize the s in $s(x)$ as standing for *self*). The second is a traditional insurance arrangement (fire insurance of a private home, damage to own vehicle, etc.) which usually involves a *deductible*. This means that the insured does not get the whole claim X covered, but only a part $r(X)$.

We have already encountered the common examples in Sect. I.4. They are:

- $r(x) = (x - b)^+$, going under the name *stop-loss reinsurance* resp. *fixed deductible* in the two settings,
- $r(x) = (1 - k)x$, called *proportional reinsurance*, resp. *proportional deductible*,
- $r(x) = x \mathbb{1}_{x>b}$, called *franchise deductible* but also relevant in reinsurance.

In the following, we use reinsurance language, but in view of the above remarks, the discussion also applies to analysis of the interplay between an insured and the company. We let u be the utility function of the cedent and u^* that of the reinsurer, w, w^* their initial wealths. The respective utility premiums H, H^* for a given risk X are then given by

$$\mathbb{E}u(w - X) = \mathbb{E}u\big(w - H - s(X)\big), \tag{4.3}$$

$$u^*(w^*) = \mathbb{E}u^*\big(w^* + H^* - r(X)\big). \tag{4.4}$$

Proposition 4.4 $H^* \ge \mathbb{E}r(X)$ *and, provided* $r(\cdot)$ *is continuous,* $H \ge \mathbb{E}r(X)$.

Proof That $H^* \ge \mathbb{E}r(X)$ follows immediately from the observations in Sect. I.2. To show $H \ge \mathbb{E}r(X)$, use the continuity of $r(\cdot)$ to choose x_0 with $r(x_0) = \mathbb{E}r(X)$, and let $f_1(x) = r(x_0) + s(x)$, $f_2(x) = r(x) + s(x) = x$. Then r being non-decreasing implies that $f_1(x) \geq f_2(x)$ for $x < x_0$ and $f_1(x) \leq f_2(x)$ for $x \geq x_0$. The conclusion of Corollary 2.8 then means $\mathbb{E}r(X) + s(X) \preceq_{\mathrm{cx}} r(X) + s(X) = X$, which gives

$$\mathbb{E}u\big(w - \mathbb{E}r(X) - s(X)\big) \;\geq\; \mathbb{E}u(w - X) \,.$$

This implies $H \geq \mathbb{E}r(X)$.   □

4.3   Local Versus Global Reinsurance

We next consider the situation where $X = V_1 + \cdots + V_N$ is a compound sum. This is of obvious relevance in reinsurance. One possibility is that the V_i are individual claims in a particular line of business, say fire insurance of private homes, another that the V_i are the total claims in different lines of business (fire insurance of private homes, fire insurance of commercial buildings, car insurance, boat insurance, etc). As an alternative to an arrangement with retention $R = r(X)$ (*global cover*), the reinsurance company could then consider an arrangement with retention $r_i(V_i)$ of claim i. Thus the reinsurer covers $R_{\mathrm{loc}} = \sum_1^N r_i(V_i)$ and the cedent $X - R_{\mathrm{loc}} = \sum_1^N s_i(V_i)$. We refer to this as *local cover*. The following observation shows that for a given premium H and a given local cover, there is a global cover with the same expected value which is preferable to both the cedent and the reinsurer, namely

$$R = R(X) = \mathbb{E}\big[R_{\mathrm{loc}} \,\big|\, X\big]. \tag{4.5}$$

Proposition 4.5 *Let R be given by* (4.5). *Then for all fixed H,*

$$\mathbb{E}u\big(w - H - (X - R)\big) \;\geq\; \mathbb{E}u\big(w - H - (X - R_{\mathrm{loc}})\big),$$
$$\mathbb{E}u^*(w^* + H - R) \;\geq\; \mathbb{E}u^*\big(w + H - R_{\mathrm{loc}}\big).$$

Proof Immediate from part (iv) of Proposition 4.1 with $\mathbb{E}[\cdot]$ replaced by $\mathbb{E}[\cdot \mid X]$.   □

Corollary 4.6 *The utility premiums H_{glob}, H_{glob}^* and H_{loc}, H_{loc}^* for global, resp. local, cover determined by* ·

$$\mathbb{E}u\big(w - H_{\mathrm{loc}} - (X - R_{\mathrm{loc}})\big) = \mathbb{E}u(w - X) = \mathbb{E}u\big(w - H_{\mathrm{glob}} - (X - R)\big),$$
$$\mathbb{E}u^*\big(w^* + H_{\mathrm{loc}}^* - R_{\mathrm{loc}}\big) = \mathbb{E}u^*(w^*) = \mathbb{E}u^*\big(w^* + H_{\mathrm{glob}}^* - R_{\mathrm{glob}}\big)$$

satisfy $H_{\mathrm{glob}} \geq H_{\mathrm{loc}}$ and $H_{\mathrm{glob}}^* \geq H_{\mathrm{loc}}^*$.

Proof Take $H = 0$ in Proposition 4.5 and note that if Z_1, Z_2 are risks with $\mathbb{E}u(w + Z_1) \geq \mathbb{E}u(w + Z_2)$, then the corresponding utility premiums H_1, H_2 satisfy $H_1 \geq H_2$. □

Now assume the reinsurer applies the premium principle $H(Z)$. We can then ask whether the global cover R in (4.5) or local cover is preferable. The cedent will prefer the global cover if

$$H(R) \leq \sum_{i=1}^{N} H\big(r_i(V_i)\big). \tag{4.6}$$

The reasons for this is that (4.6) ensures that not only is the premium to be paid with global cover smaller but also the global cover gives better expected utility, i.e.

$$\mathbb{E}u\big(w - H(R) - (X - R)\big) \geq \mathbb{E}u\Big(w - \sum H\big(r_i(V_i)\big) - \big(X - \sum r_i(V_i)\big)\Big)$$

(as follows from (4.6) combined with Proposition 4.5).

Consider first the expected value principle $H(X) = (1 + \eta)\mathbb{E}X$. Then the premiums charged for the local cover and the global cover are the same, so (4.6) holds and the global cover is preferable because of its better expected utility.

Consider next the variance principle $H(Z) = \mathbb{E}Z + \eta \operatorname{Var} Z$. For independent risks Z_1, Z_2, we have $H(Z_1 + Z_2) = H(Z_1) + H(Z_2)$ and hence

$$\sum_{i=1}^{N} H\big(r_i(V_i)\big) = H\Big(\sum_{i=1}^{N} r_i(V_i)\Big) = \mathbb{E}R_{\text{loc}} + \eta \operatorname{Var}\Big(\sum_{i=1}^{N} r_i(V_i)\Big)$$

$$\geq \mathbb{E}R_{\text{loc}} + \eta \operatorname{Var} \mathbb{E}\Big[\sum_{i=1}^{N} r_i(V_i) \,\big|\, X\Big]$$

$$= \mathbb{E}r(X) + \eta \operatorname{Var}\big(R\big) = H\big(r(X)\big).$$

Thus again (4.6) holds and the global cover is preferable.

4.4   Optimal Compensation Functions

We are concerned with comparing two insurance arrangements given by r_0, s_0 and r, s so that $r_0(X)$, $r(X)$ are the compensations to the cedent/the insured in the two situations. It is assumed throughout that both r and s are non-decreasing functions, that $r(x) + s(x) = x$ (and similarly for r_0, s_0), and that the insurer applies the expected value principle so that the premium H is the same when $\mathbb{E}r_0(X) = \mathbb{E}r(X)$ or, equivalently, when $\mathbb{E}s_0(X) = \mathbb{E}s(X)$. The utility function of the insured is denoted by u and that of the insurer by u^*; the initial wealths are w, w^*. Both u and u^* are assumed to be non-decreasing concave functions.

Many of the results are concerned with finding optimal solutions in terms of utility. Since expected values are fixed, this can be approached by finding minimal elements in the convex ordering. The common route to the various results follow the following outline:

Proposition 4.7 *Assume that* $\mathbb{E}s_0(X) = \mathbb{E}s(X)$ *and that there exists an* x_0 *such that* $s_0(x) \geq s(x)$ *for* $x < x_0$ *and* $s_0(x) \leq s(x)$ *for* $x > x_0$. *Then* $s_0(X) \preceq_{cx} s(X)$ *and* $r_0(X) \succeq_{cx} r(X)$. *In particular, for all* H *we have*

$$\mathbb{E}u\big(w - H - s_0(X)\big) \geq \mathbb{E}u\big(w - H - s(X)\big), \tag{4.7}$$

$$\mathbb{E}u^*\big(w^* + H - r_0(X)\big) \leq \mathbb{E}u^*\big(w^* + H - r(X)\big). \tag{4.8}$$

The result shows that the insurer's preferences do not coincide with those of the insured; the intuitive reason is that they both fear the tail of the risk. In particular, if s_0 satisfies the condition for any other s, then it is a majorant for the insured in utility, but a minorant for the insurer!

Proof That $s_0(x) \preceq_{cx} s(x)$ follows immediately from Corollary 2.8. Since $r(x) + s(x) = x$ and similarly for r_0, we also have $r_0(x) \leq r(x)$ for $x < x_0$ and $r_0(x) \geq r(x)$ for $x > x_0$. One more application of Corollary 2.8 gives $r_0(X) \succeq_{cx} r(X)$.   □

The first main result of this section shows that stop-loss compensation is optimal for the insured (in the case of reinsurance, add 'in the class of global reinsurance arrangements'):

Theorem 4.8 *Consider stop-loss compensation,* $r_0(x) = (x - b)^+$, *and any other compensation function* $r(x)$ *with* $\mathbb{E}r(X) = \mathbb{E}r_0(X) = \mathbb{E}(X - b)^+$. *Then* $s_0(X) \preceq_{cx} s(X)$ *where* $s_0(x) = x - r_0(x) = x \wedge b$, $s(x) = x - r(x)$. *In particular, if the premium* H *is the same in the two situations, then stop-loss gives the better utility for the cedent, i.e.* (4.7) *holds, and the lower utility for the reinsurer, i.e.* (4.8) *holds.*

Note also that given $r(x)$, one can always choose b such that $\mathbb{E}r_0(X) = \mathbb{E}r(X)$ when $X \geq 0$ a.s. This follows since $\mathbb{E}(X - b)^+$ is decreasing in b with limits $\mathbb{E}X$ and 0 at $b = 0$, resp. $b = \infty$.

Proof We show that the conditions of Proposition 4.7 are satisfied. Indeed, for $x < b$

$$s_0(x) = x \wedge b = x \geq s(x).$$

For $x > b$, $s(x) - s_0(x) = s(x) - b$ is non-decreasing and the limit as $x \to \infty$ cannot be ≤ 0 since $s(x) \leq s_0(x)$ for all x is impossible (unless $s = s_0$) because the expected values are the same. The claim now follows by graphical inspection.   □

Theorem 4.9 *Let* $r(X)$ *be the insured part of a claim* X. *Consider first risk compensation,* $r_0(x) = x \wedge b$, *and any other compensation function* $r(x)$ *with*

$\mathbb{E}r(X) = \mathbb{E}r_0(X) = \mathbb{E}[X \wedge b]$. Then $r_0(X) \preceq_{cx} r(X)$ and $s(X) \preceq_{cx} s_0(X)$. In particular, if the premium H is the same in the two situations, then first risk gives the lower utility for the cedent, i.e.

$$\mathbb{E}u\big(w - H - s_0(X)\big) \le \mathbb{E}u\big(w - H - s(X)\big),$$

and the higher utility for the reinsurer, i.e.

$$\mathbb{E}u^*\big(w^* + H - r_0(X)\big) \ge \mathbb{E}u^*\big(w^* - H - r(X)\big).$$

Proof Just note that s_0, s are the r_0, r of Theorem 4.8 and r_0, r are the s_0, s.   □

A compensation function r is called a *Vajda function* if $r(x)/x$ is non-decreasing; this is a natural requirement in many contexts. Then for the reinsurer, proportional reinsurance $s(x) = bx$ is preferable to any other arrangement with a Vajda compensation function:

Theorem 4.10 Let $r(X)$ be the insured part of a claim $X \ge 0$. Consider proportional compensation, $r_0(x) = (1 - b)x$, and any other Vajda compensation function $r(x)$ with $\mathbb{E}r(X) = \mathbb{E}r_0(X) = (1 - b)\mathbb{E}X$. Then $r_0(X) \preceq_{cx} r(X)$. In particular, if the premium H is the same in the two situations, then proportional compensation gives the better utility for the insurer,

$$\mathbb{E}u^*\big(w^* + H - r_0(X)\big) \ge \mathbb{E}u^*\big(w^* + H - r(X)\big).$$

Remark 4.11 A summary of Theorems 4.8, 4.10 is: first risk compensation ($r(x) = x \wedge b$) is always optimal for the insurer, but in comparison with a Vajda compensation function $r(x)$, proportional reinsurance does at least as well as $r(x)$, though not quite as good as first risk.   ◇

Proof Assume that x_1 satisfies $r(x_1) \le r_0(x_1)$. Since r is Vajda, we have for $0 < x \le x_1$ that

$$r(x) = x\frac{r(x)}{x} \le x\frac{r(x_1)}{x_1} \le x\frac{r_0(x_1)}{x_1} = x(1 - b) = r_0(x).$$

Hence $r(x) \le r_0(x)$ for $x < x_0$ and $r(x) \ge r_0(x)$ for $x > x_0$ where $x_0 = \sup\{x_1 : r(x_1) \le r_0(x_1)\}$. Now appeal to Proposition 4.7.   □

Exercises

4.1 Show that (4.6) holds for the standard deviation principle.

4.2 Show that first risk compensation $r(x) = x \wedge b$ is not a Vajda function.

4.3 Show that if $0 \le s(x) \le x$, $\mathbb{E}(X \wedge K)^2 = \mathbb{E}s(X)^2$, then $\mathbb{E}s(X) \le \mathbb{E}(X \wedge K)$. Hint: choose $\widetilde{K}$ with $\mathbb{E}s(X) = \mathbb{E}(X \wedge \widetilde{K})$, show that $X \wedge \widetilde{K} \preceq_{cx} s(X)$ and that $X \wedge K \le X \wedge \widetilde{K}$ a.s.

Notes
The books Kaas et al. [97] and Albrecher et al. [3] contain an extensive amount of material related to the discussion of this section.

5 Applications to Ruin Theory

Consider as in Chap. IV the Cramér–Lundberg model for the reserve of an insurance company, with fixed rates β of arrival of claims and c of premium inflow. Given two claim size distributions $B^{(1)}$, $B^{(2)}$, we may ask which one carries the larger risk in the sense of larger values of the ruin probability $\psi^{(i)}(u)$ for a fixed value of β.

Proposition 5.1 *If $B^{(1)} \preceq_{\mathrm{st}} B^{(2)}$, then $\psi^{(1)}(u) \le \psi^{(2)}(u)$ for all u.*

Proof Let for $i = 1, 2$ $S^{(i)}(t)$ be the claim surplus processes and $\tau^{(1)}(u) = \inf\{t : S^{(i)}(t) > u\}$ the ruin times. According to the a.s. characterization III) of stochastic ordering, we can assume that $S^{(1)}(t) \le S^{(2)}(t)$ for all t. In terms of the time to ruin, this implies $\tau^{(1)}(u) \ge \tau^{(2)}(u)$ for all u so that $\{\tau^{(1)}(u) < \infty\} \subseteq \{\tau^{(2)}(u) < \infty\}$. Taking probabilities, the proof is complete.   □

Of course, Proposition 5.1 is quite trivial, and a particular weakness is that we cannot compare the risks of claim size distributions with the same mean: if $B^{(1)} \preceq_{\mathrm{st}} B^{(2)}$ and $\mu_{B^{(1)}} = \mu_{B^{(2)}}$, then $B^{(1)} = B^{(2)}$. Here convex ordering is useful:

Proposition 5.2 *If $B^{(1)} \preceq_{\mathrm{cx}} B^{(2)}$, then $\psi^{(1)}(u) \le \psi^{(2)}(u)$ for all u.*

Proof Since the means of $B^{(1)}$, $B^{(2)}$ are equal, say to μ, and the stop-loss transforms are ordered, the integrated tails satisfy

$$\overline{B_0^{(1)}}(x) = \frac{1}{\mu} \int_x^\infty \overline{B^{(1)}}(y)\,dy \le \frac{1}{\mu} \int_x^\infty \overline{B^{(2)}}(y)\,dy = \overline{B_0^{(2)}}(x),$$

i.e., $B_0^{(1)} \preceq_{\mathrm{st}} B_0^{(2)}$, which implies the same order relation for all convolution powers. Hence by the compound geometric sum representation in Corollary IV.3.2, we have with $\rho = \beta\mu/c$ that

$$\psi^{(1)}(u) = (1-\rho) \sum_{n=1}^\infty \rho^n \overline{B_0^{(1)*n}}(u) \le (1-\rho) \sum_{n=1}^\infty \rho^n \overline{B_0^{(2)*n}}(u) = \psi^{(2)}(u).$$

$$(5.9)$$

□

A general picture that emerges from these results is that (in a rough formulation) *increased variation in B increases the risk* (assuming that we fix the mean). The problem is to specify what 'variation' means. A first attempt would of course be to identify 'variation' with variance. The diffusion approximation in Sect. IV.8 certainly supports this view: noting that, with fixed mean/drift, larger variance is paramount to larger second moment, it is seen that (cf. Theorem IV.8.3) in the diffusion limit, larger variance of the claim size leads to larger ruin probabilities. Here is one more (trivial!) result of the same flavor:

Corollary 5.3 *Let D refer to the distribution degenerate at μ_B. Then $\psi^{(D)}(u) \leq \psi^{(B)}(u)$ for all u.*

Proof Just note that $D \preceq_{cx} B$.   □

A partial converse to Proposition 5.2 is the following:

Proposition 5.4 *Assume $\mu_{B^{(1)}} = \mu_{B^{(2)}}$ and $\psi^{(1)}(u) \leq \psi^{(2)}(u)$ for all u, c and β. Then $B^{(1)} \preceq_{cx} B^{(2)}$.*

Proof We use what is called a light traffic approximation in queueing theory and which in insurance risk amounts to letting the premium rate c go to ∞. Let the claim size distribution be fixed at B for the moment, and recall the geometric sum representation (5.9). Here $\rho \to 0$ as $c \to \infty$ and hence for fixed u

$$\psi(u) \sim \rho \overline{B}_0(u) = \frac{\beta}{c} \int_u^\infty \overline{B}(y)\,dy = \frac{\beta}{c} \pi_B(u),$$

where π_B is the stop-loss transform. Applying this to $B^{(1)}$, $B^{(2)}$ gives the ordering $\pi_{B^{(1)}}(u) \leq \pi_{B^{(2)}}(u)$ for all u of the stop-loss transforms, so that property II) of Theorem 2.4 holds.   □

6 Maximizing the Adjustment Coefficient

We consider the problem of maximizing the adjustment coefficient γ in the Cramér–Lundberg model with fixed arrival rate β and fixed premium rate c. This is basically just a problem of maximizing a function of one variable and therefore somewhat more elementary than minimizing the ruin probability. We first note:

Proposition 6.1 *If $B^{(1)} \preceq_{cx} B^{(2)}$, then the corresponding adjustment coefficients satisfy $\gamma_2 \leq \gamma_1$.*

Proof Recall that $\kappa_i(\alpha) = \beta(\widehat{B}^{(i)}[\alpha] - 1) - c\alpha$ and that $\kappa_i(\gamma_i) = 0$. But $B^{(1)} \preceq_{cx} B^{(2)}$ implies that $\kappa_1 \leq \kappa_2$, in particular that $\kappa_2(\gamma_1) \geq \kappa_1(\gamma_1) = 0$, which by graphical inspection is only possible if $\gamma_2 \leq \gamma_1$.   □

In the following, we consider a set-up where the part $r_\zeta(V)$ paid by the reinsurer and $s_\zeta(V) = V - r_\zeta(V)$ depend on a parameter ζ taking values in some interval $[\underline{\zeta}, \overline{\zeta}]$, such that $\zeta = \underline{\zeta}$ corresponds to the case $s_\zeta(V) = 0$ of full reinsurance (everything is covered by the reinsurer) and the case $s_{\overline{\zeta}}(V) = V$ to no reinsurance. The examples to be studied are proportional reinsurance, where we write $\zeta = b$ so that $s_b(V) = bV$, $r_b(V) = (1 - b)V$ and the interval $[\underline{\zeta}, \overline{\zeta}]$ becomes $[0, 1]$, and excess-of-loss reinsurance, where we write $\zeta = a$ so that $s_a(V) = V \wedge a$,

$r_a(V) = (V - a)^+$ and $[\underline{\zeta}, \overline{\zeta}] = [0, \infty]$. We further assume that both the cedent and the reinsurer use the expected value principle with loadings η_s, resp. η_r, and that reinsurance is non-cheap, i.e. $\eta_r > \eta_s$. Thus, the claims surplus process is

$$S(t; \zeta) = \sum_{i=1}^{N(t)} s_\zeta(V_i) - ct + (1 + \eta_r)\beta t \, \mathbb{E} r_\zeta(V),$$

where $c = (1 + \eta_s)\beta\mu_B$. The drift $\mathbb{E}S(1; \zeta)$ is

$$\beta\mathbb{E}s_\zeta(V) - (1 + \eta_s)\beta\mu_B + (1 + \eta_r)\beta\mathbb{E}r_\zeta(V) = \beta[\eta_s\mathbb{E}r_\zeta(V) - \eta_s\mu_B] \qquad (6.1)$$

and the adjustment coefficient $\gamma(\zeta)$ is the positive zero of

$$\beta[\mathbb{E}e^{\gamma s_\zeta(V)} - 1] - c\gamma + (1 + \eta_r)\beta \, \mathbb{E}r_\zeta(V)\gamma \qquad (6.2)$$

provided such a solution exists.

We note that with non-cheap reinsurance, full reinsurance will lead to the positive value $(\eta_r - \eta_s)\beta\mu_B$ of (6.1) corresponding to ruin probability $\equiv 1$ and thus cannot be optimal.

Proposition 6.2 *Consider proportional reinsurance with retention $b \in [0, 1]$. Assume that the cedent and the reinsurer both use the expected value principle with safety loadings η_s, resp. η_r, where $\eta_s < \eta_r$. Let $\gamma(b)$ be the adjustment coefficient of the cedent for a given b. Assume further that a solution h of $\widehat{B}'[h] = (1 + \eta_r)\mu_B$ exists and define*

$$b_0 = \frac{h(\eta_r - \eta_s)\mu_B}{h(1 + \eta_r) - \widehat{B}[h] + 1}.$$

Then $\gamma^* = \sup_{b \in [0, 1]} \gamma(b)$ equals h/b_0 provided $b_0 \leq 1$.

Proof The rate of cash inflow to the cedent is the premium rate $(1 + \eta_s)\beta\mu_B$ minus the rate $(1 + \eta_r)(1 - b)\beta\mu_B$ of payment to the reinsurer, in total $[b(1 + \eta_r) - (\eta_r - \eta_s)]\beta\mu_B$. Thus his claim surplus process is

$$\sum_{i=n1}^{N(t)} bV_i - [b(1 + \eta_r) - (\eta_r - \eta_s)]\beta\mu_B \cdot t$$

and $\gamma(b)$ solves $\Psi(b, \gamma(b)) = 0$, where

$$\Psi(b, \gamma) = \beta(\widehat{B}[b\gamma] - 1) - [b(1 + \eta_r) - (\eta_r - \eta_s)]\beta\mu_B \cdot \gamma.$$

Thinking of Ψ as defined on the extended domain $b \in [0, \infty)$, $\gamma \in \mathbb{R}$, we have

$$0 = \frac{d}{db}\Psi(b, \gamma(b)) = \Psi_b(b, \gamma(b)) + \Psi_\gamma(b, \gamma(b))\gamma'(b)$$

(partial derivatives). A solution $b^\#$ of $\gamma'(b^\#) = 0$ must therefore satisfy

$$0 = \Psi_b(b^\#, \gamma^\#) = \beta\gamma^\# \widehat{B}'[b^\# \gamma^\#] - (1 + \eta_r)\beta\mu_B\gamma^\#,$$

where $\gamma^\# = \gamma(b^\#)$. This is the same as $\widehat{B}'[b^\# \gamma^\#] = (1 + \eta_r)\mu_B$, i.e. $b^\# \gamma^\# = h$. Using $\Psi(b^\#, \gamma^\#) = 0$ then gives

$$0 = \beta(\widehat{B}[h] - 1) - \left[h(1 + \eta_r) - (\eta_r - \eta_s)\gamma^\#\right]\beta\mu_B,$$

$$\gamma^\# = \frac{h(1 + \eta_r) - \widehat{B}[h] + 1}{(\eta_r - \eta_s)\mu_B},$$

$$b^\# = \frac{h}{\gamma^\#} = \frac{h(\eta_r - \eta_s)\mu_B}{h(1 + \eta_r) - \widehat{B}[h] + 1}.$$

□

We next consider excess-of-loss reinsurance with retention level a. If again the reinsurer uses the expected value principle with loading $\eta_r > \eta_s$, the claim surplus process is

$$S(t; a) = \sum_{i=1}^{N(t)} V_i \wedge a - ct + (1 + \eta_r)t\beta\mathbb{E}(V - a)^+.$$

For the following, note that

$$\mathbb{E}e^{\alpha(V \wedge a)} = 1 + \alpha\int_0^a e^{\alpha x}\overline{B}(x)\,dx, \quad \mathbb{E}(V - a)^+ = \int_a^\infty \overline{B}(x)\,dx,$$

using integration by parts. Since the claims are bounded, an adjustment coefficient $\gamma(a) > 0$ will exist as long as

$$0 > \mathbb{E}S(1; a) = \beta\left(\mu_B - \int_a^\infty \overline{B}(x)\,dx\right) - \beta\mu_B(1 + \eta_s) + (1 + \eta_r)\beta\int_a^\infty \overline{B}(x)\,dx,$$

i.e. $a > a_0$ where a_0 is the solution of

$$\int_{a_0}^\infty \overline{B}(x)\,dx = \frac{\eta_s}{\eta_r}\mu_B; \tag{6.3}$$

it easily follows from $\eta_s < \eta_r$ that there is a unique solution of this equation in $(0, \infty)$. At a_0, the drift is 0 and so obviously $\gamma(a_0) = 0$. The limit $a = \infty$ corresponds to no reinsurance and we write then $\gamma(\infty)$ for the adjustment coefficient. For $a_0 < a < \infty$, $\gamma(a)$ is computed as the non-zero solution of $\kappa(\gamma; a) = 0$, where $\kappa(\theta; a) = \log \mathbb{E} e^{\theta S(1;a)}$, or equivalently as the solution of

$$0 = \frac{\kappa(\gamma; a)}{\gamma\beta} = \int_0^a e^{\alpha x}\overline{B}(x)\,dx - (1+\eta_s)\mu_B + (1+\eta_r)\int_a^\infty \overline{B}(x)\,dx. \qquad (6.4)$$

Proposition 6.3 *Consider excess-of-loss reinsurance with retention level* $a \in [a_0, \infty]$, $\eta_s < \eta_r$, $c = (1+\eta_s)\beta\mu_B$, *and assume that either* (a) *B is light-tailed and* $\gamma(\infty)$ *exists and is strictly positive, or* (b) *B is heavy-tailed with* $\widehat{B}[\theta] = \infty$ *for all* $\theta > 0$. *Then* $\gamma(a)$ *attains its maximum* $\gamma(a^*)$ *at some* a^* *in the of interior* (a_0, ∞). *Further, the equation*

$$0 = \beta \int_0^{\log(1+\eta_r)/\gamma} e^{\gamma x}\overline{B}(x)\,dx - c + (1+\eta_r)\beta \int_{\log(1+\eta_r)/\gamma}^\infty \overline{B}(x)\,dx \qquad (6.5)$$

has at least one solution in $(0, \infty)$, *and* γ^* *is the largest such solution. The optimal retention level* a^* *equals* $\log(1+\eta_r)/\gamma^*$.

Proof The Lundberg equation is

$$0 = \beta\gamma \int_0^a e^{\gamma x}\overline{B}(x)\,dx - \left(c - (1+\eta_r)\beta \int_a^\infty \overline{B}(x)\,dx\right)\gamma \qquad (6.6)$$

$$= \beta \int_0^a e^{\gamma x}\overline{B}(x)\,dx - c + (1+\eta_r)\beta \int_a^\infty \overline{B}(x)\,dx, \qquad (6.7)$$

where $\gamma = \gamma(a)$ for ease of notation. Differentiation w.r.t. a gives

$$0 = \beta e^{\gamma a}\overline{B}(a) + \beta\gamma' \int_0^a x e^{\gamma x}\overline{B}(x)\,dx - (1+\eta_r)\beta\overline{B}(a),$$

$$\gamma'(a) = \frac{(1+\eta_r)\overline{B}(a) - e^{\gamma a}\overline{B}(a)}{\int_0^a x e^{\gamma x}\overline{B}(x)\,dx}. \qquad (6.8)$$

Equating this to 0 and dividing by $\overline{B}(a)$ gives $0 = 1 + \eta_r - e^{\gamma^* a^*}$, so that a^* and γ^* are indeed connected as claimed. We can then rewrite (6.7) as

$$0 = \beta \int_0^{\log(1+\eta_r)/\gamma^*} e^{\gamma^* x}\overline{B}(x)\,dx - c + (1+\eta_r)\beta \int_{\log(1+\eta_r)/\gamma^*}^\infty \overline{B}(x)\,dx$$

so that γ^* indeed solves (6.5) provided the maximum is not attained at one of the boundary points a_0 or ∞.

To exclude this, consider first the behavior at the left boundary a_0. Applying monotone convergence to $\mathbb{E}e^{\theta(V \wedge a)}$ and $\mathbb{E}(V - a)^+$ shows that $\kappa(\theta; a) \to \kappa(\theta; a_0)$ as $a \downarrow a_0$. Here $\kappa(\theta; a_0) > 0$ for any $\theta > 0$ since $\kappa'(a_0) = 0$ by construction. Inspection of the graphs gives that $\gamma(a) \leq \theta$ for a sufficiently close to a_0. Letting $\theta \downarrow 0$ gives $\gamma(a) \to 0$, $a \downarrow a_0$. Since $\gamma(a) > 0$ in the interior, a_0 therefore cannot be a maximizer.

At the right boundary ∞, we get similarly that $\kappa(\theta; a) \to \kappa(\theta)$ as $a \uparrow \infty$. In case (b), we therefore have $\kappa(\gamma^*/2; a) > 0$ for all large a which implies $\gamma(a) < \gamma^*/2$ so that the maximum cannot be attained at $a = \infty$. In case (a), we get $\gamma(a) \to \gamma(\infty) > 0$. The expression (6.8) then shows that $\kappa'(a) < 0$ and hence $\kappa(a) > \gamma(\infty)$ for all large a, and so again ∞ cannot be a maximizer. $\qquad\qquad\square$

Notes

For a general perspective on reinsurance and maximization of the adjustment coefficient, see Centeno and Guerra [47] and references there. Proposition 6.2 is from Hald and Schmidli [85], whereas Proposition 6.3 may not be in the literature in precisely this form.

Chapter IX: Extreme Value Theory

1 Introduction

Predicting the typical order of minima or maxima is important in a number of applications. In agriculture, we may want to say something about the maximal content of a pesticide in a series of samples to check if the series passes the standards. The order of the maximal sea level during the next 500 years is important for dimensioning dikes. The expected size of the maximal earthquake in the next 50 years plays a role when choosing between different types of construction of houses. A sports fan is keen to know how much the 100 m record can be expected to improve during the next 10 years, i.e. what will be the minimum time recorded. And so on.

The study of such extremes is the topic of this chapter. Mathematically, we consider a sequence $X_1, X_2, \ldots$ of random variables and are interested in properties of

$$M_n = \max(X_1, \ldots, X_n).$$

Note that it is sufficient to consider maxima instead of minima because of the simple relation

$$\min(X_1, \ldots, X_n) = -\max(-X_1, \ldots, -X_n).$$

Throughout the chapter, the X_i will be i.i.d. with common distribution F. The exact distribution of M_n is then easily available: the c.d.f. of M_n is

$$\mathbb{P}(M_n \leq x) = \mathbb{P}(X_1 \leq x, \ldots, X_n \leq x) = F^n(x).$$

Our concern will be to obtain more summary information, for example the growth rate or asymptotic distribution of M_n. It is important to distinguish between the theory of extreme values and that of rare events studied elsewhere in this book. Extreme value theory, the topic of this chapter, is about the *typical* behavior of the

© Springer Nature Switzerland AG 2020
S. Asmussen, M. Steffensen, *Risk and Insurance*, Probability Theory
and Stochastic Modelling 96, https://doi.org/10.1007/978-3-030-35176-2_9

extremes as defined by the maxima and minima, not about calculating probabilities of order say 10^{-5} or less. For example, when studying the risk of flooding of dikes in the next 50 years we will get qualified answers to questions such as what is the median or the 90% quantile of the maximal water level, but not to what is the risk of a catastrophe, which would normally only occur in a time span of hundreds or thousands of years.

Exercises

1.1 Assuming F has a density f, find the density of M_n and the joint density of M_n and second largest X, i.e. $\max\{X_k : k \leq n, \ X_k < M_n\}$.

Notes

Some main textbook references for extreme value theory are Leadbetter, Lindgren and Rootzén [106], Resnick [148], Embrechts, Klüppelberg and Mikosch [66], and de Haan and Ferreira [72]. Virtually all the results and proofs of Sects. 2–4 in this chapter can be found in these references.

2   Elementary Examples and Considerations

The development of a general theory of asymptotic properties of M_n is the main topic of this chapter and is implemented in Sects. 3–4. As an introduction, some examples are derived in this section using only elementary tools.

The topic is concerned with convergence of sequences of the form $\mathbb{P}(M_n \leq u_n)$. Namely, if we want to show that $\Phi_n(M_n)$ has a limit in distribution, the relevant choice is $u_n = \Phi_n^{-1}(x)$ for continuity points x of the limit. To this end, the following simple observation is extremely useful (note that any limit in $(0, \infty)$ can be written as $e^{-\tau}$).

Proposition 2.1 $\mathbb{P}(M_n \leq u_n) \to e^{-\tau} \quad \Longleftrightarrow \quad n\overline{F}(u_n) \to \tau.$

Proof Just note that

$$\log \mathbb{P}(M_n \leq u_n) \ = \ \log\left(1 - \overline{F}(u_n)\right)^n \ \sim \ -n\overline{F}(u_n)$$

if $\overline{F}(u_n) \to 0$.   □

The case of a standard uniform F will play a special role, so we introduce a special notation for it: if $U_1, \ldots, U_n$ are i.i.d. uniform$(0, 1)$, we write $M_n^U = \max(U_1, \ldots, U_n)$.

Example 2.2 Consider the standard uniform case $F(u) = u$, $0 < u < 1$. Then $\overline{F}(u) = 1 - u$ for $0 < u < 1$, and to get a limit τ of $n\overline{F}(u_n)$, the choice $u_n = 1 - y/n$ will work to get the limit $\tau = y$ of $n\overline{F}(u_n)$ corresponding to the limit $e^{-\tau} = e^{-y}$ of $F^n(1 - y/n)$. This means that

$$\mathbb{P}\left(n(1 - M_n^U) \geq y\right) \ = \ \mathbb{P}(M_n^U \leq 1 - y/n) \ = \ F^n(1 - y/n) \ \to \ e^{-y} \ = \ \mathbb{P}(V \geq y),$$

where V is standard exponential, i.e., $n(1 - M_n^U) \overset{D}{\longrightarrow} V$, which we informally rewrite as $M_n^U \sim 1 - V/n$ or $M_n^U \approx 1 - V/n$.

A direct proof without reference to Proposition 2.1 is of course only a small variant:

$$\mathbb{P}(M_n^U \leq u_n) = F^n(u_n) = (1 - y/n)^n \to e^{-y}. \qquad \lozenge$$

Remark 2.3 The uniform$(0,1)$ distribution is an example of a distribution with $x^* < \infty$, where $x^* = x_F^* = \inf\{x : F(x) = 1\}$ is the upper endpoint of the support. Example 2.2 illustrates the general principle that (no matter $x^* = \infty$ or $x^* < \infty$) one has

$$M_n \overset{\mathbb{P}}{\to} x^*. \qquad (2.1)$$

This follows since for $x < x^*$ one has $F(x) < 1$ and hence

$$\mathbb{P}(M_n \notin (x, x^*]) = \mathbb{P}(M_n \leq x) = F(x)^n \to 0. \qquad \lozenge$$

Example 2.4 Let F be Pareto with $F(x) = 1 - 1/x^\alpha$ for $x \geq 1$, with $\alpha > 0$. To get a limit of $F^n(u_n)$ or equivalently of $n\overline{F}(u_n)$, we can take $u_n = yn^{1/\alpha}$ since then

$$n\overline{F}(u_n) = n \cdot \frac{1}{[yn^{1/\alpha}]^\alpha} \to \frac{1}{y^\alpha}, \qquad F_n(u_n) = \left(1 - \frac{1}{[yn^{1/\alpha}]^\alpha}\right)^n \to e^{-1/y^\alpha}.$$

This means that $M_n/n^{1/\alpha}$ converges in distribution to an r.v. Y with c.d.f. $G(y) = e^{-1/y^\alpha}$, $y > 0$. This distribution is known as the *Fréchet distribution*. $\qquad \lozenge$

Example 2.5 Let F be standard exponential, $F(x) = 1 - e^{-x}$ for $x > 0$. To get a limit of $F^n(u_n)$ or equivalently of $n\overline{F}(u_n)$, we take u_n such that $e^{-u_n} = z/n$. Then

$$n\overline{F}(u_n) = z, \qquad F^n(u_n) = \left(1 - \frac{z}{n}\right)^n \to e^{-z}.$$

This means

$$e^{-z} = \lim_{n \to \infty} \mathbb{P}(M_n \leq u_n) = \lim_{n \to \infty} \mathbb{P}(M_n \leq \log n - \log z),$$

which, taking $z = e^{-y}$, we can rewrite as

$$e^{-e^{-y}} = \lim_{n \to \infty} \mathbb{P}(M_n \leq \log n + y),$$

i.e., $M_n - \log n$ converges in distribution to an r.v. Y with c.d.f. $G(y) = e^{-e^{-y}}$, $y \in \mathbb{R}$. This distribution is known as the *Gumbel distribution*. $\qquad \lozenge$

The uniform case is in some sense the most fundamental one since the general case follows from the uniform one by transformation with the inverse c.d.f. In more detail, assume F is supported by some open real interval I, and that F is strictly increasing and continuous on I. For $u \in (0, 1)$, we define $x = F^{\leftarrow}(u)$ as the unique solution $x \in I$ of $u = F(x)$. Then $F^{\leftarrow}$ is a strictly increasing and continuous bijection $(0, 1) \to I$ [the set-up can be generalized at the cost of some technicalities, and we do this in Sect. A.8 of the Appendix]. Further, an r.v. X with distribution F can be represented as $X \stackrel{\mathscr{D}}{=} F^{\leftarrow}(U)$ with U uniform$(0, 1)$:

$$\mathbb{P}\big(F^{\leftarrow}(U) \le x\big) = \mathbb{P}\big(U \le F(x)\big) = F(x).$$

It follows that

$$M_n \stackrel{\mathscr{D}}{=} \max\big(F^{\leftarrow}(U_1), \ldots, F^{\leftarrow}(U_n)\big) = F^{\leftarrow}(M_n^U). \tag{2.2}$$

Example 2.5 (continued) In the exponential case, $I = (0, \infty)$ and $F^{\leftarrow}(u) = -\log(1 - u)$. Using $M_n^U \sim 1 - V/n$ leads to

$$M_n \approx -\log\big(1 - (1 - V/n)\big) = \log n - \log V.$$

But the c.d.f. of $-\log V$ is

$$\mathbb{P}(-\log V \le y) = \mathbb{P}(V \ge e^{-y}) = e^{-e^{-y}},$$

so $-\log V$ is Gumbel and we are back to the conclusion of Example 2.5. ◇

Example 2.4 (continued) In the Pareto case, $I = (1, \infty)$ and $F^{\leftarrow}(u) = 1/(1-u)^{1/\alpha}$. Using $M_n^U \sim 1 - V/n$ leads to

$$M_n \approx \frac{1}{(V/n)^{1/\alpha}} = \frac{n^{1/\alpha}}{V^{1/\alpha}}.$$

But the c.d.f. of $1/V^{1/\alpha}$ is

$$\mathbb{P}(1/V^{1/\alpha} \le y) = \mathbb{P}(V \ge 1/y^\alpha) = e^{-1/y^\alpha},$$

so $1/V^{1/\alpha}$ is Fréchet and we are back to the conclusion of Example 2.4. ◇

Example 2.6 Assume that F has finite support, w.l.o.g. with right endpoint $x^* = 1$ and that $F(x)$ approaches 1 at power rate as $x \uparrow 1$,

$$F(x) = 1 - (1 - x)^\alpha, \tag{2.3}$$

with $\alpha > 0$ (thus $\alpha = 1$ is just the uniform case). One gets $F^{\leftarrow}(u) = 1 - (1 - u)^{1/\alpha}$ and it follows that

$$M_n \approx 1 - \frac{V^{1/\alpha}}{n^{1/\alpha}} \quad \text{where } \mathbb{P}(V^{1/\alpha} > y) = \mathbb{P}(V > y^{\alpha}) = e^{-y^{\alpha}}. \quad (2.4)$$

$\diamond$

Remark 2.7 The distribution of the r.v. $V^{1/\alpha} > 0$ with tail $e^{-y^{\alpha}}$ is in most contexts referred to as the Weibull distribution with parameter α and indeed we use that terminology in, e.g., Sect. III.4. However, in extreme value theory the tradition is that the Weibull distribution is that of $-V^{1/\alpha} < 0$, so that the c.d.f. is $F(x) = e^{-(-x)^{\alpha}}$ for $x < 0$ and $F(x) = 1$ for $x \geq 0$ (thus, *negative Weibull* would be more precise). We follow this tradition in the present chapter. $\diamond$

Example 2.8 Let F be geometric(ρ) on $\{1, 2, \ldots\}$, $\mathbb{P}(X = n) = (1 - \rho)\rho^{n-1}$ for $n = 1, 2, \ldots$ Then $X \overset{\mathcal{D}}{=} \lceil V/\rho \rceil$, where V is standard exponential so that $M_n \sim \lceil \log n/(1 - \rho) + \Gamma/\rho \rceil$, cf. Example 2.5. This does not lead to a limit result of any of the above types and a more rigorous explanation of this is given in Proposition 3.16. $\diamond$

2.1 The Exceedance Time

In applications, what is asked for is often not the growth rate of M_n but the inverse question, properties of the time $\omega(x)$ until exceedance of x, $\omega(x) = \inf\{n : X_n > x\}$. This is a much easier question: $\omega(x)$ is the time of the first success in a series of trials with success parameter $\overline{F}(x) = 1 - F(x)$ and hence geometric on $\{1, 2, \ldots\}$ with point probabilities

$$\mathbb{P}\big(\omega(x) = n\big) = F(x)^{n-1}\overline{F}(x)$$

so that, e.g., the mean is $1/\overline{F}(x)$. In applications, $\overline{F}(x)$ is typically small, and then an asymptotic representation in terms of the exponential distribution may be more convenient:

Theorem 2.9 *As $x \to \infty$, $\overline{F}(x)\omega(x) \overset{D}{\to} V$ where V is standard exponential.*

Proof By elementary calculations:

$$\mathbb{P}\big(\overline{F}(x)\omega(x) > y\big) = \mathbb{P}\big(\omega(x) > y/\overline{F}(x)\big) = \mathbb{P}(M_{\lfloor y/\overline{F}(x)\rfloor+1} \leq x)$$

$$= \big(1 - \overline{F}(x)\big)^{\lfloor y/\overline{F}(x)\rfloor+1} = \big(1 - \overline{F}(x)\big)^{y/\overline{F}(x)+O(1)} \to e^{-y}.$$

$\square$

Example 2.10 Assume maximal water levels $X_1, X_2, \ldots$ in successive years are i.i.d. $\mathcal{N}(\mu, \sigma^2)$. We ask how many years it will take until a flood occurs, where a flood is defined as exceedance of μ by three standard deviations. This gives $\mathbb{E}\omega(x) = 1/\overline{\Phi}(3) \approx 769$ in the exact geometric as well as the asymptotic exponential distribution.   ◇

3 Convergence Results

The examples of Sect. 2 all have the feature that $(M_n - b_n)/a_n$ has a limit in distribution for suitable sequences $\{a_n\}, \{b_n\}$ and the traditional set-up of extreme value theory is indeed to look for such sequences; then one needs to consider

$$\mathbb{P}\left(\frac{M_n - b_n}{a_n} \le x\right) = \mathbb{P}(M_n \le u_n) \quad \text{where } u_n = a_n x + b_n .$$

One could motivate this by an affine scaling giving a more clear picture of the asymptotics of M_n than the exact distribution formula $\mathbb{P}(M_n \le x) = \left(1 - \overline{F}(x)\right)^n$, or the representation $M_n \overset{\mathcal{D}}{=} F^{\leftarrow}(M_n^U)$ in terms of the uniform maximum.

Also, the only limits met in Sect. 2 are Fréchet, Gumbel or Weibull. This is no coincidence, since these three are the only possible ones as stated in the following result, and therefore they go under the name of *extreme value distributions*. The result is commonly referred to as the *Fisher–Tippett theorem*, even though one could argue that a completely rigorous proof was only given later by Gnedenko. Recall that two distributions G_1, G_2 are of the same type if for the corresponding r.v.s Y_1, Y_2 it holds that $Y_1 \overset{\mathcal{D}}{=} aY_2 + b$ with $a > 0$.

Theorem 3.1 *Assume in the i.i.d. setting that $(M_n - b_n)/a_n$ has a non-degenerate limit G in distribution. Then G is of one of the types Gumbel, i.e. $G(x) = e^{-e^{-x}}$, Fréchet, i.e. $G(x) = e^{-1/x^\alpha}$ for $x > 0$, or (negative) Weibull, i.e. $G(x) = e^{-(-x)^\alpha}$ for $x < 0$.*

If, for example, the limit is Gumbel, one writes $F \in$ MDA(Gumbel). The acronym MDA is for Maximum Domain of Attraction and is similar to terminology for i.i.d. sums $S_n = X_1 + \cdots + X_n$ where the limit of $(S_n - a_n)/b_n$ could be either normal (Normal Domain of Attraction) or stable (Stable Domain of Attraction). One also talks about distributions of types I, II, resp. III.

It should be noted that the union of MDA(Gumbel), MDA(Fréchet) and MDA(Weibull) is not the whole class of distributions F on the line; exceptions are, for example, certain discrete distributions like the Poisson or geometric. However, such exceptions tend to be rather special, and in fact complete and tractable necessary and sufficient conditions for F to be in each of the three classes are known. In the present section, we prove sufficiency, which is relatively easy;

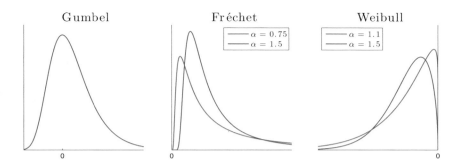

Fig. 1 The extreme value distributions—standard representation

necessity tends to be somewhat more complicated and is not treated in this book. The proof of the Fisher–Tippett theorem itself is in Sect. 4.

Plots of examples of the densities of the three extremal distributions with p.d.f.s as in Theorem 3.1 are in Fig. 1. Note that the supports are $(-\infty, \infty)$ for the Gumbel, $(0, \infty)$ for the Fréchet and $(-\infty, 0)$ for the Weibull.

For the Fréchet and Weibull, there is some flexibility in the shape determined by the parameter α. The three types can be put together into one class parametrized by a certain $\xi \in \mathbb{R}$, the *generalized extreme value distributions* with c.d.f.

$$G_\xi(x) = \exp\left\{-[1+\xi x]^{-1/\xi}\right\}. \tag{3.1}$$

The conventions are that: for $\xi = 0$ (the Gumbel case) the expression (3.1) is to be understood as the limit $\exp\{-e^{-x}\}$ as $\xi \to 0$; the support is all x such that $1+\xi x > 0$. Compared to Theorem 3.1, the Fréchet case $\xi = 1/\alpha > 0$ and the Weibull case $\xi = -1/\alpha < 0$ differ by a shift. Examples of the densities corresponding to the parametrization (3.1) are plotted in Fig. 2.

The class of all distributions of the same type (in the general $\mu + \sigma Y$ sense, not the one of the three types in extreme value theory!) is given by the set of c.d.f.s

$$G_{\xi,\mu,\sigma}(x) = \exp\left\{-\left[1+\xi\left(\frac{x-\mu}{\sigma}\right)\right]^{-1/\xi}\right\} \tag{3.2}$$

(where $1+(x-\mu)/\sigma > 0$) and statistical fitting is frequently done within this class, see further Sect. XIII.3.3. The means and variances are

$$\mu - \frac{\sigma}{\xi} + \frac{\sigma}{\xi}\Gamma(1-\xi), \quad \text{resp.} \quad \frac{\sigma^2}{\xi^2}\left(\Gamma(1-2\xi) - \Gamma(1-\xi)^2\right).$$

Before proceeding, we note:

Proposition 3.2 *Convergence in distribution of $(M_n - b_n)/a_n$ is a tail equivalence property. That is, if $(M_n - b_n)/a_n$ has limit G in distribution, then so has*

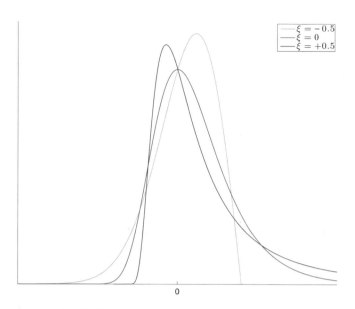

Fig. 2 The extreme value distributions—representation (3.1)

$(M_n^* - b_n)/a_n$, where $M_n^* = \max(X_1^*, \ldots, X_n^*)$ with $X_1^*, X_2^*, \ldots$ i.i.d. w.r.t. some distribution F^* such that $\overline{F}^*(t) \sim \overline{F}(t)$ as $t \to \infty$.

Proof Trivial by Proposition 2.1.   □

Exercises

3.1 Consider a Weibull r.v. Y with c.d.f. $F(x) = 1 - e^{-(-x)^\alpha}$, $x < 0$. Find a, b, ξ such that $aY + b$ has c.d.f. (3.1) and μ, σ, ξ such that Y has c.d.f. (3.2).

3.2 Find the asymptotic form of the tail $\overline{F}(x)$ and density $f(x)$ of the Fréchet distribution as $x \to \infty$.

3.1 MDA(Gumbel)

Recall that the Gumbel distribution has c.d.f. $e^{-e^{-x}}$. MDA(Gumbel) turns out to be the class of distributions F such that there exists a function $e(t)$ such that if $X \sim F$, then the conditional distribution of the overshoot $X - t$ given $X > t$ asymptotically is that of $e(t)V$, where V is a standard exponential r.v. That is, the distribution of $(X - t)/e(t)$ given $X > t$ should convergence to that of V or equivalently, one should have

$$\lim_{t\to\infty} \mathbb{P}\big((X - t)/e(t) > y \,\big|\, X > t\big) = \lim_{t\to\infty} \frac{\overline{F}(t + ye(t))}{\overline{F}(t)} = e^{-y} \qquad (3.3)$$

for all $y > 0$ (below, we require (3.3) for all $y \in \mathbb{R}$, which is usually as easy to verify).

Theorem 3.3 *Assume (3.3) holds for all $y \in \mathbb{R}$. Then $F \in MDA(Gumbel)$ and one may take $b_n = F^{\leftarrow}(1 - 1/n)$, $a_n = e(b_n)$.*

Condition (3.3) may appear rather special, but the examples below show its flexibility, and the reader will have a hard time finding a continuous F with $\overline{F}(x)$ decaying exponentially or faster such that (3.3) fails.

Proof Let $b_n = F^{\leftarrow}(1 - 1/n)$. By part 3) of Proposition A.2 in the Appendix, $n\overline{F}(b_n) \leq 1 \leq n\overline{F}(b_n-)$ and for $y > 0$

$$\liminf_{n\to\infty} \frac{n\overline{F}(b_n)}{n\overline{F}(b_n-)} \geq \liminf_{n\to\infty} \frac{\overline{F}(b_n)}{\overline{F}(b_n - ye(b_n))} = e^y.$$

Letting $y \downarrow 0$ gives $n\overline{F}(b_n) \to 1$. We then get

$$n\overline{F}(b_n + xe(b_n)) \sim n\overline{F}(b_n)e^{-x} \sim e^{-x}$$

so that $\mathbb{P}(M_n \leq b_n + xe(b_n)) \to e^{-e^{-x}}$ by Proposition 2.1. □

Remark 3.4 It can be shown that $e(x)$ may be chosen as the mean overshoot

$$\mathbb{E}[X - x \,|\, X > x] = \frac{1}{\mathbb{P}(X > x)}\mathbb{E}[X - x]^+ = \frac{1}{\overline{F}(x)}\int_x^\infty \overline{F}(t)\,dt \qquad (3.4)$$

[recall that $\mathbb{E}[X - x]^+$ also enters in Sect. VIII.2 as the stop-loss transform]. If a density f exists, then by L'Hôpital's rule the r.h.s. is asymptotically $\overline{F}(x)/f(x)$ (the inverse hazard rate) and so this is another possible choice. ◇

Example 3.5 Consider the (positive) Weibull distribution on $(0, \infty)$ with $\overline{F}(x) = e^{-x^\beta}$ and corresponding density $f(x) = \beta x^{\beta-1}e^{-x^\beta}$. Using Remark 3.4 we take $e(x) = 1/\beta x^{\beta-1}$ as trial. Now by Taylor expansion

$$\Big(x + \frac{y}{\beta x^{\beta-1}}\Big)^\beta = x^\beta\Big(1 + \frac{y}{\beta x^\beta}\Big)^\beta = x^\beta\Big(1 + \frac{y}{x^\beta} + o(1/x^\beta)\Big)$$

which easily gives (3.3).

Note that if $\beta = 1$, $e(x) \sim 1$ is immediate from the memoryless property of the exponential distribution. If $\beta < 1$, F is more heavy-tailed than the exponential distribution and $e(x) \to \infty$, whereas if $\beta > 1$, F has a lighter tail and $e(x) \to 0$.

By continuity of F, we can compute b_n as the solution of $1/n = \overline{F}(b_n) = e^{-b_n^\beta}$, which gives $b_n = \log^{1/\beta} n$. We then get $a_n = e(b_n) = \log^{1/\beta-1} n/\beta$.   ◇

Example 3.6 Example 3.5 is easily generalised to $\overline{F}(x) = e^{L(x)-dx^\beta}$ for certain smooth slowly varying functions L—at least when one understands 'easily' as referring to the general outline; in contrast, the calculations become tedious but are in principle elementary. We consider here the example $L(x) = \gamma \log x + \eta + o(1)$, which covers the gamma as well as the normal distribution (see Example 3.7). The reader may want to skip the following calculations and replace them with the direct treatment of the normal distribution in Example 3.13, which contains all the essential ideas.

Different $o(1)$ terms in the exponent of $\overline{F}(x)$ just give the same tail equivalence class so that by Proposition 3.2 we may take the $o(1)$ term to be 0. From

$$\gamma \log \left(x + \frac{y}{d\beta x^{\beta-1}}\right) = \gamma \log x + \gamma \log \left(1 + \frac{y}{d\beta x^\beta}\right) = \gamma \log x + o(1)$$

and a minor modification of the calculations in Example 3.5 (to allow for $d \neq 1$) it follows that we may take $e(x) = 1/d\beta x^{\beta-1}$.

The equation $\overline{F}(b_n) = 1/n$ gives

$$b_n = \left[\frac{1}{d}\left(\log n + \gamma \log b_n + \eta\right)\right]^{1/\beta}. \tag{3.5}$$

Thus $b_n \sim (\log n/d)^{1/\beta}$, which we first use to note that we may take

$$a_n = e\big((\log n/d)^{1/\beta}\big) = \frac{1}{d^{1/\beta}\beta \log^{1-1/\beta} n}.$$

For the final form of b_n, we need to expand (3.5) more carefully up to an $o(a_n)$ term, i.e. an $o(\log^{1/\beta-1} n)$ term. We can rewrite (3.5) as

$$\begin{aligned}
b_n &= \left(\frac{\log n}{d}\right)^{1/\beta}\left\{1 + \frac{\gamma \log b_n + \eta}{\log n}\right\}^{1/\beta} \\
&= \left(\frac{\log n}{d}\right)^{1/\beta}\left\{1 + \frac{\gamma \log b_n + \eta}{\beta \log n} + O\left(\left[\frac{\log \log n}{\log n}\right]^2\right)\right\} \\
&= \left(\frac{\log n}{d}\right)^{1/\beta}\left\{1 + \frac{\gamma \log b_n + \eta}{\beta \log n} + o\left(\frac{1}{\log n}\right)\right\}. \tag{3.6}
\end{aligned}$$

Taking logarithms gives

$$\log b_n = \frac{1}{\beta}(\log \log n - \log d) + O\left(\frac{\log \log n}{\log n}\right),$$

and substituting back into (3.6) and ignoring $o(\log^{1/\beta-1} n)$ terms gives the final expression

$$b_n = \left(\frac{\log n}{d}\right)^{1/\beta} + \frac{\gamma \log\log n}{d^{1/\beta}\beta^2 \log^{1-1/\beta} n} + \frac{\eta\beta - \gamma \log d/\beta}{d^{1/\beta}\beta^2 \log^{1-1/\beta} n}.$$

$\diamond$

Example 3.7 Recall that (up to tail equivalence) $\overline{F}(x) = e^{\gamma \log x + \eta - dx^\beta}$ in Example 3.6. For the gamma density $f(x) = x^{\alpha-1}e^{-x}/\Gamma(\alpha)$ we have $d = \beta = 1$. Further, $\overline{F}(x) \sim f(x)$ so that we may take $\gamma = \alpha - 1$, $\eta = -\log\Gamma(\alpha)$. For the standard normal distribution, $d = 1/2$, $\beta = 2$. Further, Mill's ratio formula (A.7.2) gives $\overline{\Phi}(x) = \varphi(x)/t(1 + o(1))$ and we can take $\gamma = -1$, $\eta = -\log(2\pi)/2$. $\diamond$

3.2 MDA(Fréchet)

Recall that the Fréchet(α) distribution is concentrated on $(0, \infty)$ with c.d.f. e^{-1/x^α} for $x > 0$.

Theorem 3.8 *Assume that $\overline{F}(x)$ is regularly varying at $x = \infty$, i.e. $\overline{F}(x) = L(x)/x^\alpha$ with $\alpha > 0$ and L slowly varying. Then $F \in MDA(Fréchet_\alpha)$ and one may take $a_n = F^{\leftarrow}(1 - 1/n)$, $b_n = 0$.*

Proof With $a_n = F^{\leftarrow}(1 - 1/n)$, we have for $t < 1$ that

$$\liminf_{n\to\infty} \frac{n\overline{F}(a_n)}{n\overline{F}(a_n-)} \geq \liminf_{n\to\infty} \frac{\overline{F}(a_n)}{\overline{F}(ta_n)} = t^\alpha,$$

which tends to 1 as $t \uparrow 1$ (we used here that $L(ta_n)/L(a_n) \to 1$). It thus follows from $n\overline{F}(a_n) \leq 1 \leq n\overline{F}(a_n-)$ that $n\overline{F}(a_n) \to 1$. We then get

$$n\overline{F}\big(a_n x)\big) \sim n\overline{F}(a_n)/x^\alpha \sim 1/x^\alpha$$

(using $L(a_n x)/L(a_n) \to 1$) so that $\mathbb{P}(M_n \leq a_n x) \to e^{-1/x^\alpha}$ by Proposition 2.1.
$\square$

Example 3.9 For the Pareto $\overline{F}(x) = 1/(1 + x)^\alpha$, we can take $L(x) = x^\alpha/(1 + x)^\alpha$ which has limit 1 as $x \to \infty$ and hence is slowly varying. $\diamond$

3.3 MDA(Weibull)

Recall that the (negative) Weibull(α) distribution is concentrated on $(-\infty, 0)$ with c.d.f. $e^{-(-x)^\alpha}$ for $x < 0$.

Theorem 3.10 *Assume that the right endpoint* $x^* = x_F^* = \inf\{x : F(x) = 1\}$ *is finite and that* $\overline{F}(x)$ *is regularly varying at* $x = 0$ *in the sense that* $F(x^* - 1/x) = L(x)/x^\alpha$ *with* $\alpha > 0$ *and* L *slowly varying as* $x \to \infty$. *Then* $F \in \mathrm{MDA}\big(\mathrm{Weibull}(\alpha)\big)$ *and one may take* $a_n = x^* - F^\leftarrow(1 - 1/n)$, $b_n = x^*$.

The proof is rather similar to that of Theorem 3.8 and is omitted.

Remark 3.11 The result may also be obtained as a small generalisation of Example 2.6. Assume w.l.o.g. that $x^* = 1$ and use the following variant of the regular variation condition

$$F(x) = 1 - (1 - x)^\alpha L(1 - x), \quad x < 1, \tag{3.7}$$

with $\alpha \geq 0$ and L slowly varying at $x = 0$ (thus $\alpha = 1$, $L \equiv 1$ is just the uniform case). One can then verify that

$$F^\leftarrow(u) \sim 1 - \frac{(1 - u)^{1/\alpha}}{L(1 - u)^{1/\alpha}} \quad \text{as } u \uparrow 1.$$

Since $L(V/n) \sim L(1/n)$, it follows from Example 2.6 that

$$M_n \sim F^\leftarrow(M_n^U) \sim F^\leftarrow(1 - V/n) \sim 1 - \frac{V^{1/\alpha}}{n^{1/\alpha} L(V/n)^{1/\alpha}} \sim 1 - \frac{V^{1/\alpha}}{n^{1/\alpha} L(1/n)^{1/\alpha}}$$

with V standard exponential, i.e., $n^{1/\alpha} L(1/n)^{1/\alpha}(M_n - 1) \to -V^{1/\alpha}$ where the distribution of $-V^{1/\alpha}$ is Weibull(α),

$$\mathbb{P}(-V^{1/\alpha} \leq x) = \mathbb{P}(V \geq (-x)^\alpha) = \mathrm{e}^{-(-x)^\alpha}. \qquad \diamond$$

3.4   Further Remarks and Examples

Remark 3.12 As remarked earlier, the sufficient conditions for F to be in one of the three MDAs are also necessary. The proof of this is rather easy for MDA(Fréchet) and therefore also for MDA(Weibull), since it easily follows that if the support of $X \sim F$ has a finite endpoint x_F, then $X \in \mathrm{MDA(Weibull)}$ if and only if $1/(x_F - X) \in \mathrm{MDA(Fréchet)}$. However, necessity is harder for MDA(Gumbel).

The three set of sufficient conditions for F to be in one of the three MDAs may look rather different but there is in fact a common framework in terms of the family of *generalized Pareto distributions* $G_{\xi,\beta}$ with tails of the form

$$\overline{G}_{\xi,\beta}(x) = \frac{1}{(1 + \xi x/\beta)^{1/\xi}}, \tag{3.8}$$

where $\beta > 0$ is the scale parameter and $\xi \in \mathbb{R}$ the shape parameter; as usual, (3.8) should be understood as the $\xi \to 0$ limit $\mathrm{e}^{-x/\beta}$ when $\xi = 0$. The support is $(0, \infty)$

when $\xi > 0$, $(0, \beta/|\xi|)$ when $\xi < 0$, and the whole of $\mathbb{R}$ when $\xi = 0$. Letting X^t be an r.v. distributed as the overshoot $X - t$ conditional on $X > t$, we have already seen that $F \in \text{MDA(Gumbel)}$ if and only if there exists a function $e(t)$ such that $X^t/e(t)$ has limit distribution $G_{0,\beta}$ for some $\beta > 0$. Similarly, one can prove that $F \in \text{MDA(Fréchet)}$ if and only if $X^t/e(t)$ has limit distribution $G_{\alpha,\beta}$ for some $\alpha, \beta > 0$, and in that case, one can take $e(t) = t$. That this is a sufficient condition is trivial if one knows a priori that $e(t) = t$, since regular variation of X is equivalent to X^t/t having a Pareto limit. The proof of sufficiency is, however, somewhat harder in the case of a general $e(t)$. In view of the connection between MDA(Weibull) and MDA(Fréchet) noted above, these statements immediately translate to a similar statement on a $G_{\alpha,\beta}$ with $\alpha < 0$, $\beta > 0$.   ◊

Example 3.13 The standard normal distribution has been considered in Example 3.6, but we give here a direct treatment.

We first note that the c.d.f. satisfies (3.3) with $e(t) = 1/t$. Indeed, using Mill's ratio (A.7.2) we get

$$\frac{\overline{\Phi}(t + y/t)}{\overline{\Phi}(t)} \sim \frac{t}{t + y/t} \cdot \frac{\varphi(t + y/t)}{\varphi(t)} = 1 \cdot \exp\left\{ -(t + y/t)^2/2 + t^2/2 \right\}$$

$$= \exp\{-y^2/2t^2 - y\} \sim e^{-y}.$$

Next consider the form of a_n, b_n. The equation $b_n = \Phi^{\leftarrow}(1 - 1/n)$ determining b_n is $\Phi(b_n) = 1 - 1/n$, i.e.

$$\frac{1}{n} = \frac{1}{\sqrt{2\pi}\,b_n} e^{-b_n^2/2}$$

so that

$$\frac{b_n^2}{2} = \log n - \log b_n - \log\sqrt{2\pi}.$$

This gives first $b_n \sim \sqrt{2\log n}$ and hence we can take $a_n = e(b_n) \sim 1/\sqrt{2\log n}$. Next, using $\sqrt{x + y} = \sqrt{x} + y/2\sqrt{x} + O(1/x)$, we get

$$b_n = \sqrt{2\log n - 2\log b_n - 2\log\sqrt{2\pi}}$$

$$= \sqrt{2\log n} - \frac{2\log b_n + 2\log\sqrt{2\pi}}{2\sqrt{2\log n})} + O\left(1/\log n\right)$$

$$= \sqrt{2\log n} - \frac{2 \cdot \frac{1}{2}(\log\log n + \log 2) + 2 \cdot \frac{1}{2}(\log 2 + \log \pi)}{2\sqrt{2\log n}} + O\left(1/\log n\right),$$

$$b_n = \sqrt{2\log n} - \frac{\log\log n + \log 4\pi}{2\sqrt{2\log n}} + o\left(\log\log n/\sqrt{\log n}\right). \qquad ◊$$

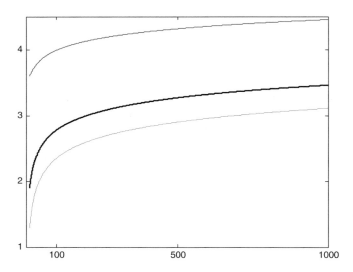

Fig. 3 Typical 95% values of normal maxima. Sample size n on the abscissa

Remark 3.14 That $a_n \to 0$ in the normal case means that M_n becomes more and more concentrated at the point b_n as $n \to \infty$. This easily leads to an impression that the band of typical values of M_n narrows markedly as n. This is however somewhat deceptive because the rate of decay of $a_n \sim 1/\sqrt{2 \log n}$ is extremely slow. An illustration of this is in Fig. 3 where the black line is the b_n function and the blue lines $b_n + z_{0.025}a_n$, resp. $b_n + z_{0.975}a_n$, where $z_{0.025}$ is the 2.5% quantile of the Gumbel distribution and $z_{0.975}$ the 97.5% quantile. That is, w.p. approximately 5% the maximum is outside the strip between the two blue lines.   ◇

Remark 3.15 A further connection between the three different types Y_G, Y_F, Y_W of extreme value distributions is that $Y_G \overset{\mathscr{D}}{=} -\log Y_F^\alpha$, $Y_W \overset{\mathscr{D}}{=} -1/Y_F$.   ◇

Typical examples of standard distributions F not in any MDA are certain discrete distributions with a rapidly decaying tail. In particular, the following result applies to the geometric and the Poisson distributions:

Proposition 3.16 *Let F be a distribution with infinite right support, $x^* = \infty$, and assume $\liminf_{x\to\infty} \overline{F}(x)/\overline{F}(x-) > 1$. Then F is not in any MDA.*

Proof Arguing *ad contradictum*, assume $(M_n - b_n)/a_n$ has a limit Y in distribution which must then necessarily be Gumbel or Fréchet, w.l.o.g. with support $[0, \infty)$ in the Fréchet case, and write $e^{-\tau} = \mathbb{P}(Y \le 1)$. Then $0 < \tau < \infty$ and we also have $a_n + b_n \to \infty$ (if this fails for a subsequence, $\mathbb{P}(M_{n_k} - b_{n_k})/a_{n_k} \le 1)$ goes to 0 rather than $e^{-\tau}$). Choose a sequence $u_k \to \infty$ with $\overline{F}(u_k)/\overline{F}(u_k-) > 1 + \delta$ for some $\delta > 0$. We can choose n_k with

$$a_{n_k} + b_{n_k} < u_k \le a_{n_k+1} + b_{n_k+1}$$

and have $n_k \overline{F}(a_{n_k} + b_{n_k}) \to e^{-\tau}$. This yields

$$1 + \delta < \liminf_{k\to\infty} \frac{\overline{F}(a_{n_k} + b_{n_k})}{\overline{F}(a_{n_k+1} + b_{n_k+1})}$$

$$= \liminf_{k\to\infty} \frac{n_k \overline{F}(a_{n_k} + b_{n_k})}{(n_k + 1)\overline{F}(a_{n_k+1} + b_{n_k+1})} = \frac{e^{-\tau}}{e^{-\tau}} = 1,$$

a contradiction.   □

Remark 3.17 Putting together the results, we obtain the following procedure for getting hands on the extreme value properties of a given distribution F:

A) Check if the right endpoint $\widehat{x}_F$ of F is finite. If no, go to B). If yes, proceed via Theorem 3.10 and check the regular variation condition to see if $F \in$ MDA(Weibull).
B) Check if F is regularly varying. If no, go to C). If yes, Theorem 3.8 gives $F \in$ MDA(Fréchet), that a_n is found by solving $\overline{F}(a_n) = 1/n$ and that $b_n = 0$.
C) If F has a density f, let $e(x) = \overline{F}(x)/f(x)$ [otherwise compute $e(x)$ as the conditional expectation of $X - x$ given $X > x$]. Check if $\overline{F}(x + ye(x))/\overline{F}(x) \to e^{-y}$. If yes, Theorem 3.3 gives $F \in$ MDA(Gumbel), that b_n is found by solving $\overline{F}(b_n) = 1/n$ and that $a_n = e(b_n)$.   ◇

Exercises

3.3 Let $0 < \beta < 1$. Show that the function $e^{-\log^\beta x}$ is slowly varying as $x \to \infty$.

3.4 Let $\beta > 1$. Find the extreme value properties of the distribution with $\overline{F}(x) = e^{-\log^\beta x}$, $x > 1$ (which MDA? a_n, b_n?).

3.5 Find the extreme value properties of the distribution with density $f(x) = e^{-x}/2 + e^{-2x}$, $x > 0$.

3.6 Show that the distribution with density $f(x) = e^{-x}/2 + xe^{-x}/2$, $x > 0$, is in MDA(Gumbel) and compute $e(t)$ up to a $1 + o(1)$ factor.

3.7 Show that the distribution with tail $\overline{F}(x) = e^{-x/\log^+ x}$ is in MDA(Gumbel) and compute $e(t)$ up to a $1 + o(1)$ factor.

3.8 Show that the distribution of $-1/V$ with V standard exponential gives an example of an $F \in$ MDA(Gumbel) with finite upper support. [Hint: $e(x) = x^2$].

3.9 What can be said about the extreme value properties of the distribution with density proportional to $(1 + \sin x/(x + 1))e^{-x}$?

3.10 The lognormal distribution with parameters μ, σ^2 is defined as the distribution of $e^{\mu+\sigma Z}$ where $Z \sim \mathcal{N}(0,1)$. Show that the lognormal distribution is in MDA(Gumbel) and find a_n, b_n in terms of the normalizing constants a_n^*, b_n^* for the standard normal case. What is $e(t)$?

3.11 Find the extreme value properties of the loggamma distribution, that is, the distribution of e^Z, where Z is gamma.

3.12 Verify the assertions of Remark 3.15.

3.13 Show that the c.d.f. $F(x) = 1 - e^{-x/(1-x)}$, $0 < x < 1$, gives another example that MDA(Gumbel) contains some distributions with bounded support.

3.14 Let $L(x) > 0$ be a differentiable function and let $M(x) = \log L(e^x)$. Show that if $M'(x) \to 0$ as $x \to \infty$ then L is slowly varying. Show thereby that $L(x) = \exp\{(\log x)^\alpha \cos(\log x)^\beta\}$ is slowly varying with $\liminf L(x) = 0$, $\limsup L(x) = \infty$ when $\alpha, \beta > 0$, $\alpha + \beta < 1$.

3.15 Formulate and prove an analogue of Proposition 3.16 for a distribution with support bounded to the right.

4 Proof of the Fisher–Tippett Theorem

Definition 4.1 A non-degenerate distribution G is called *extremal* if G is the limit in distribution of $(M_n - b_n)/a_n$ for some sequences $\{a_n\}$, $\{b_n\}$ and some distribution F such that M_n is the maximum of n i.i.d. r.v.s with distribution F.

Thus, the content of the Fisher–Tippett theorem is that an extremal distribution is also an extreme value distribution, that is, of either Gumbel, Fréchet or Weibull type.

Definition 4.2 A distribution G is called *max-stable* if $G(x) = G(a_n x + b_n)^n$ for some sequences $\{a_n\}$, $\{b_n\}$ and all x, n.

In terms of r.v.s: if Y, Y_1, Y_2, $\ldots$ are i.i.d. with common distribution G, then for each n $\max(Y_1, \ldots, Y_n)$ is of the same type as Y (cf. Sect. A.10 of the Appendix).

Proposition 4.3 *A distribution G is max-stable if and only if it is extremal.*

Proof That G is extremal means that

$$G(x) = \lim_{n\to\infty} F(a_n x + b_n)^n \qquad (4.1)$$

for suitable sequences $a_n > 0$, b_n and a suitable F. If G is max-stable, we can simply take $F = G$ and omit the limit, so G must be extremal. That an extremal G is max-stable follows by taking $s = n$ as an integer in the following lemma. $\qquad \square$

Lemma 4.4 *If G is extremal, then for any $s > 0$ there are $a(s) > 0$, $b(s)$ such that $G^s(x) = G\big(\alpha(s)x + \beta(s)\big)$ for all x.*

Proof From (4.1) it follows that

$$F(a_n x + b_n)^{\lfloor ns \rfloor} = \left[F(a_n x + b_n)^n \right]^{\lfloor ns \rfloor / n} \;\to\; G(x)^s$$

and, by passing to the subsequence $\lfloor ns \rfloor$, that $F(a_{\lfloor ns \rfloor} x + b_{\lfloor ns \rfloor})^{\lfloor ns \rfloor}$ has limit $G(x)$, i.e., the sequences $(M_{\lfloor ns \rfloor} - b_n)/a_n$ and $(M_{\lfloor ns \rfloor} - b_{\lfloor ns \rfloor})/a_{\lfloor ns \rfloor}$ have limits G^s, resp. G in distribution. Thus by the convergence-to-types theorem (Proposition A.3 in the Appendix), G^s and G are of the same type; note here and in the following that the function G^s satisfies all formal requirements to be a c.d.f. (it is non-decreasing, right continuous, and $G(-\infty) = 0$, $G(\infty) = 1$). □

Proof of Theorem 3.1 Using Lemma 4.4 in the Appendix first with $s = ts$, next with $s = s$ and finally with $s = t$, we obtain

$$G\big(\alpha(st)x + \beta(st)\big) = G^{st}(x) = [G^s(x)]^t = G^t\big(\alpha(s)x + \beta(s)\big)$$
$$= G\big(\alpha(s)\alpha(t)x + \alpha(t)\beta(s) + \beta(t)\big).$$

Thus by Lemma A.1 in the Appendix, we have

$$\alpha(st) = \alpha(s)\alpha(t), \quad \beta(st) = \alpha(t)\beta(s) + \beta(t) = \alpha(s)\beta(t) + \beta(s), \tag{4.2}$$

where in the last identity we just interchanged t and s. The expression for $\alpha(st)$ is one of the forms of the Hamel functional equation, and the only measurable solutions are powers. Thus

$$\alpha(t) = t^{-\theta} \tag{4.3}$$

for some θ. We will see that $\theta = 0$ gives the Gumbel, $\theta > 0$ the Fréchet and $\theta < 0$ the Weibull.

The Case $\theta = 0$ Letting $u = \log s$, $v = \log t$, the second identity in (4.2) becomes $g(u + v) = g(u) + g(v)$, where $g(z) = \beta(e^z)$, and using Hamel once more gives $g(z) = -cz$, $\beta(t) = -c \log t$. We then get

$$G^t(x) = G(x - c \log t), \tag{4.4}$$

and since G is non-degenerate and $G^t(x)$ is non-increasing in t, we must have $c > 0$. Further, if $G(x_0) = 1$ for some x_0, then letting $y = x_0 - c \log t$ would give $G(y) = 1$ for all t and hence $G(x) = 1$ for all x, contradicting that x is non-degenerate. Similarly, $G(x_0) = 0$ for some x_0 is impossible, so one gets $G(x) > 0$ for all x.

Now write $G(0) = \exp\{-e^{-p}\}$ and $u = -c \log t$. Then (4.4) with $x = 0$ reads

$$G(u) = \exp\{-e^{-p}t\} = \exp\{-\exp\{-u/c - p\}\} = \mathbb{P}(Y \le u/c + p),$$

where Y is Gumbel, i.e., G is of Gumbel type.

The Case $\theta > 0$ The function $\beta(s)/(1 - \alpha(s))$ is constant by the last identity in (4.2), say equal to c, and so

$$\beta(t) = \frac{\beta(s)}{1 - \alpha(s)}(1 - \alpha(t)) = c(1 - t^{-\theta}).$$

Letting $H(x) = G(x + c)$, Lemma 4.4 then takes the form

$$H^t(x) = G(t^{-\theta}(x + c) + c(1 - t^{-\theta})) = G(t^{-\theta}x + c) = H(t^{-\theta}x). \tag{4.5}$$

Letting $x = 0$ shows that $H(0) = 0$ or 1. If $H(0) = 1$, we would have $0 < H(x) < 1$ for some $x < 0$ since G and hence H is non-degenerate. But for such an x, $t^{-\theta}x$ and hence $H(t^{-\theta}x)$ is increasing in t but $H^t(x)$ is decreasing, a contradiction. Thus $H(0) = 0$.

Similarly, (4.5) gives $H^t(1) = H(t^{-\theta})$. If $H(1) = 0$ or 1, $H(x)$ would have the same constant value for all $x > 0$, contradicting non-degeneracy. Writing $\alpha = 1/\theta$, $H(1) = \exp\{-p^{-\alpha}\}$, $u = t^{-\theta}$, (4.5) with $x = 1$ then gives

$$H(u) = H^t(1) = \exp\{-p^{-\alpha}t\} = \exp\{-(pu)^{-\alpha}\},$$

i.e., H and hence G is of Fréchet type.

The Case $\theta < 0$ The proof that G is Weibull type is rather similar to the case $\theta > 0$ and is omitted. $\square$

Notes
The proofs of the Fisher–Tippett theorem in the literature are rather diverse, and in part quite long. We have followed here Resnick [148]. Another nice proof is in de Haan and Ferreira [72, pp. 6–8].

5 Records

We now change the focus somewhat and consider the record times (times of consecutive maxima)

$$U_0 = 1, \quad U_{n+1} = \inf\{k > U_n : M_k > M_{U_n}\} = \inf\{k > U_n : X_k > X_{U_n}\}.$$

We further let

$$I_j = \mathbb{1}(j \text{ is a record time}) = \sum_{n=0}^{\infty} \mathbb{1}(U_n = j).$$

For simplicity, we assume in the following that F is continuous and that $F\{\widehat{x}_F\} = \mathbb{P}(X = \widehat{x}_F) = 0$, where $\widehat{x}_F$ is the upper point in the support of F (possibly $\widehat{x}_F = \infty$); otherwise, some modifications are needed.

The assumption of continuity of F implies that the distribution of the sequences $\{I_j\}, \{U_n\}$ is independent of F (use, e.g., the transformation $V_n = F^{\leftarrow}(X_n)$ to reduce to the uniform case). In particular:

Proposition 5.1

(a) *The I_j are pairwise independent with $\mathbb{P}(I_j = 1) = 1/j$;*
(b) *The sequence $U_1, U_2, \ldots$ is a time-homogeneous Markov chain with transition probabilities*

$$\mathbb{P}(U_{n+1} = k \mid U_n = j) = \frac{j}{k(k-1)}, \quad 1 \le j < k.$$

Proof That j is a record time is equivalent to X_j being the largest among $X_1, \ldots, X_j$, and by symmetry the probability of this is $1/j$. The independence of I_j, I_k for $j < k$ follows since the knowledge that X_j is the largest among $X_1, \ldots, X_j$ does not say anything about whether X_k is the largest among $X_1, \ldots, X_k$. This shows (a).

The key step in the proof of (b) is to establish that

$$\mathbb{P}(U_1 = j_1, \ldots, U_n = j_n, X_{U_n} \in dy) = \frac{f(y)F(y)^{j_n-1}}{(j_1-1)\cdots(j_n-1)}, \qquad (5.1)$$

where f is the density of F. For this, let first $n = 1$. For $X_{U_1} \in dy$ to occur, we need $z = X_1 < y$. If also $U_1 = j_1$ is to occur, we must have $X_2 \le z, \ldots, X_{j_1-1} \le z$, and the probability of this is $F(z)^{j_1-2}$. Thus

$$\mathbb{P}(U_1 = j_1, X_{U_1} \in dy) = \int_0^y f(z) \, dz \, F(z)^{j_1-2} f(y) = \frac{f(y)F(y)^{j_1-1}}{j_1-1},$$

where we used $F'(z) = f(z)$. The argument to get from $n-1$ to n is entirely similar:

$$\mathbb{P}(U_1 = j_1, \ldots, U_n = j_n, X_{U_n} \in dy)$$

$$= \int_0^y \mathbb{P}(U_1 = j_1, \ldots, U_{n-1} = j_{n-1}, X_{U_{n-1}} \in dz) F(z)^{j_n-j_{n-1}-2} f(y)$$

$$= f(y) \int_0^y \frac{f(z)F(z)^{j_{n-1}-1} \, dz}{(j_1-1)\cdots(j_{n-1}-1)} F(z)^{j_n-j_{n-1}-1} = \frac{f(y)F(y)^{j_n-1}}{(j_1-1)\cdots(j_n-1)}.$$

Integrating (5.1) w.r.t. y we then get

$$\mathbb{P}(U_1 = j_1, \ldots, U_n = j_n) = \frac{1}{(j_1 - 1) \cdots (j_n - 1) j_n}$$

and so

$$\mathbb{P}(U_n = j_n \mid U_1 = j_1, \ldots, U_{n-1} = j_{n-1})$$

$$\frac{(j_1 - 1) \cdots (j_{n-1} - 1) j_{n-1}}{(j_1 - 1) \cdots (j_n - 1) j_n} = \frac{j_{n-1}}{j_n (j_n - 1)}.$$

Taking $j_{n-1} = j$, $j_n = k$ gives the assertion. □

Let $R_n = \sum_1^n I_j$ be the number of records before n.

Proposition 5.2

(a) $\mathbb{E} R_n = \displaystyle\sum_{j=1}^n \frac{1}{j} = \log n + \gamma + o(1)$, *where* $\gamma = 0.5772\ldots$ *is Euler's constant.*

Further,

$$\mathrm{Var}\, R_n = \sum_{j=1}^n \left[\frac{1}{j} - \frac{1}{j^2} \right] = \log n + \gamma - \frac{\pi^2}{6} + o(1).$$

(b) $R_n / \log n \overset{\text{a.s.}}{\to} 1$.

(c) $(R_n - \log n)/\sqrt{\log n}$ *has a limiting standard normal distribution.*

Proof The first expression for $\mathbb{E} R_n$ follows from

$$\mathbb{E} R_n = \sum_{j=1}^n \mathbb{P}(I_n = 1)$$

and Proposition 5.1, and the second from the definition of Euler's constant. The expression for $\mathrm{Var}\, R_n$ follows similarly from $\mathrm{Var}\, I_n = 1/j - 1/j^2$ and $\sum_1^\infty 1/j^2 = \pi^2/6$.

Let $n_k = e^{k^2}$. Then by (a) and Chebycheff's inequality,

$$\sum_{k=1}^\infty \mathbb{P}(|R_{n_k}/\mathbb{E} R_{n_k} - 1| > \epsilon) \leq \sum_{k=1}^\infty \frac{\mathrm{Var}\, R_{n_k}}{\epsilon^2 (\mathbb{E} R_{n_k})^2}$$

$$\approx \sum_{k=1}^\infty \frac{\log n_k}{\log^2 n_k} = \sum_{k=1}^\infty O(k^{-2}) < \infty$$

so that $R_{n_k}/\mathbb{E}R_{n_k} \overset{\text{a.s.}}{\to} 1$ by the Borel–Cantelli lemma and hence $R_{n_k}/\log n_k \overset{\text{a.s.}}{\to} 1$. We then get

$$\limsup_{n\to\infty} \frac{R_n}{\log n} \le \limsup_{k\to\infty} \sup_{n_k \le n \le n_{k+1}} \frac{\log n_{k+1}}{\log n} \frac{R_{n_{k+1}}}{\log n_{k+1}}$$

$$\le \limsup_{k\to\infty} \sup_{n_k \le n \le n_{k+1}} \frac{\log n_{k+1}}{\log n_k} \cdot \frac{R_{n_{k+1}}}{\log n_{k+1}} = 1,$$

where we used that R_n is increasing in n and that $\log n_{k+1}/\log n_k = (k+1)^2/k^2 \to 1$. A similar lower bound for the lim inf gives (b). For (c), use a suitable version of the CLT for independent summands with a possibly different distribution. □

Remark 5.3 Actually, the exact distribution of R_n is known: $\mathbb{P}(R_n = k) = \begin{bmatrix} n \\ k \end{bmatrix}/n!$, where $\begin{bmatrix} n \\ k \end{bmatrix}$ is a Stirling number, defined in terms of the coefficients of the nth order polynomial $\prod_0^{n-1}(x - i)$,

$$n!\binom{x}{n} = \prod_{i=0}^{n-1}(x - i) = \sum_{k=1}^{n}(-1)^{n-k}\begin{bmatrix} n \\ k \end{bmatrix}x^k, \quad x \in \mathbb{R}. \qquad \diamond$$

Corollary 5.4

(i) $\log U_n/n \overset{\text{a.s.}}{\to} 1$. *Equivalently, $U_n = e^{n(1+o(1))}$;*
(ii) *Let $\Delta_n = U_{n+1} - U_n$ be the nth interrecord time. Then $\Delta_n/U_n \overset{D}{\to} Z$, where Z has a Pareto(1) distribution, $\mathbb{P}(Z > z) = 1/(1 + z)$.*

Proof Part (i) follows from

$$\frac{n}{\log U_n} = \frac{R_{U_n}}{\log U_n} \overset{\text{a.s.}}{\to} 1.$$

Further, Proposition 5.1 gives

$$\mathbb{P}(\Delta_n > jz \mid U_n = j) = \mathbb{P}(U_{n+1} - j > jz \mid U_n = j) = \sum_{k=j(1+z)}^{\infty} \frac{j}{k(k-1)}.$$

For large j, this is of order $1/(1 + z)$, and (ii) then follows from $U_n \overset{D}{\to} \infty$. □

Remark 5.5 Combining parts (i) and (ii) shows that Δ_n is roughly of order e^n. There are quite a few statistical studies of interrecord times in sports (cf. Google!) that are not compatible with the exponential growth in Corollary 5.4, but rather predict constant order of size (e.g. the mean interrecord time has been reported as

2.4 Olympics $= 9.6$ years for the Marathon and 2.8 Olympics for the long jump). The implication is that the i.i.d. assumption is invalid in the Olympic context. ◇

Example 5.6 (The Secretary Problem)   A company has n candidates for a secretary position. They are called for interview, with the rule that the decision of appointing or not is made immediately after the interview. If no one has been accepted before the end, the last one is taken. The problem is to maximize the probability of selecting the one who is best, defined by having the highest interview score X_i. That is, one is looking for a stopping time $\tau \in \{1, \ldots, n\}$ maximizing the success probability $p(\tau) = \mathbb{P}(X_\tau = \max_1^n X_j)$.

The general theory of optimal stopping rather easily gives that the optimal τ has the form

$$\tau_c = \inf\{j \geq c : X_j > \max(X_1, \ldots, X_{j-1})\} \wedge n.$$

That is, τ_c is the first record time after c (if there is one, otherwise the last candidate is selected). So, what is the optimal choice c^* of c and what is the maximal success probability $p(\tau_{c^*})$?

The event of success is the event

$$\{I_c + \cdots + I_n = 1\} = \bigcup_{j=c}^{n} \{I_j = 1, \ I_k = 0 \text{ for } k = c, \ldots, n, \ k \neq j\} \qquad (5.2)$$

that there is exactly one record in $\{c, \ldots, n\}$. Since the events on the r.h.s. of (5.2) are mutually exclusive, it follows by independence of the I_j and $\mathbb{P}(I_j = 1) = 1/j$ that

$$p(\tau_c) = \sum_{j=c}^{n} \frac{1}{j} \prod_{k=c,\ldots,n,\,k\neq j} \left(1 - \frac{1}{k}\right) = \prod_{k=c,\ldots,n} \left(1 - \frac{1}{k}\right) \sum_{j=c}^{n} \frac{1}{j(1 - 1/j)}.$$

For large n, c, the log of the product in the last expression is approximately

$$\sum_{c}^{n} \log(1 - 1/k) \approx -\sum_{c}^{n} 1/k \approx \log c - \log n$$

whereas the sum is approximately $\sum_{c}^{n} 1/j \approx \log(n/c)$. Thus $p(\tau_c)$ is approximately $(c/n) \log(n/c)$, and since the maximum of $x \log(1/x)$ in $(0, 1)$ is attained at $x^* = 1/e$ and equals $1/e$, we obtain the classical asymptotics for the secretary problem,

$$c^* \sim \frac{n}{e}, \quad p(\tau_{c^*}) \sim \frac{1}{e}.$$

◇

Exercises

5.1 Find the distribution of the nth record height $X_{U_n} = M_{U_n}$ if F is standard exponential.

Notes

Some selected references on records are Resnick [148], Pfeifer [140], Arnold, Balakrishnan and Nagaraja [4], and Löwe [111].

Chapter X: Dependence and Further Topics in Risk Management

In an abundance of settings, one will as a first modeling attempt assume independence of the r.v.s involved. This is mathematically very convenient but extremely dangerous: in many situations where a disastrous event occurred despite its risk being calculated to be so low that it could be neglected for any practical purpose, a back mirror analysis has revealed that the reason was precisely the simultaneous occurrence of events assumed to be independent when doing the risk calculation. The study of dependence and how it may be modeled is therefore crucial, and this is the topic of this chapter.

A recurrent theme is the study of the influence of the dependence structure of a random vector $X = (X_1 \ldots X_d)$ with fixed marginals $F_i(x_i) = \mathbb{P}(X_i \leq x_i)$. The class of such X is called a *Fréchet class* and we will write $X \in \text{Fréchet}(F_1, \ldots, F_d)$.

Quite a few topics are less technical when distributions are assumed continuous. We do treat the general case for most central issues, but also sometimes assume continuity for the sake of simplicity and intuition.

On the technical side, the representation $X \overset{\mathscr{D}}{=} F_X^{\leftarrow}(U)$ with U uniform$(0, 1)$ of an r.v. with c.d.f. F_X will play a major role in large parts of the chapter. Here $F_X^{\leftarrow}$ is the generalized inverse of F_X. It is further important that if F_X is continuous, then $F_X(X)$ is uniform$(0, 1)$. A generalization, the so-called distributional transform, to distributions with atoms will also come into play. For the necessary background, see Sects. A.8–A.9 in the Appendix.

© Springer Nature Switzerland AG 2020
S. Asmussen, M. Steffensen, *Risk and Insurance*, Probability Theory and Stochastic Modelling 96, https://doi.org/10.1007/978-3-030-35176-2_10

1 Risk Measures

1.1 General Theory

In this section, we discuss approaches to quantifying the risk by means of a single number $\varrho(X)$.[1] We are loosely thinking of large values of X as 'bad' so that roughly $\varrho(X)$ should be larger the larger X is, the heavier the right tail is, etc.

The r.v. X may be, for example, the loss of an investor or the claims filed by an insured (or the whole portfolio) in a given period. Developing the concept of risk measures is largely motivated by regulations in these areas: authorities demand that companies or institutions can document their solvency and/or robustness to changes in the economic environment. A single number as 'documentation' is then more convenient than a mathematical model which, despite the potential to be more elaborate and to cover more aspects, is more difficult to comprehend, not least for people with a non-mathematical background.

Example 1.1 One popular risk measure is the VaR (Value-at-Risk) defined in terms of a suitable quantile in the distribution. In non-mathematical terms, VaR_α is the number such that the probability of losing more than VaR_α is approximately $1 - \alpha$. There are two variants, the financial VaR and the actuarial one. In insurance, one could think of X as the size of a single claim, all claims accumulated over a year, etc. Thus a large value is bad, and accordingly the actuarial VaR is defined as the α-quantile. In finance, one often uses the opposite interpretation of counting small values of X as bad so that roughly $\varrho(X)$ should be the larger the smaller X is, the heavier the left tail is, etc.; this could correspond to X being the value of an investor's portfolio. Accordingly, the financial VaR_α is defined as minus the $1 - \alpha$ quantile. See Fig. 1.

We treat here the actuarial VaR. More precisely, assuming the c.d.f. of X to be continuous and strictly increasing, for $0 < \alpha < 1$ the (left) α-quantile $v = F^\leftarrow(\alpha)$ defined as the solution of $\mathbb{P}(X \leq v) = \mathbb{P}(X < v) = \alpha$ exists and is unique, and the VaR_α at level α is defined as v. See Fig. 1 for an illustration with $\alpha = 0.95 = 95\%$, $\text{VaR}_\alpha = 4.3$. If F has jumps and/or the support has gaps, the definition is $\text{VaR}_\alpha = F^\leftarrow(\alpha)$, where $F^\leftarrow$ is the generalized inverse defined in Sect. A.8 in the Appendix. That is, $\text{VaR}_\alpha(X)$ is the smallest number satisfying

$$\mathbb{P}\big(X < \text{VaR}_\alpha(X)\big) \;\leq\; \alpha \;\leq\; \mathbb{P}\big(X \leq \text{VaR}_\alpha(X)\big). \tag{1.1}$$

---

[1] A risk measure ϱ may be defined on the class of all r.v.s X or a subclass such as those with $\mathbb{E}|X| < \infty$ or, more generally, L^p with $1 \leq p \leq \infty$.

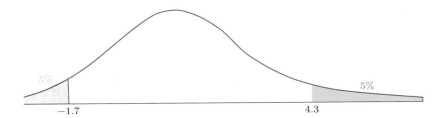

Fig. 1 Green: financial $\mathrm{VaR}_{0.95}(X) = 1.7$. Red: actuarial $\mathrm{VaR}_{0.95}(X) = 4.3$

The question arises what are the properties of the VaR risk measure and, more generally, which properties are desirable for a risk measure ϱ to have? Here are some of the main possibilities that have been suggested:

- Law invariance: if $X \overset{\mathscr{D}}{=} Y$, then $\varrho(X) = \varrho(Y)$.
- Monotonicity: if $X \leq Y$ a.s. then $\varrho(X) \leq \varrho(Y)$.
- Cash invariance: $\varrho(X + m) = \varrho(X) + m$ [also sometimes called translation invariance].
- Positive homogeneity: $\varrho(\lambda X) = \lambda\varrho(X)$ for $\lambda > 0$.
- Subadditivity: $\varrho(X + Y) \leq \varrho(X) + \varrho(Y)$.
- Convexity: $\varrho(\lambda X + (1 - \lambda)Y) \leq \lambda\varrho(X) + (1 - \lambda)\varrho(Y)$ for $0 \leq \lambda \leq 1$.
- Comonotonic additivity: If $X_1, \ldots, X_n$ are comonotone,[2] then

$$\varrho(X_1 + \cdots + X_n) = \varrho(X_1) + \cdots + \varrho(X_n).$$

Remark 1.2 When law invariance holds, the monotonicity condition $X \leq Y$ may be relaxed to stochastic order, cf. Sect. VIII.1. For risk measures defined according to the financial tradition of counting small values as bad, the monotonicity axiom becomes $X \leq Y \Rightarrow \varrho(X) \geq \varrho(Y)$ whereas cash invariance takes the form $\varrho(X + m) = \varrho(X) - m$.   ◇

Based on discussion and analysis of these concepts, the following main class of risk measures have emerged (one sometimes also encounters the smaller class of *monetary* risk measures where only monotonicity and cash invariance is required):

Definition 1.3 A risk measure is *coherent* if it is monotone, cash invariant, positive homogeneous and convex.

Law invariance seems self-evident but note that it is not assumed. Some law dependent risk measures have been suggested in connection with hedging, see also (1.6) below. Monotonicity also seems self-evident. Thus, the crucial properties characterising a coherent risk measure are positive homogeneity and convexity (or

---

[2]For the moment, just think of the case of identical marginals where comonotonicity means $X_1 = \cdots = X_n$. For the general case, see Sect. 2.

subadditivity, cf. Proposition 1.4 below). We present some motivating views on convexity in Sect. 1.4 below, but remark that positive homogeneity has not always been considered uncontroversial in the actuarial literature.

Note that independence of X, Y is not required in the definition of convexity. Comonotonic additivity is closely related to positive homogeneity. In fact, if X_1, X_2 have the same distribution and are comonotone, then $\varrho(X_1) = \varrho(X_2)$ if ϱ is law invariant and $X_1 + X_2 = 2X_1$, so comonotone additivity requires $\varrho(2X) = 2\varrho(X)$ as for positive homogeneity.

Cash invariance implies the important property

$$\varrho\big(X - \varrho(X)\big) = 0 \, ; \tag{1.2}$$

if we think of non-positive values of ϱ as being 'acceptable' or 'risk-free', this means that $\varrho(X)$ is the minimal amount that should be subtracted to make X riskless. From $\varrho(a) = \varrho(0 + a)$ we also get

$$\varrho(a) = \varrho(0) + a \, . \tag{1.3}$$

Here one may note that positive homogeneity makes it plausible that $\varrho(0) = 0$, but this is not in the list of requirements for a coherent risk measure and the argument requires additional continuity assumptions.

Proposition 1.4 *Assume ϱ is positive homogeneous. Then subadditivity and convexity are equivalent.*

Proof If ϱ is positive homogeneous and subadditive, then

$$\varrho\big(\lambda X + (1 - \lambda)Y\big) \leq \varrho(\lambda X) + \varrho\big((1 - \lambda)Y\big) = \lambda\varrho(X) + (1 - \lambda)\varrho(Y) \, ,$$

so ϱ is convex. Conversely, if ϱ is positive homogeneous and convex, then

$$\varrho(X + Y) = 2\varrho(X/2 + Y/2) \leq \varrho(X) + \varrho(Y) \, ,$$

so ϱ is subadditive.   □

1.2 Value-at-Risk and Expected Shortfall

VaR is clearly monotone and cash invariant, and hence monetary. It is also positive homogeneous. However:

Proposition 1.5 *VaR is not subadditive and hence not coherent.*

Proof The counterexamples on subadditivity are often given in the framework of discrete distributions, but to demonstrate that they have nothing to do with

Fig. 2 Densities of X (left) and $X + Y$ (right). Areas shaded in red (blue) are the ones with mass $3/4 + \epsilon$, resp. $1/4 - \epsilon$

abnormalities for this situation, we give two in the continuous set-up, even if they may be slightly longer.

In the first, let X, Y be independent with $X \overset{\mathscr{D}}{=} Y$ and assume that the common distribution is a mixture of two uniform distributions on $(-2, -1)$, resp. $(1, 2)$, with weights $1/4$ and $3/4$ for each. Then if $\alpha = 1/4 - \epsilon$ for some small ϵ, we have $\mathrm{VaR}_\alpha(X) = -1 - \epsilon_1$, $\mathrm{VaR}_\alpha(X + Y) = -\epsilon_2$ for some small $\epsilon_1, \epsilon_2 > 0$, so $\mathrm{VaR}_\alpha(X + Y) > \mathrm{VaR}_\alpha(X) + \mathrm{VaR}_\alpha(Y) = 2\,\mathrm{VaR}_\alpha(X)$, cf. Fig. 2.

For one more example, let $X_1, \ldots, X_n$ be i.i.d. potential losses and consider two portfolios A), B) such that portfolio A) consists of n copies of asset 1 and portfolio B) of 1 copy of each of assets $1, \ldots, n$. That is, portfolio A) has total loss nX_1 and portfolio B) $X_1 + \cdots + X_n$. Thus B) is much more diversified than A) and therefore should have a much smaller VaR. But take the common distribution as a mixture of two uniform distributions on $(99, 101)$, resp. $(-6, -4)$, with weights $1/21$ and $20/21$ for each, take $\alpha = 95\%$ and let $z = \mathrm{VaR}_\alpha(X_1)$. Since $\alpha < 20/21$, we have $z \in (-6, -4)$; more precisely one finds $z = -4.048$. Now $\mathbb{E}X = 0$ and hence $X_1 + \cdots + X_n$ is approximately normal$(0, n\sigma^2)$ where $\sigma^2 = \mathbb{V}\mathrm{ar}\,X = 10{,}026$. Since the 95% quantile in the normal distribution is 1.61, it follows that for large n

$$\mathrm{VaR}^A_{0.95} = n\,\mathrm{VaR}_{0.95}(X_1) = -0.4048\,n$$

$$\ll 161\sqrt{n} \approx 1.61\sqrt{n\sigma^2} \approx \mathrm{VaR}^B_{0.95},$$

a conclusion that can also be written as

$$\mathrm{VaR}_{0.95}(X_1) + \cdots + \mathrm{VaR}_{0.95}(X_n) \ll \mathrm{VaR}_{0.95}(X_1 + \cdots + X_n). \qquad \square$$

In addition to not being subadditive and hence not necessarily favoring diversification, VaR is often criticised for only taking into account the probability of a big loss, not the size of the loss. A risk measure taking this into account is the *expected shortfall* $\mathrm{ES}_\alpha(X)$ which under the minor regularity condition $\mathbb{P}(X = \mathrm{VaR}_\alpha(X)) = 0$ takes the form $\mathbb{E}[X \mid X \geq \mathrm{VaR}_\alpha(X)]$ (cf. Proposition 1.6 below). The formal mathematical definition is

$$\mathrm{ES}_\alpha(X) = \frac{1}{1-\alpha} \int_\alpha^1 \mathrm{VaR}_\lambda(X)\,d\lambda = \frac{1}{1-\alpha} \int_\alpha^1 F_X^\leftarrow(\lambda)\,d\lambda \qquad (1.4)$$

(expected shortfall is therefore often also called the *average value-at-risk*). However:

Proposition 1.6 $\mathrm{ES}_\alpha(X) = \mathrm{VaR}_\alpha(X) + \dfrac{1}{1-\alpha}\mathbb{E}\big[X - \mathrm{VaR}_\alpha(X)\big]^+$.

In particular, $\mathrm{ES}_\alpha(X) = \mathbb{E}\big[X \mid X \geq \mathrm{VaR}_\alpha(X)\big] = \mathbb{E}\big[X \mid X > \mathrm{VaR}_\alpha(X)\big]$ *when* $\mathbb{P}\big(X = \mathrm{VaR}_\alpha(X)\big) = 0$.

Proof By law invariance, we may assume $X = F_X^\leftarrow(U)$ with U uniform$(0, 1)$. The first part of the proposition then follows from

$$\mathrm{ES}_\alpha(X) - \mathrm{VaR}_\alpha(X) = \frac{1}{1-\alpha}\int_\alpha^1 \big[F_X^\leftarrow(\lambda) - \mathrm{VaR}_\alpha(X)\big]\,d\lambda$$

$$= \frac{1}{1-\alpha}\mathbb{E}\big[F_X^\leftarrow(U) - \mathrm{VaR}_\alpha(X)\big]^+ = \frac{1}{1-\alpha}\mathbb{E}\big[X - \mathrm{VaR}_\alpha(X)\big]^+.$$

For the second, note that by (1.1) we have

$$\mathbb{P}\big(X > \mathrm{VaR}_\alpha(X)\big) = \mathbb{P}\big(X \geq \mathrm{VaR}_\alpha(X)\big) = 1 - \alpha$$

when $\mathbb{P}\big(X = \mathrm{VaR}_\alpha(X)\big) = 0$. □

Proposition 1.7 *The expected shortfall in* (1.4) *is a subadditive and hence coherent risk measure.*

Proof We give the proof in the class of continuous distributions only. Let X, Y be risks and $Z = X + Y$. Let further for a given α q_X, q_Y, q_Z be the level α VaRs for X, Y, Z and define $I_X = \mathbb{1}(X > q_X)$, etc. Then

$$(1 - \alpha)\big(\mathrm{ES}_\alpha(Z) - \mathrm{ES}_\alpha(X) - \mathrm{ES}_\alpha(Y)\big) = \mathbb{E}\big[ZI_Z - XI_X - YI_Y\big]$$
$$= \mathbb{E}\big[X(I_Z - I_X) + Y(I_Z - I_Y)\big]. \tag{1.5}$$

Now $X < q_X$ when $I_X = 0$. Then also $I_Z - I_X \geq 0$ since $I_Z \geq 0$ and hence $X(I_Z - I_X) \leq q_X(I_Z - I_X)$. If $I_X = 1$, then $X > q_X$ and $I_Z - I_X \leq 0$, and hence also $X(I_Z - I_X) \leq q_X(I_Z - I_X)$. Thus

$$\mathbb{E}\big[X(I_Z - I_X)\big] \leq q_X\mathbb{E}(I_Z - I_X) = q_X\big((1 - \alpha) - (1 - \alpha)\big) = 0.$$

Together with a similar estimate for Y, we conclude that (1.5) is non-positive, which gives the assertion. □

There exist several proofs of Proposition 1.7, see Embrechts and Wang [67]. We return to one in Sect. 2.1; it is hardly shorter but may be more iluminating.

Example 1.8 Let L be a linear subspace of $L_2(\Omega, \mathscr{F}, \mathbb{P})$ such that any $X \in L$ has a normal distribution and consider for $\alpha > 50\%$ VaR_α as a risk measure on L. With z_α the α-quantile of the standard normal distribution, we then have

$$\mathrm{VaR}_\alpha(X) = \mu_X + z_\alpha\sigma_X, \quad X \in L,$$

where $\mu_X = \mathbb{E}X$, $\sigma_X^2 = \mathbb{V}\mathrm{ar}(X)$. By Cauchy–Schwarz,

$$\sigma_{X+Y}^2 = \sigma_X^2 + \sigma_Y^2 + 2\,\mathbb{C}\mathrm{ov}(X, Y) \leq \sigma_X^2 + \sigma_Y^2 + 2\sigma_X\sigma_Y = (\sigma_X + \sigma_Y)^2$$

and $z_\alpha > 0$ then yields

$$\begin{aligned}
\mathrm{VaR}_\alpha(X + Y) &= \mu_{X+Y} + z_\alpha\sigma_{X+Y} \leq \mu_X + \mu_Y + z_\alpha(\sigma_X + \sigma_Y) \\
&= \mathrm{VaR}_\alpha(X) + \mathrm{VaR}_\alpha(Y),
\end{aligned}$$

i.e., VaR_α *is a subadditive and hence coherent risk measure when restricted to* L. ◊

1.3   Further Remarks

The expectation $\mathbb{E}X$ is an obvious candidate for a risk measure, but obviously a very crude one. In particular, it considers risk as symmetric in the sense that the prospect of gains is assessed in the same way as the risk of losses, whereas the emphasis is usually on losses. Inspired by axiomatic discussions of insurance premiums and mean-variance portfolio theory, one could try to take the variance or standard deviation into account by considering

$$\varrho(X) = \mathbb{E}X + c\,\mathbb{V}\mathrm{ar}(X) \quad \text{or} \quad \varrho(X) = \mathbb{E}X + c\sqrt{\mathbb{V}\mathrm{ar}(X)}.$$

However, here even the requirement of monotonicity fails.

It has also been suggested to use expectations in an extended way by considering

$$\varrho(X) = \sup_{Q \in \mathcal{Q}} \mathbb{E}_Q X, \tag{1.6}$$

where $\mathcal{Q}$ is a suitable class of probability measures. Here, law invariance does not hold.

Example 1.9 The property $\varrho(X - \varrho(X)) = 0$ in (1.2) leads to considering precisely the risks with $\varrho(X) \leq 0$ as acceptable. Turning this around, one could start from an *acceptance set* $\mathcal{A} \subseteq \mathbb{R}$ and define

$$\varrho(X) = \inf\{m : X - m \in \mathcal{A}\}.$$

The properties of this risk measure obviously depend on the properties of $\mathcal{A}$. For example, it is convex if $\mathcal{A}$ is so.

A particular case is $\mathcal{A} = \{x : u(x) \geq u_0\}$ where u is some utility function and u_0 is some threshold value.

Concavity of u yields convexity of $\mathcal{A}$, and the definition of the risk measure reduces to

$$\varrho(X) \;=\; \inf\{m \,:\, u(X - m) \geq u_0\}\,. \qquad\qquad \Diamond$$

1.4 Quotes on Risk Diversification

The axioms of subadditivity (or convexity, cf. Proposition 1.4) are motivated by the economic principle that diversification should not increase risk. It may not be straightforward to understand, in particular for a person with a different background, and so we have collected (in alphabetical order) some of the explanations of various authors:

- Artzner, Delbaen, Ebere and Heath in [5]:
    A merger does not create extra risk.
- Biagini, Meyer-Brandis and Svindland in [101], p. 141:
    Consider the possibilities to invest in either X or Y or in a fraction $\lambda X + (1 - \lambda)Y$ of both. Favoring diversification means that the risk of the diversified investment $\lambda X + (1 - \lambda)Y$ should not exceed the risks of both X and Y, thereby accounting for the fact that the downside risk is lower in the diversified investment $\lambda X + (1 - \lambda)Y$ as compared to the most risky of X and Y. Formally this property is known as quasi-convexity of ϱ:

$$\varrho\big(\lambda X + (1 - \lambda)Y\big) \leq \max\big(\varrho(X), \varrho(Y)\big)\,.$$

If the risk measure satisfies cash invariance then it can indeed be shown that quasi-convexity is equivalent to convexity.
- Föllmer and Schied in [77], p. 159:
    Consider the collection of future outcomes that can be generated with the resources available to an investor: One investment strategy leads to X, while a second strategy leads to Y. If one *diversifies*, spending only a fraction λ on the resources and using the remaining part for the second alternative, one obtains $\lambda X + (1 - \lambda)Y$. Thus, the axiom of convexity gives a precise meaning to the idea that diversification should not increase the risk.
- Hult, Lindskog, Hammarlid and Rehn in [93]:
    [on convexity p. 161; to understand this remark, one may compare with the quote from Biagini et al. above] For example, investing a fraction of the initial capital in one stock and the remaining in another stock, rather than everything in the more risky stock, reduces the overall risk.
    [on subadditivity p. 162] This property says that diversification should be rewarded. A bank consisting of two units should be required to put away less buffer capital than the sum of the buffer capital for the two units considered as separate entities. In particular if the regulator enforces the use of a subadditive

risk measure, then it does not encourage companies to break into parts in order to reduce the buffer capital requirements.

- Klüppelberg and Stelzer in [101], p. 244:

    For a risk measure to be coherent it is necessary that the risk measure of the sum of the two risks is always below the sum of the two risk measures. Since e.g. banks estimate the VaR for each unit and add all resulting VaRs up to estimate the risk of the whole bank, the use of VaR may underestimate the true bank's VaR considerably.

- McNeil, Frey and Embrechts in [116], p. 273:

    First subadditivity reflects the idea that risk can be reduced by diversification, a time-honoured principle in finance and economics. Second, if a regulator uses a non-subadditive risk measure in determining the regulatory capital for a financial institution, that institution has an incentive to legally break up in various subsidiaries in order to reduce its regulatory capital requirements. Similarly, if the risk measure used by an organized exchange in determining the margin requirements of investors is not subadditive, an investor could reduce the margin he has to pay by opening a different account for every position in his portfolio. Finally, subadditivity makes decentralization of risk-management systems possible. Consider as an example two trading desks with positions leading to losses L_1 and L_2. Imagine that a risk manager wants to insure that $\varrho(L)$, the risk of the overall loss $L = L_1 + L_2$, is smaller than some number M. If he uses a subadditive risk measure ϱ, he may simply choose bounds M_1 and M_2 such that $M_1 + M_2 \leq M$ and impose on each of the desks the constraint that $\varrho(L_i) \leq M_i$; subadditivity of ϱ then automatically ensures that $\varrho(L) \leq M_1 + M_2 \leq M$.

- Woo in [175], p. 276:

    There should be some benefits in pooling risks. The PML [Probable Maximum Loss, p. 266] for the joint X and Y portfolios should not exceed the separate PMLs for A and B.

Exercises

1.1 As discussed in Sect. I.3, two of the most common premium rules in insurance are the variance principle $\varrho(X) = \mathbb{E}X + \beta \mathbb{V}\mathrm{ar}(X)$ and the standard deviation principle $\varrho(X) = \mathbb{E}X + \beta\sqrt{\mathbb{V}\mathrm{ar}(X)}$. Which of the axioms for risk measures are satisfied?

Notes

The fundamental paper on coherent risk measures is Artzner et al. [5], with some roots in Deprez and Gerber [59]. A somewhat related axiomatic discussion from the insurance point of view is in Wang, Young and Panjer [174]. Some main textbook references are McNeil et al. [115], Föllmer and Schied [77] and Rüschendorf [154]. The research literature is immense and will not be reviewed here.

Historically, the boom of interest in VaR was initiated by the regulations for banking in the Basel II documents, see e.g. [20], and for insurance in [92]. The typical values of α are 99.97% for Economic Capital and 95% to 99.9% for regulatory capital, whereas in insurance Solvency II [159] takes it as $\alpha = 97.5\%$. Expected shortfall came up in later versions like Basel III [21].

2 The Fréchet–Höffding Bounds. Comonotonicity

Let $X = (X_1, \ldots, X_d) \in \mathbb{R}^d$ be a random vector. Its c.d.f. is

$$F(x_1, \ldots, x_d) = \mathbb{P}(X_1 \leq x_1, \ldots, X_d \leq x_d)$$

and the marginals are

$$F_i(x_i) = \mathbb{P}(X_i \leq x_i) = \lim_{x_j \to \infty, j \neq i} F(x_1, \ldots, x_d).$$

Theorem 2.1 (Fréchet–Höffding) *For $X \in$ Fréchet$(F_1, \ldots, F_d)$,*

$$\left(\sum_{i=1}^d F_i(x_i) - (d-1) \right)^+ \leq F(x_1, \ldots, x_d) \leq \min_{i=1,\ldots,d} F_i(x_i).$$

Proof The upper bounds follows immediately since for each i

$$\mathbb{P}(X_1 \leq x_1, \ldots, X_d \leq x_d) \leq \mathbb{P}(X_i \leq x_i) = F_i(x_i).$$

For the lower one, use $F(x_1, \ldots, x_d) \geq 0$ together with

$$1 - F(x_1, \ldots, x_d) = \mathbb{P}\big(\{X_1 \leq x_1, \ldots, X_d \leq x_d\}^c\big) = \mathbb{P}\Big(\bigcap_{i=1}^d \{X_i \leq x_i\} \Big)^c$$

$$= \mathbb{P}\Big(\bigcup_{i=1}^d \{X_i > x_i\} \Big) \leq \sum_{i=1}^d \mathbb{P}(X_i > x_i) = \sum_{i=1}^d 1 - F_i(x_i).$$

$\square$

The upper Fréchet–Höffding bound can be interpreted in terms of *comonotonicity*. In the case $F_1 = \cdots = F_d$ of identical marginals, comonotonicity of $(X_1, \ldots, X_d)$ means $X_1 = \cdots = X_d$ a.s. In general, note that we have $X_i \overset{\mathscr{D}}{=} F_i^\leftarrow(U_i)$ with U_i uniform$(0, 1)$, where $F^\leftarrow$ is the generalized inverse of F. We then call $(X_1, \ldots, X_d)$ comonotonic if we can use the same U for all i, i.e. if

$$(X_1, \ldots, X_d) \overset{\mathscr{D}}{=} \big(F_1^\leftarrow(U), \ldots, F_d^\leftarrow(U) \big)$$

with U uniform$(0, 1)$. We have:

Corollary 2.2 *The following assertions are equivalent for a random vector $(X_1, \ldots, X_d)$:*

(i) $(X_1, \ldots, X_d)$ *is comonotonic;*
(ii) $(X_1, \ldots, X_d)$ *attains the upper Fréchet–Höffding bound;*

(iii) *there exists an r.v. Z and non-decreasing functions $h_1, \ldots, h_d$ such that*
$$(X_1, \ldots, X_d) \overset{\mathcal{D}}{=} \big(h_1(Z), \ldots, h_d(Z)\big);$$
(iv) *for some comonotonic set B, $X \in B$ a.s.*

[A set $B \subseteq \mathbb{R}^d$ is comonotonic if $x, y \in B$ implies that either $x_i \leq y_i$ for all i or $y_i \leq x_i$ for all i.]

Proof Since $F_1^{\leftarrow}(u) \leq x$ if and only if $u \leq F(x)$ (part 4 of Proposition A.2 in the Appendix), we have

$$\mathbb{P}\big(F_1^{\leftarrow}(U) \leq x_1, \ldots, F_d^{\leftarrow}(U) \leq x_d\big) = \mathbb{P}\big(U \leq F_1(x_1), \ldots, U \leq F_d(x_d)\big)$$
$$= \mathbb{P}\Big(U \leq \min_{i=1,\ldots,d} F_i(x_i)\Big) = \min_{i=1,\ldots,d} F_i(x_i).$$

This gives (i) $\Rightarrow$ (ii), and the converse follows since the distribution of a random vector is uniquely specified by its c.d.f.

For (i) $\Rightarrow$ (iii), we can just take $Z = U$, $h_i = F_i^{\leftarrow}$. Assume conversely that (iii) holds. Then

$$F(x_1, \ldots, x_d) = \mathbb{P}\big(h_1(Z) \leq x_1, \ldots, h_d(Z) \leq x_d\big) = \mathbb{P}(Z \in A_1, \ldots, Z \in A_d),$$

where each A_i is an interval of the form $(-\infty, v_i]$ or $(-\infty, v_i)$. Therefore for some i we have $A_i \subseteq A_j$ for all j. This gives

$$F(x_1, \ldots, x_d) = \mathbb{P}(Z \in A_i) = \min_{j=1,\ldots,d} \mathbb{P}(Z \in A_j) = \min_{j=1,\ldots,d} F_j(x_j)$$

so that (ii) holds.

We omit the proof that (iv) is equivalent to (i)–(iii). $\square$

Comonotonicity is often interpreted as 'maximal positive dependence'. This is, however, not universally true in the sense of leading to the greatest risk (for a counterexample, see Example 7.4), but certainly, with fixed marginals comonotonicity maximizes covariance (and hence correlation):

Corollary 2.3 *For any two r.v.s $X_1, X_2 \in L^2$,*

$$\mathbb{C}\mathrm{ov}(X_1, X_2) \leq \mathbb{C}\mathrm{ov}\big(F_1^{\leftarrow}(U), F_2^{\leftarrow}(U)\big)$$
$$= \int_{-\infty}^{\infty} \int_{-\infty}^{\infty} \big[\min\big(F_1(x_1), F_2(x_2)\big) - F_1(x_1)F_2(x_2)\big]\,dx_1\,dx_2.$$

Proof By Proposition A.1 in Sect. A.1, $\mathbb{C}\mathrm{ov}(X_1, X_2)$ equals

$$\int_{-\infty}^{\infty} \int_{-\infty}^{\infty} \big[F(x_1, x_2) - F_1(x_1)F_2(x_2)\big]\,dx_1\,dx_2.$$

Now just insert the upper Fréchet–Höffding bound. $\square$

The situation for the lower Fréchet–Höffding bound is only clear cut for $d = 2$ and so is the concept of 'maximal negative dependence'. Call X_1, X_2 *countermonotonic* if

$$(X_1, X_2) \overset{\mathscr{D}}{=} \left(F_1^{\leftarrow}(U), F_2^{\leftarrow}(1 - U)\right)$$

with U uniform$(0, 1)$.

Proposition 2.4 *For $d = 2$, the lower Fréchet–Höffding bound is attained for (X_1, X_2) countermonotonic. For $d > 2$ and continuous F_i, the lower Fréchet–Höffding bound is not the c.d.f. of a random vector, and hence unattainable.*

Proof The first statement follows from

$$\mathbb{P}\left(F_1^{\leftarrow}(U) \leq x_1, F_2^{\leftarrow}(1 - U) \leq x_2\right) = \mathbb{P}\left(U \leq F_1(x_1), 1 - U \leq F_2(x_2)\right)$$
$$= \mathbb{P}\left(1 - F_2(x_2) \leq U \leq F_1(x_1)\right) = \left[F_1(x_1) - \left(1 - F_2(x_2)\right)\right]^+ .$$

For the second, assume $\left(\sum_1^d F(x_i) - (d - 1)\right)^+$ was a c.d.f. If we let $U_i = F_i(X_i)$ and assume all F_i are continuous, the c.d.f. of $(U_1, \ldots, U_d)$ is then

$$\mathbb{P}\left(F_1(X_1) \leq u_1, \ldots, F_d(X_d) \leq u_d)\right) = \mathbb{P}\left(X_1 \leq F_1^{\leftarrow}(u_1), \ldots, X_d \leq F_d^{\leftarrow}(u_d)\right)$$
$$= \left[\sum_{i=1}^{d} F_i(F_i^{\leftarrow}(u_i)) - (d - 1)\right]^+ = \left[\sum_{i=1}^{d} u_i - (d - 1)\right]^+ \qquad (2.1)$$

(if continuity of F_i fails for some i, use the distributional transform in the Appendix). Thus, it suffices the consider the uniform case, assuming (2.1) is the c.d.f. of a random vector $(U_1, \ldots, U_d)$ with uniform marginals. In

$$\mathbb{P}\left((U_1, \ldots, U_d) \in (1/2, 1]^d\right) = 1 - \mathbb{P}\left(\bigcup_{i=1}^{d}\{U_i \leq 1/2\}\right),$$

we use the inclusion-exclusion formula

$$\mathbb{P}\left(\bigcup_{i=1}^{d} A_1\right) = \sum_{i=1}^{d} \mathbb{P}(A_1) - \sum_{\substack{i,j=1 \\ i \neq j}}^{d} \mathbb{P}(A_i \cap A_j) + \sum_{\substack{i,j,k=1 \\ i \neq j \neq k}}^{d} \mathbb{P}(A_i \cap A_j \cap A_k) - \cdots$$

with $A_i = \{U_i \leq 1/2\}$. For $n \geq 2$, insertion of the assumed lower Fréchet–Höffding bound gives

$$\mathbb{P}(A_{i_1} \cap \cdots \cap A_{i_n}) = \left(n/2 + (d - n) - (d - 1)\right)^+ = (1 - n/2)^+ = 0$$

(n of the U_i must be at most $1/2$ whereas there are no restrictions on the remaining $d - n$). Thus

$$\mathbb{P}\big((U_1, \ldots, U_d) \in (1/2, 1]^d\big) \;=\; 1 - d\mathbb{P}(A_1) \;=\; 1 - d/2\,,$$

which is < 0 for $d > 2$, a contradiction. □

In terms of covariances, we get the following analogue of Proposition 2.3:

Corollary 2.5 *For any two r.v.s $X_1, X_2 \in L^2$,*

$$\mathbb{C}\mathrm{ov}(X_1, X_2) \;\geq\; \mathbb{C}\mathrm{ov}\big(F_1^\leftarrow(U), F_2^\leftarrow(1 - U)\big)$$

$$= \int_{-\infty}^{\infty} \int_{-\infty}^{\infty} \big[F_1(x_1) + F_2(x_2) - 1\big) - F_1(x_1)F_2(x_2)\big]\,\mathrm{d}x_1\,\mathrm{d}x_2\,.$$

Example 2.6 For X_1, X_2 standard exponential, Corollary 2.5 gives the lower bound $-0.645\ldots$ on $\mathbb{C}\mathrm{orr}(X_1, X_2) = \mathbb{C}\mathrm{ov}(X_1, X_2)$. This follows since in the countermonotonic case,

$$\mathbb{E}[X_1 X_2] \;=\; \mathbb{E}\big(F_1^\leftarrow(U), F_2^\leftarrow(1 - U)\big] \;=\; \int_0^1 \log u \log(1 - u)\,\mathrm{d}u \;=\; 0.355\ldots$$

so the Fréchet–Höffding lower bound on the correlation is

$$\big(0.355 - (\mathbb{E}X_1)^2\big)/\mathbb{V}\mathrm{ar}\,X_1 \;=\; (0.355 - 1^2)/1 \;=\; -0.645\,. \qquad \diamond$$

Remark 2.7 Example 2.6 shows that in general $\mathbb{C}\mathrm{orr}(X_1, X_2) = -1$ cannot be attained. If X_1, X_2 have the same distribution, the upper bound on $\mathbb{C}\mathrm{ov}(X_1, X_2)$ in Proposition 2.3 is simply their common variance and so $\mathbb{C}\mathrm{orr}(X_1, X_2) = 1$ is attained for comonotonicity. More generally, it can be shown that correlation 1 is attained if and only X_1, X_2 are of the same type and -1 if and only $X_1, -X_2$ are so (this does not contradict Example 2.6 since it occurs precisely when the distribution of X_1 is symmetric). $\diamond$

2.1 Applications of Comonotonicity

The following result says that comonotonicity is a majorant in the convex ordering of sums with fixed marginals:

Theorem 2.8 *Assume that $X = (X_1, \ldots, X_d) \in \text{Fréchet}(F_1, \ldots, F_d)$, i.e. the marginals are $F_1, \ldots, F_d$. Then*

$$X_1 + \cdots + X_d \;\preceq_{\mathrm{cx}}\; F_1^\leftarrow(U) + \cdots + F_d^\leftarrow(U)$$

for U uniform$(0, 1)$.

Proof Since $F_i^{\leftarrow}(U)$ has distribution F_i, the two sums have the same mean, and so it suffices by Theorem VIII.2.4(II) to show that for any fixed t

$$\mathbb{E}\Big(\sum_{i=1}^{d} X_i - t\Big)^+ \leq \mathbb{E}\Big(\sum_{i=1}^{d} F_i^{\leftarrow}(U) - t\Big)^+. \tag{2.2}$$

We give the proof only for the case when all F_i and hence all $F_i^{\leftarrow}$ are continuous; for the general case, see [154] pp. 56–57. Then there is a unique $u_0 \in (0, 1)$ such that $\sum_1^d F_i^{\leftarrow}(u_0) = t$ and we write $v_i = F_i^{\leftarrow}(u_0)$; then $t = \sum_1^d v_i$. If $0 < u \leq u_0$ we then have $F_i^{\leftarrow}(u) \leq v_i$ whereas $F_i^{\leftarrow}(u) \geq v_i$ if $u_0 \leq u < 1$. Hence the $F_i^{\leftarrow}(u) - v_i$ are either all ≤ 0 or all ≥ 0, which gives the second equality in

$$\Big(\sum_{i=1}^{d} F_i^{\leftarrow}(u) - t\Big)^+ = \Big(\sum_{i=1}^{d} \big(F_i^{\leftarrow}(u) - v_i\big)\Big)^+ = \sum_{i=1}^{d} \Big(F_i^{\leftarrow}(u) - v_i\Big)^+ \tag{2.3}$$

for all $u \in (0, 1)$. Using this, we can bound the l.h.s. of (2.2) as desired:

$$\mathbb{E}\Big(\sum_{i=1}^{d} X_i - t\Big)^+ = \mathbb{E}\Big(\sum_{i=1}^{d} (X_i - v_i)\Big)^+ \leq \sum_{i=1}^{d} \mathbb{E}(X_i - v_i)^+$$

$$= \sum_{i=1}^{d} \mathbb{E}\big(F_i^{\leftarrow}(U) - v_i\big)^+ = \mathbb{E}\Big(\sum_{i=1}^{d} F_i^{\leftarrow}(U) - t\Big)^+,$$

where we used the subadditivity of the function $x \to x^+$ for the inequality, that $F_i^{\leftarrow}(U) \overset{\mathscr{D}}{=} X_i$ for the following equality and (2.3) for the last. $\qquad\square$

Example 2.9 We next give another proof of the subadditivity of expected shortfall (Proposition 1.7), valid also if not all $F_i^{\leftarrow}$ are continuous. Recalling $\mathrm{VaR}_\alpha(X) = F_X^{\leftarrow}(\alpha)$, we have

$$\mathrm{ES}_\alpha(X) = \mathrm{VaR}_\alpha(X) + \frac{1}{1-\alpha}\mathbb{E}\big[X - \mathrm{VaR}_\alpha(X)\big]^+ \tag{2.4}$$

$$= \frac{1}{1-\alpha}\int_\alpha^1 \mathrm{VaR}_\lambda(X)\,\mathrm{d}\lambda = \frac{1}{1-\alpha}\int_\alpha^1 F_X^{\leftarrow}(\lambda)\,\mathrm{d}\lambda$$

$$= \frac{1}{1-\alpha}\mathbb{E}\big[F_X^{\leftarrow}(U);\ U \geq \alpha\big], \tag{2.5}$$

where U is any uniform$(0, 1)$ r.v. The aim is to show $\mathrm{ES}_\alpha(Z) \leq \mathrm{ES}_\alpha(X) + \mathrm{ES}_\alpha(Y)$, where $Z = X + Y$ (and no independence or continuity assumptions on X, Y are in force).

We can assume (by extending the underlying probability space if necessary) that in addition to X, Y there are defined uniform$(0, 1)$ r.v.s V_Z, V_X, V_Y which are mutually independent and independent of X, Y. Let $D_Z = F_Z(Z-) + V_Z \Delta F(Z)$ be the distributional transform of Z (cf. Sect. A.9 of the Appendix), $B_Z = \mathbb{1}(D_Z \geq \alpha)$ and similarly for X, Y, so that D_Z, D_X, D_Y are uniform$(0, 1)$. Taking $U = D_X$ in (2.5) gives

$$\mathrm{ES}_\alpha(X) = \frac{1}{1-\alpha} \mathbb{E}\big[F_X^\leftarrow(D_X)B_X\big]$$

$$\geq \frac{1}{1-\alpha} \mathbb{E}\big[F_X^\leftarrow(D_X)B_Z\big] = \frac{1}{1-\alpha}\mathbb{E}[XB_Z],$$

where the last step used $F_X^\leftarrow(D_X) = X$ a.s. and the inequality follows from Corollary 2.3 by noting that by Corollary 2.2 (iii) $F_X^\leftarrow(D_X)$, B_X are comonotonic with $\mathbb{E}B_X = \mathbb{E}B_Z (= 1 - \alpha)$. Together with a similar estimate for Y we then get

$$\mathrm{ES}_\alpha(X) + \mathrm{ES}_\alpha(Y) \geq \frac{1}{1-\alpha}\big(\mathbb{E}[XB_Z] + \mathbb{E}[YB_Z]\big) = \frac{1}{1-\alpha}\mathbb{E}[ZB_Z]$$

$$= \frac{1}{1-\alpha}\mathbb{E}\big[F_Z^\leftarrow(D_Z)B_Z\big] = \mathrm{ES}_\alpha(Z),$$

where the last steps used similar identities as when evaluating $\mathrm{ES}_\alpha(X)$. $\diamond$

Exercises
2.1 Let U_1, U_2 be uniform$(0, 1)$. Compute the minimal attainable value of $\mathbb{C}\mathrm{orr}(U_1, U_2)$.

2.2 Let X_1, X_2 be Pareto with common tail $1/(1 + x)^\alpha$. Compute numerically the minimal attainable value of $\mathbb{C}\mathrm{orr}(X_1, X_2)$ and show its dependence on α in the form of a graph.

Notes
The Fréchet–Höffding bounds are standard. For the lower bound, a discussion of attainability in special cases and some actuarial applications are in Müller and Stoyan [126, Section 8.3].

3 Special Dependence Structures

When considering a set $X_1, \ldots, X_d$ of r.v.s with marginal distributions $F_1, \ldots, F_d$, one quite often has an idea of what are realistic assumptions on the F_i. How to specify the dependence structure, that is, the joint distribution of the random vector $X = (X_1, \ldots, X_d)$, is, however, usually less obvious. In this section, we consider a few possibilities.

3.1 The Multivariate Normal Distribution

A multivariate normal r.v. $X = X_{\mu,\Sigma} \in \mathbb{R}^d$ has distribution specified by its mean vector μ and its covariance matrix Σ. It is standard that the m.g.f. $\mathbb{E}e^{a^{\mathsf T}X}$ equals $\exp\{a^{\mathsf T}\mu + a^{\mathsf T}\Sigma a/2\}$ (in particular, the characteristic function at $t \in \mathbb{R}^d$ is $\exp\{it^{\mathsf T}\mu - t^{\mathsf T}\Sigma t/2\}$) and that the density exists when Σ is invertible and then equals

$$\frac{1}{(2\pi)^{d/2}|\Sigma|} \exp\left\{ - (x - \mu)^{\mathsf T}\Sigma^{-1}(x - \mu)/2\right\}.$$

In contrast, there is no analytic expression for the c.d.f., not even for $d = 2$.

It is sometimes convenient to represent X in terms of i.i.d. standard normals $Y_1, Y_2, \ldots$ Assume w.l.o.g. $\mu = \mathbf{0}$. The case $d = 2$ is simple. We can write

$$\Sigma = \begin{pmatrix} \sigma_1^2 & \rho\sigma_1\sigma_2 \\ \rho\sigma_1\sigma_2 & \sigma_2^2 \end{pmatrix},$$

where $\rho = \mathbb{C}\mathrm{orr}(X_1, X_2)$, take Y_1, Y_2, Y_3 independent $\mathcal{N}(0, 1)$, and let

$$X_1 = \sigma_1\left(\sqrt{1 - |\rho|}Y_1 + \sqrt{|\rho|}Y_3\right), \quad X_2 = \sigma_2\left(\sqrt{1 - |\rho|}Y_2 \pm \sqrt{|\rho|}Y_3\right),$$

where $+$ is for $\rho \geq 0$, $-$ for $\rho \leq 0$. An approach working for all values of d is to find a square root of Σ, that is, a matrix C satisfying $\Sigma = CC^{\mathsf T}$, since we can then represent $X_{0,\Sigma}$ as $CX_{0,I}$. Component by component,

$$X_i = \sum_{k=1}^{d} c_{ik}Y_k, \quad i = 1, \ldots, d, \tag{3.1}$$

where $Y_1, \ldots, Y_d$ are i.i.d. $\mathcal{N}(0, 1)$ r.v.s.

An obvious candidate for C is $\Sigma^{1/2}$, the nonnegative definite (symmetric) square root of Σ. This can be found from a diagonal form $\Sigma = B(\lambda_i)_{\mathrm{diag}}B^{\mathsf T}$ with B orthonormal by letting $\Sigma^{1/2} = B(\lambda_i^{1/2})_{\mathrm{diag}}B^{\mathsf T}$. Another much used approach is to use Cholesky factorization to compute C as lower triangular.

3.2 Normal Mixtures

The multivariate t distribution is defined as the distribution of

$$\left(\frac{Y_1}{\sqrt{A}}, \ldots, \frac{Y_d}{\sqrt{A}}\right),$$

where $Y_1, \ldots, Y_d$ are i.i.d. standard normals, and A an independent χ_f^2/f r.v. ($f =$ degrees of freedom); note that A is the same in all components!

A first generalization is to take $\boldsymbol{Y} = (Y_1 \ldots Y_d)^{\mathsf{T}}$ multivariate normal $(\boldsymbol{0}, \boldsymbol{\Sigma})$. Going one step further, a *normal mixture* is defined as the distribution of

$$\boldsymbol{X} = \boldsymbol{\mu}(W) + \sqrt{W}\,\boldsymbol{Z} \qquad \text{where} \quad \boldsymbol{Z} \sim \mathcal{N}(\boldsymbol{0}, \boldsymbol{\Sigma}), \tag{3.2}$$

W is a $(0, \infty)$-valued r.v. and $\boldsymbol{\mu}(\cdot)$ an $\mathbb{R}^d$-valued function.

If $d = 1$ and W only attains a finite set $\sigma_1^2, \ldots, \sigma_n^2$ of values, say w.p. θ_i for σ_i^2, we obtain a distribution on $\mathbb{R}$ with density of the form

$$\sum_{i=1}^{n} \theta_i \frac{1}{\sqrt{2\pi\sigma_i^2}} \exp\{-(x - \mu_i)^2/2\sigma_i^2\}.$$

An extremely important example for $d > 1$ is a *multivariate generalized hyperbolic distribution* which corresponds to $\boldsymbol{\mu}(w) = \boldsymbol{\mu} + w\boldsymbol{\gamma}$ and W having density

$$\frac{\chi^{-\lambda/2}\psi^{\lambda/2}}{2K_\lambda(\chi^{1/2}\psi^{\lambda/2})}\, w^{\lambda-1} \exp\{-(\chi/w + \psi w)/2\} \tag{3.3}$$

at $w > 0$, where K_λ denotes a modified Bessel function of the third kind. The multivariate generalized hyperbolic density comes out as

$$c\,\frac{K_{\lambda-d/2}\big(q(\boldsymbol{x})\big) \exp\left\{(\boldsymbol{x} - \boldsymbol{\mu})^{\mathsf{T}}\boldsymbol{\Sigma}^{-1}\boldsymbol{\gamma}\right\}}{q(\boldsymbol{x})^{d/2-\lambda}}, \tag{3.4}$$

where $q(\boldsymbol{x})^2 = \chi + (\boldsymbol{x} - \boldsymbol{\mu})^{\mathsf{T}}\boldsymbol{\Sigma}^{-1}(\boldsymbol{x} - \boldsymbol{\mu})$, $c = \dfrac{\chi^{-\lambda/2}\psi^{\lambda/2}(\psi + \boldsymbol{\gamma}^{\mathsf{T}}\boldsymbol{\Sigma}^{-1}\boldsymbol{\gamma})^{d/2-\lambda}}{(2\pi)^{d/2}|\boldsymbol{\Sigma}|^{1/2}K_\lambda(\chi^{1/2}\psi^{1/2})}$.

The multivariate generalized hyperbolic distribution has become quite popular in applications because of its flexibility to fit many different types of data.

The density (3.3) is known as the *generalized inverse Gaussian* (GIG) density, and for $\lambda = 2$ just as the inverse Gaussian (IG) density. The terminology comes from the IG distribution being the distribution of the time of first passage of Brownian motion with drift $\psi^{1/2}$ to level $\chi^{1/2}$.

3.3 Spherical and Elliptical Distributions

For vectors $\boldsymbol{a}, \boldsymbol{b} \in \mathbb{R}^d$, the usual scalar product is $\boldsymbol{a}^{\mathsf{T}}\boldsymbol{b}$ and the norm (2-norm) $\|\boldsymbol{a}\|_2$ of $\boldsymbol{a}$ is $(\boldsymbol{a}^{\mathsf{T}}\boldsymbol{a})^{1/2}$. That is, if $\boldsymbol{a} = (a_i)_{i=1,\ldots,d}$, $\boldsymbol{b} = (b_i)_{i=1,\ldots,d}$, then

$$\boldsymbol{a}^{\mathsf{T}}\boldsymbol{b} = a_1 b_1 + \cdots + a_d b_d, \quad \|\boldsymbol{a}\|_2^2 = a_1^2 + \cdots + a_d^2.$$

The unit sphere in $\mathbb{R}^d$ is denoted $\mathcal{B}_2$ and consists of all vectors with $\|a\|_2 = 1$ (another much used notation is $\mathcal{S}^{d-1}$).

A matrix U is *orthonormal* if U is invertible with $U^{-1} = U^\mathsf{T}$. That $a = Ub$ for some orthonormal U defines an equivalence relation on $\mathbb{R}^d$, and the equivalence class of a is precisely the set of vectors b with $\|b\|_2 = \|a\|_2$, that is, the set $\|a\|\mathcal{B}_2$.

In the following, S will denote a random vector uniformly distributed on $\mathcal{B}_2$. For the formal definition, one may, for example, take this distribution as the distribution of $Y/\|Y\|_2$, where $Y \sim N_d(\mathbf{0}, I_d)$, or equivalently, as the conditional distribution of Y given $\|Y\|_2 = 1$. Since $UY \stackrel{\mathcal{D}}{=} Y$ for U orthonormal this implies $US \stackrel{\mathcal{D}}{=} S$.

A random vector X is said to have a *spherical distribution* if $X \stackrel{\mathcal{D}}{=} RS$, where $R \geq 0$ is a one-dimensional r.v. independent of S (for example, the $N_d(\mathbf{0}, I_d)$ case corresponds to R being χ^2 with d degrees of freedom).

Proposition 3.1 *The distribution of X is spherical if and only if either of the following two equivalent properties hold:*

(a) $UX \stackrel{\mathcal{D}}{=} X$ *for all orthogonal U;*
(b) *for any a, the distribution of $a^\mathsf{T}X$ depends only on $\|a\|_2$.*

Proof That (a) holds if X is spherical follows immediately from $US \stackrel{\mathcal{D}}{=} S$. That (a)$\Rightarrow$(b) follows since if $\|b\|_2 = \|a\|_2$, then $b = Ua$ for some orthonormal U and hence

$$b^\mathsf{T}X = a^\mathsf{T}U^\mathsf{T}X \stackrel{\mathcal{D}}{=} a^\mathsf{T}X$$

since U^T is orthogonal. We omit the proof that (b) implies (a).   □

A random vector X is said to have an *elliptical distribution* if $X \stackrel{\mathcal{D}}{=} \mu + AZ$, where $\mu \in \mathbb{R}^d$, Z has a spherical distribution on $\mathbb{R}^k$ for some k (possibly with $k \neq d$) and A is an $d \times k$ matrix. This class has also received much attention.

Exercises
3.1 Verify that $b = Ua$ for some orthonormal U if $\|b\|_2 = \|a\|_2$.

3.4   Multivariate Regular Variation

Recall that an r.v. $X \in (0, \infty)$ is said to be regularly varying (RV) if $\mathbb{P}(X > t) \sim L(t)/t^\alpha$ as $t \to \infty$ with $\alpha \geq 0$ and L slowly varying. The number α is the *index*; we exclude the case $\alpha = 0$ (for example, $\mathbb{P}(X > t) \sim 1/\log t$) so necessarily $\alpha > 0$.

Proposition 3.2 *A one-dimensional r.v. X is RV with index α if and only if $\mathbb{P}(X > zt)/\mathbb{P}(X > t) \to z^{-\alpha}$ for all $t > 0$.*

Proof The 'only if' part follows immediately from the definition of a slowly varying function. For 'if', simply take $L(t) = t^\alpha \mathbb{P}(X > t)$ and note that then

$$\frac{L(zt)}{L(t)} = \frac{(zt)^\alpha \mathbb{P}(X > zt)}{t^\alpha \mathbb{P}(X > t)} \to z^\alpha z^{-\alpha} = 1. \qquad \square$$

The characterization in Proposition 3.2 may be slightly less intuitive than the definition of RV, but generalizes more readily to multidimensions. To this end, one notes that if $\|x\| = \|x\|_1 = |x_1| + \cdots + |x_d|$ is the 1-norm on $\mathbb{R}^d$, then $x \to (\|x\|, \boldsymbol{\theta})$ is a bijection of $\mathbb{R}^d \setminus \{\mathbf{0}\}$ onto $(0, \infty) \times \mathcal{B}_1$, where $\boldsymbol{\theta} = x/\|x\|$ and $\mathcal{B}_1 = \{x : \|x\| = 1\}$ is the unit sphere. The component $\|x\|$ represents 'size' and $\boldsymbol{\theta}$ 'direction', analogous to the use of polar coordinates for $d = 2$. With these preliminaries, we define a random vector $X = (X_1, \ldots, X_d)$ to be multivariate regularly varying (MRV) with index $\alpha \geq 0$ if there exists a probability measure μ on $\mathcal{B}_1$ such that

$$\frac{\mathbb{P}(\|X\| > zt, \boldsymbol{\Theta} \in A)}{\mathbb{P}(\|X\| > t)} \to z^{-\alpha} \mu(A) \qquad (3.5)$$

for all continuity sets A of μ where $\boldsymbol{\Theta} = X/\|X\|$. One should note that this is an asymptotic relation and thus the distribution of $\|X\|$ together with μ does not uniquely determine the distribution of X.

Taking $A = \mathcal{B}_1$, it is seen that $\|X\|_1 = |X_1| + \cdots + |X_d|$ is RV. That is,

$$\mathbb{P}(|X_1| + \cdots + |X_d| > t) = \frac{L(t)}{t^\alpha} \qquad (3.6)$$

for some slowly varying L. Further, (3.5) with $z = 1$ shows that $X/\|X\|$ converges in distribution to μ conditionally upon $\|X\| > t$.

Example 3.3 Let $d = 1$. Then $\mathcal{B}_1 = \{1, -1\}$ and $\|x\| = |x|$. Letting $\mu_1 = \mu\{1\}$ and $\mu_{-1} = \mu\{-1\}$, $\boldsymbol{\Theta} = X/\|X\| = 1$ simply means $X > 0$ and so for $t > 0$ the definition yields

$$\mathbb{P}(X > t) = \mathbb{P}(|X| > t, \boldsymbol{\Theta} = 1) \sim \mu_1 \mathbb{P}(|X| > t) \sim \mu_1 \frac{L(t)}{t^\alpha},$$

Similarly, $\boldsymbol{\Theta} = -1$ means $X < 0$ and so

$$\mathbb{P}(X < -t) = \mathbb{P}(|X| > t, \boldsymbol{\Theta} = -1) \sim \mu_{-1} \mathbb{P}(|X| > t) \sim \mu_{-1} \frac{L(t)}{t^\alpha},$$

i.e., both the right and left tail of X are RV with the same α and the slowly varying functions only differ by a constant. Thus even with $d = 1$, the definition (3.5) of RV extends the one we have used so far by taking not only the left tail but also the right into account (the two tails have the same α and the slowly varying functions differ asymptotically only by constants). $\diamond$

Example 3.4 Let $d = 2$ and let $X_1, X_2 \geq 0$ be independent with common tail $L_0(t)/t^\alpha$. Then the tail of $\|X\| = X_1 + X_2$ is RV with tail $L(t)/t^\alpha$ where $L(t) \sim 2L_0(t)$. Further, if $X_1 + X_2 > t$ then asymptotically either $X_1 > t$ or $X_2 > t$, w.p. 1/2 for each and the other X_i remains of order $O(1)$, cf. Proposition III.2.1 and the following discussion and Exercise III.2.3, i.e., asymptotically Θ is concentrated on the points $(1, 0)$ and $(0, 1)$, which are both in $\mathcal{B}_1$. Thus (3.5) holds where μ is the probability measure on $\mathcal{B}_1$ giving mass 1/2 to each of $(1, 0)$ and $(0, 1)$.   ◇

There are several equivalent definitions of MRV. One of the most frequently encountered is in terms of the so-called *limit measure* or *exponent measure* ν. Here 'limit' refers to convergence in the space of measures on $\mathbb{R}^d \setminus \{0\}$ that are finite away from 0, i.e. $\mu\{x : \|x\| > \epsilon\} < \infty$ for all $\epsilon > 0$, and $\nu_n \to \nu$ means that $\nu_n(C) \to \nu(C)$ for all continuity sets C of ν such that there exists an $\epsilon > 0$ such that $\epsilon < \|x\| < \epsilon^{-1}$ for all $x \in C$ (an abstract formulation is vague convergence in a suitable one-point compactification of $\mathbb{R}^d \setminus \{0\}$). The characterization then means convergence in this sense of the distribution of X/t when properly normalized, written as

$$a(t)\mathbb{P}(X/t \in \cdot) \to \nu \qquad (3.7)$$

for some suitable function $a(t)$, with ν non-degenerate.

Proposition 3.5 *If X is MRV in the sense of (3.5), then (3.7) holds and one can take*

$$a(t) = \frac{1}{\mathbb{P}(\|X\| > t)} . \qquad (3.8)$$

Further, ν is a product measure in polar coordinates $(\|x\|, \theta)$ and subject to (3.8), ν can be expressed in terms of α, μ by means of

$$\nu\{\|x\| \in dz, \theta \in d\theta\} = \alpha z^{-\alpha-1} dz \times \mu(d\theta) . \qquad (3.9)$$

Conversely, if (3.7) holds then $a(t)$ is RV for some $\alpha > 0$ and (3.5) holds with

$$\mu(B) = \frac{\nu\{\|x\| > 1, \theta \in B\}}{\nu\{\|x\| > 1\}} . \qquad (3.10)$$

Proof Since the mapping $x \to (\|x\|, \theta)$ is a bicontinuous bijection of $\mathbb{R}^d \setminus \{0\}$ onto $(0, \infty) \times \mathcal{B}_1$, the conclusion of (3.5) can be rewritten as $\nu(t) \to \nu$ where $a(t) = 1/\mathbb{P}(\|X\| > t)$, $\nu(t)$ is the distribution of the pair $(\|X\|, \Theta)$ and ν is given by (3.9). This proves the first part of the proposition.

For the second, consider $a(t)\mathbb{P}(X \in tB)$ and $a(tc)\mathbb{P}(X \in tcB)$ to find that $a(ct)/a(t) \to \nu(B)/\nu(cB) \in (0, \infty)$ showing that a is RV for some $\alpha > 0$ (cf.

Exercise 3.2). We also get

$$v(cB) = c^{-\alpha}v(B) \qquad \forall c > 0 \tag{3.11}$$

(note that $\alpha = 0$ is ruled out because then we would have $\mu(cB) = \mu(B)$ and finiteness of μ away from 0 would imply that μ is trivially 0). We then get

$$z^\alpha \lim_{t\to\infty} \frac{\mathbb{P}(\|X\| > zt, \boldsymbol{\Theta} \in B)}{\mathbb{P}(\|X\| > t)}$$

$$= \lim_{t\to\infty} \frac{a(zt)\mathbb{P}(\|X\| > zt, \boldsymbol{\Theta} \in B)}{a(t)\mathbb{P}(\|X\| > t)} = \frac{v\{\|x\| > 1, \boldsymbol{\theta} \in B\}}{v\{\|x\| > 1\}},$$

so that indeed (3.5) holds with μ given by (3.10). □

A simple application of Proposition 3.5 follows. Recall that $\boldsymbol{a}\cdot\boldsymbol{x} = a_1x_1 + \cdots + a_dx_d$.

Corollary 3.6 *If X is MRV as in (3.5), then*

$$\mathbb{P}(\boldsymbol{a}\cdot\boldsymbol{X} > t) \sim w(\boldsymbol{a})\frac{L(t)}{t^\alpha} \qquad \text{where} \quad w(\boldsymbol{a}) = \int_{\mathcal{B}_1} [(\boldsymbol{a}\cdot\boldsymbol{\theta})^+]^\alpha \mu(d\boldsymbol{\theta})$$

for all $\boldsymbol{a} \in \mathbb{R}^d$, where L is as in (3.6). In particular, each component X_i is RV. More precisely

$$\mathbb{P}(X_i > t) \sim w_i\frac{L(t)}{t^\alpha} \qquad \text{where} \quad w_i = w(\mathbf{1}_i) = \int_{\mathcal{B}_1} \theta_i^{+\alpha} \mu(d\boldsymbol{\theta}).$$

Proof First observe that by (3.7), (3.8) we have

$$\mathbb{P}(\boldsymbol{a}\cdot\boldsymbol{X} > t) \sim \frac{1}{a(t)}v\{\boldsymbol{a}\cdot\boldsymbol{x} > 1\} \sim \frac{L(t)}{t^\alpha}v\{\boldsymbol{a}\cdot\boldsymbol{x} > 1\}.$$

But $\int_0^\infty \mathbb{1}_{z>b}\alpha z^{-\alpha-1}dz = b^{-\alpha}$ for $b > 0$, and so

$$v(\boldsymbol{a}\cdot\boldsymbol{x} > 1) = \int_{\mathcal{B}_1} d\boldsymbol{\theta} \int_0^\infty \mathbb{1}_{z\boldsymbol{\theta}\cdot\boldsymbol{a}>1}\alpha z^{-\alpha-1}dz = w(\boldsymbol{a}). \qquad □$$

Exercises

3.2 Show the following strengthening of Proposition 3.2: X is RV for some α if and only if $\mathbb{P}(X > zt)/\mathbb{P}(X > t)$ has a limit $h(z)$ for all z.

3.3 Show that $\|x\|_2 \le \|x\|_1 \le d\|x\|_2$. *Hint*: it suffices to consider the case where all $x_i \ge 0$ and $\|x\|_1 = 1$. It may be useful to consider r.v.s Y_1, Y_2 taking values in $\{x_1, \ldots, x_d\}$ with probabilities $1/d, \ldots, 1/d$ for Y_1 and $x_1, \ldots, x_d$ for Y_2.

3.4 If $X_1, X_2 \geq 0$ are independent and X_1 subexponential, then $X_1 - X_2$ has the same right tail as X_1, cf. Exercise III.2.4. For the case where both X_1 and X_2 are RV, retrieve this result via Example 3.4 and Proposition 3.6.

Notes
Resnick [149] gives a detailed textbook exposition of multivariate regular variation. Another alternative characterization of MRV is in terms of linear combinations, stating that (analogous to the Cramér–Wold device for the multivariate normal distribution) except for a quite special case MRV is equivalent to RV of $a \cdot X$ for all a, cf. Boman and Lindskog [29] and the references there.

4 Copulas

A d-dimensional *copula* C is defined as a probability distribution on $(0, 1)^d$ with uniform marginals, that is, as the distribution of $(U_1, \ldots, U_d)$ where the U_i are uniform$(0, 1)$ but not necessarily independent.

If a random vector $X = (X_1, \ldots, X_d)$ with multivariate c.d.f. F_X has *continuous* marginals $F_1, \ldots, F_d$ where $F_i(x_i) = \mathbb{P}(X_i \leq x_i)$, then the random vector $\big(F_1(X_1), \ldots, F_d(X_d)\big)$ has uniform marginals and hence its distribution C is a copula satisfying

$$F_X(x_1, \ldots, x_d) = C\big(F_1(x_1), \ldots, F_d(x_d)\big). \tag{4.1}$$

The continuity assumption is not necessary for the existence of C:

Theorem 4.1 (Sklar) *For any random vector $X = (X_1, \ldots, X_d)$ there exists a copula with the property* (4.1). *If the marginals F_i are continuous then C is unique.*

Proof As noted before (4.1), the result is elementary in the case of continuous marginals. In the general case, replace $F_i(X_i)$ with the distributional transform $D_i = F_i(X_i-) + V_i \Delta F_i(X_i)$, where V_i is an independent uniform$(0, 1)$ r.v. and $\Delta F_i(x) = F_i(x) - F_i(x-) = \mathbb{P}(X_i = x)$. As shown in Proposition A.1, D_i has a uniform$(0, 1)$ distribution and satisfies $X = F_i^{\leftarrow}(D_i)$ a.s. This allows the argument for the continuous case to go through immediately.   □

In the continuous case, such a C is called *the* copula of X (otherwise *a* copula of X). Copulas are basically a way to represent dependence structure, in the same way as the c.d.f. is a way to represent a distribution, and a copula of a random vector is seldom of intrinsic interest. In the present authors' opinion, there is nothing deep in the concept *per se* and also Sklar's theorem is of course elementary at least in the continuous case. These points are also stressed in Mikosch [121], and somewhat contrasts with parts of the literature which easily leave the impression of an area deeper than it is!

Remark 4.2 Copulas are nonstructural models in the sense that there is a lack of classes that on the one hand are characterized by a finite number of parameters so that statistical estimation is feasible, and on the other are flexible enough to describe a broad range of dependence structure. The way they are used is therefore most often as scenarios: for given marginals, experiment with different copulas (typically by Monte Carlo) and see how sensitive the quantity of interest is to the particular choice of copula. This approach is today widely used in areas such as multivariate loss distributions (loss from different lines of business), credit risk where losses from different debtors are highly dependent, market risk (multivariate asset returns), and risk aggregation where one combines market risk, credit risk, operational risk, etc. See in particular McNeil et al. [115].

As a concrete example, assume we want to infer what could be typical values of $\theta = \mathbb{P}(X_1 + \cdots + X_d > x)$ where x is some fixed large number and the distribution F_i of each X_i is lognormal with known or estimated parameters μ_i, σ_i^2. One then selects M copulas $C_1, \ldots, C_M$ and for copula m, one generates R i.i.d. replicates with the given marginals by simulating $(U_{1;rm}, \ldots, U_{d;rm})$ for $r = 1, \ldots, R$. One then lets

$$(X_{1;rm}, \ldots, X_{d;rm}) = \left(F_1^{\leftarrow}(U_{1;rm}), \ldots, F_d^{\leftarrow}(U_{d;rm})\right),$$

$$\widehat{\theta}_m = \frac{1}{R} \sum_{r=1}^{R} \mathbb{1}(X_{1;rm} + \cdots + X_{d;rm} > x)$$

and inspects the variability in the set $\{\widehat{\theta}_1, \ldots, \widehat{\theta}_M\}$.   ◇

The simplest example of a copula is the *independence copula*, corresponding to $U_1, \ldots, U_d$ being independent so that $C(u_1, \ldots, u_d) = u_1 \ldots u_n$. The concept of comonotonicity encountered in Sect. 2 leads to calling the case $U_1 = \cdots = U_d$ the *comonotonic copula*; here $C(u_1, \ldots, u_d) = \min(u_1, \ldots, u_d)$. We can construct a random vector with comonotonic copula and marginals $F_1, \ldots, F_d$ by letting $X_i = F_i^{-1}(U)$ with the *same* uniform r.v. U for different i. If $d = 2$, the *countermonotonic copula* is defined as the distribution of $(U, 1 - U)$ with U uniform$(0, 1)$. Intuitively, comonotonicity is the strongest positive dependence possible and countermonotonicity the strongest negative one when $d = 2$ (but there are counterexamples showing that life is not quite as simple as this; see in particular Remark 7.4 below).

A *Gaussian copula* is the copula of a multivariate normal X (cf. Exercise 4.1) and a *t copula* is the copula of a multivariate t random vector. Beyond these simple examples, there is an abundance of specific parametric copulas, see the standard textbooks by Joe [96], Nelsen [130] and McNeil et al. [115]. Some main examples follow.

4.1 Archimedean Copulas

An Archimedean copula is defined by having form

$$C(u_1, \ldots, u_d) = \phi^{-1}\left(\phi(u_1) + \cdots + \phi(u_d)\right), \qquad (4.2)$$

where ϕ is called the *generator* ϕ and ϕ^{-1} is its functional inverse. Obviously, certain conditions on ϕ (or equivalently on ϕ^{-1}) must hold for (4.2) to be a copula, and the most basic case is the following:

Proposition 4.3 *Let $Z > 0$ be an r.v., $\psi(s) = \mathbb{E}e^{-sZ}$ its Laplace transform and let $V_1, \ldots, V_d$ be independent standard exponential r.v.s and independent of Z. Then the copula of $(\psi(V_1/Z), \ldots, \psi(V_d/Z))$ is given by (4.2) with $\phi = \psi^{-1}$.*

Proof Using that ψ is continuous and strictly decreasing in the first step, we get

$$\mathbb{P}(\psi(V_1/Z) \leq u_1, \ldots, \psi(V_d/Z) \leq u_d)$$
$$= \mathbb{P}(V_1/Z \geq \phi(u_1), \ldots, V_d/Z \geq \phi(u_d))$$
$$= \mathbb{E}\,\mathbb{P}(V_1 \geq Z\phi(u_1), \ldots, V_d \geq Z\phi(u_d) \mid Z)$$
$$= \mathbb{E}[\exp\{-Z\phi(u_1)\} \cdots \exp\{-Z\phi(u_d)\}]$$
$$= \mathbb{E}\exp\{-Z(\phi(u_1) + \cdots + \phi(u_d))\}$$
$$= \psi(\phi(u_1) + \cdots + \phi(u_d)),$$

which is the same as (4.2).   □

Some principal examples are the *Clayton copula*, where $\phi(u) = (u^{-\alpha} - 1)/\alpha$ for some $\alpha > 0$; the *Frank copula*, where

$$\phi(u) = -\log \frac{e^{\alpha u} - 1}{e^{\alpha} - 1}$$

for some $\alpha \in \mathbb{R}$; and the *Gumbel* or *Gumbel–Hougaard copula*, where $\phi(u) = [-\log u]^{\alpha}$ for some $\alpha \geq 1$.

Remark 4.4 From an analytical point of view, the main feature of the Laplace transform $\psi(s) = \mathbb{E}e^{-sZ}$ of a non-negative r.v. Z is that it is *completely monotone*, meaning that the derivatives alternate in sign: $\psi \geq 0$, $\psi' \leq 0$, $\psi'' \geq 0$, $\psi''' \leq 0$ and so on. Conversely, *Bernstein's theorem* says that any such function is the Laplace transform $\int_0^{\infty} e^{-sz} \mu(dz)$ of a non-negative measure. The assumption $\phi^{-1}(s) = \mathbb{E}e^{-sZ}$ of Proposition 4.3 is not quite necessary for (4.2) to define a copula. For example, complete monotonicity can be weakened somewhat for a given d (cf. [154, pp. 8–9]). Also one can allow $\phi^{-1}(\infty) \neq 1$ after due modifications.   ◊

Remark 4.5 The definition implies immediately that r.v.s with an Archimedean copula are *exchangeable*. This may be highly natural in some applications and absolutely not in others!   ◊

4.2 Extreme Value Copulas

Extreme value copulas are connected to multivariate extreme value theory, i.e. to the multivariate c.d.f. F^n of the componentwise maximum M_n of n i.i.d. copies of a random vector $X = (X_1, \ldots, X_d)$ with c.d.f. $F : \mathbb{R}^d \to \mathbb{R}$. A copula C is called an extreme value copula if for some F it is the limit of the copula of $(M_n - b_n)/a_n$ for suitable sequences a_n, b_n.

Proposition 4.6 *A copula C is an extreme value copula if and only if*

$$C(u_1, \ldots, u_d) = C(u_1^{1/m}, \ldots, u_d^{1/m})^m \tag{4.3}$$

for all $m = 1, 2, \ldots$ and all $(u_1, \ldots, u_d) \in (0, 1)^d$.

Proof Let C_F be the copula of F and C_{M_n} that of M_n. Then C_{M_n} is also the copula of $(M_n - b_n)/a_n$ and since the c.d.f. of the ith component of M_n is F_i^n, we have

$$C_{M_n}(F_1(x_1)^n, \ldots, F_d(x_d)^n) = F(x_1, \ldots, x_d)^n = C_F(F_1(x_1), \ldots, F_d(x_d))^n.$$

If C is an extreme value copula, it follows by replacing u_i by $F_i(x_i)^n$ that

$$C(u_1, \ldots, u_d) = \lim_{n \to \infty} C_F(u_1^{1/n}, \ldots, u_d^{1/n})^n,$$

which in turn leads to

$$
\begin{aligned}
C(u_1, \ldots, u_d) &= \lim_{k \to \infty} C_F(u_1^{1/mk}, \ldots, u_d^{1/mk})^{mk} \\
&= \lim_{k \to \infty} \left[C_F((u_1^{1/m})^{1/k}, \ldots, (u_d^{1/m})^{1/k})^k \right]^m \\
&= C(u_1^{1/m}, \ldots, u_d^{1/m})^m,
\end{aligned}
$$

so that (4.3) holds. Conversely, (4.3) implies that we can just take $F = C$. $\qquad\square$

It can be shown that the functional equation (4.3) implies that an extreme value copula C can be represented in terms of the so-called *Pickands dependence function*, a certain convex function A (say). For $d = 2$ this means

$$C(u_1, u_2) = \exp\left\{ -(|\log u_1| + |\log u_2|) A\left(\frac{|\log u_1|}{|\log u_1| + |\log u_2|}\right) \right\}, \tag{4.4}$$

where $s \vee (1 - s) \le A(s) \le 1$ for $0 \le s \le 1$. Taking $A(s) = (s^\theta + (1-s)^\theta)^{1/\theta}$ gives the Gumbel copula mentioned above, which for a general d takes the form

$$C(u_1, \ldots, u_d) = \exp\left\{ -(|\log u_1|^\theta - \cdots - |\log u_1|^\theta)^{1/\theta} \right\}. \tag{4.5}$$

4.3 Frailty Copulas

The idea is to create dependence via an unobservable random vector $\mathbf{Z} = (Z_1, \ldots, Z_d) \in (0, \infty)^d$. More precisely, we let $H_1, \ldots, H_d$ be any set of c.d.f.s and take $X_1, \ldots, X_d$ to be conditionally independent given $\mathbf{Z} = (z_1, \ldots, z_d)$ with conditional c.d.f. $H_i^{z_i}$ for X_i. Thus

$$F(x_1, \ldots, x_d) = \int_{(0,\infty)^d} H_1^{z_1}(x_1) \cdots H_d^{z_d}(x_d) \, G(\mathrm{d}z_1, \ldots, \mathrm{d}z_d), \qquad (4.6)$$

where G is the distribution of $\mathbf{Z}$. The copula of such an F is called a frailty copula.

Proposition 4.7 *The copula of* (4.6) *is given by*

$$C(u_1, \ldots, u_d) = \psi\big(\psi_i^{-1}(u_1), \ldots, \psi_d^{-1}(u_d)\big), \qquad (4.7)$$

where $\psi_i(s) = \mathbb{E}e^{-sZ_i}$ *is the Laplace transform of* Z_i *and* $\psi(s_1, \ldots, s_d) = \mathbb{E}\exp\{-s_1 Z_1 - \cdots - s_d Z_d\}$ *the multivariate Laplace transform of* $\mathbf{Z}$.

Proof Let ρ be the r.h.s. of (4.7) evaluated at $u_1 = F_1(x_1), \ldots, u_d = F_d(x_d)$. We then have to show that $\rho = F(x_1, \ldots, x_d)$. But $F_i(x)$ equals $\psi_i\big(-\log H_i(x)\big)$ (take all $x_i = x$ in (4.6) and all other $x_j = \infty$) and so

$$\rho = \psi\big(-\log H_1(x), \ldots, -\log H_d(x)\big)$$
$$= \mathbb{E}\big\{\log H_1(x)Z_1 + \cdots + \log H_d(x)Z_d\big\} = \mathbb{E}\big[H_1^{Z_1}(x_1) \cdots H_d^{Z_d}(x_d)\big],$$

which is the same as $F(x_1, \ldots, x_d)$. $\qquad\square$

Remark 4.8 If $\mathbf{Z}$ is comonotonic and we write $\mathbf{Z} = Z_1$, then $\psi(s_1, \ldots, s_d) = \mathbb{E}\exp\{-(s_1 + \cdots + s_d)Z\}$ and therefore (4.7) takes the form (4.2), i.e., an Archimedean copula given as in Proposition 4.3 (which is a quite general form in this class) is a special case of a frailty copula. $\diamond$

4.4 Further Examples

Example 4.9 (The Farlie–Gumbel–Morgenstern Copula) This corresponds for $d = 2$ to

$$C(u_1, u_2) = u_1 u_2 \big(1 + \epsilon(1 - u_1)(1 - u_2)\big)$$

with $\epsilon \in [-1, 1]$ (the form is more complicated when $d > 2$). $\diamond$

Fig. 3 Scatter-plots of six copulas

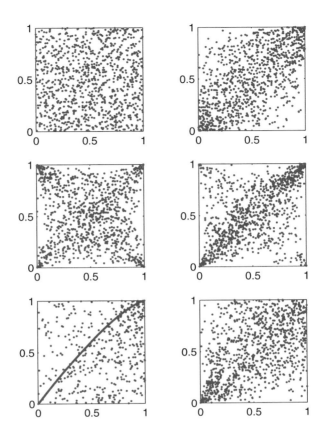

Example 4.10 A common way to illustrate the dependence structure of a bivariate copula is via a scatter-plot, that is, a plot of R simulated values (necessarily $(0, 1)^2$-valued).

Six such scatter-plots with $R = 1,000$ are given in Fig. 3. In lexicographic order, the first is from the independence copula (the uniform distribution on $(0, 1)^2$), the second from a Gaussian copula with correlation 0.6, the third from a t copula with $f = 1$ and $\rho = 0$, the fourth from a t copula with $f = 1$ and $\rho = 0.6$. The fifth is from the Marshall–Olkin exponential copula, defined as the copula of $X_1 = T_1 \wedge T_{12}$, $X_2 = T_2 \wedge T_{12}$, where T_1, T_2, T_{12} are independent exponentials with rates $\lambda_1 = 1$, $\lambda_2 = 3$, $\lambda_{12} = 5$; the singular component of course corresponds to the case $X_1 = X_2$, which arises if $T_{12} < T_1 \wedge T_2$. Finally, the sixth is from the Clayton copula with $\alpha = 1$. A huge number of scatter-plots for further copulas is given in Nelsen [130].

Singular copulas are concentrated on suitable lines or curves, blue in Fig. 4. Here the left panel corresponds to the comonotonic copula concentrated on the diagonal

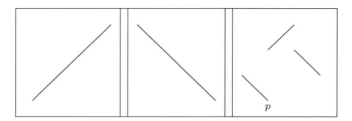

Fig. 4 Three singular copulas

$u_1 = u_2$, the middle is the countermonotonic copula concentrated on $u_2 = 1 - u_1$ and the third is an example of a so-called *shuffle*. It is one of the other candidates for the copula of the comonotonic vector with $\mathbb{P}(X_1 = X_2 = 0) = p$, $\mathbb{P}(X_1 = X_2 = 1) = 1 - p$ since in this case the only value of $F(x_1, x_2)$ which is $\notin \{0, 1\}$ is $F(0, 0) = p$ so that the only requirement for this copula is $C(p, p) = p$. It thereby in particular gives an example of non-uniqueness in the discrete case.   ◇

Example 4.11 Given a copula C, its *tail copula* is defined by the c.d.f. $C(1 - u_1, \ldots, 1 - u_d)$. In terms of r.v.s, it corresponds to replacing $(U_1, \ldots, U_d)$ by $(1 - U_1, \ldots, 1 - U_d)$. It serves the purpose of switching the dependence in the right and left tail. For example, the Clayton copula exhibits a strong dependence in the left tail but not the right, cf. the lower right part of Fig. 4, but for its tail copula it is the other way round.   ◇

Exercises
4.1 Let X be multivariate normal $\mathcal{N}(\mu, \Sigma)$. Show that the copula of X is independent of μ and only depends on Σ through the correlations $\mathbb{C}\mathrm{orr}(X_i, X_j)$, $i \neq j$.

4.2 Let $X_1, \ldots, X_p$ be continuous r.v.s and $h_1, \ldots, h_p$ strictly increasing functions. Show that $X_1, \ldots, X_p$ and $h_1(X_1), \ldots, h_p(X_p)$ generate the same copula.

4.3 Show that the tail copula of the Marshall–Olkin bivariate exponential distribution is given by $C(u_1, u_2) = (u_1^{1-\alpha_1} u_2) \wedge (u_1 u_2^{1-\alpha_2})$, where $\alpha_i = \lambda_{12}/(\lambda_i + \lambda_{12})$.

4.4 Show that the Farlie–Gumbel–Morgenstern copula depends only on ϵ but not F_1, F_2.

4.5 Show that the Clayton copula approaches comonotonicity as $\alpha \to \infty$ and independence as $\alpha \to 0$. Show more generally that the dependence is increasing in α in the sense that the c.d.f. is nondecreasing in α.

5 Pearson, Kendall and Spearman

5.1 Linear (Pearson) Correlation

The most naive measure of dependence is that of the standard correlation $\mathbb{C}\mathrm{orr}(X, Y)$, as defined by

$$\mathbb{C}\mathrm{orr}(X, Y) = \frac{\mathbb{C}\mathrm{ov}(X, Y)}{\sqrt{\mathbb{V}\mathrm{ar}(X)\,\mathbb{V}\mathrm{ar}(Y)}} \tag{5.1}$$

and taking a value in $[-1, 1]$. For a given set F_X, F_Y of marginals, ± 1 need not be attained; the maximal and minimal values correspond to comonotonicity, resp. countermonotonicity, i.e.

$$\mathbb{C}\mathrm{orr}\left(F_X^{\leftarrow}(U), F_Y^{\leftarrow}(1 - U)\right) \le \mathbb{C}\mathrm{orr}(X, Y) \le \mathbb{C}\mathrm{orr}\left(F_X^{\leftarrow}(U), F_Y^{\leftarrow}(U)\right),$$

cf. Sect. 2. The term 'correlation' is often refined to 'linear correlation' due to invariance under linear and affine transformations,

$$\mathbb{C}\mathrm{orr}(X, Y) = \mathbb{C}\mathrm{orr}(aX + b, cY + d) \quad \text{when } ac > 0.$$

Taking the linear correlation as a quantitative measure of dependence is partly motivated by the Gaussian case where X, Y are independent if and only if $\mathbb{C}\mathrm{orr}(X, Y) = 0$ [the 'only if' holds in general]. However, in different settings the situation may be quite different since $\mathbb{C}\mathrm{ov}(X, Y)$ may be zero or close to zero even if there is a strong functional relation and hence dependence between X, Y. An extremely simple example is $Y = X^2$ with $X \sim \mathcal{N}(0, 1)$. Some further ones follow.

Example 5.1 A mathematically simple example is the uniform distribution on a circle $(x - a)^2 + (y - b)^2 = r^2$, cf. Fig. 5a; a representation is for example $X = a + r \cos 2\pi U$, $Y = b + r \sin 2\pi U$, where U is uniform$(0, 1)$.

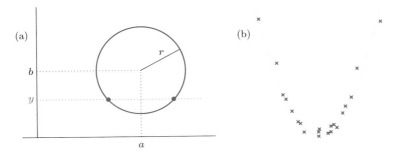

Fig. 5 Dependent but linearly uncorrelated data

That $\mathbb{C}\mathrm{ov}(X, Y) = 0$ may be intuitively obvious here but can more pedantically be verified by noting that

$$\mathbb{E}(XY) \;=\; \mathbb{E}\,\mathbb{E}[XY \mid Y] \;=\; \mathbb{E}\big[Y\mathbb{E}[X \mid Y]\big] \;=\; \mathbb{E}[Y \cdot a] \;=\; ab \;=\; \mathbb{E}X\,\mathbb{E}Y$$

($\mathbb{E}[X \mid Y] = a$ since for any $y \in (b-r, b+r)$ there are two symmetric equiprobable values of x, cf. again the figure).

A related example often met in the literature is $Y \approx X^2$ and X symmetric as for the data in Fig. 5b; for example, $X = V_1$, $Y = X^2 + V_2$, where V_1, V_2 are independent and normal $(0, \sigma_X^2)$, resp. $(0, \sigma_Y^2)$. The reason why $\mathbb{C}\mathrm{ov}(X, Y)$ (empirically) is close to zero is again that $\mathbb{E}(X \mid Y) = \mathbb{E}Y$ (here equal to 0). $\diamond$

Example 5.2 An important example is stochastic volatility models in finance. There one often assumes that the log-return X_{n+1} at time $n+1$ is given as $X_{n+1} = A_n\epsilon_{n+1}$, where the ϵ_k are i.i.d. $\mathcal{N}(0, 1)$ and the A_n adapted w.r.t. the filtration $\mathcal{F}_n = \sigma(\epsilon_k : k \leq n)$ (i.e. A_n is $\mathcal{F}_n$-measurable). Thus A_n is the volatility at time $n + 1$. Here the A_n are typically strongly dependent and hence so are the X_n. But

$$\mathbb{E}X_{n+1} \;=\; \mathbb{E}\,\mathbb{E}[X_{n+1} \mid \mathcal{F}_n] \;=\; \mathbb{E}\big[A_n\mathbb{E}[\epsilon_{n+1} \mid \mathcal{F}_n]\big] \;=\; \mathbb{E}[A_n \cdot 0] \;=\; 0\,,$$

$$\mathbb{E}(X_{n+k}X_n) \;=\; \mathbb{E}\,\mathbb{E}\big[X_{n+k}X_n \mid \mathcal{F}_{n+k-1}\big]$$
$$=\; \mathbb{E}\big[X_nA_{n+k-1}\mathbb{E}[\epsilon_{n+k} \mid \mathcal{F}_{n+k-1}]\big] \;=\; \mathbb{E}\big[X_nA_{n+k-1} \cdot 0\big] \;=\; 0$$

for all n and all $k \neq 0$ and so $\mathbb{C}\mathrm{ov}(X_{n+k}, X_n) = 0$.

Assuming in addition that the A_n form a stationary sequence, similar calculations yield

$$\mathbb{C}\mathrm{ov}(X_{n+k}^2, X_n^2) = \mathbb{C}\mathrm{ov}(A_{n+k-1}^2, A_{n-1}^2)\,. \tag{5.2}$$

The implication, seen also in empirical studies, is that whereas the X_n appear uncorrelated, the opposite is true for the X_n^2. Note that both the A_n and the A_n^2 are typically highly dependent (and correlated). For example, a popular model is a log-AR model where $A_n = \mathrm{e}^{R_n}$ with $R_{n+1} = \beta R_n + \delta_{n+1}$. $\diamond$

One further reason that correlation is an inadequate measure of dependence is the fact that it is not preserved by monotone transformations. That is, if f, g are increasing functions, then $\mathbb{C}\mathrm{orr}\big(f(X), g(Y)\big)$ can be very different from $\mathbb{C}\mathrm{orr}(X, Y)$.

In finance and the actuarial sciences, positive dependence (large values of X tend to go with large values of Y) is usually much more interesting than negative dependence (large values of X tend to go with small values of Y and vice versa). There are many possible rigorous definitions of positive dependence, but most of them imply the desirable property

$$\mathbb{C}\mathrm{ov}\big(f(X), g(Y)\big) \geq 0 \tag{5.3}$$

when f, g are non-decreasing functions.

Exercises
5.1 Verify (5.2).

5.2 Kendall's Tau and Spearman's Rho

Correlation as defined by (5.1) often goes under the name *Pearson correlation*. However, a couple of further one-dimensional quantifications of the dependence between X and Y, associated with the names of Spearman and Kendall, also carry the name correlation and will be consider next. A common feature is that they are both $[-1, 1]$-valued and that (in contrast to the Pearson correlation) all values in $[-1, 1]$ can be attained for a given set of marginals, with 1 being attained for (X, Y) comonotonic and -1 for countermonotonicity (and 0 for independence, as should be).

To avoid certain technicalities associated with ties, we assume in the following that the distributions of X, Y are both continuous.

For the *Kendall rank correlation coefficient* or *Kendall's* τ, we define two pairs $(x, y), (x^*, y^*) \in \mathbb{R}^2$ as *concordant* if either $x < x^*$ and $y < y^*$, or $x > x^*$ and $y > y^*$, and as *discordant* otherwise, cf. Fig. 6, and consider an independent copy (X^*, Y^*) of (X, Y). Kendall's τ is then defined as the difference of the probabilities that (X, Y) and (X^*, Y^*) are concordant and that they are discordant, i.e. as

$$
\begin{aligned}
\rho_\tau &= \mathbb{P}\big((X_1 - X_1^*)(Y_1 - Y_1^*) > 0\big) - \mathbb{P}\big((X_1 - X_1^*)(Y_1 - Y_1^*) < 0\big) \\
&= \mathbb{E}\,\text{sign}\big[(X_1 - X_1^*)(Y_1 - Y_1^*)\big].
\end{aligned}
$$

If X, Y are in some sense positive dependent, one should then observe a positive value of τ, and similarly one would expect a negative one in the case of negative dependence. Obviously $-1 \le \rho_\tau \le 1$.

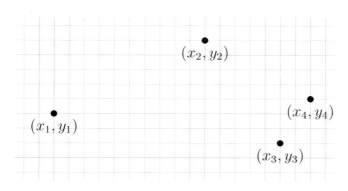

Fig. 6 Concordant pairs: 12, 14, 34; discordant 13, 23, 24

Spearman's rank correlation coefficient or *Spearman's* ρ is defined as

$$\rho_S(X, Y) = \operatorname{Corr}\big(F_X(X), F_Y(Y)\big).$$

The connection to ranks is in the general setting loosely that since $F_X(X)$ is uniform$(0, 1)$, the value u observed when $X = x$ measures on a scale from 0 to 1 where on the line X is located. This statement may be most easily understood in terms of the empirical version that we will introduce in a moment in Sect. 5.3. However, we first note the following alternative expression:

Proposition 5.3 *For X, Y continuous,*

$$\rho_S(X, Y) = 1 - 6\,\mathbb{E}\big[\big(F_X(X) - F_Y(Y)\big)^2\big]. \tag{5.4}$$

Proof Since $U_X = F_X(X), U_Y = F_Y(Y)$ are uniform, we have

$$\mathbb{E}U_X = \mathbb{E}U_Y = \frac{1}{2}, \quad \mathbb{E}U_X^2 = \mathbb{E}U_Y^2 = \frac{1}{3}, \quad \operatorname{Var} U_X = \operatorname{Var} U_Y = \frac{1}{12}.$$

Using the formula $xy = \big(x^2 + y^2 - (x - y)^2\big)/2$, we therefore get

$$
\begin{aligned}
\operatorname{Corr}(U_X, U_y Y) &= \frac{\mathbb{E}(U_X U_Y) - \mathbb{E}U_X\,\mathbb{E}U_Y}{(\operatorname{Var} U_X \cdot \operatorname{Var} U_Y)^{1/2}} = 12\big[\mathbb{E}(U_X, U_Y) - (1/2)^2\big] \\
&= 12\big[\mathbb{E}U_X^2/2 + \mathbb{E}U_Y^2/2 - \mathbb{E}(U_X - U_Y)^2/2 - 1/4\big] \\
&= 12\big[1/3 - \mathbb{E}(U_X - U_Y)^2/2 - 1/4\big] \\
&= 1 - 6\mathbb{E}(U_X - U_Y)^2/2
\end{aligned}
$$

as asserted. $\square$

In the bivariate normal distribution with $\operatorname{Corr}(X_1, X_2) = \rho$,

$$\rho_\tau(X_1, X_2) = \frac{2}{\pi} \arcsin(\rho), \quad \rho_S(X_1, X_2) = \frac{6}{\pi} \arcsin(\rho/2)$$

(the expression for ρ_τ is more generally valid for elliptical distributions).

5.3 Empirical Versions

Consider a set of data $(x_1, y_1), \ldots, (x_n, y_n)$, assumed i.i.d. outcomes of a random pair distributed as (X, Y). What are the natural estimators for ρ, ρ_τ, ρ_S and what are their asymptotic properties? In particular, how does one test for independence?

The natural estimators are obviously the empirical counterparts. For ρ, this is

$$\widehat{\rho} = \frac{S_{xy}}{S_x S_y} = \frac{\sum_{i=1}^{n}(x_i - \overline{x})(y_i - \overline{y})}{\left[\sum_{i=1}^{n}(x_i - \overline{x})^2 \sum_{i=1}^{n}(y_i - \overline{y})^2\right]^{1/2}},$$

where

$$s_x^2 = \frac{1}{n-1}\sum_{i=1}^{n}(x_i - \overline{x})^2, \quad s_{xy} = \frac{1}{n-1}\sum_{i=1}^{n}(x_i - \overline{x})(y_i - \overline{y}), \quad \overline{x} = \frac{1}{n}\sum_{i=1}^{n}x_i$$

(and similarly for $\overline{y}, s_y^2$). The test for $\rho = 0$ rejects for large absolute values of $\widehat{\rho}/\sqrt{(1 - \widehat{\rho}^2)/(n - 2)}$ which under the null hypothesis has a t distribution with $n-2$ degrees of freedom.

The empirical counterpart of Kendall's rank correlation coefficient is

$$\widehat{\rho}_\tau = \frac{C - D}{\binom{n}{2}} = 1 - \frac{4D}{n(n - 1)},$$

where C is the number of concordant pairs and D the number of discordant pairs. Since the denominator in the definition is the number of all pairs, we have $-1 \leq \widehat{\rho}_\tau \leq 1$. If X, Y are independent, $\widehat{\rho}_\tau$ is asymptotically normal with mean 0 and variance $2(2n + 5)/9n(n - 1)$, and this can be used to test independence.

For the empirical version of Spearman's rho, we use the r.h.s. of (5.4) with X replaced by the data and F_X by the empirical c.d.f. $\widehat{F}_X(x) = \sum_{1}^{n} \mathbb{1}(x_i \leq x)/n$ (and similarly for the Y component). Now $n\widehat{F}_X(x_i)$ has a simple interpretation as the rank of x_i (i.e. the rank is 1 if x_i is the smallest among the x_j, it is n if x_i is the largest, etc). If d_i is the difference between the rank of x_i and y_i, the empirical analogue of $\mathbb{E}(F_X(X) - F_Y(Y))^2$ is therefore

$$\sum_{i=1}^{n}\frac{1}{n}\left(F_X(x_i) - F_Y(y_i)\right)^2 = \frac{1}{n^3}\sum_{i=1}^{n}d_i^2.$$

The estimator traditionally used is

$$\widehat{\rho}_S = 1 - \frac{6}{n(n^2 - 1)}\sum_{i=1}^{n}d_i^2;$$

the replacement of one n^2 by $n^2 - 1$ is due to issues such as the empirical mean of $\widehat{F}_X$ not being exactly $1/2$ but

$$\frac{1}{n}(1/n + 2/n + \cdots + n/n) = \frac{1}{2}(1 + 1/n).$$

The test for independence rejects for large absolute values of $\widehat{\rho}_S \sqrt{(n-2)/(1 - \widehat{\rho}_S^2)}$, which asymptotically follows a t-distribution with $n - 2$ degrees of freedom.

6 Further Dependence Concepts

6.1 Association

We call a random vector $X \in \mathbb{R}^d$ *associated* if $\mathrm{Cov}\left(f(X), g(X)\right) \geq 0$ for all non-decreasing $f, g : \mathbb{R}^d \to \mathbb{R}$ such that the expectations are well-defined.

Proposition 6.1

(a) *Any univariate r.v. $X \in \mathbb{R}$ is associated;*
(b) *If $X = (X_1, \ldots, X_d)$ is associated, then so is $(X_{i_1}, \ldots, X_{i_k})$ whenever $1 \leq i_1 < i_2 < \ldots < i_k \leq d$;*
(c) *If X and Y are both associated, then so is the concatenation $(X\ Y)$;*
(d) *If X is associated, then so is $\left(f_1(X_1) \ \ldots \ f_d(X_d)\right)$ whenever $f_1, \ldots, f_d$ are non-decreasing;*
(e) *If $X_1, \ldots, X_d$ are independent, then X is associated.*

Proof We show only part (a) and leave the (easy) parts (b)–(e) as Exercise 6.1. There are various proofs around, all of them slightly tricky. We use here a coupling argument.

Let $f, g : \mathbb{R} \to \mathbb{R}$ be non-decreasing and let X' be independent of X with $X' \overset{\mathscr{D}}{=} X$. Then

$$
\begin{aligned}
0 &\leq \mathbb{E}\left[\left(f(X) - f(X')\right)\left(g(X) - g(X')\right)\right] \\
&= \mathbb{E}\left[f(X)g(X)\right] - \mathbb{E}\left[f(X)g(X')\right] - \mathbb{E}\left[f(X')g(X)\right] + \mathbb{E}\left[f(X')g(X')\right] \\
&= 2\mathbb{E}\left[f(X)g(X)\right] - 2\mathbb{E}f(X) \cdot \mathbb{E}g(X) \\
&= 2\,\mathrm{Cov}\left(f(X), g(X)\right),
\end{aligned}
$$

where in the first step we used that either $f(X) - f(X') \geq 0$ and $g(X) - g(X') \geq 0$ at the same time (namely, when $X \geq X'$) or $f(X) - f(X') \leq 0$ and $g(X) - g(X') \leq 0$ at the same time (namely, when $X \leq X'$). $\square$

The inequalities $\mathbb{E}[f(X)g(X)] \geq \mathbb{E}f(X)\,\mathbb{E}g(X)$ and $\mathbb{E}[f(X)h(X)] \leq \mathbb{E}f(X)\,\mathbb{E}h(X)$ for f, g non-decreasing and h non-increasing that follow from part (a) sometimes go under the name *Chebycheff's covariance inequality*.

6.2 Tail Dependence

In many contexts, one is interested in the extent to which very large values of two r.v.s X, Y tend to occur together. A commonly used measure for this is the *tail dependence coefficient* $\lambda(Y|X)$ which for $X \overset{\mathcal{D}}{=} Y$ is defined by

$$\lambda(Y|X) = \lim_{x \to \infty} \mathbb{P}\big(Y > x \mid X > x\big).$$

When $X \overset{\mathcal{D}}{=} Y$ fails, for example when X, Y are of a different order of magnitude, this definition leads to undesired features such as that symmetry (meaning $\lambda(Y|X) = \lambda(X|Y)$) may fail, that one may have $\lambda(aY|X) \neq \lambda(Y|X)$, etc. Therefore (taking the distributions of X, Y to be continuous for simplicity) the general definition is

$$\lambda(Y|X) = \lim_{u \uparrow 1} \mathbb{P}\big(F_Y(Y) > u \mid F_X(X) > u\big). \tag{6.1}$$

If in particular $\lambda(Y|X) = 0$ we call X, Y *tail independent*.

Proposition 6.2 *The tail dependence coefficient depends only on the copula $C_{X,Y}$ of X, Y and not the marginals. More precisely,*

$$\lambda(Y|X) = \lim_{u \uparrow 1} \frac{1 - 2u + C_{X,Y}(u, u)}{1 - u}. \tag{6.2}$$

Further, $\lambda(X|Y) = \lambda(Y|X)$.

Proof Let $U_1 = F_X(X)$, $U_2 = F_Y(Y)$. Then

$$\mathbb{P}(U_1 > u, U_2 > u) = 1 - \mathbb{P}(U_1 \leq u \text{ or } U_2 \leq u)$$
$$= 1 - \big[\mathbb{P}(U_1 \leq u) + (U_2 \leq u) - \mathbb{P}(U_1 \leq u, U_2 \leq u)\big]$$
$$= 1 - 2u + C_{X,Y}(u, u).$$

Inserting this into (6.1), (6.2) follows immediately, and $\lambda(X|Y) = \lambda(Y|X)$ is clear from (6.2) and $C_{X,Y} = C_{Y,X}$ on the diagonal. $\qquad\qquad\square$

Proposition 6.3 *Assume that (X, Y) has a bivariate Gaussian distribution with correlation $\rho \in (-1, 1)$. Then $\lambda(Y|X) = 0$.*

Proof Since $\lambda(Y|X)$ depends only on the copula, we may assume that both marginals are $\mathcal{N}(0, 1)$. Given $X > x$, X is close to x (cf. Remark A.1 in the Appendix) and the conditional distribution of Y given $X = x$ is $\mathcal{N}(\rho x, 1 - \rho^2)$. Thus up to the first order of approximation

$$\lambda(Y|X) \approx \lim_{x \to \infty} \mathbb{P}(Y > x | X = x) = \lim_{x \to \infty} \overline{\Phi}\big((1 - \rho)x/(1 - \rho^2)^{1/2}\big) = 0.$$

The rigorous verification is left as Exercise 6.4.   □

In view of the properties of the conditional distribution of Y given $X = x$ that were used, Proposition 6.3 is somewhat counterintuitive and indicates that $\lambda(Y|X)$ is only a quite rough measure of tail dependence. In the multivariate regularly varying case:

Proposition 6.4 *Let $X = (X_1 \, X_2)$ be MRV with parameters α, L, μ so that*

$$\mathbb{P}\big(|X_1| + |X_2| > x\big) = L(x)/x^\alpha, \quad \mathbb{P}(X_i > x) \sim w_i L(x)/x^\alpha$$

for $i = 1, 2$ where $w_i = \int_{\mathcal{B}_1} \theta_i^{+\alpha} \mu(d\theta_1, d\theta_2)$, cf. Proposition 3.6. Then

$$\lambda(X_1|X_2) = \lambda(X_2|X_1) = \int_{\mathcal{B}_1^+} \min\big(\theta_1^\alpha/w_1, \theta_2^\alpha/w_2\big) \mu(d\theta_1, d\theta_2)$$

where $\mathcal{B}_1^+ = \big\{(\theta_1, \theta_2) \in \mathcal{B}_1 : \theta_1 > 0, \theta_2 > 0\big\}$.

Proof Let $d_i = w_i^{1/\alpha}$, $d = d_1 + d_2$. Then X_1/d_1, X_2/d_2 both have asymptotic tails $L(x)/x^\alpha$ and so

$$\lambda(X_1|X_2) = \lambda(X_1/d_1|X_2/d_2) = \lim_{x \to \infty} \frac{\mathbb{P}\big(X_1/d_1 > x, \, X_2/d_2 > x\big)}{\mathbb{P}(X_2/d_2 > x)}$$

$$= \lim_{x \to \infty} \frac{\mathbb{P}\big(X_1/d_1 > x/d, \, X_2/d_2 > x/d\big)}{\mathbb{P}(X_2/d_2 > x/d)}$$

$$= \lim_{x \to \infty} \frac{\mathbb{P}\big(X_1 > a_1 x, \, X_2 > a_2 x\big)}{d^\alpha L(x)/x^\alpha}, \tag{6.3}$$

where $a_1 = d_1/d$, $a_2 = d_2/d$.

We give two arguments for estimating the numerator in (6.3), one using the angular measure μ alone and the other the limit measure ν in Proposition 3.5 together with the connection to μ given in (3.9).

The argument based on μ alone is illustrated in Fig. 7 where the green shaded area is the region $X_1 > a_1 x$, $X_2 > a_2 x$. A point $z = (z_1, z_2)$ with $\|z\| = z_1 + z_2 = y$ is in this region precisely when $\theta = (\theta_1, \theta_2) = z/y$ in $\mathcal{B}_1$ is between the two red

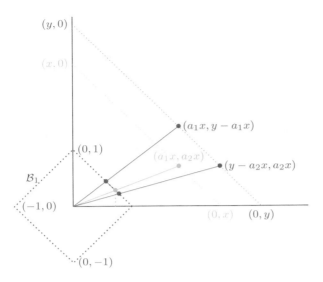

Fig. 7 $d = 2$

dotted points on $\mathcal{B}_1$, which means $\theta_1 > a_1x/y$, $\theta_2 > a_2x/y$. It follows that

$$\mathbb{P}\big(X_1 > a_1x,\ X_2 > a_2x\big)$$

$$= \int_x^\infty \mathbb{P}\big(\|X\| \in \mathrm{d}y\big) \int_{\mathcal{B}_1^+} \mathbb{1}(\theta_1 > a_1x/y,\ \theta_2 > a_2x)\, \mu(\mathrm{d}\theta_1, \mathrm{d}\theta_2)$$

$$= \int_{\mathcal{B}_1^+} \mu(\mathrm{d}\theta_1, \mathrm{d}\theta_2) \int_x^\infty \mathbb{1}(\theta_1 > a_1x/y,\ \theta_2 > a_2x/y)\, \mathbb{P}\big(\|X\| \in \mathrm{d}y\big)$$

$$= \int_{\mathcal{B}_1^+} \mu(\mathrm{d}\theta_1, \mathrm{d}\theta_2) \int_{(\max(a_1x/\theta_1,a_2x/\theta_2)}^\infty \mathbb{P}\big(\|X\| \in \mathrm{d}y\big)$$

$$= \int_{\mathcal{B}_1^+} \frac{L\big(\max(a_1x/\theta_1, a_2x/\theta_2)\big)}{[(\max(a_1x/\theta_1, a_2x/\theta_2)]^\alpha}\, \mu(\mathrm{d}\theta_1, \mathrm{d}\theta_2)$$

$$\sim \frac{L(x)}{x^\alpha} \int_{\mathcal{B}_1^+} \big[\min(\theta_1/a_1, \theta_2/a_2)\big]^\alpha\, \mu(\mathrm{d}\theta_1, \mathrm{d}\theta_2)$$

$$= \frac{d^\alpha L(x)}{x^\alpha} \int_{\mathcal{B}_1^+} \min(\theta_1^\alpha/w_1, \theta_2^\alpha/w_2)\, \mu(\mathrm{d}\theta_1, \mathrm{d}\theta_2)\,.$$

Combining with (6.3) completes the proof.

In the second argument, we use Proposition 3.5 to infer that

$$\frac{x^\alpha}{L(x)}\mathbb{P}(X_1 > a_1 x,\ X_2 > a_2 x) \ \to\ \nu\{x_1, x_2 : x_1 > a_1, x_2 > a_2\}$$

$$= \nu\{r, \boldsymbol{\theta} : \theta_1 > a_1/r, \theta_2 > a_2/r\} \ = \ \nu\{r, \boldsymbol{\theta} : r > a_1/\theta_1 \vee a_2/\theta_2\}$$

$$= \int_{\mathcal{B}_1^+} \frac{1}{(a_1/\theta_1 \vee a_2/\theta_2)^\alpha}\mu(\mathrm{d}\theta_1, \mathrm{d}\theta_2) \ = \ \int_{\mathcal{B}_1^+} \left(\theta_1/a_1 \wedge \theta_2/a_2\right)^\alpha \mu(\mathrm{d}\theta_1, \mathrm{d}\theta_2)\,,$$

where the third equality used (3.9). The rest of the algebra is the same.   □

Exercises

6.1 Show parts (b)–(e) of Proposition 6.1.

6.2 Compute $\lambda(Y|X)$ if $X = N_X + N$, $y = N_Y + N$ with N, N_X, NY independent and Poisson with rates μ, μ_X, μ_Y.

6.3 Compute the tail dependence coefficient for the Marshall–Olkin bivariate exponential distribution.

6.4 Give the details of a rigorous proof that $\lambda(Y|X) = 0$ in the bivariate normal distribution with $|\rho| \neq 1$.

6.5 Show that $\lambda(Y|X) = 2 - 2\lim_{t\downarrow 0} \phi'(t)/\phi''(2t)$ for an Archimedean copula.

Notes
As noted in connection with Proposition 6.3, $\lambda(Y|X)$ is a rather crude measure of tail dependence and accordingly, various more refined concepts have been developed. One is the *residual tail dependence coefficient* q_0 taking values in [0, 1]. It is again a copula concept and with uniform marginals, it has the form

$$\mathbb{P}(U_1 > y, U_2 > y) \ \sim\ \ell(1-y)(1-y)^{1/q_0}\,, \quad y \uparrow 1\,,$$

where ℓ is slowly varying. Independence corresponds to $q_0 = 1/2$, and in the bivariate Gaussian case, $q_0 = (1 + \rho)/2$. A somewhat related concept is *hidden regular variation*, developed for MRV where there is a great variety of dependence structures leading to extremal independence $\lambda(X_2 | X_1) = 0$, that is, the angular measure μ being a two point distribution degenerate at $(0, 1)$ and $(1, 0)$. In this case, one may gain some further asymptotic information by considering the interior of the cone $\mathbb{R}_+^2$. That is, (3.7) may still hold for B bounded away from the axis and a faster growing $a(t)$. This theory, surveyed in Resnick [149, Sect. 9.4], is quite similar with the main difference that the angular measure need not be finite since it is defined on a non-compact set. See also de Haan and Zhou [73].

7   Tails of Sums of Dependent Risks

We now consider the question of giving estimates for $\mathbb{P}(S > x)$ where $S = X_1 + \cdots + X_d$ with the components of the vector $(X_1 \ldots X_d)$ possibly dependent and having marginals $F_1, \ldots, F_d$, i.e., $F_i(x) = \mathbb{P}(X_i \leq x)$ and $(X_1, \ldots, X_d) \in$

Fréchet($F_1, \ldots, F_d$). For simplicity, we will assume that the F_i are continuous and strictly increasing.

7.1 Bounds

Rüschendorf [154] refers to the following inequality as one of the 'standard bounds'.

Proposition 7.1

$$\mathbb{P}(X_1 + \cdots + X_d > x) \leq \inf_{x_1 + \cdots + x_d = x} \left(\overline{F}_1(x_1) + \cdots + \overline{F}_d(x_d) \right).$$

Proof If $S = X_1 + \cdots + X_d > x = x_1 + \cdots + x_d$, then at least one X_i must exceed x_i (otherwise $S \leq x$). Hence

$$\mathbb{P}(S > x) \leq \mathbb{P}\left(\bigcup_{i=1}^{n} \{X_i > x_i\} \right) \leq \sum_{i=1}^{d} \mathbb{P}(X_i > x_i) = \sum_{i=1}^{d} \overline{F}(x_i).$$

Taking the inf gives the result. □

It is remarkable that despite the simplicity of its proof, the inequality in Proposition 7.1 is sharp, i.e. attained for a certain dependence structure. We verify this for $d = 2$.

Proposition 7.2 *Given F_1, F_2 and x, there exists a random vector (X_1, X_2) with marginals F_1, F_2 such that*

$$\mathbb{P}(X_1 + X_2 > x) = \inf_{x_1 + x_2 = x} \left(\overline{F}_1(x_1) + \overline{F}_2(x_2) \right). \tag{7.1}$$

Proof Denote the r.h.s. of (7.1) by p^*. Then for x_1 such that $\overline{F}_1(x_1) \leq p^*$

$$x_1 + \overline{F}_2^{\leftarrow}\left(p^* - \overline{F}_1(x_1)\right) \geq x_1 + \overline{F}_2^{\leftarrow}\left(\overline{F}_2(x - x_1)\right)$$
$$= x_1 + (x - x_1) = x, \tag{7.2}$$

where the first step used that $p^* \leq \overline{F}_1(x_1) + \overline{F}_2(x - x_1)$ and that $\overline{F}_2^{\leftarrow}$ is non-increasing, and the second the assumed regularity properties of F_2. Now let U_1 be uniform$(0, 1)$ and let $U_2 = p^* - U_1$ for $U_1 \leq p^*$, $U_2 = U_1$ for $U_1 > p^*$. Then U_2 is again uniform$(0, 1)$ since clearly $\mathbb{P}(U_2 \in A) = \mathbb{P}(U_1 \in A)$ for $A \subseteq (p^*, 1)$, whereas also

$$\mathbb{P}(U_2 \in A) = \mathbb{P}(U_2 \in A, \ U_2 \leq p^*) = \mathbb{P}(p^* - U_1 \in A, \ U_1 \leq p^*)$$
$$= \mathbb{P}(U_1 \in A, \ U_1 \leq p^*) = \mathbb{P}(U_1 \in A)$$

for $A \subseteq (0, p^*]$. Hence $X_1 = \overline{F}_1^{\leftarrow}(U_1)$ and $X_2 = \overline{F}_1^{\leftarrow}(U_2)$ have marginals F_1, F_2. On the set $U_1 \leq p^*$, we have

$$U_1 = \overline{F}_1(X_1) \leq p^*, \quad X_2 = \overline{F}_2^{\leftarrow}\big(p^* - \overline{F}_1(X_1)\big),$$

and hence $X_1 + X_2 \geq x$ by (7.2). Hence $\mathbb{P}(X_1 + X_2 > x) \geq \mathbb{P}(U_1 \leq p^*) = p^*$, and by the definition of p^* and Proposition 7.1 equality must hold.   □

Proposition 7.3 *Assume $F_1 = F_2 = F$ and that F has a density f that is non-increasing on some interval $[a, \infty)$. Then $\mathbb{P}(X_1 + X_2 > x) \leq 2\overline{F}(x/2)$ for all large x.*

Proof The value $2\overline{F}(x/2)$ of the inf in Proposition 7.1 is obviously attained for $x_1 = x_2 = x/2$. The assumption ensures that $\overline{F}$ is convex on $[a, \infty)$. Thus for $x_1 + x_2 = x$, $x_2 \geq x_1 \geq a$ we have $\overline{F}(x_1) + \overline{F}(x_2) \geq 2\overline{F}(x/2)$. For $x_1 + x_2 = x$, $x_1 < a$ we have $\overline{F}(x_1) + \overline{F}(x_2) \geq \overline{F}(a)$. Hence the inf equals $\overline{F}(x/2)$ when x is so large that $2\overline{F}(x/2) \leq \overline{F}(a)$.   □

Remark 7.4 The value $2\overline{F}(x/2)$ is to be compared with the smaller $\overline{F}(x/2)$ one gets in the comonotonic case, and gives a key example that comonotonicity is not necessarily the dependence structure leading to the maximal risk.

A further illustration of this comes from F being regular varying, $\overline{F}(x) = L(x)/x^\alpha$. Here $\mathbb{P}(X_1 + X_2 > x) = \overline{F}(x/2) \sim 2^\alpha L(x)/x^\alpha$ in the comonotonic case, whereas if X_1, X_2 are independent, then by Proposition III.2.4

$$\mathbb{P}(X_1 + X_2 > x) \sim 2\overline{F}(x) = 2L(x)/x^\alpha.$$

Thus if $\alpha < 1$ so that $2^\alpha < 2$, the asymptotic order of $\mathbb{P}(X_1 + X_2 > x)$ is larger with independence than with comonotonicity, again contrary to the naive first guess.

With independence and regular variation, we have $\mathbb{P}(X_1 + X_2 > x) \sim 2\overline{F}(x)$. This is of the same rough order of magnitude $1/x^\alpha$ as the standard bound and as in the comonotonic case. With lighter tails than for regular variation, the situation is, however, typically the opposite. For example, if F is standard exponential, then $\mathbb{P}(X_1 + X_2 > x) \sim x e^{-x}$ in the independent case, whereas $\mathbb{P}(X_1 + X_2 > x) = \overline{F}(x/2) = e^{-x/2}$ is of larger magnitude in the comonotonic case (though still not attaining the upper bound $2e^{-x/2}$!). See also Example III.1.3 and Proposition 7.8 below.   ◇

7.2   A Copula Representation

To indicate how copulas can come into play to evaluate $\mathbb{P}(S > x)$, we present the following example:

Proposition 7.5 *If the copula C of $X_1 \geq 0$, $X_2 \geq 0$ is absolutely continuous, then*

$$\frac{\mathbb{P}(X_1 + X_2 > x)}{\overline{F}_1(x)} = 1 + \int_0^x \frac{1 - C_1\big(F_1(z), F_2(x - z)\big)}{\overline{F}_1(x)} \, F_1(\mathrm{d}z),$$

where $C_1(u_1, u_2) = \dfrac{\partial C(u_1, u_2)}{\partial u_1}$.

For the proof we need:

Lemma 7.6 *If U_1, U_2 are uniform$(0, 1)$ with absolutely continuous copula C, then*
$\mathbb{P}(U_2 \leq u_2 \mid U_1 = u) = C_1(u_1, u_2)$.

Proof Since $\mathbb{P}(u_1 \leq U_1 < u_1 + h)$, we get

$$\mathbb{P}(U_2 \leq u_2 \mid U_1 = u) = \lim_{h \downarrow 0} \frac{\mathbb{P}(U_2 \leq u_2, \, u_1 \leq U_1 < u_1 + h)}{\mathbb{P}(u_1 \leq U_1 < u_1 + h)}$$

$$= \lim_{h \downarrow 0} \frac{C(u_1 + h, u_2) - C(u_1, u_2)}{h} = C_1(u_1, u_2).$$

$\square$

Proof of Proposition 7.5 From Lemma 7.6, we get

$$\mathbb{P}(X_1 + X_2 \leq x \mid X_1 = z) = \mathbb{P}\big(F_2(X_2) \leq F_2(x - z) \,\big|\, F_1(X_1) = F_1(z)\big)$$

$$= C_1\big(F_1(z), F_2(x - z)\big).$$

The result then follows from

$$\frac{\mathbb{P}(X_1 + X_2 > x)}{\overline{F}_1(x)} = 1 + \frac{F_1(x) - \mathbb{P}(X_1 + X_2 \leq x)}{\overline{F}_1(x)}$$

$$= 1 + \int_{-\infty}^x \frac{1 - \mathbb{P}(X_1 + X_2 \leq x \mid X_2 = z)}{\overline{F}_1(x)} \, \overline{F}_1(\mathrm{d}z).$$

$\square$

7.3   Asymptotics

For simplicity, we will concentrate on the case $d = 2$ and positive r.v.s X_1, X_2 with continuous marginals F_1, F_2 such that F_1 is more heavy-tailed than F_2, more precisely in the sense that

$$c = \lim_{x \to \infty} \frac{\overline{F}_2(x)}{\overline{F}_1(x)} \tag{7.3}$$

exists and is in $[0, 1]$. Tail dependence as measured by $\lambda(Y|X)$ defined by (6.1) is a less relevant concept since it is scale-free whereas $\mathbb{P}(X_1 + X_2 > x)$ is changed if one of the r.v.s is scaled. We shall therefore instead work with

$$\widehat{\lambda} = \lim_{x \to \infty} \mathbb{P}(X_2 > x | X_1 > x). \tag{7.4}$$

A key question is asymptotic comparison of the tails of S and X_1: does $\mathbb{P}(S > x)/\overline{F}_1(x)$ go to 1, to a constant in $(1, \infty)$, or to ∞? A relatively simple result in this frame is the following:

Proposition 7.7 *Assume that F_1 is regularly varying and that $\widehat{\lambda} = 0$. Then $\mathbb{P}(X_1 + X_2 > x) \sim (1 + c)\overline{F}_1(x)$.*

This is the same conclusion as in Proposition III.2.10, with independence relaxed to $\widehat{\lambda} = 0$. The proof is given later.

Another simple case is the following:

Proposition 7.8 *Assume that F_1 is in MDA(Gumbel) and that*

$$m = \inf_{a>0} \liminf_{x \to \infty} \mathbb{P}(X_2 > ae(x) \mid X_1 > x) > 0, \tag{7.5}$$

where $e(\cdot)$ is the mean excess function. Then $\mathbb{P}(S > x)/\overline{F}_1(x) \to \infty$.

Remark 7.9 The result covers a fairly broad spectrum of situations with lighter than regularly varying marginals. In fact, MDA(Gumbel) gave the main examples of such marginals in Chap. IX and one had $e(x) = cx/\log x$ for the lognormal case and typically $e(x) = cx^{1-\beta}$ with $\beta > 0$ otherwise. When $0 < \beta < 1$, (7.5) essentially says that X_2 has to grow with X_1 but the rate could be slower. When $\beta > 1$, (7.5) roughly can only fail if X_2 becomes small as X_1 becomes large, a quite atypical type of behavior.   ◇

Proof of Proposition 7.8 The assumption $F_1 \in$ MDA(Gumbel) implies that $e(x)$ is self-neglecting, i.e. that $e(x + ae(x))/e(x) \to 1$ for all $a \in \mathbb{R}$, cf. Exercise 7.1. Thus for a given $a > 0$, we have $e(x) \le 2e(x - ae(x))$ for all large x. Hence

$$\liminf_{x \to \infty} \frac{\mathbb{P}(S > x)}{\overline{F}_1(x)} \ge \liminf_{x \to \infty} \frac{\mathbb{P}(X_1 > x - ae(x),\ X_2 > ae(x))}{\overline{F}_1(x)}$$

$$= \liminf_{x \to \infty} \frac{\overline{F}_1(x - ae(x))\mathbb{P}(X_2 > ae(x) \mid X_1 > x - ae(x))}{\overline{F}_1(x)}$$

$$\ge e^a \liminf_{x \to \infty} \mathbb{P}(X_2 > 2ae(x - ae(x)) \mid X_1 > x - ae(x)) \ge me^a,$$

where we used that $\overline{F}(x - ae(x))/\overline{F}(x) \to e^a$ in MDA(Gumbel). Letting $a \to \infty$ completes the proof. □

In the regularly varying case (i.e., $F_1 \in$ MDA(Fréchet)) one cannot have $\mathbb{P}(S > x)/\overline{F}_1(x) \to \infty$. More precisely:

Proposition 7.10 *Assume that F_1 is regularly varying, i.e. $\overline{F}_1(x) = L(x)/x^\alpha$. Then*

$$
\limsup_{x\to\infty} \frac{\mathbb{P}(S > x)}{\overline{F}_1(x)}
$$

$$
\leq \begin{cases} \left(\widehat{\lambda}^{1/(\alpha+1)} + (1+c-2\widehat{\lambda})^{1/(\alpha+1)}\right)^{\alpha+1} & 0 \leq \widehat{\lambda} \leq (1+c)/3, \\ 2^\alpha(1+c-\widehat{\lambda}) & (1+c)/3 \leq \widehat{\lambda} \leq 1. \end{cases}
$$

Proof A straightforward extension of the proof of Proposition III.2.4, to be given in a moment below, gives

$$
\limsup_{x\to\infty} \frac{\mathbb{P}(S > x)}{\overline{F}_1(x)} \leq \frac{1+c-2\widehat{\lambda}}{(1-\delta)^\alpha} + \frac{\widehat{\lambda}}{\delta^\alpha} \tag{7.6}
$$

for $0 < \delta < 1/2$. Here the first term on the r.h.s. increases with δ and the second decreases. Easy calculus shows that the minimizer δ^* of the r.h.s. is given by

$$
\delta^* = \begin{cases} \dfrac{1}{1 + \left((1+c)/3 - 2\right)^{1/(\alpha+1)}} & 0 \leq \widehat{\lambda} \leq (1+c)/3, \\ 1/2 & (1+c)/3 \leq \widehat{\lambda} \leq 1. \end{cases}
$$

Inserting this into (7.6) gives the result.

For the proof of (7.6), note as in the proof of Proposition III.2.4 that $\{S > x\} \subseteq A_1 \cup A_2 \cup A_3$, where

$$
A_1 = \{X_1 > (1-\delta)x\}, \quad A_2 = \{X_2 > (1-\delta)x\}, \quad A_3 = \{X_1 > \delta x, X_2 > \delta x\},
$$

for $0 < \delta < 1/2$. For such δ,

$$
\mathbb{P}\big((A_1 \cup A_2) \cap A_3\big) \geq \mathbb{P}(A_2 \cap A_3) \geq \mathbb{P}(A_2 \cap A_1).
$$

Hence

$$
\begin{aligned}
\mathbb{P}(S > x) &\leq \mathbb{P}(A_1 \cup A_2) + \mathbb{P}A_3 - \mathbb{P}\big((A_1 \cup A_2) \cap A_3\big) \\
&\leq \mathbb{P}A_1 + \mathbb{P}A_2 + \mathbb{P}A_3 - 2\mathbb{P}(A_1 \cap A_2) \\
&= \overline{F}_1\big((1-\delta)x\big) + \overline{F}_2\big((1-\delta)x\big) + \mathbb{P}(X_1 > \delta x, X_2 > \delta x) \\
&\quad - 2\mathbb{P}\big(X_1 > (1-\delta)x, X_2 > (1-\delta)x\big),
\end{aligned}
$$

so that the lim sup in the proposition is bounded by the lim sup of

$$(1 - 2\widehat{\lambda}) \frac{\overline{F}_1((1 - \delta)x)}{\overline{F}_1(x)} + \frac{\overline{F}_2((1 - \delta)x)}{\overline{F}_1(x)} + \frac{\overline{F}_1(\delta x)}{\overline{F}_1(x)} \mathbb{P}(X_2 > \delta x \mid X_1 > \delta x),$$

which equals

$$(1 - 2\widehat{\lambda}) \frac{1}{(1 - \delta)^\alpha} + \frac{c}{(1 - \delta)^\alpha} + \frac{\widehat{\lambda}}{\delta^\alpha}$$

as asserted.   □

Proof of Proposition 7.7 That $1 + c$ is an asymptotic upper bound for $\mathbb{P}(S > x)/\overline{F}_1(x)$ follows immediately by inserting $\widehat{\lambda} = 0$ in the bound of Proposition 7.10. That it is also a lower bound follows from

$$\begin{aligned}
\mathbb{P}(S > x) &\geq \mathbb{P}\big(\max(X_1, X_2) > x\big) \\
&= \mathbb{P}(X_1 > x) + \mathbb{P}(X_2 > x) - \mathbb{P}(X_1 > x, X_2 > x) \\
&= \mathbb{P}(X_1 > x) + \mathbb{P}(X_2 > x) - \mathbb{P}(X_1 > x)\mathbb{P}(X_2 > x \mid X_1 > x) \\
&\sim \overline{F}_1(x)(1 + c - \widehat{\lambda}) = \overline{F}_1(x)(1 + c).
\end{aligned}$$

□

Exercises
7.1 Let F be a distribution and $e(x) > 0$ a function such that $\overline{F}(x + ae(x)) \sim e^{-a}\overline{F}(x)$ for all $a \in \mathbb{R}$. Show that $e(x + ae(x)) \sim e(x)$.

Notes
Some key sources for the material of this section are Rüschendorf [154] and Asmussen, Albrecher and Kortschak [9]. There is, however, an extensive research literature.

8   Dependence Orderings

We consider multivariate r.v.s $X = (X_1 \ldots X_d)$, $Y = (Y_1 \ldots Y_d)$, and discuss the question of quantifying questions such as when the components of Y are more dependent than those of X. From the point of view of actuarial or financial applications, the main case is positive dependence: if one X_i is larger than typical, then so are some or all of the X_j with $j \neq i$.

Intuitively, one expects that positive dependence will increase the risk. One main example to quantify this will be to compare sums $X_1 + \cdots + X_d$ and $Y_1 + \cdots + Y_d$ in terms of tail probabilities or convex ordering when (in appropriate definitions) the components of Y are more dependent than those of X. Similar issues are comparison of $\max(X_1, \ldots, X_d)$ and $\max(Y_1, \ldots, Y_d)$, and of the corresponding minima. By far, the most interesting case is X, Y being in the same Fréchet class, that is, having the same marginals $F_i(x_i) = \mathbb{P}(X_i \le x_i) = \mathbb{P}(Y_i \le x_i)$.

8.1 Orthant Dependencies and Orderings

A random vector $X = (X_1, \ldots, X_d)$ is said to be *positive upper orthant dependent* if

$$\mathbb{P}(X_1 > x_1, \ldots, X_d > x_d) \ge \mathbb{P}(X_1 > x_1) \cdots \mathbb{P}(X_d > x_d) = \overline{F}_1(x_1) \cdots \overline{F}_d(x_d),$$

where $F_1, \ldots, F_d$ are the marginals, and *positive lower orthant dependent* if

$$\mathbb{P}(X_1 \le x_1, \ldots, X_d \le x_d) \ge \mathbb{P}(X_1 \le x_1) \cdots \mathbb{P}(X_d \le x_d) = F_1(x_1) \cdots F_d(x_d).$$

Similarly, X is said to be smaller than Y in upper orthant ordering, written $X \le_{\mathrm{uo}} Y$, if

$$\mathbb{P}(X_1 > x_1, \ldots, X_d > x_d) \le \mathbb{P}(Y_1 > x_1, \ldots, Y_d > x_d) \tag{8.1}$$

and smaller than Y in lower orthant ordering, $X \le_{\mathrm{lo}} Y$, if

$$\mathbb{P}(X_1 \le x_1, \ldots, X_d \le x_d) \le \mathbb{P}(Y_1 \le x_1, \ldots, Y_d \le x_d). \tag{8.2}$$

If both $X \le_{\mathrm{uo}} Y$ and $X \le_{\mathrm{lo}} Y$, then X is said to be smaller than Y in concordance ordering, $X \le_{\mathrm{c}} Y$. An easy limit argument shows that $>$ in (8.1) may be replaced by $\ge$, and similarly for (8.2). This gives immediately that

$$X \le_{\mathrm{uo}} Y \iff -Y \le_{\mathrm{lo}} -X. \tag{8.3}$$

If $X^\perp$ denotes a random vector with the same marginals $F_1, \ldots, F_d$ but independent components, then positive upper orthant dependence of X is immediately seen to be the same as $X^\perp \le_{\mathrm{uo}} X$. That is, large values of one X_i somehow stimulates large values of the other X_j. Similar remarks apply to lower orthant ordering.

In terms of maxima and extrema, we immediately have:

Proposition 8.1 *If $X \le_{\mathrm{lo}} Y$, then $Y_i \preceq_{\mathrm{st}} X_i$, $i = 1, \ldots, d$, and*

$$\max(Y_1, \ldots, Y_d) \preceq_{\mathrm{st}} \max(X_1, \ldots, X_d).$$

If instead $X \le_{\mathrm{uo}} Y$, then $X_i \preceq_{\mathrm{st}} Y_i$ for all i and

$$\min(X_1, \ldots, X_d) \preceq_{\mathrm{st}} \min(Y_1, \ldots, Y_d).$$

Note that the stochastic ordering inequality for the maximum goes the other way than $X \leq_{lo} Y$. Intuitively, one can understand this as more positive dependence meaning fewer sources of risk (see also Remark 8.13 below).

Proof Assuming $X \leq_{lo} Y$ and letting $x_j \to \infty$ for the $j \neq i$ in (8.2) gives $\mathbb{P}(X_i \leq x_i) \leq \mathbb{P}(Y_i \leq x_i)$, implying $Y_i \preceq_{st} X_i$. Taking instead $x_1 = \cdots = x_d = x$ gives

$$\mathbb{P}\big(\max(X_1, \ldots, X_d) \leq x\big) \;\leq\; \mathbb{P}\big(\max(Y_1, \ldots, Y_d) \leq x\big)$$

and the assertion on ordering of maxima. The proofs for the case $X \leq_{uo} Y$ are similar, or one may appeal to (8.3).   □

Corollary 8.2 *If $X \leq_c Y$, then $Y_i \overset{\mathcal{D}}{=} X_i$ for all i and*

$$\min(X_1, \ldots, X_d) \preceq_{st} \min(Y_1, \ldots, Y_d) \leq \max(Y_1, \ldots, Y_d) \preceq_{st} \max(X_1, \ldots, X_d).$$

Proposition 8.3 $X \leq_{lo} Y$, *resp.* $X \leq_{uo} Y$ *if and only if*

$$\mathbb{E} f_1(X_1) \cdots f_d(X_d) \;\leq\; \mathbb{E} f_1(Y_1) \cdots f_d(Y_d)$$

for all set $f_1, \ldots, f_d$ of non-increasing, resp. non-decreasing functions.

Proof Just note that $\mathbb{1}(X_1 \leq x_1, \ldots, X_d \leq x_d) = \mathbb{1}(X_1 \leq x_1) \cdots \mathbb{1}(X_d \leq x_d)$ and use standard approximation arguments.   □

Proposition 8.4 *Assume for $d = 2$ that $X = (X_1, X_2)$ and $Y = (Y_1, Y_2)$ have the same marginals F_1, F_2. Then*

$$X \leq_{lo} Y \iff X \leq_{uo} Y \iff X \leq_c Y. \tag{8.4}$$

If one of these equivalent statements hold, then $\rho(X_1, X_2) \leq \rho(Y_1, Y_2)$, where ρ is any of the Pearson, Kendall or Spearman correlations from Sect. 5.

When X and Y have the same marginals, a further equivalent ordering is the supermodular one to be introduced in Sect. 8.2. This and all equivalences in (8.4) only hold for $d = 2$!

Proof The first equivalence in (8.4) follows from

$$\mathbb{P}(X_1 > x_1, X_2 > x_2) = 1 - F_1(x_1) - F_2(x_2) + \mathbb{P}(X_1 \leq x_1, X_2 \leq x_2),$$
$$\mathbb{P}(Y_1 > x_1, Y_2 > x_2) = 1 - F_1(x_1) - F_2(x_2) + \mathbb{P}(Y_1 \leq x_1, Y_2 \leq x_2),$$

and the second is then clear.

For the ordering of the Pearson correlation coefficient, use Proposition 8.3 to conclude that $\mathbb{E}X_1X_2 \le \mathbb{E}Y_1Y_2$. For Spearman, just use the definition:

$$\mathrm{Cov}\left(F_1(X_1), F_2(X_2)\right) = \mathbb{E}\left[F_1(X_1)F_2(X_2)\right] - \mathbb{E}F_1(X_1)\,\mathbb{E}F_2(X_2)$$

$$= \mathbb{E}\left[F_1(X_1)F_2(X_2)\right] - \mathbb{E}F_1(Y_1)\,\mathbb{E}F_2(Y_2)$$

$$\le \mathbb{E}\left[F_1(Y_1)F_2(Y_2)\right] - \mathbb{E}F_1(Y_1)\,\mathbb{E}F_2(Y_2) = \mathrm{Cov}\left(F_1(Y_1), F_2(Y_2)\right),$$

where the inequality follows from Proposition 8.3. The case of Kendall's τ is deferred to Sect. 8.2. ☐

Orthant dependence induces higher variability of sums in terms of convex order, at least for $d = 2$ (for the case $d > 2$, see Proposition 8.12 below):

Proposition 8.5 *Assume* $X = (X_1, X_2)$ *is positive lower orthant dependent. Then* $X_1^{\perp} + X_2^{\perp} \preceq_{cx} X_1 + X_2$. *More generally, if* $X \preceq_{lo} Y$, *then* $X_1 + X_2 \preceq_{cx} Y_1 + Y_2$.

Proof Since $\mathbb{E}[X_1^{\perp} + X_2^{\perp}] = \mathbb{E}[X_1 + X_2]$, it suffices to show $\mathbb{E}(b - X_1^{\perp} - X_2^{\perp})^+ \le \mathbb{E}(b - X_1 + X_2)^+$. But

$$\int \mathbb{1}\{x_1 \le t, x_2 \le b - t\}\,dt = \int \mathbb{1}\{x_1 \le t \le b - x_2\}\,dt = (b - x_1 - x_2)^+$$

and so

$$\mathbb{E}(b - X_1 + X_2)^+ = \int \mathbb{P}(X_1 \le t, X_2 \le b - t)\,dt$$

$$\ge \int \mathbb{P}(X_1^{\perp} \le t, X_2^{\perp} \le b - t)\,dt = \mathbb{E}(b - X_1^{\perp} - X_2^{\perp})^+.$$

The proof of the last statement is similar. ☐

Orthant dependence can improve the bounds in Proposition 7.1:

Proposition 8.6 *If X is lower orthant dependent, then*

$$\mathbb{P}(X_1 + \cdots + X_d > x) \le 1 - \sup_{x_1 + \cdots x_d = x} F_1(x_1) \cdots F_d(x_d). \tag{8.5}$$

Proof Just note that

$$\mathbb{P}(X_1 + \cdots + X_d \le x) \ge F_1(x_1) \cdots F_d(x_d)$$

whenever $x_1 + \cdots x_d = x$, take the sup and subtract both sides from 1. ☐

Remark 8.7 Rüschendorf [154, p. 93] refers to (8.5) as a substantial improvement of Proposition 7.1. However, for $d = 2$ and identical marginals $F_1 = F_2 = F$, (8.5) can be written

$$\mathbb{P}(X_1 + X_2 > x) \leq \inf_{x_1+x_2=x} \left[\overline{F}(x_1) + \overline{F}(x_2) - \overline{F}(x_1)\overline{F}(x_2) \right].$$

If F is standard exponential, this gives

$$\mathbb{P}(X_1 + X_2 > x) \leq \inf_{x_1+x_2=x} \left[e^{-x_1} + e^{-x_2} - e^{-x} \right] = 2e^{-x/2} - e^{-x},$$

where the last step follows by Proposition 7.3. For large x, this is only marginally smaller than the bound $2e^{-x/2}$ provided by Proposition 7.1. This does not of course exclude a substantial difference in other situations!   ◊

Exercises
8.1 Show that if $X = (X_1, X_2)$ is associated, then X is upper orthant dependent.

8.2 The Supermodular Ordering

For vectors $x = (x_1 \ldots x_d)$, $y = (y_1 \ldots y_d) \in \mathbb{R}^d$, define the pointwise maximum and minimum as the vectors

$$x \wedge y = (x_1 \wedge y_1 \ldots x_d \wedge y_d), \quad x \vee y = (x_1 \vee y_1 \ldots x_d \vee y_d).$$

A function $f : \mathbb{R}^d \to \mathbb{R}$ is called *supermodular*, written $f \in \mathcal{SM}$, if

$$f(x) + f(y) \leq f(x \wedge y) + f(x \vee y) \tag{8.6}$$

for all $x, y \in \mathbb{R}^d$ (note that this holds with equality for any function f if $d = 1$, so the concept is only interesting for $d > 1$).

Example 8.8 For $d = 2$, some examples of supermodular functions of $x = (x_1, x_2)$ are $x_1 x_2$, $(x_1 + x_2)^2$, $\min(x_1, x_2)$, $\mathbb{1}(x_1 > z_1, x_2 > z_2)$, $f(x_1 - x_2)$, where f is concave and continuous. Examples where the inequality in (8.6) is reversed (i.e. $-f$ is supermodular) are $|x_1 - x_2|^p$ for $p > 1$, $\max(x_1, x_2)$, $f(x_1 - x_2)$ where f is convex and continuous.   ◊

Proposition 8.9 *Let $f : \mathbb{R}^d \to \mathbb{R}$. Then*

(a) *$f \in \mathcal{SM}$ if and only if f is supermodular in any two components, that is, if $f(x_1, \ldots, x_d)$ is a supermodular function of x_i, x_j for $i \neq j$ and fixed values of the x_k with $k \neq i, j$;*

(b) *if f is twice differentiable, then $f \in \mathcal{SM}$ if and only if*

$$f_{ij} = \frac{\partial^2 f}{\partial x_i \partial x_j} \geq 0 \quad \text{for } i \neq j.$$

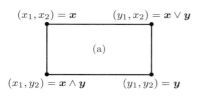

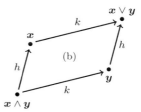

Fig. 8 (a) $d = 2$; (b) $d > 2$

Proof The 'only if' part of (a) is trivial. The 'if' part follows from (b) if f is twice differentiable; the general case is omitted.

For (b), we take first $d = 2$. Assume $f \in \mathcal{SM}$. Then for any x_1, x_2 we have

$$f(x_1 + \epsilon, x_2 + \delta) + f(x_1, x_2) \ge f(x_1 + \epsilon, x_2) + f(x_1, x_2 + \delta),$$

which implies $f_2(x_1 + \epsilon, x_2) \ge f_2(x_1, x_2)$, where $f_2 = \partial f / \partial x_2$. This in turn is sufficient for $f_{12} \ge 0$.

Assume conversely $f_{12} \ge 0$. The inequality (8.6) is trivial if x, y are totally ordered, that is, if $x \wedge y$ equals either x or y. Otherwise, assume e.g. $x_1 \le y_1$, $x_2 \ge y_2$. Then (cf. Fig. 8a)

$$
\begin{aligned}
f(x \wedge y) &+ f(x \vee y) - f(x) - f(y) \\
&= f(x_1, y_2) + f(y_1, x_2) - f(x_1, x_2) - f(y_1, y_2) \\
&= \bigl(f(y_1, x_2) - f(y_1, y_2)\bigr) - \bigl(f(x_1, x_2) - f(x_1, y_2)\bigr) \\
&= \int_{y_2}^{x_2} f_2(y_1, y)\, dy - \int_{y_2}^{x_2} f_2(x_1, y)\, dy \\
&= \int_{y_2}^{x_2} dy \int_{x_1}^{y_1} f_{12}(x, y)\, dx \ge 0.
\end{aligned}
$$

Now consider a general d. The 'only if' statement of (b) then follows from what was shown for $d = 2$. For the 'if' part, let $h = x - x \wedge y$, $k = y - x \wedge y$. Then clearly $h_i, k_j \ge 0$ for all i, j. Applying the identity $x_i \vee y_i + x_i \wedge y_i = x_i + y_i$ for each i yields $x \vee y + x \wedge y = x + y$. In particular, $x \vee y = x + k = y + h$, so that the points $x \wedge y, x, y, x \vee y$ form a parallelogram, cf. Fig. 8b. Further, $x_i \vee y_i = x_i + k_i = y_i + h_i$ shows that h_i and k_i cannot be positive at the same time, i.e. $h_i k_i = 0$. It follows that

$$
\begin{aligned}
f(x \wedge y) &+ f(x \vee y) - f(x) - f(y) \\
&= f(x \wedge y + k + h) - f(x \wedge y + k) + \bigl(f(x \wedge y) - f(x \wedge y + h)\bigr)
\end{aligned}
$$

$$= \int_0^1 \frac{d}{ds}\big[f(\boldsymbol{x} \wedge \boldsymbol{y} + \boldsymbol{k} + s\boldsymbol{h}) - f(\boldsymbol{x} \wedge \boldsymbol{y} + s\boldsymbol{h})\big]\,ds$$

$$= \int_0^1 ds \int_0^1 \frac{d}{ds}\frac{d}{dt} f(\boldsymbol{x} \wedge \boldsymbol{y} + t\boldsymbol{k} + s\boldsymbol{h})\,dt$$

$$= \int_0^1 ds \int_0^1 \sum_{i,j=1}^{d} h_i k_j f_{ij}(\boldsymbol{x} \wedge \boldsymbol{y} + t\boldsymbol{k} + s\boldsymbol{h})\,dtt$$

$$= \int_0^1 ds \int_0^1 \sum_{i \ne j} h_i k_j f_{ij}(\boldsymbol{x} \wedge \boldsymbol{y} + t\boldsymbol{k} + s\boldsymbol{h})\,dt \ \ge\ 0.$$

$\square$

If in Fig. 8 we think of the four points as possible outcomes of an r.v. $\boldsymbol{Z} = (Z_1, Z_2)$, strong dependence of Z_1 and Z_2 should mean that the points $\boldsymbol{x} \wedge \boldsymbol{y}$ and $\boldsymbol{x} \vee \boldsymbol{y}$ have a substantially larger pooled likelihood than the points $\boldsymbol{x}$ and $\boldsymbol{y}$. This motivates us to define a dependence ordering, called the *supermodular ordering* and denoted $\boldsymbol{X} \le_{\mathrm{sm}} \boldsymbol{Y}$, by letting $\boldsymbol{X} \le_{\mathrm{sm}} \boldsymbol{Y}$ mean $\mathbb{E}f(\boldsymbol{X}) \le \mathbb{E}f(\boldsymbol{Y})$ for all $f \in \mathcal{SM}$. Some main properties of this ordering are:

Proposition 8.10 *Assume $\boldsymbol{X} \le_{\mathrm{sm}} \boldsymbol{Y}$. Then:*

(a) $\boldsymbol{X} \le_{\mathrm{lo}} \boldsymbol{Y}$, $\boldsymbol{X} \le_{\mathrm{uo}} \boldsymbol{Y}$ *and* $\boldsymbol{X} \le_{\mathrm{c}} \boldsymbol{Y}$;
(b) $X_i \overset{\mathscr{D}}{=} Y_i$ *for all i;*
(c) $\big(g_1(X_1) \ldots g_d(X_d)\big) \le_{\mathrm{sm}} \big(g_1(Y_1) \ldots g_d(Y_d)\big)$ *whenever $g_1,\ldots,g_d$ are non-decreasing;*
(d) $\mathbb{C}\mathrm{ov}(X_i, X_j) \le \mathbb{C}\mathrm{ov}(Y_i, Y_j)$ *for $i \ne j$. More generally,* $\mathbb{C}\mathrm{ov}\big(g_i(X_i), g_j(X_j)\big)$ *whenever g_i, g_j are non-decreasing.*

Proof For (a), just check that for fixed $x_1, \ldots, x_d$ the functions mapping $x_1^*, \ldots, x_d^*$ into

$$\mathbb{1}(x_1^* \le x_1, \ldots, x_d^* \le x_d), \quad \mathbb{1}(x_1^* > x_1, \ldots, x_d^* > x_d)$$

are supermodular.

Any function depending only on one x_i is supermodular, and so $\mathbb{E}h(X_i) \le \mathbb{E}h(Y_i)$ and $-\mathbb{E}h(X_i) \le -\mathbb{E}h(Y_i)$ for all $h : \mathbb{R} \to \mathbb{R}$, implying $\mathbb{E}h(X_i) = \mathbb{E}h(Y_i)$ and (b). For (c), just note that if $f \in \mathcal{SM}$, then $f\big(g_1(\cdot), \ldots, g_d(\cdot)\big) \in \mathcal{SM}$ under the assumptions of (c). The first assertions of (d) follows since $\boldsymbol{x} \to x_i x_j$ is supermodular, and the second assertion then follows by combining with (c). $\square$

Proposition 8.11 *Assume $d = 2$, that $\boldsymbol{X}$, $\boldsymbol{Y}$ have identical marginals F_1, F_2, and that any of $\boldsymbol{X} \le_{\mathrm{lo}} \boldsymbol{Y}$, $\boldsymbol{X} \le_{\mathrm{uo}} \boldsymbol{Y}$, $\boldsymbol{X} \le_{\mathrm{c}} \boldsymbol{Y}$ holds. Then $\boldsymbol{X} \le_{\mathrm{sm}} \boldsymbol{Y}$.*

Proof Let $f \in C^2$ be supermodular and define

$$\widetilde{f}(x_1, x_2) = f(x_1, x_2) - f(0, x_2) - f(x_1, 0) + f(0, 0).$$

Define $A_{++} = \{X_1 > 0, X_2 > 0\}$, $A_{-+} = \{X_1 < 0, X_2 > 0\}$, $\Omega_{+-} = \{x_1 > 0, x_2 > 0\}$, $\Omega_{--} = \{x_1 < 0, x_2 < 0\}$, etc. We shall first show that

$$\mathbb{E}\big[\widetilde{f}(X_1, X_2); A_{++}\big] = \int_{\Omega_{++}} \big(F^X(z_1, z_2) + B_{++}(z_1, z_2)\big) f_{12}(z_1, z_2)\, dz_1\, dz_2$$

(8.7)

and similarly for the other sign combinations, where

$$B_{++}(z_1, z_2) = 1 - F_1(z_1) - F_2(z_2), \quad B_{--}(z_1, z_2) = 0,$$

$$B_{+-}(z_1, z_2) = -F_2(z_2), \quad B_{-+}(z_1, z_2) = -F_1(z_1).$$

To this end, note that by elementary calculus,

$$\widetilde{f}(x_1, x_2) = \int_0^{x_2} \int_0^{x_1} f_{12}(z_1, z_2)\, dz_1\, dz_2$$

for all z_1, z_2, no matter the sign of x_1, x_2 in view of the usual convention $\int_a^b f = -\int_b^a f$. This gives

$$\mathbb{E}\big[\widetilde{f}(X_1, X_2); A_{++}\big] = \int_{\Omega_{++}} \mathbb{P}(X_1 > z_1, X_2 > z_2) f_{12}(z_1, z_2)\, dz_1\, dz_2.$$

The claim (8.7) then follows from

$$1 = \mathbb{P}(X_1 > z_1, X_2 > z_2) + F_1(z_1) + F_2(z_2) - F^X(z_1, z_2).$$

For, say, the $-+$ case, we get similarly that

$$\mathbb{E}\big[\widetilde{f}(X_1, X_2); A_{-+}\big] = -\int_{\Omega_{-+}} \mathbb{P}(X_1 \le z_1, X_2 > z_2) f_{12}(z_1, z_2)\, dz_1\, dz_2$$

$$= -\int_{\Omega_{-+}} \big(F_1(z_1) - F(z_1, z_2)\big) f_{12}(z_1, z_2)\, dz_1\, dz_2$$

and the asserted version of (8.7). The $+-$ and $--$ cases are similar.

To prove $\mathbb{E}f(X_1, X_2) \le \mathbb{E}f(Y_1, Y_2)$, now just note that

$$\mathbb{E}f(Y_1, Y_2) - \mathbb{E}f(X_1, X_2) = \mathbb{E}\widetilde{f}(Y_1, Y_2) - \widetilde{\mathbb{E}}f(X_1, X_2)$$

$$= \int_{\mathbb{R}^2} \left[F^Y(z_1, z_2) - F^X(z_1, z_2) \right) f_{12}(z_1, z_2) \, \mathrm{d}z_1 \, \mathrm{d}z_2 \ge 0,$$

where the two first step used the identical marginals, the second also (8.7), and the last $X \le_{\mathrm{lo}} Y$ together with $f_{12} \ge 0$.   □

Proof of Proposition 8.4, Completed It remains to show the ordering of Kendall's τ. We have for an independent copy (X_1^*, X_2^*) of (X_1, X_2) and continuous marginals that

$$\tau(X_1, X_2) = \mathbb{P}\big((X_1 - X_1^*)(X_2 - X_2^*) > 0\big) - \mathbb{P}\big((X_1 - X_1^*)(X_2 - X_2^*) < 0\big)$$

$$= 2\,\mathbb{P}\big((X_1 - X_1^*)(X_2 - X_2^*) > 0\big) - 1 = 4\mathbb{P}(X_1 \le X_1^*, X_2 \le X_2^*) - 1$$

$$= 4\,\mathbb{E}F_X(X_1, X_2) - 1 \le 4\,\mathbb{E}F_Y(X_1, X_2) - 1$$

$$\le 4\,\mathbb{E}F_Y(Y_1, Y_2) - 1 = \tau(Y_1, Y_2),$$

where the first inequality used $X \le_{\mathrm{lo}} Y$ and the second that $h(y_1, y_2) = (F_Y(y_1, y_2)$ is supermodular (if $g_Y(y_1, y_2)$ is the density of (Y_1, Y_2), then $h_{12} = g_Y \ge 0$).   □

Proposition 8.12 *If $X \le_{\mathrm{sm}} Y$, then*

$$X_1 + \cdots + X_d \preceq_{\mathrm{cx}} Y_1 + \cdots + Y_d.$$

Proof Just note that if $g \in C^2$ is convex and we let $f(x_1, \dots, x_d) = g(x_1 + \cdots + x_d)$, then $f_{ij}(x_1, \dots, x_d) = g''(x_1 + \cdots + x_d) \ge 0$ so that $f \in \mathscr{S}M$, and hence

$$\mathbb{E}g(X_1 + \cdots + X_d) \le \mathbb{E}g(Y_1 + \cdots + Y_d).$$ □

Exercises
8.2 Check (8.7) for the $+-$ and $--$ cases.

8.3 The Multivariate Normal Distribution

We consider a random vector $X = (X_1 \ \dots \ X_d)$ with an $N_d(\mathbf{0}, \mathbf{R})$ distribution, where $\mathbf{R} = (r_{ij})_{i,j=1,\dots,d}$.

The intuition is of course that increasing correlations increases dependence and thereby risk, as illustrated by the following simple calculations:

Example 8.13 Let $d = 2$ and X_1, X_2 be standard normal with covariance $\rho \ge 0$. Then $S = X_1 + X_2 \sim N(0, 2(1 + \rho))$ and accordingly, the tails are heavier the larger ρ is. Consider instead $M = \max(X_1, X_2)$. Using the representation

$$X_1 = \sqrt{1-\rho}\, U_0 + \sqrt{\rho}\, U_1, \quad X_2 = \sqrt{1-\rho}\, U_0 + \sqrt{\rho}\, U_2$$

with U_0, U_1, U_2 i.i.d. standard normal, we get

$$M = \sqrt{1-\rho}\, U_0 + \sqrt{\rho}\, M^* \qquad \text{where } M^* = \max(U_1, U_2).$$

Thus, as ρ increases, more and more weight is shifted from U_0 to the stochastically larger M^*. $\diamond$

These considerations motivate the following result:

Theorem 8.14 *Let* $X = (X_1 \ldots X_d) \sim \mathcal{N}_d(0, R^X)$, $Y = (Y_1 \ldots Y_d) \sim \mathcal{N}_d(0, R^Y)$, *where the diagonal elements of* R^X, R^Y *agree and* $R^X \leq R^Y$. *Then* $X \leq_{\mathrm{sm}} Y$ *and hence also* $X \leq_{\mathrm{lo}} Y$, $X \leq_{\mathrm{uo}} Y$, $X \leq_{\mathrm{c}} Y$.

Lemma 8.15 *Let* $h : \mathbb{R}^d \to \mathbb{R}$ *be* C^2 *with* $|h(x)| = \mathrm{o}(e^{a\|x\|^2})$ *for all* $a > 0$ *and similarly for the partial derivatives* h_i, h_{ij}. *Assume that* $h_{ij}(x) \geq 0$ *for some* $i \neq j$ *and all* x. *Then*

$$\frac{\partial}{\partial r_{ij}} \mathbb{E} h(X) \geq 0.$$

Proof With $\varphi(x)$ the density of X, we get

$$\frac{\partial}{\partial r_{ij}} \mathbb{E} h(X) = \frac{\partial}{\partial r_{ij}} \int_{\mathbb{R}^d} h(x)\varphi(x)\,\mathrm{d}x = \int_{\mathbb{R}^d} h(x)\frac{\partial}{\partial r_{ij}}\varphi(x)\,\mathrm{d}x$$

$$= \int_{\mathbb{R}^d} h(x)\frac{\partial^2}{\partial x_i \partial x_j}\varphi(x)\,\mathrm{d}x = -\int_{\mathbb{R}^d} h_i(x)\frac{\partial}{\partial x_j}\varphi(x)\,\mathrm{d}x$$

$$= \int_{\mathbb{R}^d} h_{ij}(x)\varphi(x)\,\mathrm{d}x \geq 0,$$

where the third step used the PDE for φ given in Lemma A.2 of the Appendix and the following two integrations by parts are justified by the boundary assumptions on the h_i, h_{ij}. $\square$

Proof of Theorem 8.14 We proceed in three steps A), B), C).

A) Assume $d = 2$, let h be the indicator function of the orthant $[z_1, \infty) \times [z_2, \infty)$ and for $i = 1, 2$, define $f_{i,n}(x) = 1$ for $x \geq z_i$ and $f_{i,n}(x) = e^{-n(x-x_i)^4}$ for $x \leq z_i$. Letting $h_n(x_1, x_2) = f_{1,n}(x_1) f_{2,n}(x_2)$, we have $h_n \downarrow h$. Also the boundary conditions in Lemma 8.15 are trivial to verify, and hence

$$\mathbb{P}(X_1 \geq z_1, X_2 \geq z_2) = \lim_{n \to \infty} \mathbb{E}h_n(X_1, X_2)$$

$$\leq \lim_{n \to \infty} \mathbb{E}h_n(Y_1, Y_2) = \mathbb{P}(Y_1 \geq z_1, Y_2 \geq z_2).$$

Thus $X \leq_c Y$ and hence (since $d = 2$) $X \leq_{sm} Y$.

B) We assume here that covariances only disagree for one pair $i \neq j$, say $i = 1$, $j = 2$, i.e., $r_{12}^X = r_{21}^X < r_{21}^Y = r_{12}^Y$ while all other r_{ij}^X and r_{ij}^Y coincide. In particular, the marginal distributions of components $3, \ldots, d$ coincide for X and Y, whereas the formulas in Sect. A.7.1 of the Appendix show that the conditional distributions of components 1,2 given components $3, \ldots, d$ have the same mean but covariances of the form

$$\begin{pmatrix} r_{11} & r_{12}^X \\ r_{21}^X & r_{22} \end{pmatrix} - S \quad \text{resp.} \quad \begin{pmatrix} r_{11} & r_{12}^Y \\ r_{21}^Y & r_{22} \end{pmatrix} - S$$

with the same S. Step A) therefore shows that these conditional distributions are supermodularly ordered. Using the plausible (and easily verified) fact that $\leq_{sm}$ is closed under mixtures then gives the assertion.

C) Let $f \in C^2$ be supermodular. The result of B) shows that

$$\frac{\partial}{\partial r_{ij}} \mathbb{E}f(X) \Big|_{R^X = R^*} \geq 0 \tag{8.8}$$

for any positive definite covariance matrix R^* and all $i \neq j$ [note that this argument applies irrespective of whether f satisfies the boundary conditions in Lemma 8.15 or not!]. With $Z(t) \sim \mathcal{N}_d(0, R(t))$, where $R(t) = t R^Y + (1-t) R^X$ is obviously positive definite when R^X and R^Y are so, we get by the chain rule that

$$\mathbb{E}f(Y) - \mathbb{E}f(X) = \int_0^1 \frac{d}{dt} \mathbb{E}f(Z(t)) \, dt$$

$$= \int_0^1 \sum_{i,j=1}^d \frac{\partial}{\partial r_{ij}} \mathbb{E}f(Z(t)) \cdot \frac{d}{dt} \left(t r_{ij}^Y + (1-t) r_{ij}^X \right) dt$$

$$= \int_0^1 \sum_{i,j=1}^d \frac{\partial}{\partial r_{ij}} \mathbb{E}f(Z(t)) \cdot \left(r_{ij}^Y - r_{ij}^X \right) dt,$$

which is non-negative by (8.8) with $R^* = R(t)$ and the assumption $R^X \leq R^Y$. $\qquad\square$

Exercises

8.3 Verify the assertions of Example 8.8.

Notes

Good references for dependence orderings are Chapter 3 of Müller and Stoyan [126] and Section 6.1 of Rüschendorf [154]. A further main result is that any random vector X is supermodularly dominated by its comonotonic version $\left(F_1^{\leftarrow}(U), \ldots, F_d^{\leftarrow}(U)\right)$. Much attention has also been given to an increasing supermodular order $X \prec_{\text{ism}} Y$, defined by $\mathbb{E}f(X) \leq \mathbb{E}f(Y)$ for non-decreasing supermodular f. It holds that $X \leq_{\text{sm}} Y$ if and only if $X \prec_{\text{ism}} Y$ and X, Y have the same marginals or even just $\mathbb{E}X_i = \mathbb{E}Y_i$ for all i.

Theorem 8.14 on the multivariate normal distribution is close to Slepian's inequality, to which we return in Sect. XIII.5.

Chapter XI: Stochastic Control in Non-Life Insurance

1 Introduction

In most of the insurance models considered so far, strategies have been static: premiums, arrangements for premiums, reinsurance, dividends or investment, etc. have been fixed once and for all at $t = 0$. The underlying philosophy is not necessarily to follow that same strategy forever, but to get a feeling of its effects in the long run. Stochastic control deals with the more realistic situation of dynamic strategies that are updated along the way for some optimality purpose. The updating may depend on how relevant quantities (for example reserves or the economic environment) evolve in time.

The more precise formulation requires setting up a model and defining an optimality criterion. Consider first the modelling aspect. Here by far the most popular model for the uncontrolled process X is a diffusion

$$\mathrm{d}X(t) = \mu\big(X(t)\big)\,\mathrm{d}t + \sigma\big(X(t)\big)\,\mathrm{d}W(t), \tag{1.1}$$

or a more general solution of a SDE of the form

$$\mathrm{d}X(t) = \mu\big(t, X(t)\big)\,\mathrm{d}t + \sigma\big(t, X(t)\big)\,\mathrm{d}W(t). \tag{1.2}$$

In finance, X could be geometric Brownian motion

$$\mathrm{d}X(t) = aX(t)\,\mathrm{d}t + bX(t)\,\mathrm{d}W(t).$$

In non-life insurance, a natural choice would be the diffusion approximation

$$\mathrm{d}X(t) = (c - \beta\mu_B)\,\mathrm{d}t + \sqrt{\beta\mu_B^{(2)}}\,\mathrm{d}W(t)$$

© Springer Nature Switzerland AG 2020
S. Asmussen, M. Steffensen, *Risk and Insurance*, Probability Theory
and Stochastic Modelling 96, https://doi.org/10.1007/978-3-030-35176-2_11

of a Cramér–Lundberg model with premium rate c, Poisson rate β and claim size distribution B with first and second moment μ_B, $\mu_B^{(2)}$, cf. Sect. IV.8; in this setting, one usually assumes 0 to be an absorbing barrier for X. Extensions could involve an additional $-L\,\mathrm{d}t$ term with L a fixed rate of paying dividends or liabilities (administration costs, rent, etc), and an additional $rX(t)\,\mathrm{d}t$ term if the reserve is invested at risk-free interest rate r (further variants will show up in the following).

In (1.1), (1.2) the drift and the variance may depend on a parameter u such that $\mu(t, x) = \mu(t, x, u)$, $\sigma^2(t, x) = \sigma^2(t, x, u)$, and u is fixed (static) and may therefore be suppressed in the notation. In the controlled setting, the decision-maker may be able to adapt u dynamically to strategies depending on t, $X(t)$ or other relevant information. We shall only consider *feedback* or *Markovian* controls, meaning that the u used at time t is a function of t and the state x of the controlled process at that time. In the problems we deal with, we will only encounter the diffusion form (1.1) and not (1.2). Further, the form of the optimization problems we consider (see below) motivate that an optimal u cannot depend on t, so we take the $u = u(x)$ independent of t. Formally, we write $U = (u(x))_{x\in I}$, where I is the relevant interval in which X lives, and the controlled process X^U then evolves as

$$\mathrm{d}X^U(t) = \mu\big(X^U(t), u(X^U(t))\big)\,\mathrm{d}t + \sigma\big(X^U(t), u(X^U(t))\big)\,\mathrm{d}W(t)\,. \qquad (1.3)$$

The strategy U is restricted to be chosen in a set $\mathscr{U}$ of admissible strategies. One condition on $\mathscr{U}$ that we always tacitly assume is that (1.3) admits a strong solution. Further restrictions imposed on $u(x)$ for a fixed x are denoted $u(x) \in \mathscr{U}(x)$.

Example 1.1 Consider proportional reinsurance with retention u and assume that both the cedent and the reinsurer use the expected value principle with safety loadings η_s, resp. η_r. This means that the cedent will receive premiums at rate $\beta\mu_B(1+\eta_s)$, whereas his payment rates are $u\beta\mu_B$ for claims and $(1-u)\beta\mu_B(1+\eta_r)$ for premiums to the reinsurer. The variance constant for his claim payments is $u^2\beta\mu_B^{(2)}$. Thus the uncontrolled process is

$$\begin{aligned}
\mathrm{d}X(t) &= \beta\mu_B\big(1 + \eta_s - u - (1-u)(1+\eta_r)\big)\,\mathrm{d}t + u\sqrt{\beta\mu_B^{(2)}}\,\mathrm{d}W(t) \\
&= \beta\mu_B\big(\eta_s - (1-u)\eta_r\big)\,\mathrm{d}t + u\sqrt{\beta\mu_B^{(2)}}\,\mathrm{d}W(t)\,.
\end{aligned}$$

In the controlled setting, $u = u(x)$ is steered by the decision-maker to depend on $x = X(t)$ in a suitable way. The set $\mathscr{U}$ of admissible strategies is defined by the requirement $0 \le u(x) \le 1$. $\Diamond$

Example 1.2 In the case of dividend payments at a continuous rate, the control can be defined as the function giving the rate $u(x)$ at level x of the reserve, so that

$$\mathrm{d}X^U(t) = \big(c - \beta\mu_B - u(X^U(t))\big)\,\mathrm{d}t + \sqrt{\beta\mu_B^{(2)}}\,\mathrm{d}W(t)\,.$$

Example 1.3 A classical example of control problems in finance is the consumption-investment model where an individual invests his wealth at a financial market in order to finance consumption during his course of life. The individual has access to the Black–Scholes market with price dynamics given by

$$dB(t) = rB(t), \qquad dS(t) = \alpha S(t)\,dt + \sigma S(t)\,dW(t),$$

where $S(0) = s_0$ and B is the bank account with $B(0) = 1$. In our formulation, where we do not allow dependence of u on t, let the individual invest the proportion $u = u(x)$ of his wealth x in stocks; we can also say that he holds the amount ux in stocks, he owns the number ux/S of stocks, and he gains $ux/S \cdot dS$ from his stock investments. Further, he invests the proportion $1 - u$ of x in bonds, i.e. he holds the amount $(1 - u)x$ in bonds, he owns the number $(1 - u)x/B$ of bonds, and he gains $(1 - u)x/B \cdot dB$ from his bond investment. All in all, the dynamics of his wealth follows the SDE

$$
\begin{aligned}
dX^U(t) &= u(X^U(t))X^U(t)\frac{dS(t)}{S(t)} + \big(1 - u(X^U(t))X^U(t)\big)\frac{dB(t)}{B(t)} \\
&\quad - c(X^U(t))\,dt \\
&= \big(r + u(X^U(t))(\alpha - r)\big)X^U(t)\,dt + u(X^U(t))\sigma X^U(t)\,dW(t) \qquad (1.4) \\
&\quad - c(X^U(t))\,dt, \qquad (1.5)
\end{aligned}
$$

where $c(\cdot)$ is the consumption rate. This SDE decomposes the wealth dynamics into one part (1.4) containing all capital gains, and another part (1.5) containing the cash flow in and out of X.   ◇

The optimality criterion may have several forms depending on the context. Some of the most popular ones are to minimize the ruin probability

$$\psi^U(x) = \mathbb{P}\big(\tau^U(x) < \infty\big) \quad \text{where } \tau^U(x) = \inf\{t > 0 : X^U(t) < 0\} \qquad (1.6)$$

or, in the context of dividend payments as in Example 1.2, to maximize the expected total amount of discounted dividends

$$\mathbb{E}_x \int_0^\infty e^{-rt}\, u(X^U(t))\,dt = \mathbb{E}_x \int_0^{\tau^U(x)} e^{-rt}\, u(X^U(t))\,dt,$$

or, in the consumption-investment problem in Example 1.3 to maximize

$$\int_0^\infty e^{-dt}\, v\big(c(X^U(t))\big)\,dt$$

for some utility function v and some subjective discount rate d.

More formally, one considers a payout $\mathcal{J}^U$ of strategy U and defines the *value function* V^U and the *optimal value function* $V(x)$ by

$$V^U(x) = \mathbb{E}_x \mathcal{J}^U, \qquad V(x) = \sup_{U \in \mathcal{U}} V^U(x).$$

The main goal of stochastic control is to find an optimal strategy $U^* = (u^*(x))_{x \in I}$, i.e. a U^* satisfying $V^{U^*}(x) = V(x)$. Also the optimal value function $V(x)$ is of some interest, though often to a lesser extent.

Remark 1.4 Our aim here is to treat a number of basic examples as smoothly as possible and our formulation is therefore not quite the most general one occurring in the literature. In particular, we have imposed some restrictions on the form of the strategy U. One is the feedback assumption. It often seems intuitively obvious that U must be feedback, but there are subtle mathematical difficulties inherent in this question. There are also many problems where it would be reasonable to let u depend on t in addition to x; one example is the consumption-investment problem with a finite time horizon T where one expects an optimal strategy to increase consumption as t becomes closer to T, or where at time t the individual earns income at rate $a(t)$, giving an added term $a(t) \, dt$ in the SDE that thereby takes the more general form (1.2).

Another restriction is the form (1.3), excluding controls that are not absolutely continuous. One example of such is lump sum payments, another is a control that pays out at a boundary. For example, for dividend payments a general formulation of this is to let $U = (U(t))_{t \geq 0}$, where $U(t)$ is the total amount of dividend payments before t. Then instead

$$dX^U(t) = (c - \beta \mu_B) \, dt - dU(t) + \sqrt{\beta \mu_B^{(2)}} \, dW(t)$$

and the problem is to choose U optimally in the set of adapted non-decreasing D-functions. Obvious requirements on $\mathcal{U}$ are that $\Delta U(t) \leq X(t)$ and that $dU(t) = 0$ for $t > \tau^U(x)$ (the ruin time, cf. (1.6)).   ◊

Notes
Some standard textbooks on stochastic control are Davis [53], Fleming and Rishel [74], Fleming and Soner [75] and Øksendal and Sulem [136].

Schmidli [155] is the basic textbook reference for stochastic control in insurance up to 2008; updates are in Asmussen and Albrecher [8] from 2010 and Azcue and Muler [17] from 2014. Specific results given in this chapter without citations are in these books and references there. The research literature is booming and we will not attempt to survey it here. However, quite a few papers deal with variants/combinations of topics treated here.

In more general formulations of stochastic control, the assumption of U having feedback form is often relaxed to the control used at time t having the form $u(t)$ for some adapted process $\{u(t)\}_{t \geq 0}$ rather than just $u(X^U(t))$. It is often counterintuitive that an optimal control would not be feedback and certainly, we do not know of realistic examples of this. However, some care is needed in the rigorous mathematical verification of this, cf. [8] p. 447.

2   Minimizing the Ruin Probability

We now consider the optimal choice of strategy for minimizing $\psi^U(x)$ as given by (1.6), or equivalently to maximize the survival probability $\phi(u) = 1 - \psi(u)$. This has the appealing simplifying feature that the quantity to be optimized can be computed explicitly in the diffusion setting and that the problem of finding a whole strategy that is optimal becomes local, that is, reduces to the standard problem of maximizing a function of one variable, namely the ratio between the drift and variance.

2.1   General Theory

Consider a diffusion (1.1) on $[0, \infty)$ with $\mu(t, x) \equiv \mu(x)$, $\sigma^2(t, x) \equiv \sigma^2(x)$ independent of t, and assume that a strong solution exists for any initial value $x_0 > 0$. Define $\kappa(x) = \mu(x)/\sigma^2(x)$, $\tau = \inf\{t > 0 : X(t) = 0\}$, $\psi(x) = \mathbb{P}_x(\tau < \infty)$. Define further the scale function S by

$$S(x) = \int_0^x s(y)\,dy, \quad S(\infty) = \int_0^\infty s(y)\,dy, \tag{2.1}$$

where $s(y) = \exp\left\{\int_0^y -2\kappa(z)\,dz\right\}$.

Proposition 2.1 *Assume that the drift $\mu(x)$ and the variance $\sigma^2(x)$ of X are measurable and satisfy*

$$\sup_{0 \le x \le A} \mu(x) < \infty, \quad \sup_{0 \le x \le A} \sigma^2(x) < \infty, \quad \inf_{0 \le x \le A} \sigma^2(x) > 0 \tag{2.2}$$

for all $A < \infty$. If

$$S(\infty) < \infty, \tag{2.3}$$

then $0 < \psi(x) < 1$ for all $x > 0$ and

$$\psi(x) = 1 - \frac{S(x)}{S(\infty)}. \tag{2.4}$$

Conversely, if (2.3) fails, then $\psi(u) = 1$ for all $u > 0$.

Proof Define $\tau(\ell) = \inf\{t > 0 : X(t) = 0 \text{ or } X(t) = \ell\}$. Then for $0 < x < \ell$, we have by Lemma A.1 of the Appendix that

$$\mathbb{P}_x\big(X(\tau(\ell)) = \ell\big) = \frac{S(x)}{S(\ell)}.$$

Now just use that the events $\{X(\tau(\ell)) = \ell\}$ decrease to $\{\tau = \infty\}$ as $\ell \uparrow \infty$, so by monotone convergence

$$1 - \psi(x) = \lim_{\ell \uparrow \infty} \frac{S(x)}{S(\ell)} = \frac{S(x)}{S(\infty)},$$

from which the result immediately follows in both cases.   □

The assumption of $\sigma^2(x)$ being bounded below on bounded intervals is crucial. A particularly unpleasant case from the point of view of ruin theory is $\sigma^2(x) \to 0$ as $x \to 0$. Then it may happen, for example, that $\psi(x)$, as defined above as the probability of actually hitting 0, is zero for all $x > 0$ but that nevertheless $X(t) \overset{\text{a.s.}}{\to} 0$ (the problem leads into the complicated area of boundary classification of diffusions, see e.g. Breiman [33] or Karlin and Taylor [99, p. 226]). In control problems, $\sigma^2(x) \to 0$ can be interpreted as freezing the business more and more as the reserve gets small. This is appealing from the point of view of avoiding ruin, but $X(t) \overset{\text{a.s.}}{\to} 0$ may be less so from different angles.

Proposition 2.2 *For a given function κ on $[\ell_d, \ell_u]$ and a given $\delta \in [\ell_d, \ell_u]$, let*

$$\varphi(\kappa) = \frac{\displaystyle\int_{\ell_d}^{\delta} \exp\left\{-2\int_{\ell_d}^{y} \kappa(z)\,dz\right\} dy}{\displaystyle\int_{\ell_d}^{\ell_u} \exp\left\{-2\int_{\ell_d}^{y} \kappa(z)\,dz\right\} dy}.$$

Then $\kappa_0 \le \kappa_1$ implies $\varphi(\kappa_0) \le \varphi(\kappa_1)$.

Proof Define

$$\xi = \kappa_1 - \kappa_0, \quad \kappa_t = \kappa_0 + t\xi = (1 - t)\kappa_0 + t\kappa_1, \quad g(t) = \varphi(\kappa_t).$$

Then

$$g'(t) = \lim_{h \downarrow 0} \frac{\varphi(\kappa_{t+h}) - \varphi(\kappa_t)}{h} = \lim_{h \downarrow 0} \frac{\varphi(\kappa_t + h\xi) - \varphi(\kappa_t)}{h} = \varphi_\xi(\kappa_t),$$

$$\varphi(\kappa_1) - \varphi(\kappa_0) = g(1) - g(0) = \int_0^1 g'(t)\,dt = \int_0^1 \varphi_\xi(\kappa_t)\,dt,$$

where φ_ξ denotes the directional derivative. Thus the result will follow if we can show $\varphi_\xi(\kappa_t) \geq 0$ for all t. Define for a fixed t

$$H(x) = \exp\left\{-2\int_{\ell_d}^x \kappa_t(y)\,dy\right\}, \quad K(x) = \int_{\ell_d}^x \xi(y)\,dy,$$

$$A(\delta) = \int_{\ell_d}^\delta H(x)\,dx, \quad B(\delta) = \int_{\ell_d}^\delta H(x)K(x)\,dx.$$

Then $g(t) = \varphi(\kappa_t) = A(\delta)/A(\ell_u)$ and

$$
\begin{aligned}
\varphi(\kappa_t + h\xi) &= \frac{\int_{\ell_d}^\delta \exp\left\{-2\int_{\ell_d}^x \kappa_t(y)\,dy - 2h\int_{\ell_d}^x \xi(y)\,dy\right\}dx}{\int_{\ell_d}^{\ell_u} \exp\left\{-2\int_{\ell_d}^x \kappa_t(y)\,dy - 2h\int_{\ell_d}^x \xi(y)\,dy\right\}dx} \\
&\approx \frac{\int_{\ell_d}^\delta H(x)\big(1 - 2hK(x)\big)\,dx}{\int_{\ell_d}^{\ell_u} H(x)\big(1 - 2hK(x)\big)\,dx} = \frac{A(\delta) - 2hB(\delta)}{A(\ell_u) - 2hB(\ell_u)} \\
&= \varphi(\kappa_t)\frac{1 - 2hB(\delta)/A(\delta)}{1 - 2hB(\ell_u)/A(\ell_u)} \\
&\approx \varphi(\kappa_t)\big(1 - 2hB(\delta)/A(\delta)\big)\big(1 + 2hB(\ell_u)/A(\ell_u)\big),
\end{aligned}
$$

which gives $\varphi_\xi(\kappa_t) = 2\varphi(\kappa_t)\big[C(\ell_u) - C(\delta)\big]$ where $C(\delta) = B(\delta)/A(\delta)$. Thus it suffices for $\varphi_\xi(\kappa_t) \geq 0$ that C is non-decreasing, which follows since $H \geq 0$ implies

$$C'(\delta) = \frac{A(\delta)B'(\delta) - A'(\delta)B(\delta)}{A(\delta)^2} = \frac{H(\delta)\big[A(\delta)K(\delta) - B(\delta)\big]}{A(\delta)^2} \geq 0$$

where we used that $\xi \geq 0$ and $\delta \geq x$ entails

$$K(\delta) \geq K(x), \quad A(\delta)K(\delta) = \int_{\ell_d}^\delta H(x)K(\delta)\,dx \geq \int_{\ell_d}^\delta H(x)K(x)\,dx = B(\delta).$$

$\square$

Corollary 2.3 *Consider two diffusions X_0, X_1 with κ-functions satisfying $\kappa_0 \leq \kappa_1$. Then it holds for the corresponding ruin probabilities that $\psi_1(x) \leq \psi_0(x)$ for all x.*

Proof By the same monotone convergence argument as in the proof of Proposition 2.1.

$\square$

Notes
Proposition 2.2, as well as the validity of the underlying Lemma A.1 of the Appendix in the generality stated there, originates from Pestien and Sudderth [138]. The present proof of Proposition 2.2, coming from [11], is different. That minimizing the ruin probability means maximizing κ is traditionally proved in the insurance risk literature via the HJB equation (see Sect. 3), but this approach seems less direct to us and leads to slightly stronger conditions. Reference [138] was exploited by Bäuerle and Bayraktar [41] in a rather similar vein as in this section.

2.2   Reinsurance Control

Consider a general reinsurance situation with generic claim Y having distribution B, the cedent paying $s(Y)$ of Y and the reinsurer $r(Y)$ where $s(Y) + r(Y) = Y$, cf. Sect. I.4. Assuming that both the cedent and the reinsurer calculate premiums by the expected value principle with loadings η_s, resp. η_r, the cedent then receives premiums at rate $(1 + \eta_s)\beta\mu_B$, pays claims at rate $\beta\mathbb{E}s(Y)$ and pays reinsurance premium at rate $(1 + \eta_r)\beta\mathbb{E}r(V)$. Thus his governing SDE becomes

$$
\mathrm{d}X(t) = \big((1 + \eta_s)\beta\mu_B - \beta\mathbb{E}s(Y) - (1 + \eta_r)\beta\mathbb{E}r(Y)\big)\,\mathrm{d}t + \sqrt{\beta\mathbb{E}s(Y)^2}\,\mathrm{d}W(t)
$$

$$
= \big(\eta_s\beta\mu_B - \eta_r\beta\mathbb{E}r(Y)\big)\,\mathrm{d}t + \sqrt{\beta\mathbb{E}s(Y)^2}\,\mathrm{d}W(t)\,.
$$

In a control situation where $s = s^u$, $r = r^u$ depend on a parameter u, the optimal choice is therefore to let u be independent of the level $X^U(t)$ and t, and to be chosen as the maximizer of

$$
\frac{\mu_B - q\mu_r(u)}{\mu_{s,2}(u)} \tag{2.5}
$$

(the β cancels in μ/σ^2) where $q = \eta_r/\eta_s$, $\mu_r(u) = \mathbb{E}r^u(Y)$, $\mu_{s,2}(u) = \mathbb{E}s^u(Y)^2$. We impose the natural condition $q > 1$ of non-cheap reinsurance. Maximizing (2.5) amounts to analyzing the sign variation of the derivative, or equivalently of

$$
h(u) = -q\mu'_r(u)\mu_{s,2}(u) - \big(\mu_B - q\mu_r(u)\big)\mu'_{s,2}(u)\,. \tag{2.6}
$$

Consider first proportional reinsurance where $s^u(Y) = uY$, $r^u(Y) = (1 - u)Y$, $u \in [0, 1]$.

Proposition 2.4 *In proportional reinsurance with $q = \eta_r/\eta_s \geq 2$, the ruin probability is minimized by taking no reinsurance, i.e. $u = 1$. If $q \leq 2$, there is a unique minimum at $u = 2(1 - 1/q)$.*

Proof Since $\mu_r(u) = (1 - u)\mu_B$, $\mu_{s,2}(u) = u^2\mu_B^{(2)}$, we have

$$
h(u) = \mu_B\mu_B^{(2)}\big[qu^2 - 2\big(1 - q(1 - u)\big)u\big] = \mu_B\mu_B^{(2)}u\big[-qu + 2(q - 1)\big]
$$

$$
= -\mu_B\mu_B^{(2)}qu[u - \breve{u}]\,,
$$

where $\breve{u} = 2(1 - 1/q)$. This shows that κ as function of $u \in [0, \infty)$ attains a unique maximum at $u = \breve{u}$ and is increasing on the interval $[0, \breve{u})$. But $\breve{u} \in (0, 1]$ if $q \geq 2$ whereas $\breve{u} \geq 1$ if $q \leq 2$. From these observations the result follows.   □

Consider next excess-of-loss reinsurance where the control π is the retention level and $s^u(Y) = Y \wedge u$, $r^u(Y) = (Y - u)^+$. The following result shows that the control problem depends on the full distribution B of the claim Y and not just the two first moments as for proportional reinsurance:

Proposition 2.5 *In excess-of-loss reinsurance, the optimal retention level u^* is in $(0, \infty)$ and solves*

$$q \int_0^u x\overline{B}(x)\, dx \;=\; \left(\mu_B - q \int_u^\infty \overline{B}(x)\, dx \right) u. \tag{2.7}$$

Proof We have

$$\mu_r(u) \;=\; \int_u^\infty \overline{B}(x)\, dx, \qquad \mu_{s,2}(u) \;=\; 2 \int_0^u x\overline{B}(x)\, dx$$

which, leaving out an unimportant factor of 2, gives that

$$h(u) \;=\; q\overline{B}(u) \int_0^u x\overline{B}(x)\, dx - \left(\mu_B - q \int_u^\infty \overline{B}(x)\, dx \right) u\overline{B}(u) \;=\; \overline{B}(u)h_0(u),$$

where

$$h_0(u) \;=\; q \int_0^u x\overline{B}(x)\, dx - \left(\mu_B - q \int_u^\infty \overline{B}(x)\, dx \right) u.$$

Condition (2.7) then follows from $\overline{B}(u) \neq 0$.

It remains to show that (2.7) necessarily has at least one solution in $(0, \infty)$. This follows since (1) both sides are continuous and 0 at $u = 0$, (2) the l.h.s. has derivative 0 at $u = 0$ and the r.h.s. $\mu_B(1 - q) < 0$ so l.h.s. > r.h.s. for small u, (3) l.h.s. < r.h.s. for large u since the l.h.s. has the finite limit $q\mu_B^{(2)}/2$ as $u \to \infty$ whereas the r.h.s. goes to ∞ at rate $u\mu_B$.   □

Example 2.6 Assume B is exponential, for ease of notation with unit rate. Then (2.7) means that

$$q(1 - e^{-u} - ue^{-u}) \;=\; (1 - qe^{-u})u,$$

i.e. that u^* solves the fixed-point problem $u^* = q(1 - e^{-u^*})$. Using $q > 1$ then easily gives a direct proof of uniqueness of the solution. In fact, one has an explicit expression for u^* in terms of the Lambert W function $\mathcal{W}$ defined as the inverse of the function te^t, i.e. by $\mathcal{W}(x)e^{\mathcal{W}(x)} = x$. Namely, rewriting $u^* = q(1 - e^{-u^*})$ as

$$u^* = q(1 - e^{-q}e^{q-u^*}), \qquad q - u^* = qe^{-q}e^{q-u^*}$$

gives $u^* = q + \mathcal{W}(-qe^{-q})$.   ◇

Example 2.7 Assume B is Pareto on $(1, \infty)$ with finite second moment, i.e. $\overline{B}(x) = x^{-\alpha}$ with $\alpha > 2$. Then for $u > 1$

$$\int_0^u x\overline{B}(x)\,\mathrm{d}x = \int_0^1 x\,\mathrm{d}x + \int_1^u \frac{1}{x^{\alpha-1}}\,\mathrm{d}x = \frac{1}{2} + \frac{1}{\alpha-2}\left(1 - \frac{1}{u^{\alpha-2}}\right)$$

$$= \frac{\alpha}{2(\alpha-2)} - \frac{u^{2-\alpha}}{\alpha-2}.$$

Similarly but easier,

$$\mu_B = \int_0^\infty \overline{B}(x)\,\mathrm{d}x = \frac{\alpha}{\alpha-1} \quad \text{and for } u > 1, \int_u^\infty \overline{B}(x)\,\mathrm{d}x = \frac{u^{1-\alpha}}{\alpha-1}.$$

See further Exercises 2.1, 2.2.   ◊

2.3   Investment Control

The starting point for the model is two stochastic differential equations

$$\mathrm{d}X(t) = a_1\,\mathrm{d}t + a_2\,\mathrm{d}W^0(t), \quad \mathrm{d}S(t) = b_1 S(t)\,\mathrm{d}t + b_2 S(t)\,\mathrm{d}W^1(t),$$

where W^0, W^1 are independent standard Brownian motions and $X^0(0) = x$. Here X describes the evolution of the reserve (surplus) of the company without investment and S the price process of a risky asset. Thus, X is Brownian motion with drift and S is geometric Brownian motion. It is now assumed that the company is free to invest an amount $u(x)$ in the risky asset at reserve level x; here $u(x)$ can exceed x, i.e. it is possible to use additional sources (or borrow money) for the investment. Thus in the presence of investment, the reserve evolves according to

$$\mathrm{d}X^U(t) = a_1\,\mathrm{d}t + a_2\,\mathrm{d}W^0(t) + b_1 u(X^U(t))\,\mathrm{d}t + b_2 u(X^U(t))\,\mathrm{d}W^1(t)$$

$$= (a_1 + b_1 u(X^U(t)))\,\mathrm{d}t + \sqrt{a_2^2 + b_2^2 u(X^U(t))^2}\,\mathrm{d}W(t),$$

where

$$W = \frac{a_2}{\sqrt{a_2^2 + b_2^2 u(X^U(t))^2}}\,W^0 + \frac{b_2 u(X^U(t))}{\sqrt{a_2^2 + b_2^2 u(X^U(t))^2}}\,W^1.$$

Here indeed W is Brownian motion, as follows e.g. from adaptedness of $u(X^U(t))$ and a theorem of Lévy stating that it suffices that the quadratic variation be t.

For the optimization, we impose the conditions $b_1 > 0$, $u(x) \geq 0$ (if say $b_1 < 0$, $u(x) \geq 0$, then the investment decreases the gain in terms of the drift but increases the risk in terms of the variance). We have $\kappa = (a_1 + b_1 u)/(a_2^2 + b_2^2 u^2)$, and maximizing κ amounts to analyzing the sign variation of the derivative, i.e. of

$$(a_2^2 + b_2^2 u^2)b_1 - 2(a_1 + b_1 u)b_2^2 u = -b_1 b_2^2 u^2 - 2a_1 b_2^2 u + a_2^2 b_1 . \tag{2.8}$$

The optimal choice is therefore to let $u^*(x) \equiv u^*$ be independent of the level x and chosen as the positive root

$$u^* = \frac{1}{b_1 b_2}\sqrt{a_1^2 b_2^2 + a_2^2 b_1^2} - \frac{a_1}{b_1}$$

of the quadratic (2.8). Thus (somewhat surprisingly) no matter how large the current capital x, it is optimal to always invest the constant amount u^* of money into the risky asset for minimizing the probability of ruin.

2.4 Premium Control

Consider a Bayesian situation where the N potential customers have rates $A_1, \ldots, A_N$ of generating claims, assumed i.i.d. with common distribution G, and where customer i will insure at the company if $A_i > bu$ where u is the offered premium and b is a suitable constant. In the plain Cramér–Lundberg model, the ruin probability will then be minimized by taking $u = \infty$, since then no customers insure and the reserve just stays constant so that the ruin probability is zero. A more interesting situation arises if the company has to pay liabilities at rate $L > 0$. If N is large, claim arrivals will approximately be governed by a Poisson process with rate $n(u)\alpha(u)$, where

$$n(u) = N\overline{G}(bu), \qquad \alpha(u) = \mathbb{E}[A \mid A > bu], \tag{2.9}$$

and we get in the diffusion approximation that

$$dX^U(t) = n(u)(u - \alpha(u)\mu_B)\,dt - L\,dt + \sqrt{n(u)\alpha(u)\mu_B^{(2)}}\,dW(t) .$$

The functional form of $\alpha(u)$ reflects adverse selection: since $\alpha(u)$ grows at rate at least bu, one has typically that the higher the premium one charges, the less desirable (to the company) is the average customer. In the traditional empirical Bayes model without selection, the average arrival rate would instead of $\alpha(u)$ simply be $\mathbb{E}A = \alpha(0)$, i.e., the adverse selection correction depends on the premium policy. We get

$$\kappa = \frac{n(u)(u - \alpha(u)\mu_B) - L}{n(u)\alpha(u)\mu_B^{(2)}} = \frac{u}{\alpha(u)\mu_B^{(2)}} - \frac{\mu_B}{\mu_B^{(2)}} - \frac{L}{n(u)\alpha(u)\mu_B^{(2)}} .$$

In view of (2.9), the maximization problem therefore in an essential way depends on the whole distribution G of the A_i.

Motivated by Example II.1.2, we take G to be exponential, say with rate δ. Recall that the Lambert W function $\mathcal{W}$ is defined by $\mathcal{W}(x)e^{\mathcal{W}(x)} = x$.

Proposition 2.8 *For the premium optimization problem with $\mathbb{P}(A > x) = e^{-\delta x}$, it holds that:*

(i) κ *attains a unique maximum at $u^* = \mathcal{W}(N/L\delta b)/\delta b$;*
(ii) *If $\kappa^{u^*} \le 0$ then $\psi^*(x) = 1$ for all x. Otherwise $\psi^*(x) < 1$ for all $x > 0$.*

Proof We have $n(u) = Ne^{-\delta cu}$, $\alpha(u) = 1/\delta + bu$,

$$\kappa = \frac{u}{(1/\delta + bu)2/\delta^2} - \frac{\delta}{2} - \frac{Le^{\delta bu}}{N(1/\delta + bu)2/\delta^2},$$

$$\frac{2N}{\delta^2}\kappa = N\frac{\delta u}{1 + \delta bu} - \frac{Le^{\delta bu}}{1 + \delta bu} - \frac{N}{\delta} = \frac{N}{b}\left(1 - \frac{1}{1 + \delta bu}\right) - \frac{Le^{\delta cu}}{1 + \delta bu} - \frac{N}{\delta}$$

$$= -h(q) + \frac{N}{b} - \frac{N}{\delta},$$

where $q = 1 + \delta bu$, $h(q) = (N/b + L\delta e^{q-1})/q$. Thus maximization of κ w.r.t. π is equivalent to the minimization of h w.r.t. q, and we get

$$h'(q) = \frac{qL\delta e^{q-1} - N/b - L\delta e^{q-1}}{q^2}.$$

Thus $h'(q) = 0$ means $(q - 1)L\delta e^{q-1} = N/b$, i.e. $q - 1 = \mathcal{W}(N/L\delta b)$ corresponding to $u = u^*$. Since $q > 1$, we also see that $h'(q)$ and hence κ' is < 0 for $u < u^*$ and > 0 for $u > u^*$ so that indeed the sign variation is the correct for a maximum. From this part (i) follows. Part (ii) is obvious.   □

Exercises
2.1 Find u^* when $\alpha = 3$ in Example 2.7.

2.2 Explain that the assumption $\alpha > 2$ is not minimal in Example 2.7, and set up the equation determining u^* when $\alpha = 2$. What about $\alpha = 1$?

Notes
Section 2.4 on premium as the control is a simple case of Asmussen, Christensen and Taksar [10].

3   The Hamilton–Jacobi–Bellman Equation

The analysis in the preceding section on how to minimize the ruin probability had the simplifying feature that the value function was explicitly available. This is, however, not the typical case in stochastic control problems where one instead proceeds via the so-called *Hamilton–Jacobi–Bellman* (HJB) *equation* given as (3.4) below. It is an ODE that determines the value function but, more importantly, also allows us to identify the optimal control. The rigorous treatment of the topic is, however, highly technical and we will here give an introduction which is informal but still allows us to derive the solutions in some key examples in non-life insurance. This is then followed up in Chap. XII with life insurance examples, worked out with some of the details omitted in this chapter.

The setting is as in Sect. 1 a controlled diffusion

$$dX^U(t) = \mu\big(X^U(t), u(X^U(t))\big)\, dt + \sigma\big(X^U(t), u(X^U(t))\big)\, dW(t) \tag{3.1}$$

with payout J^U of strategy U. The value function $V^U(x)$ and the optimal value function $V(x)$ are defined by

$$V^U(x) = \mathbb{E}_x J^U, \qquad V(x) = \sup_{U \in \mathcal{U}} V^U(x).$$

The horizon of the control problem is denoted T. We will make the simplifying assumptions that T is an exit time (as, for example, the time of ruin) and that J^U has the form

$$e^{-\delta T} r_T(X^U(T)) + \int_0^T e^{-\delta s} r\big(X^U(s), u(X^U(s))\big)\, ds. \tag{3.2}$$

The restriction that U be in the set $\mathcal{U}$ of admissible strategies will (in addition to existence of strong solutions, etc.) mean that $u(x)$ belongs to some set $\mathcal{U}(x)$ for all x. Important technical tools are to 1) extend the value function $V(t, x)$ to depend on t by considering the remaining payout after t,

$$V(t, x) = \tag{3.3}$$

$$\sup_{u \in \mathcal{U}(t)} \mathbb{E}_{x,t}\Big[e^{-\delta(T-t)} r_T(X^U(T)) + \int_t^T e^{-\delta(s-t)} r\big(X^U(s), u(X^U(s))\big)\, ds \Big],$$

where $\mathbb{E}_{x,t}$ denotes $\mathbb{E}[\,\cdot\mid X^U(t) = x,\ t < T]$, and 2) to involve generators. More precisely, $\mathcal{L}^u$ will be the generator of the (time-homogeneous) Markov process according to which X^U evolves when control $u(x) \equiv u$ is used. That is, $\mathcal{L}^u$ is the differential operator

$$\mathcal{L}^u f(x) = \mu(x, u) f'(x) + \frac{1}{2}\sigma^2(x, u) f''(x).$$

The importance of generators is that, among other properties,

$$\mathbb{E}_{x,t} f(X^U(t + h)) = f(x) + h\, \mathcal{L}^{u(x)} f(x) + o(h)$$

under suitable conditions on f.

Note that the assumption that T be an exit time ensures that the value function $V(t, x)$ only depends on $T - t$, not on t itself. One is eventually interested in $V(x) = V(0, x)$.

Theorem 3.1 *Under suitable regularity assumptions, the optimal value function $V(\cdot)$ is the solution of*

$$0 = \sup_{u \in \mathcal{U}(x)} \left[\mathcal{L}^u V(x) - \delta V(x) + r(x, u) \right]. \tag{3.4}$$

Remark 3.2 The 'regularity conditions' for the HJB equation (3.4) include that $V(\cdot)$ is in the domain of $\mathcal{L}^u$ for all u (we do not specify the remaining conditions and the reader should not take the proof below as more than a heuristic justification). The argument suggests that the maximizer u^* (when it exists) is the optimal control when $X(t) = x$. However, to establish that the solution of (3.4) indeed solves the control problem, more work is needed. Another complication is that it is not a priori clear whether the HJB equation has a unique solution. If it has, then one usually needs to prove separately (in a so-called *verification step*) that the obtained solution is indeed the value function of the optimal control problem. This can be done by either justifying all steps of the derivation of the HJB derivation rigorously, or by proving that the solution of the HJB equation dominates all other possible value functions; such a procedure often involves martingale arguments and (extensions of) Itô's formula. The second possibility is usually the more feasible one.

If the solution of the HJB equation is not unique, then the stochastic control problem can become very difficult. This effect can for instance occur if the value function is not as regular as the HJB equation would ask for. In that case one can often still work with either *weak solutions* or so-called *viscosity solutions*, see the textbook references at the end of Sect. 1. ◊

Proof of Theorem 3.1 Let U be a control such that u is used as control in $[t, t + h)$ and the optimal control is used in $[t + h, T)$. Then $V(x)$ has two parts, the contribution from $[t, t + h)$ and the contribution from $[t + h, T)$. This gives

$$V(x) \geq r(x, u)h + o(h) + e^{-\delta h} \mathbb{E}_{x,t} V(X^U(t + h))$$
$$= r(x, u)h - \delta h V(x) + \mathcal{L}^u V(x)h + V(x) + o(h),$$

which shows that (3.4) holds with $=$ replaced by $\geq$. To see that the sup is actually 0, choose a control u such that the above scheme gives an expected reward of at least $V(x) - \epsilon$. The same calculation then gives

$$V(x) - \epsilon \leq \mathcal{L}^u V(x) + r(x, u) + V(x) - \delta V(x).$$

Let $\epsilon \downarrow 0$. □

The application of the HJB equation typically follows this pattern: first for a fixed x express the minimizer $u^*(x)$ as a function

$$\xi\big(x, V(x), V'(x), V''(x)\big)$$

of x, $V(x)$, $V'(x)$, $V''(x)$. Then substitute this back into the HJB equation to get a second-order ODE

$$
\begin{aligned}
0 = {} & \mu\big(x, \xi\big(x, V(x), V'(x), V''(x)\big)\big)V'(x) \\
& + \frac{1}{2}\sigma^2\big(x, \xi\big(x, V(x), V'(x), V''(x)\big)\big)V''(x) \\
& + r\big(x, \xi\big(x, V(x), V'(x), V''(x)\big)\big) - \delta V(x))
\end{aligned}
$$

for V.

As may become clear from the above remarks, giving a rigorous and systematic treatment of stochastic control theory in insurance is outside the scope of this book. In the sequel, we shall rather consider some particular examples to get the flavor of the topic, putting more emphasis on how to arrive at the form of the results than to make all details completely rigorous.

Example 3.3 Ruin probability minimization as studied in Sect. 2 gives a simple example of how the HJB formalism works. To fit into the framework of (3.2), we consider maximization of the survival probability $1 - \psi^U(x)$ rather than minimization of $\psi^U(x)$ itself. This means that $T = \tau(x) \wedge \infty$, $\delta = 0$, $r_T(X^U(T)) = \mathbb{1}(T = \infty)$, $r(X^U(s), u(s)) = 0$, and that $\mu(x, u)$ and $\sigma^2(x, u)$ do not depend on x. The value function is then $1 - \psi^U(x)$ and the HJB equation reduces to

$$0 = \sup_u \mathscr{L}^u V(x) = \sup_u \left[\mu(u)V'(x) + \frac{1}{2}\sigma^2(u)V''(x)\right].$$

Assuming the solution $u^*(x)$ of this optimization problem exists, we thus have

$$0 = \mu(u^*(x))V'(x) + \frac{1}{2}\sigma^2(u^*(x))V''(x)$$

and the most general solution of this ODE has the form $C_1 + C_2 S(x)$, where S is the scale function given by

$$s(y) = \exp\left\{-2\int_a^y \mu(u^*(z))/2\sigma^2(u^*(z))\,dz\right\}, \quad S(x) = \int_b^x s(y)\,dy$$

(this is in fact one of the ways to arrive at the expression for $\psi(x)$ in Theorem 2.1!). Since by properties of Brownian motion ruin happens immediately starting from

$x = 0$, we have $C_1 = -C_2$ and we may w.l.o.g. assume $V(x) = 1 - S(x)$. This gives

$$V'(x) = -s(x), \quad V''(x) = 2s(x)\mu(u^*(x))/\sigma^2(u^*(x)).$$

Assuming the maximum is attained in the interior and can be obtained by differentiation, it follows by writing $m(u)$, $s_2(u)$ for the derivatives of $\mu(u)$, $\sigma^2(u)$ w.r.t. u that

$$
\begin{aligned}
0 &= m(u^*)V'(x) + \frac{1}{2}s_2(u^*)V''(x) \\
&= -\frac{s(x)}{\sigma^2(u^*(x))}\left[m(u^*(x))\sigma^2(u^*(x)) - \mu(u^*(x))s_2(u^*(x))\right] \\
&= -s(x)\frac{d}{du}\frac{\mu(u)}{\sigma^2(u)}\bigg|_{u=u^*(x)}.
\end{aligned}
$$

Since $s(x) \neq 0$, we are thereby back to the first-order condition of Sect. 2 (of course, modifications are needed if the sup is attained on the boundary as we will see examples of in the following Sect. 4).   ◇

4   Optimal Dividends

We consider only the case of restricted dividend payment. That is, the admissible strategies are defined by $0 \le u(x) \le u_0$ for some $u_0 < \infty$. We have

$$\mathrm{d}X^U(t) = \left(a - u(X^U(t))\right)\mathrm{d}t + b\,\mathrm{d}W(t). \tag{4.1}$$

Thus $\mathscr{L}^u f = (a-u)f' + bf''/2$, $r(x, u) = u$, and the HJB equation takes the form

$$
0 = \sup_{0 \le u \le u_0}\left[(a-u)V'(x) + bV''(x)/2 - \delta V(x) + u\right] \tag{4.2}
$$

$$
= aV'(x) + bV''(x)/2 - \delta V(x) + \sup_{0 \le u \le u_0}\left[u(1 - V'(x))\right]. \tag{4.3}
$$

The maximizer is therefore $u^*(x) = 0$ if $V'(x) > 1$ and $u^*(x) = u_0$ if $V'(x) < 1$, and thus

$$
0 = aV'(x) + bV''(x)/2 - \delta V(x) \qquad\qquad \text{if } V'(x) > 1, \tag{4.4}
$$

$$
0 = (a - u_0)V'(x) + bV''(x)/2 - \delta V(x) + u_0 \qquad \text{if } V'(x) < 1. \tag{4.5}
$$

Obviously, it cannot be optimal to take $u(x) = 0$ for all $x > 0$, so there must be an x with $V'(x) < 1$. The structure of the sets where $V'(x) > 1$, resp. $V'(x) < 1$, is not clear at this stage, but the standard formula for the solutions of second-order homogeneous linear ODEs immediately give the following lemma:

Lemma 4.1 *Let $\theta_1, -\theta_2$ denote the positive, resp. negative roots of the quadratic $bz^2/2 + az - \delta$ and $\theta_3, -\theta_4$ the similar roots of $bz^2/2 + (a - u_0)z - \delta$. Then $C_1 e^{\theta_1 x} + C_2 e^{-\theta_2 x}$ solves (4.4) and $C_3 e^{\theta_3 x} + C_4 e^{-\theta_4 x} + u_0/\delta$ solves (4.5). Furthermore, any solutions have this form.*

Define the constant A by

$$A = \frac{u_0}{\delta} - \frac{1}{\theta_4}.$$

The arguments to follow proceed by first finding a C^2 trial solution f to the HJB equation and next use Itô to verify that $f = V$. It is clear that the value function must satisfy $0 \leq V(x) \leq \int_0^\infty e^{-\delta t} u_0 \, dt$ and (by properties of Brownian motion) $V(0) = 0$, so we can limit the search to f that are bounded with $f(0) = 0$. We will first investigate whether it may occur that $u^*(x) = u_0$ for all x, that is, if it is always optimal to pay out dividends at the maximal allowed rate. The answer is:

Theorem 4.2 *Assume $A \leq 0$, i.e. $u_0 \leq \delta/\theta_4$. Then there exists a unique optimal strategy given by $u^*(x) = u_0$ for all x. For this strategy,*

$$V(x) = \frac{u_0}{\delta}(1 - e^{-\theta_4 x}). \tag{4.6}$$

Proof By Lemma 4.1, our trial f must have the form $C_4 e^{-\theta_4 x} + u_0/\delta$ ($C_3 \neq 0$ is excluded since we need f to be bounded with $f \geq 0$). From $f(0) = 0$ it then follows that $f(x)$ must be given by the r.h.s. of (4.6). Clearly this f is concave, C^2 and a solution to the HJB equation since the condition on u_0 ensures that $f'(0) \leq 1$ and hence $f'(x) < 1$ for $x > 0$.

Having thus motivated our trial solution $u^*(x) = u_0$ for all x, we proceed to verify that it is indeed optimal. Let U be some arbitrary strategy. Then, with τ the ruin time and $x = X(0) = X^U(0)$, $u(s) = u(X^U(s))$,

$$e^{-\delta(\tau \wedge t)} f(X^U(\tau \wedge t))$$

$$= f(x) + \int_0^{\tau \wedge t} e^{-\delta s} b \, dW(s)$$

$$+ \int_0^{\tau \wedge t} e^{-\delta s} \left[-\delta f(X^U(s)) + (a - u(s)) f(X^U(s)) + \frac{1}{2} f''(X^U(s)) \right] ds$$

$$\leq f(x) + \int_0^{\tau \wedge t} e^{-\delta s} b \, dW(s) - \int_0^{\tau \wedge t} e^{-\delta s} u(s) \, ds ,$$

where the application of Itô in the first step used that f is C^2 and bounded, and the last step that f solves (4.2). Further equality holds if $U = U^*$. Boundedness of f ensures that we can take expectations to get

$$f(x) \geq e^{-\delta(\tau \wedge t)} f(X^U(\tau \wedge t)) + \mathbb{E}_x \int_0^{\tau \wedge t} e^{-\delta s} u(s) \, ds$$

and let $t \to \infty$ to get $f(x) \geq 0 + V^U(x)$ (here we used in addition $X^U(\tau) = 0$ and monotone convergence). Again equality holds if $U = U^*$, completing the proof.   □

It remains to settle the case $u_0 > \delta/\theta_4$ which turns out to be more intriguing.

Theorem 4.3 *Assume $A > 0$, i.e. $u_0 > \delta/\theta_4$. Then there exists a unique optimal strategy given by $u^*(x) = u_0$ for $x > x_0$ and $u^*(x) = 0$ for all $x < x_0$ where*

$$x_0 = \frac{1}{\theta_1 + \theta_2} \log \frac{1 + A\theta_2}{1 - A\theta_1} > 0. \tag{4.7}$$

For this strategy,

$$V(x) = \begin{cases} K_1(e^{\theta_1 x} - e^{-\theta_2 x}) & \text{for } 0 \leq x \leq x_0, \\ u_0/\delta - K_2 e^{-\theta_4 x} & \text{for } x > x_0, \end{cases} \tag{4.8}$$

where the values of K_1, K_2 follow by inserting (4.7) in (4.11), (4.12) below.

Proof Motivated by what we found for $u_0 \leq \delta/\theta_4$, we take the trial solution f to be C^2 and concave, implying that there should be some $x_0 > 0$ such that $f'(x) \leq 1$ for $x \geq x_0$ and $f'(x) \geq 1$ for $x \leq x_0$; the analysis in the proof above shows that we cannot have $x_0 = 0$. Lemma 4.1 then gives that $f(x)$ should be given by the r.h.s. of (4.8) with $K_1 = C_1$ (that $C_2 = -C_1$ follows from $f(0) = 0$) and $K_2 = -C_4$ (that $C_3 = 0$ follows again from f being bounded with $f \geq 0$).
For the C^2 property it is necessary that

$$f(x_0-) = f(x_0+), \quad f'(x_0-) = 1 = f'(x_0+), \tag{4.9}$$

which gives the equations

$$K_1(e^{\theta_1 x_0} - e^{-\theta_2 x_0}) = u_0/\delta - K_2 e^{-\theta_4 x_0}, \tag{4.10}$$

$$K_1(\theta_1 e^{\theta_1 x_0} + \theta_2 e^{-\theta_2 x_0}) = 1, \tag{4.11}$$

$$K_2 \theta_4 e^{-\theta_4 x_0} = 1. \tag{4.12}$$

This argument is known as the principle of *smooth fit*. Inserting (4.12) in (4.10), the r.h.s. reduces to A, and dividing (4.10) by (4.11) then gives

$$\frac{e^{\theta_1 x_0} - e^{-\theta_2 x_0}}{\theta_1 e^{\theta_1 x_0} + \theta_2 e^{-\theta_2 x_0}} = A, \qquad \text{i.e.} \quad (1 - A\theta_1)e^{\theta_1 x_0} = (1 + A\theta_2)e^{-\theta_2 x_0}.$$

From this (4.7) follows provided we can show $A\theta_1 < 1$, i.e.

$$\frac{u_0}{\delta} < \frac{1}{\theta_1} + \frac{1}{\theta_4}. \tag{4.13}$$

But $\sqrt{r^2 + s} - r < s/2r$ for $r, s > 0$, and so

$$\theta_1 = \frac{\sqrt{a^2 + 4b\delta} - a}{2b} < \frac{\delta}{a}, \quad \frac{1}{\theta_1} + \frac{1}{\theta_4} > \frac{a}{\delta} \geq \frac{u_0}{\delta}$$

when $u_0 \leq a$, showing (4.13). For $u_0 > a$, use that similarly $\theta_4 < \delta/(u_0 - a)$.

The f we have found is thus bounded and non-negative with $f(0) = 0$. It also solves the HJB equation and is C^2, except possibly at $x = x_0$. However, the C^2 property there follows by letting $x \uparrow x_0$ in (4.4) and $x \downarrow x_0$ in (4.5), giving

$$a \cdot 1 + bf''(x_0-) = \delta f(1) = (a - u_0) \cdot 1 + bf''(x_0+) + u_0$$

so that indeed $f''(x_0-) = f''(x_0+)$. From these properties of f one gets $f = V$ exactly as in the proof of Theorem 4.2.   □

Notes

The case $0 \leq u \leq \infty$ of unbounded dividends is intrinsically more difficult and requires a different formulation as discussed in Remark 1.4. It turns out that the optimal strategy is to pay out nothing if the reserve is $< x_0$ for a certain $x_0 > 0$ and so much at x_0 that x_0 acts as a reflecting upper barrier. This is an instance of so-called *impulse control*.

5   Control Problems for the Cramér–Lundberg Model

The structure of the controlled diffusions we have been working with so far have most often been motivated from the diffusion approximation of a Cramér–Lundberg risk process

$$R(t) = x + ct - \sum_{i=1}^{N(t)} Y_i \tag{5.1}$$

(note that claims in this chapter are denoted $Y_1, Y_2, \ldots$ and not $V_1, V_2, \ldots$ as in Chap. IV). It is therefore natural to try to look at the same problems directly in the model (5.1). This turns out to be substantially more difficult. One can see this already by considering the first control problem we looked at for diffusions, ruin probability minimization: in that setting there was an explicit expression for the ruin probability, in the Cramér–Lundberg model there is only so in special cases and hardly any at all if the premium rate c, the Poisson arrival rate β and the distribution B of the claims are allowed to vary with the current level of R.

5.1 The Generator and the HJB Equation

Given the role the generator played in the diffusion setting, the first step in stochastic control theory for the Cramér–Lundberg process is to look at its analogue there. That is, we are looking for an operator $\mathscr{L}$ acting on a suitable space of functions such that

$$\mathbb{E}\big[f(R(t+h)) \,\big|\, R(t) = x\big] \;=\; f(x) + h\,\mathscr{L}f(x) + \mathrm{o}(1)\,. \tag{5.2}$$

We won't specify the maximal class of functions for which one can make sense of $\mathscr{L}f$, but at least:

Proposition 5.1 *Assume f is bounded with $f(x) = 0$ for $x < 0$ and differentiable on $[0, \infty)$. Then (5.2) holds with*

$$\mathscr{L}f(x) \;=\; cf'(x) + \beta\Big[\int_0^x f(x-y)\,B(\mathrm{d}y) - f(x)\Big]\,.$$

Proof We follow the usual approach of splitting up according to no claim arrivals in $(t, t + h]$, precisely one, or at least two. Up to $\mathrm{o}(h)$ terms, these events have probabilities $1 - \beta h$, βh and 0. In the first case, R simply grows to $x + ch$, in the second it makes a downward jump of size say y occurring w.p. $B(\mathrm{d}y)$. This gives

$$\mathbb{E}\big[f(R(t+h)) \,\big|\, R(t) = x\big]$$

$$= (1 - \beta h)f(x + ch) + \beta h \int_0^x f(x-y)\,B(\mathrm{d}y) + \mathrm{o}(h)$$

$$= (1 - \beta h)\big(f(x) + hcf'(x)\big) + \beta h \int_0^x f(x-y)\,B(\mathrm{d}y) + \mathrm{o}(h)\,,$$

where in the first step we used that $f(x) = 0$ for $x \le 0$. From this the result easily follows.   □

We restrict ourselves again to a time horizon T that is an exit time, to controls $U = \left(c^U(x), \beta^U(x), B^U(dy; x)\right)_{x \geq 0}$ of feedback type, and to pay-outs of the form

$$e^{-\delta T} r_T(R^U(T)) + \int_0^T e^{-\delta s} r\left(R^U(s), u(R^U(s))\right) ds \ .$$

Precisely the same heuristics as for the diffusion models in the proof of Theorem 3.1 then gives:

Theorem 5.2 *Under suitable regularity assumptions, the optimal value function $V(\cdot)$ is the solution of*

$$0 = \sup_{u \in \mathscr{U}(x)} \left[\mathscr{L}^u V(x) - \delta V(x) + r(x, u) \right]. \tag{5.3}$$

Unfortunately, the cases where the HJB equation (5.3) can be solved explicitly are much more rare than in the diffusion setting. In general, one has to resort to numerical methods, see Remark 5.5 below for examples. We shall not present examples of such calculations here but refer to Schmidli [155]. Instead, we shall present a couple of the (rare!) instances where some relevant information can indeed be extracted from the HJB equation.

5.2 Ruin Probability Minimization

The problem of finding an optimal control minimizing the ruin probability turns out to be substantially more complicated for the Cramér–Lundberg model than for the diffusion case. In particular, the form of B plays a more crucial role, and in most examples where $u^*(x) \equiv u^*$ was found to be level-independent for diffusions, this is no longer so for the Cramér–Lundberg model (asymptotic considerations may, however, sometimes mitigate these difficulties as exemplified below).

Considering the equivalent problem of survival probability maximization, we have $\delta = 0$, $r(x, u) \equiv 0$ and $r_T(R^U(T)) = \mathbb{1}(T = \infty)$, and the value function is the survival probability. In the rest of this subsection, we consider reinsurance control. Then if $s(y; u)$ is the part of a claim of size y covered by the cedent if the control is u and $c(u)$ the net rate of premiums (premiums from policy holders minus premium to the reinsurer), the HJB equation (5.3) takes the form

$$0 = \sup_{0 \leq u \leq 1} \left\{ c(u) f'(x) + \beta \left[\int_0^\infty f(x - s(y; u)) B(dy) - f(y) \right] \right\}. \tag{5.4}$$

This at best specifies the value function up to a constant. However, the optimal survival probability must be strictly positive since it is at least the optimal survival probability for the uncontrolled process. Hence one can specify a (dummy) non-zero boundary value at 0, and we take $f(0) = 1$. Then (with quite some effort) (5.4) can be shown to have a unique, strictly increasing and continuously differentiable solution. Further, after appropriate scaling this solution is indeed the maximal survival probability of the controlled process.

As a specific example we consider, as in Sect. 2.2, ruin probability minimization in proportional reinsurance with both the cedent and the reinsurer applying the expected value principle with loadings η_s, η_r. Then

$$\mathscr{U}(x) = \{u : 0 \le u \le 1\}, \quad \beta^U(x) \equiv \beta, \quad B^U(dy; x) = B(dy/u(x))$$

and, as in Example 1.1,

$$c(u) = \beta \mu_B \big[\eta_s - (1 - u)\eta_r\big].$$

We assume for ease of exposition that $\beta = 1$, $\mu_B = 1$ (this just amounts to a scaling of time and space). The HJB equation (5.3) can then be written as

$$0 = \sup_{0 \le u \le 1} H(x, u) \tag{5.5}$$

$$\text{where } H(x, u) = (\eta_s + (u - 1)\eta_r) f'(x) + \mathbb{E} f(x - uY) - f(x). \tag{5.6}$$

Since $H(x, u)$ is continuous in both variables, compactness of $[0, 1]$ gives that the supremum is attained for all x and so

$$0 = H(x, u^*(x)) = (\eta_s + (u^*(x) - 1)\eta_r) f'(x) + \mathbb{E} f(x - u^*(x)Y) - f(x). \tag{5.7}$$

In the following, we write $\mathbb{P}_x^*$ for the governing probability measure of R when started from $R(0) = x$ and developing according to the optimal strategy $(u^*(z))_{z \ge 0}$. We also for convenience assume that B has infinite support.

Proposition 5.3 *In the setting of maximizing the survival probability via dynamical proportional reinsurance, the value function V is strictly increasing with $V(0) > 0$. Furthermore, V' is continuous with $V'(x) > 0$ for all $x \ge 0$.*

Proof Since the strategy with $u(x) \equiv 1$ (no reinsurance) has positive survival probability, it is clear that $V(0) > 0$. Let $\sigma(y) = \inf\{t : R(t) = y\}$. Then $\mathbb{P}_x^*(\tau = \infty, \sigma(y) = \infty) = 0$ for $0 \le x < y$. Indeed, if $u(z) = 0$ for Lebesgue a.a. z (full reinsurance), then R decreases monotonically from x to 0 at rate $\eta_s - \eta_r$. If on the other hand $\{z \le y : u(z) > 0\}$ has positive Lebesgue measure, then so has $F = \{z \le y : u(z) > \epsilon\}$ for some $\epsilon > 0$. Then a claim arriving from any $z \in F$ has probability $\overline{B}(z/u(z)) \ge \overline{B}(y/\epsilon) > 0$ for causing ruin (recall that we assumed $\overline{B}(v) > 0$ for all v). Since there is an infinity of such arrivals when $\sigma(y) = \infty$, a geometric trials argument therefore gives $\tau < \infty$ a.s.

For $x < y$, we now have

$$
\begin{aligned}
\mathbb{P}^*_x(\tau = \infty) &= \mathbb{P}^*_x(\tau = \infty, \sigma(y) = \infty) + \mathbb{P}^*_x(\sigma(y) < \infty)\mathbb{P}^*_y(\tau = \infty) \\
&< 0 + \mathbb{P}^*_y(\tau = \infty)
\end{aligned}
$$

(obviously, $\mathbb{P}^*_x(\sigma(y) = \infty) > \mathbb{P}^*_x(\tau > \sigma(y)) > 0$). This gives the strict increase of V and hence also that $V'(x) \geq 0$. Since trivially $\mathbb{E}f(x - uY) \leq f(x)$, we therefore must have $\eta_s + (u^*(x) - 1)\eta_r \geq 0$, i.e. $u^*(x) \geq u_0$ where $u_0 = 1 - \eta_s/\eta_r > 0$. Then $u^*(x) > 0$ gives $\mathbb{E}f(x - u^*(x)Y) < f(x)$ so indeed $u^*(x) > u_0$ and $V'(x) > 0$. The continuity of V' now follows from (5.7) since the coefficient to $V'(x)$ is bounded away from 0 and everything else is continuous. □

We shall use this to show a marked difference from the diffusion case, that it is optimal to purchase no reinsurance for any initial capital x below some positive level:

Proposition 5.4 *In ruin probability minimization for the Cramér–Lundberg model with proportional reinsurance, there exists an $x_0 > 0$ such that $u^*(x) = 1$ for $0 \leq x \leq x_0$.*

Proof Recalling that $f(z) = f'(z) = 0$ for $z < 0$, we have

$$
\frac{d}{du}\mathbb{E}f(x - uY) = \frac{d}{du}\mathbb{E}\left[-f(0) + \int_0^{x-uY} f'(y)\,dy \right] = -\mathbb{E}\left[Yf'(x - uY)\right].
$$

By choosing $x_0 > 0$ small enough, we can obtain

$$
\mathbb{E}[Y; Y \leq x_0/u_0)] < \eta_r/4, \quad 3f'(0)/4 \leq f'(y) \leq 2f'(0)
$$

for $y \leq x_0$. Then for $x \leq x_0$ and $u \geq u_0$, we have

$$
\mathbb{E}\left[Yf'(x - uV)\right] \leq 2f'(0)\mathbb{E}[Y; Y \leq x_0/u_0)] \leq \eta_r f'(0)/2.
$$

Hence with $H(x, u)$ as in (5.6) we have

$$
\frac{d}{du}H(x, u) = \eta_r f\check{S}(x) - \mathbb{E}\left[Yf'(x - uV)\right] > \eta_r f'(0)/4 > 0
$$

for such x and u, i.e. $H(x, u)$ is strictly increasing in $u \in [u_0, 1]$. Recalling from above that $u^*(x) \geq u_0$, this gives the assertion. □

Remark 5.5 For many purposes, it is useful to rewrite the HJB equation (5.3) as

$$
f'(x) = \inf_{u_0 \leq u \leq 1} \left\{ \frac{\beta}{c(u)}\left[f(x) - \int_0^\infty f(x - yu)\,B(dy) \right] \right\}. \tag{5.8}
$$

In particular, this provides algorithms for computing a numerical approximation $\widehat{f}$ to f. The simplest is an Euler scheme, which in the setting of proportional reinsurance with $\beta = \mu_B = 1$ works a follows. First note that (5.8) means

$$f'(x) = \inf_{u_0 \le u \le 1} \left\{ \frac{1}{\eta_s + (u-1)\eta_r} \left[f(x) - \int_0^{x/u} f(x - yu) \, B(\mathrm{d}y) \right] \right\}. \tag{5.9}$$

Then choose a step length $h > 0$, let $\widehat{f}_0 = 1$ and

$$\widehat{f}_{(k+1)h} = \widehat{f}_{kh} + \inf_{u_0 \le u \le 1} \left\{ \frac{h}{\eta_s + (u-1)\eta_r} \left[\widehat{f}_{kh} - I(h, k, u) \right] \right\},$$

where $I(h, k, u)$ is some approximation to $\int_0^{x/u} f(x - yu) \, B(\mathrm{d}y)$, e.g.

$$\sum_{i=1}^{k} \widehat{f}_{(k-i)h} \left[B(ih/u) - B((i-1)h/u) \right].$$

Of course, the inf has to be evaluated by some sort of discretization. A more sophisticated algorithm starts by noting that

$$f(x) - \int_0^{x/u} f(x - uy) \, B(\mathrm{d}y) = \overline{B}(x/u) + \int_0^x f'(z) \overline{B}((x - z)/u) \, \mathrm{d}z.$$

This means that one can rewrite (5.9) as $\mathcal{R}w = w$, where $w(x) = f'(x)$ and $\mathcal{R}$ is the operator

$$\mathcal{R}w(x) = \inf_{u_0 \le u \le 1} \left\{ \frac{1}{\eta_s + (u-1)\eta_r} \left[\overline{B}(x/u) + \int_0^x w(z) \overline{B}((x - z)/u) \, \mathrm{d}z \right] \right\}.$$

One then solves the fixed-point problem $\mathcal{R}w = w$ numerically by iteration, starting from some suitable w_0 and letting $w_{n+1} = \mathcal{R}w_n$.   ◇

Even if one has to approximate the solution to the HJB equation numerically in practical examples, asymptotic results for $x \to \infty$ can, however, be obtained. In particular, if a strictly positive solution γ^* to the adjustment equation

$$\inf_{u \in [0,1]} \left\{ \beta \left(\widehat{B}[u\gamma] - 1 \right) - c(u)\gamma \right\} = 0 \tag{5.10}$$

exists, then one can show that under some mild additional assumptions the Cramér–Lundberg approximation

$$\lim_{x \to \infty} e^{\gamma^* x} (1 - V(x)) = C \tag{5.11}$$

holds for the optimally controlled process R^{U^*} with some constant $C > 0$. If moreover the value u^* for which the infimum in (5.10) is attained is unique, then $\lim_{x\to\infty} u^*(x)$ exists, i.e. for increasing initial capital x the optimal strategy converges to a constant reinsurance fraction.

If on the other hand B is regularly varying with $\overline{B}(x) = L(x)/x^\alpha$, then $1 - V(x) \sim C\overline{B}_0(x)$, where B_0 is the integrated tail and C some explicit constant. Further, again $\lim_{x\to\infty} u^*(x)$ exists. For subexponential but lighter tails of B like lognormal or Weibull, available results are not as explicit, but in that case the optimal strategy can be shown to satisfy $\limsup_{x\to\infty} u^*(x) = u_0 = \inf\{u : c(u) > 0\}$, i.e. for large x one tries to reinsure as much as still possible without resulting in negative drift. The intuitive reason is that (in contrast to regularly varying tails) proportional reinsurance makes the tail of the distributions smaller and so it is more attractive to purchase more reinsurance.

5.3   Optimal Investment

We next consider the optimal investment problem for the Cramér–Lundberg model where the company has the possibility to dynamically invest an amount of $u(x)$ into a financial asset that is modeled by geometric Brownian motion $M(t)$ with

$$\mathrm{d}M(t) = b_1 M(t)\,\mathrm{d}t + b_2 M(t)\,\mathrm{d}W(t).$$

Taking $c = 1$, the controlled process then satisfies

$$\mathrm{d}R^U(t) = \big(1 + u(X^U(t))b_1\big)\mathrm{d}t + u(X^U(t))b_2\,\mathrm{d}W(t) - \mathrm{d}\sum_{i=1}^{N(t)} Y_i.$$

The goal is again to find the minimal ruin probability $\psi^*(x)$ of R^U. In the present model, the HJB equation $0 = \sup_{u \geq 0} \mathcal{L}^u V(x)$ translates into

$$0 = \sup_{u \geq 0}\left[(1 + b_1 u)\,V'(x) + \frac{u^2 b_2^2}{2}\,V''(x) + \beta\left(\int_0^x V(x - y)B(\mathrm{d}y) - V(x)\right)\right].$$
$$(5.12)$$

Note that $u^*(x) \to 0$ as $x \to 0$ (otherwise path properties of Brownian motion imply immediate ruin so the investment will lead to $1 - \psi^*(0) = V(0) = 0$, which cannot be optimal). Thus we obtain a boundary condition $V'(0) = \beta V(0)$. The second boundary condition is again $\lim_{x\to\infty} V(x) = 1$.

Since the sup in (5.12) is not attained when $V''(x) \geq 0$, we assume $V''(x) < 0$. Then the sup is attained for

$$u^*(x) = -\frac{b_1 V'(x)}{b_2^2 V''(x)}$$

and plugging this into the HJB equation one gets

$$V'(x) - \frac{b_1^2}{2b_2^2} \frac{V'(x)^2}{V''(x)} + \beta \left(\int_0^x V(x-y) B(\mathrm{d}y) - V(x) \right) = 0. \qquad (5.13)$$

It is now considerably more difficult than in the diffusion case to solve this equation and retrieve further information about the optimal strategy, but remarkably, this is possible asymptotically for $u^*(x)$ and $\psi^*(x) = 1 - V(x)$ as $x \to \infty$: for a light-tailed claim size distribution B, if the adjustment coefficient γ_I exists as the positive solution of

$$\beta \big(\widehat{B}[r] - 1 \big) - r = \frac{b_1^2}{2b_2^2}, \qquad (5.14)$$

then $\psi^*(x) \leq e^{-\gamma_I x}$ and (under a mild additional condition) the Cramér–Lundberg approximation $\lim_{x \to \infty} e^{\gamma_I x} \psi^*(x) = C$ holds for some constant C. Without investment, the r.h.s. of (5.14) is zero, so clearly $\gamma_I > \gamma$, and hence optimal investment can substantially decrease the probability of ruin. Furthermore $\lim_{x \to \infty} u^*(x) = b_1/(b_2^2 \gamma_I)$, so asymptotically the optimal strategy is to invest a constant amount into the risky asset, which is somewhat surprising at first sight.

On the other hand, for a heavy-tailed B (i.e. $\widehat{B}[r] = \infty$ for all $r > 0$), one can show that $u^*(x)$ is unbounded. If the hazard rate of B tends to zero, then the optimal strategy converges and $\lim_{x \to \infty} u^*(x) = \infty$. A quite pleasant result is that for $B, B_0 \in \mathscr{S}$ the optimal investment strategy leads to

$$\psi^*(x) \sim \frac{2\beta b_2^2}{b_1^2} \int_x^\infty \frac{1}{\int_0^y \frac{1}{1-B(z)} \mathrm{d}z} \, \mathrm{d}y, \qquad x \to \infty. \qquad (5.15)$$

If further $\overline{B}(x) = L(x)/x^\alpha$ is regularly varying, then (5.15) translates into

$$\psi^*(x) \sim \frac{2\beta b_2^2 (\alpha + 1)}{b_1^2 \alpha} \overline{B}(x) \quad \text{and} \quad u^*(x) \sim \frac{b_1}{b_2^2 (\alpha + 1)} x, \qquad x \to \infty.$$

Accordingly, for $x \to \infty$ it is optimal to invest the constant fraction $b_1/(b_2^2(\alpha + 1))$ of the surplus into the risky asset.

5.4 *Optimal Dividends*

As in Sect. 4, we consider the case of restricted dividend payments discounted by δ. That is, the admissible strategies are defined by $0 \le u(t) \le u_0$ for some $u_0 < \infty$. The HJB equation takes the form $\sup_{0 \le u \le u_0} H(x, u)$, where

$$H(x, u) = (c - u)V'(x) + \beta \left[\int_0^x V(x - y) B(dy) - V(x) \right] - \delta V(x) + u .$$

We find therefore that the maximizer is $u^*(x) = 0$ if $V'(x) > 1$ and $u^*(x) = u_0$ if $V'(x) < 1$. This is the same as in Sect. 4, and an explicit solution is in fact possible provided one assumes B to be exponential, say $B(dy) = \alpha e^{-\alpha y} dy$. In that case,

$$\beta \frac{d}{dx} \int_0^x V(x - y) B(dy) = \beta \alpha \frac{d}{dx} e^{-\alpha x} \int_0^x V(y) e^{\alpha y} dy$$

$$= \beta \alpha V(x) - \beta \alpha^2 e^{-\alpha x} \int_0^x V(y) e^{\alpha y} dy$$

$$= \beta \alpha V(x) + \alpha \big[(c - u)V'(x) - \beta V(x) - \delta V(x) + u \big],$$

$$\frac{\partial}{\partial x} H(x, u) = (c - u)V''(x) - \beta V'(x) - \delta V(x)$$

$$+ \beta \alpha V(x) + \alpha \big[(c - u)V'(x) - \beta V(x) - \delta V(x) + u \big]$$

and it follows after simple rearrangements that

$$0 = \begin{cases} cV''(x) - (\beta + \delta - \alpha c)V'(x) - \alpha \delta V(x) & V'(x) > 1, \\ (c - u_0)V''(x) - (\beta + \delta - \alpha(c - u_0))V'(x) - \alpha \delta V(x) + \alpha u_0 & V'(x) < 1. \end{cases}$$

The rest of the analysis now follows the pattern of Sect. 4, with the roots of the two associated quadratics playing a major role to get piecewise solutions and by applying the principle of smooth fit. Again, the values of the parameters (now $\alpha, \beta, \delta, c, u_0$) play a role and for example, one needs to proceed somewhat differently in the cases $c > u_0$, $c = u_0$ and $c < u_0$.

Notes
For a derivation of the missing steps and a detailed discussion of the above results the reader is referred to Schmidli [155, Ch. IV].

6 Examples Involving Game Theory

The stochastic control problems we have considered so far involve only optimality concerns of one insurance company. A more general setting is to consider a market of two or more companies, typically each aiming at maximizing its profit and/or position in the market. This type of question can naturally be formulated in terms of concepts from game theory. The literature on this topic in an insurance context is, however, more sparse and more recent than that on control problems, and we shall here just sketch a few examples which hopefully can at least give some indication of how game theory can come in.

For simplicity, we restrict to two companies I_1, I_2 (players) such that company j has access to a set $\mathcal{U}_j$ of strategies U_j. Assume that there is a single payoff function $\mathcal{J}^{U_1, U_2}$ defined for $(U_1, U_2) \in \mathcal{U}_1 \times \mathcal{U}_2$. Company I_1 is trying to maximize the expected payoff $\mathbb{E}\, \mathcal{J}^{U_1, U_2}$ while simultaneously I_2 is trying to minimize the same quantity.

Much of game theory is concerned with discussing pairs U_1^*, U_2^* of strategies that are unilaterally optimal for both players in the sense that none of them has an incentive to deviate from their U_j^*. The most classical concept is that of *Nash equilibrium*, defined by the requirement

$$\max_{U_1 \in \mathcal{U}_1} \mathbb{E}\, \mathcal{J}^{U_1, U_2^*} = \mathbb{E}\, \mathcal{J}^{U_1^*, U_2^*} = \min_{U_2 \in \mathcal{U}_2} \mathbb{E}\, \mathcal{J}^{U_1^*, U_2}. \tag{6.1}$$

Behind this is the implicit assumption of an oligopoly game where the players make their decisions simultaneously. In a *Stackelberg game* on the other hand, there is a leader, say I_2, making his decision first and the follower (in this case I_1) then makes his, knowing the strategy of the leader. More precisely, a Stackelberg equilibrium is defined as a strategy pair (U_1^*, U_2^*) satisfying

$$U_1^* = \widehat{U}_1(U_2^*) \quad \text{and} \quad \mathcal{J}^{U_1^*, U_2^*} \le \mathcal{J}^{\widehat{U}_1(U_2), U_2} \text{ for all } U_2, \tag{6.2}$$

where $\widehat{U}_1(U_2)$ is the follower's best strategy given the leader uses U_2.

Our examples will involve two companies competing on their insurance business: they try to optimize their strategy in such a way that their surplus rises faster than that of the opponent. If one company has a faster increasing surplus than the surplus of its competitor, then this company will eventually dominate the market. More precisely, we will assume that, with given strategies u_1, u_2 fixed in time, the reserve of company j develops according to

$$\mathrm{d}X_j(t) = \mu_j(x, u_1, u_2)\, \mathrm{d}t + \sigma_j(x, u_1, u_2)\, \mathrm{d}W_j(t), \tag{6.3}$$

where W_1, W_2 are independent Brownian motions and that the comparisons of surpluses is in terms of $\Delta(t) = X_1(t) - X_2(t)$ such that

$$\mathcal{J}^{U_1, U_2} = \mathbb{1}(\tau = \ell_u) \quad \text{where } \tau = \inf\{t > 0: \Delta(t) \notin (\ell_d, \ell_u)\}$$

with (ℓ_d, ℓ_u) some interval containing $x_1 - x_2$ (thus τ is the exit time). We have $\Delta(0) = x_1 - x_2$ and

$$\mathrm{d}\Delta(t) = \mu(x, u_1, u_2)\, \mathrm{d}t + \sigma(x, u_1, u_2)\, \mathrm{d}W(t), \tag{6.4}$$

where W is another Brownian motion and

$$\mu(\cdot) = \mu_1(\cdot) - \mu_2(\cdot), \qquad \sigma^2(\cdot) = \sigma_1^2(\cdot) + \sigma_2^2(\cdot).$$

This set-up is a special case of what goes under the name of a *differential game*.

Scale function arguments similar to those of Sect. 2 now give the following result, showing that the concern of I_1 is to maximize $\kappa = \mu/\sigma^2$ and that of I_1 to minimize κ:

Proposition 6.1

(i) $\mathbb{P}_\delta(\Delta(\tau) = \ell_u) = \dfrac{S(\delta)}{S(\ell_u)}$ *where*

$$s(y) = \exp\left\{\int_{\ell_d}^y -2\kappa(z)\,dz\right\}, \quad S(x) = \int_{\ell_d}^x s(y)\,dy.$$

(ii) $\underset{u_1}{\operatorname{argmax}}\, \mathbb{P}_\delta(\Delta(\tau) = \ell_u) = \underset{u_1}{\operatorname{argmax}} \dfrac{\mu(x, u_1, u_2)}{\sigma^2(x, u_1, u_2)},$

$$\underset{u_2}{\operatorname{argmax}}\, \mathbb{P}_\delta(\Delta(\tau) = \ell_d) = \underset{u_2}{\operatorname{argmin}}\, \mathbb{P}_\delta(\Delta(\tau) = \ell_u) = \underset{u_2}{\operatorname{argmin}} \dfrac{\mu(x, u_1, u_2)}{\sigma^2(x, u_1, u_2)}.$$

6.1 An Optimality Property of Stop-Loss Reinsurance

Now consider reinsurance where $\mathscr{U}_1 = \mathscr{U}_2$ is the set of all possible reinsurance arrangements as specified by a pair r, s of functions with $0 \le r(v) \le v$ and $r(v) + s(v) = v$. Following strategy U_j, company j will choose arrangement $u_j(x) = s_j(\cdot; x)$ when $\Delta(t) = x$, both companies (cedents) apply the expected value principle with loadings η_1, η_2 and the reinsurer applies the expected value principle with loading η. Assuming in addition that the reserves earn interest at rate r, this gives in the diffusion approximation that

$$\mu(x, u_1, u_2) = (1 + \eta_1)\mathbb{E}Y - (1 + \eta_2)\mathbb{E}Y - (1 + \eta)\big(\mathbb{E}r_1(Y) - \mathbb{E}r_2(Y)\big) + rx,$$

$$\sigma^2(x, u_1, u_2) = \beta_1 \mathbb{E}s_1(Y)^2 + \beta_2 \mathbb{E}s_2(Y)^2.$$

The following result shows that when looking for a Nash equilibrium, both companies may restrict their strategies to $u \in \mathscr{U}^{XL}$, the set of all excess-of-loss arrangements where $r(Y) = (Y - K)^+$, $s(Y) = \min(Y, K)$ for some retention level $K > 0$.

Proposition 6.2 *For any* $u_1^*, u_2^*,$

$$\operatorname*{argmax}_{u_1} \mathbb{P}_\delta(\Delta(\tau) = \ell_u) = \operatorname*{argmax}_{u_1 \in \mathscr{U}^{XL}} \frac{\mu(x, u_1, u_2^*)}{\sigma^2(x, u_1, u_2^*)},$$

$$\operatorname*{argmax}_{u_2} \mathbb{P}_\delta(\Delta(\tau) = \ell_d) = \operatorname*{argmin}_{u_2} \mathbb{P}_\delta(\Delta(\tau) = \ell_u) = \operatorname*{argmin}_{u_2 \in \mathscr{U}^{XL}} \frac{\mu(x, u_1^*, u_2)}{\sigma^2(x, u_1^*, u_2)}.$$

Proof Since $\mathbb{E}[Y \wedge K]^2$ is continuous in K and increases monotonically from 0 to $\mathbb{E}Y^2$, $s_1(Y) < Y$ implies that it is possible to choose K with $\mathbb{E}[Y \wedge K]^2 = \mathbb{E}s_1(Y)^2$, cf. Exercise VIII.4.3. Replacing $s_1(Y)$ with $Y \wedge K$ leaves the denominator in κ unchanged but decreases $\mathbb{E}r_1(Y)$, cf. Exercise VIII.4.3. Hence it increases μ_1 and therefore κ. The proof of the second statement is similar.   □

6.2   Premiums Based on Customer Preferences

We next consider a situation where I_1, I_2 compete by offering different premiums p_1, p_2 for the same product. In Sect. 2.4 the alternatives for a customer when offered a premium p by a single company I was to either insure with I or not. In the present two-company situation, the choice is instead whether to insure with I_1 or I_2. The reason that the customer just doesn't choose say I_2 if $p_2 < p_1$ is modeled by a personal preference parameter $v \in [0, 1]$, such that he will incur a cost cv by insuring with I_1 and $(1 - c)v$ by insuring with I_2. In economics, such customer behavior is often referred to as *market friction*. In behavioral economics, it is further common to attribute individuals a discounting factor $d > r$, where r is the risk-free interest rate. This expresses that an individual's time horizon is shorter than what rational financial considerations would lead to.

Given p_1, p_2, v, c, d, r, the customer can now set up a calculation leading to his choice of company. Let w_0 be his initial wealth and $w_j(t)$ his wealth at time t when insuring with I_j. Choosing I_1, he will incur a one-time cost of cv. Thus $w_1(0) = w_0 - cv$, $dw_1(t) = (rw_1(t) - p_1) dt$ so that he will calculate his overall discounted wealth as

$$v_1 = \int_0^\infty e^{-dt} dw_1(t) = \frac{r(w_0 - cv) - p_1}{d - r}.$$

Choosing I_2 will result in v_2 given by the same expression with v, p_1 replaced by $1 - v$, p_2, and so I_1 will be preferred if $p_1 - p_2 < rc(1 - 2v)$, i.e. if

$$v < v_0 = v_0(p_1, p_2) = \frac{1}{2}\left(1 - \frac{p_1 - p_2}{rc}\right)$$

and I_2 otherwise

We take again a Bayesian view, this time representing the preference parameters of the N customers as i.i.d. outcomes of a continuous r.v. Z (whereas the claim rates are the same α for all customers). Thus the portfolio sizes will be

$$n_1(p_1, p_2) = N\mathbb{P}(Z < z_0), \quad n_2(p_1, p_2) = N\mathbb{P}(Z > z_0)$$

so that the diffusion coefficients of Δ become

$$\mu(x, p_1, p_2) = \mu_0(p_1, p_2) + rx \qquad \text{where}$$
$$\mu_0(p_1, p_2) = N\mathbb{P}(Z < z_0)(p_1 - \beta\mu_B) - N\mathbb{P}(Z > z_0)(p_2 - \beta\mu_B),$$
$$\sigma^2(p_1, p_2) = N\mathbb{P}(Z < z_0)\beta\mu_B^{(2)} + N\mathbb{P}(Z > z_0)\beta\mu_B^{(2)} = N\beta\mu_B^{(2)}.$$

From the fact that $\sigma^2(p_1, p_2)$ and rx do not depend on (p_1, p_2), we then get from Proposition 6.1 that:

Proposition 6.3 *The strategy with premiums p_1^*, p_2^* not dependent on the running reserve difference is a Nash equilibrium (p_1^*, p_2^*) provided*

$$\sup_{p_1} \mu_0(p_1, p_2^*) = \inf_{p_2} \mu_0(p_1^*, p_2) = \mu_0(p_1^*, p_2^*). \tag{6.5}$$

This holds if p_1^*, p_2^* satisfy the first-order conditions

$$0 = \frac{\partial}{\partial p_1}\mu_0(p_1, p_2)\Big|_{p_1=p_1^*, p_2=p_2^*} = \frac{\partial}{\partial p_2}\mu_0(p_1, p_2)\Big|_{p_1=p_1^*, p_2=p_2^*} \tag{6.6}$$

together with the second-order conditions

$$0 > \frac{\partial^2}{\partial p_1^2}\mu_0(p_1, p_2)\Big|_{p_1=p_1^*, p_2=p_2^*}, \quad 0 < \frac{\partial^2}{\partial p_2^2}\mu_0(p_1, p_2)\Big|_{p_1=p_1^*, p_2=p_2^*}. \tag{6.7}$$

Notes
The reinsurance problem considered in Sect. 6.1 is from Taksar and Zheng [170], who also consider the case where the reinsurer uses the variance principle rather than the expected value principle. Section 6.2 is from Asmussen, Christensen and Thøgersen [11], where it is also verified that the conditions of Proposition 6.3 hold in a wide class of parameters if V is beta distributed. In a companion paper [12], Stackelberg equilibria are studied in the setting of different deductibles for I_1, I_2.

Further relevant recent references on game theory in insurance include Dutang, Albrecher and Loisel [64], Emms [68], Pantelous and Passalidou [137]; more complete lists are in [11] and [12].

Chapter XII: Stochastic Control in Life Insurance

Stochastic control first appeared in life insurance mathematics as an application of the so-called linear regulator. The idea of linear regulation was part of the origin of stochastic control theory developed in the 1950s and 1960s and has frequently been applied in engineering since then. During the 1990s, the linear regulator appeared in what we could call insurance engineering. The idea is to control the surplus of a pension fund via dividends to policy holders. In Sect. VI.4, we have modeled the surplus of an insurance contract. If we aggregate over a portfolio of contracts, we can study the surplus at portfolio level. In that case, it is appealing to approximate event risk by diffusive risk, and this is exactly where we start out in Sect. 1. However, it is actually possible to work with the same ideas but keep the modelling of event risk at an individual level and this is how we continue in Sect. 2. The area of personal finance where one considers the financial decision making of an individual is related. We present in Sect. 3 the classical consumption-investment problem where stochastic control theory was originally introduced to financial economics, around 1970. In the subsequent three Sects. 4–6, we study extensions to uncertain lifetime, access to the insurance market, and generalizations to multistate risk models. Finally, in Sect. 7, we merge some of the ideas and formalize the control problem of the pension fund with the objectives of the policy holders.

We stress that the key difference between what we have done so far in the life insurance context is not the underlying modelling aspects. The key difference is the ultimate objective of the study. So far, we have concentrated on valuation aspects. In Sect. VII.8, we did model behavior of the policy holder, but even there the final objective was to properly evaluate the obligations to the policy holder, even in the presence of behavioral options. In this chapter on stochastic control in life insurance, the ultimate purpose is not the valuation. The ultimate result is a decision to make. We can think of a situation where many different decisions lead to the same value. This makes the valuation in itself insufficient as an object of study in order to choose among these different decision processes. We need to introduce further objectives

© Springer Nature Switzerland AG 2020
S. Asmussen, M. Steffensen, *Risk and Insurance*, Probability Theory
and Stochastic Modelling 96, https://doi.org/10.1007/978-3-030-35176-2_12

under which we can order the different decisions such that we can say, in a precise way, that certain decisions are preferred over other decisions, although they may cost the same.

1 The Diffusion Approximation

In this section, we go back and study the original linear regulator appearing in stochastic control. The mathematics is similar to that occurring in engineering. The wording is different since the context is pension economics. As mentioned in the introduction of this chapter, we approximate the surplus studied in Sect. VI.4 by a diffusion process.

The surplus accumulates by a stochastic process of accumulated surplus contributions, C, and capital gains from investments. From the surplus, redistributions to the policy holders in terms of accumulated dividends, D, are withdrawn. In Chap. VI, we modeled the process of dividends similarly to the underlying payment process B and the process of surplus contributions C, i.e. driven by an underlying finite-state Markov process. At portfolio level, we choose to approximate the portfolio risk by a diffusion process. So, the idea is to model the surplus contribution process by

$$\mathrm{d}C(t) \ = \ c(t)\,\mathrm{d}t + \sigma(t)\,\mathrm{d}W(t)$$

for deterministic functions c and σ. In addition to the systematic surplus contributions, the capital gains appear. In this section, we disregard stochastic capital gains and control of capital gains via investment decisions and focus on cash flow control. Thus, we assume that deterministic interest is earned on the surplus and the only control process of the pension fund is the dividend distribution, which we model as

$$\mathrm{d}D(t) \ = \ \delta(t)\,\mathrm{d}t.$$

Then the dynamics of the surplus can be written as

$$\mathrm{d}X(t) \ = \ \big(rX(t) + c(t) - \delta(t)\big)\,\mathrm{d}t + \sigma(t)\,\mathrm{d}W(t), \quad X(0) \ = \ x_0,$$

and δ is the control process.

We need to specify preferences that we can use to rank different dividend strategies. We introduce a disutility process U such that $U(t)$ contains accumulated disutility until time t. We model U as an absolutely continuous process with rate $u\big(t, \delta(t), X(t)\big)$, i.e.

$$\mathrm{d}U(t) \ = \ u\big(t, \delta(t), X(t)\big)\,\mathrm{d}t .$$

Thus, at any point in time, disutility is lost from a certain position of the dividend rate δ and a certain position of the surplus itself X. Now, the pension fund chooses a dividend strategy in order to minimize future expected accumulated disutility, i.e.

$$\inf_{\delta} \mathbb{E}\left[\int_0^n e^{-\rho t}\, dU(t)\right],$$

according to some time horizon n, possibly infinity. The discount factor $e^{-\rho t}$ relates disutility earned at different time points and makes them addable via the integral. A positive ρ means that the pension fund is more concerned with disutility earned sooner.

In the methodology called dynamic programming, the optimization problem is embedded in a continuum of problems which are then connected via a so-called Hamilton–Jacobi–Bellman (HJB) equation. For a given surplus x at time t, we consider the problem of minimizing expected disutility from time t and onwards by optimal dividend decisions over that period. We introduce the so-called *value function*

$$V(t, X(t)) = \inf_{\delta} \mathbb{E}\left[\int_t^n e^{-\rho(s-t)}\, dU(s) \,\Big|\, X(t)\right]. \tag{1.1}$$

The preferences are now modeled by specification of u. Throughout this section, u takes a quadratic form which is exactly the objective that leads to a solution in terms of a so-called linear regulation. We introduce

$$u(t, \delta, x) = \frac{1}{2}p(t)(\delta - a(t))^2 + \frac{1}{2}q(t)x^2, \tag{1.2}$$

where $p > 0$ and $q > 0$. This criterion punishes quadratic deviations of the present dividend rate from a dividend target rate a and deviations of the surplus from 0. Such a disutility criterion reflects a trade-off between policy holders preferring stability of dividends, relative to a, over non-stability, and the pension fund preferring stability of the surplus relative to 0. One can extend the objective here to include a target capital $\widehat{x}$ such that we instead punish quadratic deviation of the surplus away from $\widehat{x}$ via a term $q(t)(x - \widehat{x}(t))^2/2$. One can even introduce a cross-term in the form $\chi(t)(\delta - a'(t))(x - \widehat{x}'(t))$ for some functions a', $\widehat{x}'$, and χ. However, we choose not to go to this level of generality. One can specify preferences that fully disregard the objective of the pension fund by setting $q = 0$. The value function is then minimized to 0 via $\delta(t) = a(t)$. The deterministic functions p and q balance off these preferences for stability of dividends and surplus, respectively.

We now prove a verification theorem that characterizes the solution to the optimization problems in terms of the corresponding HJB equation. Similar to the characterization of values in Chap. VI we present only the verification theorem. For each case, a solution to the HJB equation is verified to be, indeed, equal to the value function of the optimization problem. This is an ideal sufficiency result about the value function since we consider cases only where we can characterize a solution to the HJB equation essentially explicitly. In other cases, of which we consider none, where one cannot find directly a solution to the HJB equation, one may establish the result that the value function solves the HJB equation. This necessity result about the value function typically requires differentiability and integrability conditions that are very difficult, if possible at all, to prove upfront.

Here and throughout this chapter, we

Theorem 1.1 *Assume that there exists a sufficiently integrable function* $\overline{V}(t, x)$ *such that* $\overline{V}(n, x) = 0$ *and*

$$\overline{V}_t(t, x) - \rho\overline{V}(t, x)$$
$$= \sup_{\delta} \left\{ -\overline{V}_x(t, x)(rx + c(t) - \delta) - \frac{1}{2}\overline{V}_{xx}(t, x)\sigma^2(t) - u(t, \delta, x) \right\}.$$
$$\tag{1.3}$$

Then $\overline{V}$ *is indeed the value function introduced in* (1.1) *and the optimal control is the realization of the supremum, i.e.*

$$V = \overline{V},$$
$$\delta^* = \arg\sup_{\delta} \left(\overline{V}_x(t, x)\delta - u(t, \delta, x) \right). \tag{1.4}$$

Proof Consider a specific strategy δ. Expanding

$$d\left(e^{-\rho s}\overline{V}(s, X(s))\right) = e^{-\rho s}\left(-\rho\overline{V}(s, X(s))\,ds + d\overline{V}(s, X(s))\right)$$

by Itô and integrating over (t, n), we get

$$e^{-\rho t}\overline{V}(t, X(t)) = -\int_t^n d\left(e^{-\rho s}\overline{V}(s, X(s))\right) + e^{-\rho n}\overline{V}(n, X(n))$$
$$= -\int_t^n e^{-\rho s}\left(-\rho\overline{V}(s, X(s))\,ds + \overline{V}_s(s, X(s))\,ds + \overline{V}_x(x, X(s))\,dX(s)\right.$$
$$\left. + \frac{1}{2}\overline{V}_{xx}(s, X(s))\sigma^2(s)\,ds\right) + e^{-\rho n}\overline{V}(n, X(n)).$$

First, consider an arbitrary strategy δ. For this strategy we know by (1.3) that

$$\overline{V}_t(t, X(t)) - \rho\overline{V}(t, X(t))$$
$$\geq -\overline{V}_x(t, X(t))(rX(t) + c(t) - \delta(t))$$
$$- \frac{1}{2}\overline{V}_{xx}(t, X(t))\sigma^2(t) - u(t, \delta(t), X(t)),$$

such that

$$e^{-\rho t}\overline{V}(t, X(t)) - e^{-\rho n}\overline{V}(n, X(n))$$

$$= -\int_t^n e^{-\rho s}\left[\left(-\rho\overline{V}(s, X(s)) + \overline{V}_s(s, X(s))\right)ds + \overline{V}_x(s, X(s))\,dX(s)\right.$$

$$\left. + \frac{1}{2}\overline{V}_{xx}(s, X(s))\sigma^2(s)\,ds\right]$$

$$\leq \int_t^n e^{-\rho s}\left[u(s, \delta(s), X(s))\,ds - \overline{V}_x(s, X(s))\sigma(s)\,dW(s)\right].$$

Replacing $\overline{V}(n, X(n))$ by 0 in accordance with the terminal condition, taking conditional expectation on both sides, and multiplying by $e^{\rho t}$ yields

$$\overline{V}(t, x) \leq \mathbb{E}_{t,x}\left[\int_t^n e^{-\rho(s-t)}u(s, \delta(s), X(s))\,ds\right].$$

Since the strategy was chosen arbitrarily, we have that

$$\overline{V}(t, X(t)) \leq \inf_\delta \mathbb{E}_{t,x}\left[\int_t^n e^{-\rho(s-t)}u(s, \delta(s), X(s))\,ds\right]. \tag{1.5}$$

Now we consider the specific strategy attaining the supremum of (1.3), specified in (1.4). For this strategy we have that

$$\overline{V}_t(t, X(t)) - \rho\overline{V}(t, X(t))$$

$$= -\overline{V}_x(t, X(t))\left(rX(t) + c(t) - \delta^*\right) - \frac{1}{2}\overline{V}_{xx}(t, x)\sigma^2(t) - u(t, \delta^*, X(t)),$$

such that, with the same arguments as above, we get

$$\overline{V}(t, x) = \mathbb{E}_{t,x}\left[\int_t^n e^{-\rho(s-t)}u(s, \delta^*, X(s))\,ds\right]$$

$$\geq \inf_\delta \mathbb{E}_{t,x}\left[\int_t^n e^{-\rho(s-t)}u(s, \delta(s), X(s))\,ds\right], \tag{1.6}$$

where, in the last inequality, we have used that the infimum over strategies yields a smaller conditional expectation than the specific strategy realizing the supremum in (1.3), δ^*. By combining (1.5) and (1.6), we have that

$$\overline{V}(t, x) = \inf_\delta \mathbb{E}_{t,x}\left[\int_t^n e^{-\rho(s-t)}u(s, \delta(s), X(s))\right],$$

and the infimum is obtained by the strategy realizing the supremum. $\square$

We are now going to solve the HJB equation (1.3). In our search for a function $\overline{V}$, we simply call our candidate function V. First, we calculate the so-called first-order conditions by differentiating the inner part of the supremum with respect to δ and setting to zero. We get that

$$0 = V_x(t, x) - p(t)(\delta - a(t)).$$

There are different routes to take here that all lead to the same conclusion. We choose to propose already now a candidate for the value function and calculate the optimal strategy candidates that conform with this proposal. We guess that the value function is given by

$$V(t, x) = \frac{1}{2} f(t)(x - g(t))^2 + \frac{1}{2} h(t), \tag{1.7}$$

such that the derivatives are

$$V_t(t, x) = \frac{1}{2} f'(t)(x - g(t))^2 - g'(t) f(t)(x - g(t)) + \frac{1}{2} h'(t),$$

$$V_x(t, x) = f(t)(x - g(t)), \qquad V_{xx}(t, x) = f(t).$$

The guess (1.7) is an example of a value function candidate that inherits the quadratic structure of the disutility function. Such an inherited functional structure is typically the starting point for guessing a solution. Under the assumption that our guess on the value function is correct, we can plug the derivatives of the value function into the first-order conditions to get

$$0 = f(t)(x - g(t)) - p(t)(\delta - a(t))$$

$$\Rightarrow \quad \delta(t, x) = a(t) + \frac{f(t)(x - g(t))}{p(t)}. \tag{1.8}$$

Note that when solving for δ we immediately realize that δ becomes a function of t, x and we call also this function δ. However, in the arguments above and below one has to distinguish carefully between the point δ, the function $\delta(t, x)$ and the process $\delta(t)$ which is defined by $\delta(t, X(t))$.

In order to be sure that we actually find the supremum when differentiating and setting to zero, we also take the second derivative of the inner part of the supremum with respect to δ. We find the second-order derivative to be negative, $-p(t) < 0$, to confirm that we actually are at a supremum. Now, we are ready to plug the guess on the value function, its derivatives, and the strategy we have constructed from them into the HJB equation. Omitting the argument t in $f = f(t)$ for simplicity and similarly for a, c, f', g, g', h, p, q, we get

$$\frac{1}{2}f'(x-g)^2 - g'f(x-g) + \frac{1}{2}h' - \rho\left(\frac{1}{2}f(x-g)^2 + \frac{1}{2}h\right)$$

$$= -f(x-g)\left(r(x-g) + rg + c - a - \frac{f(x-g)}{p}\right)$$

$$- \frac{1}{2}f\sigma^2 - \frac{1}{2}p\!\left(\frac{f(x-g)}{p}\right)^2 - \frac{1}{2}qx^2.$$

The side condition reads

$$\frac{1}{2}f(n)(x-g(n))^2 + \frac{1}{2}h(n) = 0,$$

from where we derive the side conditions for f and h, respectively,

$$f(n) = h(n) = 0.$$

Apart from the last term $\frac{1}{2}q(t)x^2$, we see that x always appears in terms of $(x - g(t))^2$ or $x - g(t)$. For the last term we rewrite

$$\frac{1}{2}q(t)x^2 = \frac{1}{2}q(t)(x-g(t))^2 + q(t)g(t)(x-g(t)) + \frac{1}{2}q(t)g^2(t),$$

so that we can now start collecting terms with $(x - g(t))^2$, terms with $x - g(t)$, and terms where x does not appear, respectively. Collecting terms with $(x - g(t))^2$ and finally dividing by $(x - g(t))^2$ gives

$$f'(t) = (\rho - 2r)f(t) + \frac{f^2(t)}{p(t)} - q(t), \qquad f(n) = 0.$$

This is a so-called Riccati equation which, in general, has to be solved numerically.
  Collecting terms with $x - g(t)$ and dividing by $x - g(t)$ now gives

$$g'(t) = \left(r + \frac{q(t)}{f(t)}\right)g(t) + c(t) - a(t), \qquad g(n) = 0,$$

where the side condition for g is chosen for simplicity at our discretion. This is a linear differential equation with solution

$$g(t) = \int_t^n e^{-\int_t^s (r + q(\tau)/f(\tau))\, d\tau}\left(a(s) - c(s)\right) ds.$$

Finally, collecting the residual terms where x does not appear gives

$$h'(t) = \rho h(t) - f(t)\sigma^2 - q(t)g^2(t), \qquad h(n) = 0.$$

This is also a linear differential equation with solution

$$h(t) = \int_t^n e^{-\int_t^s \rho} \big(\sigma^2 f(s) + q(s)g^2(s)\big)\,\mathrm{d}s.$$

Once we have identified f, g, and h, such that our guess solves the HJB equation, we are done. We are then typically not interested in actually calculating V but just the control δ. But this is just given by (1.8) with the function f and g plugged in. Note that we then never really have to actually calculate h. The linearity of (1.8) has given the linear regulator its name.

2 Finite-State Markov Process Linear Regulation

In Sect. 1, the dynamics of the surplus were approximated by a diffusion process and the dividends were taken to be of absolutely continuous form. In this section, we return to the more fundamental structure of the surplus and show how one can even work with linear regulation in that case. The surplus is modeled as in Sect. VI.4. As in Sect. 1, however, we disregard stochastic investment gains and focus exclusively on the dividend payout.

Thus, the surplus dynamics are given by

$$\mathrm{d}X(t) = rX(t)\,\mathrm{d}t + \mathrm{d}(C - D)(t)\,, \qquad X(0) = x_0,$$

with

$$\mathrm{d}C(t) = c^{Z(t)}(t)\,\mathrm{d}t + \sum_{k:k\neq Z(t-)} c^{Z(t-)k}(t)\,\mathrm{d}N^k(t)\,,$$

$$\mathrm{d}D(t) = \delta(t)\,\mathrm{d}t + \sum_{k:k\neq Z(t-)} \delta^k(t)\,\mathrm{d}N^k(t)\,.$$

For mathematical convenience, we disregard lump sum payments at deterministic time points. However, including such would not destroy the idea implemented here. The control processes are $\delta, \delta^0, \ldots, \delta^J$.

We now need to specify preferences such that we can rank different dividend strategies. As in the previous section, we now introduce a process U of accumulated disutilities. However, due to the structure of C and D, we allow for lump sum disutilities at the discontinuities of C and D. Thus, inheriting the structure of the involved payment processes, U is now assumed to have the dynamics

$$\mathrm{d}U(t) = u^{Z(t)}\big(t, \delta(t), X(t)\big)\,\mathrm{d}t + \sum_{k:k\neq Z(t-)} u^{Z(t-)k}\big(t, \delta^k(t), X(t-)\big)\,\mathrm{d}N^k(t).$$

Now, the pension fund chooses a dividend strategy in order to minimize future expected accumulated disutility, i.e.

$$\inf_{\delta,\delta^0,\dots,\delta^J} \mathbb{E}\left[\int_0^n e^{-\rho t}\, dU(t)\right],$$

according to some time horizon n, possibly infinity with the same interpretation of ρ as in Sect. 1. The value function becomes

$$V^{Z(t)}(t, X(t)) = \inf_{\delta,\delta^0,\dots,\delta^J} \mathbb{E}\left[\int_t^n e^{-\rho(s-t)}\, dU(s)\,\Big|\, Z(t), X(t)\right]. \tag{2.1}$$

The preferences are now modeled by specification of u^j and u^{jk}. Generalizing the quadratic disutility functions introduced in Sect. 1, we form the coefficients of U as

$$u^j(t, \delta, x) = \frac{1}{2}p^j(t)(\delta - a^j(t))^2 + \frac{1}{2}q^j(t)x^2,$$

$$u^{jk}(t, \delta^k, x) = \frac{1}{2}p^{jk}(t)(\delta^k - a^{jk}(t))^2 + \frac{1}{2}q^{jk}(t)x^2,$$

where $p^j, q^j, p^{jk}, q^{jk} > 0$. Compared to Sect. 1, there are some differences: First, there are now two coefficients corresponding to disutility rates and the lump sum disutilities upon transitions of Z. Second, for each type of dividend payment, we allow the target to be state dependent. Third, the weights that balance off preferences for stable dividend payments, relative to a^Z and a^{Zk}, towards a stable surplus, relative to zero, are also allowed to be state dependent. The comments in Sect. 1 apply in the sense that we could have a target for x different from zero and we could have cross-terms mixing preferences for dividend rates and surplus. However, for notational convenience, we disregard that level of generality.

We now prove a verification theorem that characterizes the solution to the optimization problems in terms of 'its' HJB equation.

Theorem 2.1 *Assume that there exists a sufficiently integrable function $\overline{V}^j(t, x)$ such that $\overline{V}^j(n, x) = 0$,*

$$\overline{V}_t^j(t, x) - \rho\overline{V}^j(t, x) = \sup_{\delta,\delta^0,\dots,\delta^J}\Big\{-\overline{V}_x^j(t, x)(rx + c^j(t) - \delta) - u^j(t, \delta, x)$$

$$- \sum_{k:k\neq j} \mu^{jk}(t)R^{jk}(t, \delta^k, x)\Big\}, \tag{2.2}$$

where

$$R^{jk}(t, \delta^k, x) = u^{jk}(t, \delta^k, x) + \overline{V}^k(t, x + c^{jk}(t) - \delta^k) - \overline{V}^j(t, x).$$

Then $V = \overline{V}$ and the optimal controls are given by

$$\delta^* = \arg\sup_{\delta} \left\{ \overline{V}_x^j(t, x)\delta - u^j(t, \delta, x) \right\}, \tag{2.3}$$

$$\delta^{k*} = \arg\sup_{\delta^k} \left\{ -R^{jk}(t, \delta^k, x) \right\}.$$

Proof Consider a specific strategy $\delta, \delta^k, \forall k$. Given a function $\overline{V}^j(t, x)$ we get by first expanding

$$d\big(e^{-\rho t}\overline{V}^{Z(t)}(t, X(t))\big) = e^{-\rho t}\big(-\rho\overline{V}^{Z(t)}(t, X(t))\,dt + d\overline{V}^{Z(t)}(t, X(t))\big)$$

by Itô and next integrating over (t, n) that

$$e^{-\rho t}\overline{V}^{Z(t)}(t, X(t)) - e^{-\rho n}\overline{V}^{Z(n)}(n, X(n))$$

$$= -\int_t^n e^{-\rho s}\Big(-\rho\overline{V}^{Z(t)}(s, X(s))\,ds + \overline{V}_s^{Z(t)}(s, X(s))\,ds$$

$$+ \overline{V}_x^{Z(t)}(s, X(s))\big(rX(s) + c^{Z(s)}(s) - \delta(s)\big)\,ds$$

$$+ \sum_k \Big[\overline{V}^k\big(s, X(s-) + c^{Z(s-)k}(s) - \delta^k(s)\big)$$

$$- \overline{V}^{Z(s-)}(s-, X(s-))\Big]\,dN^k(s)\Big).$$

First, consider an arbitrary strategy specified by δ and the δ^k, $k = 0, \ldots, J$. For this strategy, we know by (2.2) that

$$\overline{V}_t^{Z(t)}(t, X(t)) - \rho\overline{V}^{Z(t)}(t, X(t))$$

$$\geq -\overline{V}_x^{Z(t)}(t, X(t))\big(rX(t) + c^{Z(t)}(t) - \delta(t)\big) - u^{Z(t)}(t, \delta(t), X(t))$$

$$- \sum_{k:k\neq Z(t)} \mu^{Z(t)k}(t)R^{Z(t)k}\big(t, \delta^k(t), X(t)\big),$$

such that

$$e^{-\rho t}\overline{V}^{Z(t)}(t, X(t)) - e^{-\rho n}\overline{V}^{Z(n)}(n, X(n))$$

$$\leq \int_t^n e^{-\rho s}\Big[dU(s) - \sum_{k:k\neq Z(s-)} R^{Z(s-)k}(s, \delta^k(s), X(s-))\,dM^k(s)\Big].$$

Replacing $\overline{V}^{Z(n)}(n, X(n))$ by 0 in accordance with the terminal condition, taking expectation on both sides, and multiplying by $e^{\rho t}$ yields

$$\overline{V}^j(t, x) \le \mathbb{E}_{t,x,j}\left[\int_t^n e^{-\rho(s-t)}dU(s)\right].$$

Since the strategy was chosen arbitrarily, we have that

$$\overline{V}^j(t, x) \le \inf_{\delta,\delta^0,\dots,\delta^J} \mathbb{E}_{t,x,j}\left[\int_t^n e^{-\rho(s-t)}\,dU(s)\right]. \tag{2.4}$$

Now we consider the specific strategy attaining the supremum of (2.2), specified in (2.3). For this strategy we have that

$$\overline{V}_t^{Z(t)}(t, X(t)) - \rho\overline{V}^{Z(t)}(t, X(t))$$
$$= -\overline{V}_x^{Z(t)}(t, X(t))\big(rX(t) + c^{Z(t)}(t) - \delta(t)\big) - u^{Z(t)}\big(t, \delta(t), X(t)\big)$$
$$- \sum_{k: k \ne Z(t)} \mu^{Z(t)k}(t) R^{Z(t)k}\big(t, \delta^k(t), X(t)\big),$$

such that, with the same arguments as above, we have

$$\overline{V}^j(t, x) = \mathbb{E}_{t,x,j}\left[\int_t^n e^{-\rho(s-t)}dU(s)\right]$$
$$\ge \inf_{\delta,\delta^0,\dots,\delta^J} \mathbb{E}_{t,x,j}\left[\int_t^n e^{-\rho(s-t)}dU(s)\right], \tag{2.5}$$

where, in the last inequality, we have used that the infimum over strategies yields a smaller conditional expectation than the specific strategy attaining the supremum in (2.2). By combining (2.4) and (2.5) we have that

$$\overline{V}^j(t, x) = \inf_{\delta,\delta^0,\dots,\delta^J} \mathbb{E}_{t,x,j}\left[\int_t^n e^{-\rho(s-t)}dU(s)\right],$$

and the infimum is obtained by the strategy realizing the supremum.   □

We are now going to solve the HJB equation. As before we simply call our object of study V instead of $\overline{V}$. First we calculate the so-called first-order conditions by differentiating the inner part of the supremum with respect to δ and δ^k, respectively, and setting equal to zero. We get the first-order conditions,

$$0 = V_x^j(t, x) - p^j(t)(\delta - a^j(t)),$$
$$0 = p^{jk}(\delta^k - a^{jk}(t)) - q^{jk}(t)(x + c^{jk}(t) - \delta^k) - V_x^k\big(t, x + c^{jk}(t) - \delta^k\big).$$

We follow the route in Sect. 1 and guess that the value function inherits the quadratic structure of the disutility functions such that

$$V^j(t, x) = \frac{1}{2} f^j(t)(x - g^j(t))^2 + \frac{1}{2} h^j(t).$$

The derivatives are given by

$$V_t^j(t, x) = \frac{1}{2} f^{j\prime}(t)(x - g^j(t))^2 - g^{j\prime} f^j(t)(x - g^j(t)) + \frac{1}{2} h^{j\prime}(t),$$

$$V_x^j(t, x) = f^j(t)(x - g^j(t)), \qquad V_{xx}^j(t, x) = f^j(t).$$

Under the assumption that our guess on the value function is correct, we can plug the derivatives of the value function into the first-order conditions to get

$$0 = f^j(t)(x - g^j(t)) - p^j(t)(\delta - a^j(t))$$

$$\Rightarrow \delta^j(t, x) = a^j(t) + \frac{f^j(t)(x - g^j(t))}{p^j(t)},$$

$$0 = p^{jk}(t)(\delta^k - a^{jk}(t)) - q^{jk}(t)(x + c^{jk}(t) - \delta^k)$$

$$\qquad - f^k(t)\left(x + c^{jk}(t) - \delta^k - g^k(t)\right)$$

$$\Rightarrow \delta^{jk}(t, x) = \frac{p^{jk}}{S^{jk}(t)} a^{jk}(t) + \frac{q^{jk}(t)}{S^{jk}(t)}(x + c^{jk}(t))$$

$$\qquad + \frac{f^k(t)}{S^{jk}(t)}\left(x + c^{jk}(t) - g^k(t)\right),$$

with

$$S^{jk}(t) = p^{jk}(t) + q^{jk}(t) + f^k(t).$$

Now, we are ready to plug the guess on the value function, its derivatives, and the strategy we have constructed from them into the HJB equation, giving that

$$\frac{1}{2} f^{j\prime}(t)\left(x - g^j(t)\right)^2 - g^{j\prime} f^j(t)(x - g^j(t)) + \frac{1}{2} h^{j\prime}(t)$$

$$\qquad - \rho\left(\frac{1}{2} f^j(t)(x - g^j(t))^2 + \frac{1}{2} h^j(t)\right)$$

equals

$$- f^j(t)(x - g^j(t))\left(r(x - g^j(t)) + rg^j(t) + c^j(t) - a^j(t) - \frac{f^j(t)(x - g^j(t))}{p^j(t)}\right)$$

$$\qquad - \frac{1}{2} p^j(t)\left(\frac{f^j(t)\left(x - g^j(t)\right)}{p^j(t)}\right)^2 - \frac{1}{2} q^j(t)x^2$$

$$- \sum_{k:k \neq j} \mu^{jk}(t) \left[\frac{1}{2} p^{jk}(t) \left(\delta^k - a^{jk}(t) \right)^2 + \frac{1}{2} q^{jk}(t) x^2 \right.$$

$$\left. + \frac{1}{2} f^k(t) \left(x - g^k(t) \right)^2 + \frac{1}{2} h^k(t) - \left(\frac{1}{2} f^j(t)(x - g^j(t))^2 + \frac{1}{2} h^j(t) \right) \right].$$

The boundary condition reads

$$\frac{1}{2} f^j(n)(x - g^j(n))^2 + \frac{1}{2} h^j(n) = 0,$$

from where we derive the boundary conditions for f and h, respectively,

$$f^j(n) = h^j(n) = 0.$$

We can now do some extensive rewriting in order to get ODEs for f and g, respectively. For f we get $f(n) = 0$ and

$$f^{j\prime}(t) = (\rho - 2r) f^j(t) + \frac{f^j(t)^2}{p^j(t)} - q^j(t) - \sum_{k:k \neq j} \mu^{jk}(t) R^{f;jk}(t)$$

with

$$R^{f;jk}(t) = \left(\frac{p^{jk}(t)}{S^{jk}(t)} \left(q^{jk}(t) + f^k(t) \right) - f^j(t) \right).$$

This is a J-dimensional Riccati equation.

For g we get

$$g^{j\prime}(t) = \left(r + \frac{q^j(t)}{f^j(t)} \right) g^j(t) + c^j(t) - a^j(t) - \sum_{k:k \neq j} \mu^{jk}(t) R^{g;jk}(t),$$

with

$$R^{g;jk}(t) = \frac{p^{jk}(t)}{S^{jk}(t)} \left[\frac{q^{jk}(t) + f^k(t)}{f^j(t)} \left(a^{jk}(t) - c^{jk}(t) - g^j(t) \right) + \frac{f^k(t)}{f^j(t)} g^k(t) \right].$$

Since h does not appear in the control, we disregard further specification of h here.

Sections 1 and 2 present the idea of linear regulation of pension funds based on preferences for stability expressed through quadratic disutility. Section 1 handled the diffusion approximation which was the starting point for the early development of stochastic control theory in general and applications of linear regulation in engineering in particular. In Sect. 2 we showed that the fundamental idea of linear regulation holds far beyond the diffusion model, including a situation where the lifetime uncertainty is modeled on individual policy level.

3   The Consumption-Investment Problem

In the preceding sections, we worked with the linear regulation which is an engineering approach to pension fund decision making. We now take a completely different point of view and a completely new family of preferences. We are going to work with maximization of utility for an individual who consumes and invests in a financial market. We start out with the so-called Merton's problem, which has become the starting point for generalizations in many different directions since its introduction around 1970. It is also the starting point for us in the subsequent sections introducing lifetime uncertainty to the model. We consider an individual investing his wealth in a financial market and earning income at the labor market in order to finance consumption during his course of life. We assume that the individual has access to the Black–Scholes market with price dynamics given by

$$dB(t) = r B(t)\, dt\,, \quad B(0) = 1,$$

$$dS(t) = \alpha S(t)\, dt + \sigma S(t)\, dW(t)\,, \quad S(0) = s_0.$$

The individual earns income at rate a and consumes at rate c over a time horizon n. If the individual invests the proportion π of his wealth X in stocks, we can also say that he holds the amount πX in stocks, he owns the number $\frac{\pi X}{S}$ of stocks, and he gains $\frac{\pi X}{S}\, dS$ from his stock investments. Further, he invests the proportion $1 - \pi$ of X in bonds, i.e. he holds the amount $(1 - \pi) X$ in bonds, he owns the number $\frac{(1-\pi)X}{B}$ of bonds, and he gains $\frac{(1-\pi)X}{B}\, dB$ from his bond investment. Note that, since the interest rate is deterministic, a bond position is equivalent to holding money in the bank account. All in all, the dynamics of his wealth follows the following SDE,

$$
\begin{aligned}
dX(t) &= \pi(t)X(t)\frac{dS(t)}{S(t)} + (1 - \pi(t))X(t)\frac{dB(t)}{B(t)} + a(t)dt - c(t)\, dt \\
&= \big(r + \pi(t)(\alpha - r)\big)X(t)\, dt + \pi(t)\sigma X(t)\, dW(t) \tag{3.1} \\
&\quad + a(t)\, dt - c(t)\, dt. \tag{3.2}
\end{aligned}
$$

This differential equation decomposes the wealth dynamics into one part (3.1) containing all capital gains, and another part (3.2) containing the cash flow in and out of X. We assume that the initial wealth of the individual is x_0.

The objective of the individual is to generate as much utility as possible from consumption. Since the development of the wealth is uncertain due to uncertain capital gains, we consider the utility of consumption that the individual expects to obtain. We assume that the investor's preferences can be modeled via the power function such that utility is measured via

$$u(\cdot) = \frac{1}{1 - \gamma}(\cdot)^{1-\gamma}.$$

For utility of consumption c we have the coefficient of relative risk aversion given by

$$-\frac{cu''(c)}{u'(c)} = \gamma \frac{cc^{-\gamma-1}}{c^{-\gamma}} = \gamma,$$

and we speak, simply, of γ as the risk aversion. The power utility function essentially defines the class of utility functions where the coefficient of relative risk aversion is constant, also called CRRA utility (Constant Relative Risk Aversion, cf. Sect. I.2). We can now write the expected future utility in the following way:

$$\mathbb{E}\left[\int_0^n e^{-\rho t} w^{\gamma}(t) \frac{1}{1-\gamma} c^{1-\gamma}(t)\, dt + e^{-\rho n} \Delta W^{\gamma}(n) \frac{1}{1-\gamma} X^{1-\gamma}(n)\right]. \qquad (3.3)$$

The first term measures utility from consumption over $(0, n)$ whereas the last term measures utility from saving $X(n)$ until time n. This may be saved for the individual himself to generate utility of consumption at or after time n or for someone else to generate utility of wealth at or after time n. Utility of consumption $u(c(t))$ at time t is weighted with a utility discount factor $e^{-\rho t}$ which is a personal time preference for consuming now rather than later (for positive ρ). The parameter ρ is also spoken of as the impatience rate. Correspondingly, the utility of terminal wealth is discounted by $e^{-\rho n}$. Further, time preferences are allowed through the weight functions w and ΔW which are, without loss of generality, taken to the power γ. One could, of course, integrate the utility discount factor into the time weights w and ΔW but we choose to keep it there to be able to immediately spell out the solution to the canonical cases where w or ΔW are 0 or 1, respectively. Though canonical, the weights play important roles when introducing life event risk later but we choose to introduce them already here to prepare the reader for the type of solutions that arise from them.

The optimization problem is formalized by the dynamics in (3.1), (3.2) and the objective

$$\sup_{\pi,c} \mathbb{E}\left[\int_0^n e^{-\rho t} w^{\gamma}(t) \frac{1}{1-\gamma} c^{1-\gamma}(t)\, dt + e^{-\rho n} \Delta W^{\gamma}(n) \frac{1}{1-\gamma} X^{1-\gamma}(n)\right].$$

In the methodology called dynamic programming, the optimization problem is embedded in a continuum of problems which are then connected through the Hamilton–Jacobi–Bellman (HJB) equation. For a given wealth x at time t, we consider the problem of generating utility from time t and onwards by making optimal investment and consumption decisions over that period. We introduce the value function

$$V(t, x) = \sup_{\pi,c} \mathbb{E}_{t,x}\left[\int_t^n e^{-\rho(s-t)} w^{\gamma}(s) \frac{1}{1-\gamma} c^{1-\gamma}(s)\, ds\right.$$

$$\left. + e^{-\rho(n-t)} \Delta W^{\gamma}(n) \frac{1}{1-\gamma} X^{1-\gamma}(n)\right]. \qquad (3.4)$$

We now prove a verification theorem that characterizes the solution to the optimization problem in terms of 'its' HJB equation.

Theorem 3.1 *Assume that there exists a sufficiently integrable function* $\overline{V}(t, x)$ *such that*

$$\overline{V}_t(t, x) - \rho \overline{V}(t, x) = \inf_{\pi, c} \left\{ -\overline{V}_x(t, x)[(r + \pi(\alpha - r))x + a(t) - c] \right. \tag{3.5}$$

$$\left. -\frac{1}{2}\overline{V}_{xx}(t, x)\pi^2\sigma^2 x^2 - w^\gamma(t)\frac{1}{1-\gamma}c^{1-\gamma} \right\},$$

$$\overline{V}(n, x) = \Delta W^\gamma(n)\frac{1}{1-\gamma}x^{1-\gamma}.$$

Then $V = \overline{V}$ *and the optimal controls are given by*

$$c^* = \arg\inf_c \left\{ \overline{V}_x(t, x)c - w^\gamma(t)\frac{1}{1-\gamma}c^{1-\gamma} \right\}, \tag{3.6}$$

$$\pi^* = \arg\inf_\pi \left\{ -\overline{V}_x(t, x)\pi(\alpha - r)x - \frac{1}{2}\overline{V}_{xx}(t, x)\pi^2\sigma^2 x^2 \right\}.$$

Proof Consider a specific strategy (π, c). Given a function $\overline{V}(t, x)$, we have by Itô that

$$d\left(e^{-\rho t}\overline{V}(t, X(t))\right) = e^{-\rho t}\left(-\rho\overline{V}(t, X(t))\, dt + d\overline{V}(t, X(t))\right)$$

$$= e^{-\rho t}\left(-\rho\overline{V}(t, X(t))\, dt + \overline{V}_t(t, X(t))\, dt + \overline{V}_x(t, X(t))\, dX(t)\right.$$

$$\left. + \frac{1}{2}\overline{V}_{xx}(t, X(t))\pi^2(t)\sigma^2 X^2(t)\, dt\right).$$

By integrating over (t, n) we get

$$e^{-\rho t}\overline{V}(t, X(t)) = -\int_t^n d\left(e^{-\rho s}\overline{V}(s, X(s))\right) + e^{-\rho n}\overline{V}(n, X(n))$$

$$= -\int_t^n e^{-\rho s}\left(-\rho\overline{V}(s, X(s))\, ds + \overline{V}_s(s, X(s))\, ds + \overline{V}_x(s, X(s))\, dX(s)\right.$$

$$\left. + \frac{1}{2}\overline{V}_{xx}(s, X(s))\pi^2(s)\sigma^2 X^2(s)\, ds\right) + e^{-\rho n}\overline{V}(n, X(n)).$$

First, consider an arbitrary strategy (π, c). For this strategy we know by (3.5) that

$$\overline{V}_t(t, X(t)) - \rho\overline{V}(t, X(t))$$

$$\leq -\overline{V}_x(t, X(t))\left[(r + \pi(t)(\alpha - r))X(t) + a(t) - c(t)\right]$$

$$-\frac{1}{2}\overline{V}_{xx}(t, x)\pi^2(t)\sigma^2 X(t)^2 - w^\gamma(t)\frac{1}{1-\gamma}c^{1-\gamma}(t),$$

so that

$$e^{-\rho t}\overline{V}(t, X(t))$$

$$= -\int_t^n e^{-\rho s}\left[-\rho\overline{V}(s, X(s))\,ds + \overline{V}_s(s, X(s))\,ds + \overline{V}_x(s, X(s))dX(s)\right.$$

$$\left. +\frac{1}{2}\overline{V}_{xx}(s, X(s))\pi^2(s)\sigma^2 X^2(s)\,ds\right] + e^{-\rho n}\overline{V}(n, X(n))$$

$$\geq \int_t^n e^{-\rho s}\left[w^\gamma(s)\frac{1}{1-\gamma}c^{1-\gamma}(s)\,ds - \overline{V}_x s, X(s))\pi(s)X(s)\sigma\,dW(s)\right]$$

$$+ e^{-\rho n}\overline{V}(n, X(n)).$$

Replacing $\overline{V}(n, X(n))$ by $\Delta W^\gamma(n)\frac{1}{1-\gamma}X(n)^{1-\gamma}$ in accordance with the terminal condition, taking the expectation on both sides, and multiplying by $e^{\rho t}$ yields

$$\overline{V}(t, x) \geq \mathbb{E}_{t,x}\left[\int_t^n e^{-\rho(s-t)}w^\gamma(s)\frac{1}{1-\gamma}c^{1-\gamma}(s)\,ds\right.$$

$$\left. + e^{-\rho(n-t)}\Delta W^\gamma(n)\frac{1}{1-\gamma}X(n)^{1-\gamma}\right].$$

Since the strategy was chosen arbitrarily, we have that

$$\overline{V}(t, x) \geq \sup_{\pi,c}\mathbb{E}_{t,x}\left[\int_t^n e^{-\rho(s-t)}w^\gamma(s)\frac{1}{1-\gamma}c^{1-\gamma}(s)\,ds\right.$$

$$\left. + e^{-\rho(n-t)}\Delta W^\gamma(n)\frac{1}{1-\gamma}X(n)^{1-\gamma}\right]. \qquad (3.7)$$

Now we consider the specific strategy realizing the infimum of (3.5), specified in (3.6). For this strategy we have that

$$\overline{V}_t(t, X(t)) - \rho\overline{V}(t, X(t))$$

$$= -\overline{V}_x(t, X(t))\left[(r + \pi(\alpha - r))X(t) + a(t) - c(t)\right]$$

$$- \frac{1}{2}\overline{V}_{xx}(t, x)\pi^2(t)\sigma^2 X(t)^2 - w^\gamma(t)\frac{1}{1-\gamma}c^{1-\gamma}(t),$$

so that, with the same arguments as above, we have

$$\overline{V}(t, x) = \mathbb{E}_{t,x}\left[\int_t^n e^{-\rho(s-t)}w^\gamma(s)\frac{1}{1-\gamma}c^{1-\gamma}(s)\,ds\right.$$

$$\left. + e^{-\rho(n-t)}\Delta W^\gamma(n)\frac{1}{1-\gamma}X(n)^{1-\gamma}\right]$$

$$\leq \sup_{\pi,c} \mathbb{E}_{t,x} \left[\int_t^n e^{-\rho(s-t)} w^\gamma(s) \frac{1}{1-\gamma} c^{1-\gamma}(s) \, ds \right.$$

$$\left. + e^{-\rho(n-t)} \Delta W^\gamma(n) \frac{1}{1-\gamma} X(n)^{1-\gamma} \right], \qquad (3.8)$$

where, in the last inequality, we have used that the supremum over strategies yields a higher conditional expectation than the specific strategy realizing the infimum in (3.5). By combining (3.7) and (3.8) we have that

$$\overline{V}(t,x) = \sup_{\pi,c} \mathbb{E}_{t,x} \left[\int_t^n e^{-\rho(s-t)} w^\gamma(s) \frac{1}{1-\gamma} c^{1-\gamma}(s) \, ds \right.$$

$$\left. + e^{-\rho(n-t)} \Delta W^\gamma(n) \frac{1}{1-\gamma} X(n)^{1-\gamma} \right],$$

and the supremum is obtained by the strategy realizing the infimum.   □

We are now going to solve the HJB equation. As before, we simply call our object of study V instead of $\overline{V}$. First, we calculate the first-order conditions by differentiating the inner part of the infimum with respect to c and π, respectively, and setting to zero. We get that

$$0 = V_x(t,x) - w^\gamma(t) c^{-\gamma},$$

$$0 = -V_x(t,x)(\alpha - r)x - V_{xx}(t,x) \pi \sigma^2 x^2.$$

There are different routes to take here that all lead to the same conclusion. We choose to propose already now a candidate for the value function and calculate the optimal strategy candidates that conform with this proposal. We guess that the value function is given by

$$V(t,x) = \frac{1}{1-\gamma} f(t)^\gamma (x + g(t))^{1-\gamma}, \qquad (3.9)$$

so that the derivatives are given by

$$V_t(t,x) = \frac{\gamma}{1-\gamma} \left(\frac{x+g(t)}{f(t)} \right)^{1-\gamma} f'(t) + \left(\frac{x+g(t)}{f(t)} \right)^{-\gamma} g'(t),$$

$$V_x(t,x) = \left(\frac{x+g(t)}{f(t)} \right)^{-\gamma}, \qquad V_{xx}(t,x) = -\gamma \left(\frac{x+g(t)}{f(t)} \right)^{-\gamma-1} \frac{1}{f(t)}.$$

Under the assumption that our guess on the value function is correct, we can plug the derivatives of the value function into the first-order conditions to get

$$0 = \left(\frac{x + g(t)}{f(t)}\right)^{-\gamma} - w^{\gamma}(t)c^{-\gamma} \Rightarrow c = w(t)\frac{x + g(t)}{f(t)}, \tag{3.10}$$

$$0 = -\left(\frac{x + g(t)}{f(t)}\right)^{-\gamma}(\alpha - r)x + \gamma\left(\frac{x + g(t)}{f(t)}\right)^{-\gamma-1}\frac{1}{f(t)}\pi\sigma^2 x^2$$

$$\Rightarrow \pi = \frac{1}{\gamma}\frac{\alpha - r}{\sigma^2}\frac{x + g(t)}{x}. \tag{3.11}$$

In order to be sure that we are actually finding the infimum when differentiating and setting to zero, we also take the second derivative of the inner part of the infimum with respect to c and π. We find that the second-order derivatives are positive,

$$w^{\gamma}(t)\gamma c^{-\gamma-1} > 0, \qquad \gamma\left(\frac{x + g(t)}{f(t)}\right)^{-\gamma-1}\frac{1}{f(t)}\sigma^2 x^2 > 0,$$

so that we are actually at an infimum.

Now, we are ready to plug the guess on the value function, its derivatives, and the strategy we have constructed from them into the HJB equation. After multiplication by $\left(\frac{x+g(t)}{f(t)}\right)^{\gamma}$ and addition and subtraction of $rg(t)$ on the right-hand side, we get, suppressing here the dependence of a, f, g, and w on t,

$$\frac{\gamma}{1 - \gamma}\frac{x + g}{f}f' + g' - \rho\frac{1}{1 - \gamma}\frac{x + g}{f}f$$

$$= -r\frac{x + g}{f}f + rg - a - \frac{1}{\gamma}\left(\frac{\alpha - r}{\sigma}\right)^2\frac{x + g}{f}f + w\frac{x + g}{f}$$

$$+ \frac{1}{2}\frac{1}{\gamma}\left(\frac{\alpha - r}{\sigma}\right)^2\frac{x + g}{f}f - w\frac{1}{1 - \gamma}\frac{x + g}{f}. \tag{3.12}$$

The boundary condition reads

$$\frac{1}{1 - \gamma}f(n)^{\gamma}(x + g(n))^{1-\gamma} = \Delta W^{\gamma}(n)\frac{1}{1 - \gamma}x^{1-\gamma}, \tag{3.13}$$

from where we derive the boundary conditions for f and g, respectively,

$$f(n) = \Delta W(n), \quad g(n) = 0.$$

We can now collect all terms in (3.12) without $\frac{x+g}{f}$ to form a differential equation for g with boundary condition,

$$g'(t) = rg(t) - a(t), \qquad g(n) = 0. \tag{3.14}$$

This linear ODE has the solution

$$g(t) = \int_t^n e^{-\int_t^s r} a(s)\, ds \,. \tag{3.15}$$

After getting rid of all terms without $\frac{x+g}{f}$ by the differential equation (3.14), we divide the rest of (3.12) by $\frac{\gamma}{1-\gamma} \frac{x+g}{f}$ to get the following ODE for f with boundary condition,

$$f'(t) = \tilde{r} f(t) - w(t), \qquad f(n) = \Delta W, \tag{3.16}$$

where we have introduced

$$\tilde{r} = \frac{1}{\gamma}\rho + \left(1 - \frac{1}{\gamma}\right)r + \left(1 - \frac{1}{\gamma}\right)\frac{1}{2}\frac{1}{\gamma}\theta^2, \tag{3.17}$$

reminding the reader that θ equals $(\alpha - r)/\sigma$. Note that $\tilde{r}$ is a weighted average of ρ and r with weights $\frac{1}{\gamma}$ and $1 - \frac{1}{\gamma}$ respectively, plus a term coming from access to the stock market, $\left(1 - \frac{1}{\gamma}\right)\frac{1}{2}\frac{1}{\gamma}\theta^2$. This linear ODE for f has the solution

$$f(t) = \int_t^n e^{-\int_t^s \tilde{r}} w(s)\, ds + e^{-\int_t^n \tilde{r}} \Delta W(n)\,. \tag{3.18}$$

Note that g equals the present value of future labor income. Similar expressions play a prominent role when we consider generalized versions of the problem studied later. It is spoken of as the human wealth in contrast to X, which is the financial wealth. The notion of human wealth used in personal finance is different from the same phrase used in human resource management in business economics. The total wealth $X(t) + g(t)$ is the total assets of the individual, including his future labor earnings.

We have now completed our search for an optimal solution since this is fully uncovered by the functions f and g. However, we can gain further insight by studying the dynamics of the optimally controlled process X and the dynamics of the optimal controls c and π. The dynamics of the total wealth is given by

$$
\begin{aligned}
d\big(X(t) &+ g(t)\big) \\
&= \big(r + \pi(t)(\alpha - r)\big)X(t)\, dt + \pi(t)\sigma X(t)\, dW(t) \\
&\quad + a(t)\, dt - c(t)\, dt + rg(t)\, dt - a(t)\, dt \\
&= \left(r + \frac{1}{\gamma}\theta^2 - \frac{w(t)}{f(t)}\right)\big(X(t) + g(t)\big)\, dt + \frac{1}{\gamma}\theta\big(X(t) + g(t)\big)\, dW(t)\,.
\end{aligned}
$$

Thus, we see that the total wealth process $X + g$ is a geometric Brownian motion with starting point $X(0) + g(0)$. Therefore, if $X(0) + g(0) > 0$ then $X + g > 0$, meaning that we know that $X > -g$. We do not know that $X > 0$, which is sometimes introduced as an extra constraint spoken of as a no borrowing constraint. With $X > -g$ we do actually allow the individual to borrow against future labor income. Allowing for borrowing against future labor income is a natural problem formulation that gives access to the explicit solutions above. Borrowing constraints considerably complicates the problem and one has to resort to, in general, numerical optimization algorithms.

We are also going to study the dynamics of the strategies for c and π in the case where w is constant. The dynamics of c are given by

$$
\begin{aligned}
\mathrm{d}c(t, X(t)) &= \frac{f(t)\mathrm{d}\big(X(t) + g(t)\big) - \big(X(t) + g(t)\big)\mathrm{d}f(t)}{f^2(t)}\,w \\
&= \frac{w}{f^2(t)}\bigg[\Big(r + \frac{1}{\gamma}\theta^2 - \frac{w}{f(t)}\Big)\big(X(t) + g(t)\big)\,\mathrm{d}t f(t) \\
&\quad + \frac{1}{\gamma}\theta\big(X(t) + g(t)\big)\,\mathrm{d}W(t)f(t) - \big(X(t) + g(t)\big)\big(\tilde{r}f(t) - w(t)\big)\,\mathrm{d}t\bigg] \\
&= \Big(r - \tilde{r} + \frac{1}{\gamma}\theta^2\Big)c(t, X(t))\,\mathrm{d}t + \frac{1}{\gamma}\theta c(t, X(t))\,\mathrm{d}W(t). \quad (3.19)
\end{aligned}
$$

Plugging in $\tilde{r}$ we can write the drift coefficient as

$$
\frac{r - \rho}{\gamma} + \frac{1}{2\gamma}\Big(\frac{1}{\gamma} + 1\Big)\theta^2. \quad (3.20)
$$

We see that the growth rate of the consumption rate consists of $\frac{r-\rho}{\gamma}$, the excess of the interest rate r over the subjective utility discount rate ρ weighted with $\frac{1}{\gamma}$, plus a term coming from access to the stock market, $\frac{1}{2\gamma}\big(\frac{1}{\gamma} + 1\big)\theta^2$. Note that in the special case $\rho = r$, the growth rate of consumption comes from risky investments only.

Finally, we also derive the dynamics of the stock proportion

$$
\begin{aligned}
&\mathrm{d}\pi(t, X(t)) \\
&= \frac{\alpha - r}{\gamma\sigma^2}\bigg(\frac{X(t)\big(\mathrm{d}X(t) + \mathrm{d}g(t)\big) - \big(X(t) + g(t)\big)\mathrm{d}X(t)}{X^2(t)} + \frac{g(t)}{X^3(t)}\big(\mathrm{d}X(t)\big)^2\bigg) \\
&= \frac{\alpha - r}{\gamma\sigma^2 X^2(t)}\bigg(-X(t)a(t)\,\mathrm{d}t - g(t)a(t)\mathrm{d}t + g(t)w(t)\frac{X(t) + g(t)}{f(t)}\,\mathrm{d}t \\
&\quad - g(t)\pi(t)(\alpha - r)X(t)\,\mathrm{d}t - g(t)\pi(t)\sigma X(t)\,\mathrm{d}W(t)\bigg) \\
&\quad + \frac{\alpha - r}{\gamma\sigma^2 X(t)}g(t)\pi^2(t)\sigma^2\,\mathrm{d}t.
\end{aligned}
$$

After this follows a series of rearrangements where we make extensive use of the relations

$$\frac{X(t) + g(t)}{X(t)} = 1 + \frac{g(t)}{X(t)} = \pi(t, X(t))\gamma \frac{\sigma^2}{\alpha - r},$$

such that, e.g.,

$$\frac{X(t) + g(t)}{X(t)} \frac{g(t)}{X(t)} = \pi^2(t, X(t))\gamma^2 \frac{\sigma^4}{(\alpha - r)^2} - \pi(t, X(t))\gamma \frac{\sigma^2}{\alpha - r},$$

and

$$\frac{1}{\gamma} \frac{\alpha - r}{\sigma^2} \frac{g(t)}{X(t)} \pi^2(t, X(t))\sigma^2 = \pi^3(t, X(t))\sigma^2 - \frac{1}{\gamma}(\alpha - r)\pi^2(t, X(t)).$$

Finally, we arrive at

$$\begin{aligned}
d\pi(t) = &\left[\left(\frac{w(t)}{f(t)} - \frac{a(t)}{g(t)} \right) \left(\gamma \frac{\sigma^2 \pi(t)}{\alpha - r} - 1 \right) + \left(\frac{\alpha - r}{\sigma} \right)^2 \frac{1}{\gamma} \right. \\
&\left. - \left(1 + \frac{1}{\gamma} \right) (\alpha - r)\pi(t) + \pi^2(t)\sigma^2 \right] \pi(t) \, dt \\
&+ \left(\frac{\alpha - r}{\sigma^2} \frac{1}{\gamma} - \pi(t) \right) \pi(t)\sigma \, dW(t).
\end{aligned}$$

This can be studied to understand the profile of π.

4 Uncertain Lifetime

In this section, we generalize the consumption-investment problem to a situation with uncertain lifetime. We build up the section so that it progressively takes into account the ideas one might naturally get, even when they lead to dead ends. First, we think of the dynamics of wealth as unchanged and we represent the dynamics of wealth until death of the individual. These dynamics are

$$dX(t) = \left(r + \pi(t)(\alpha - r)\right)X(t) \, dt + \pi(t)\sigma X(t) \, dW(t) + a(t) \, dt - c(t) \, dt. \quad (4.1)$$

We now generalize the objective function of the individual. Since utility is obtained from consumption only as long as the individual is alive, it appears natural to multiply the utility of the consumption rate and the utility of terminal wealth by the survival indicator

$$V(t, X(t)) = \sup_{\pi, c} \mathbb{E}\left[\int_t^n e^{-\int_t^s \rho} w^\gamma(s) I(s) u(c(s))\, ds\right.$$
$$\left. + e^{-\int_t^n \rho} \Delta W^\gamma(n) I(n) U(X(n)) \,\middle|\, X(t)\right].$$

We then tacitly define this value function as a conditional expectation also conditioning upon survival until time t. An important question arises: What happens to the financial wealth upon death? If also upon death, utility is gained from wealth, we can generalize the value function to

$$V(t, X(t))$$
$$= \sup_{\pi, c} \mathbb{E}\left[\int_t^n e^{-\int_t^s \rho} \left[w^\gamma(s) I(s) u(c(s))\, ds + w^{01}(s)^\gamma u^{01}(X(s-))\, dN(s)\right]\right.$$
$$\left. + e^{-\int_t^n \rho} \Delta W^\gamma(n) I(n) U(X(n)) \,\middle|\, X(t)\right].$$

The appearance of $X(s-)$ in the utility function upon death reflects that the wealth is consumed upon death. More realistically, the wealth is consumed continuously by the dependants after the death of the individual and the term represents the indirect utility that the dependants gain from that. The consumption of wealth upon death does not appear in the wealth dynamics (4.1), which represents the dynamics during survival only. Upon death, the wealth is considered as consumed or reallocated to future consumption by the dependants and we therefore set it to zero where it remains after the death. So a more complete description of the wealth dynamics is then

$$dX(t) = I(t)\left[(r + \pi(t)(\alpha - r))X(t)\, dt + \pi(t)\sigma X(t)\, dW(t)\right.$$
$$\left. + a(t)\, dt - c(t)\, dt\right] - X(t-)\, dN(t).$$

If we make the usual assumption that the lifetime distribution is independent of the strategy, we can take plain expectation over the lifetime distribution, leaving the expectation operator to work over diffusive capital gains risk only

$$V(t, X(t)) = \sup_{\pi, c} \mathbb{E}\left[\int_t^n e^{-\int_t^s \rho + \mu} \left\{w^\gamma(s) u(c(s)) + \mu(s) w^{01}(s)^\gamma u^{01}(X(s))\right\} ds\right.$$
$$\left. + e^{-\int_t^n \rho + \mu} \Delta W^\gamma(n) U(X(n)) \,\middle|\, X(t)\right]. \qquad (4.2)$$

The HJB equation related to this objective reads, here written in terms of V itself rather than the candidate function $\overline{V}$,

$$V_t(t, x) - \rho V(t, x)$$

$$= \inf_{\pi, c} \Big\{ - V_x(t, x)\big[(r + \pi(\alpha - r))x + a(t) - c\big] - w^\gamma(t)u(c)$$

$$- \frac{1}{2}V_{xx}(t, x)\pi^2\sigma^2 x^2 - \mu(t)\big(w^{01}(t)^\gamma u^{01}(x) - V(t, x)\big)\Big\},$$

$$V(n, x) = \Delta W^\gamma(n)U(x).$$

We are not going to verify this HJB equation. Before proceeding any further, let us note that the term $\mu(t)V(t, x)$ on the right-hand side can be collected with the term $\rho V(t, x)$ on the l.h.s. Thus, this particular term is taken care of by replacing ρ by $\rho + \mu$ in the calculations of the consumption-investment problem without lifetime uncertainty. Then, the main difference compared to (3.5) is the term $\mu(t)w^{01}(t)^\gamma u^{01}(x)$. We now consider the power utility case where all three utility functions are of power form with the same risk aversion, i.e.

$$u(\cdot) = u^{01}(\cdot) = U(\cdot) = \frac{1}{1 - \gamma}(\cdot)^{1-\gamma}.$$

The consequences of the extra term $\mu(t)w^{01}(t)^\gamma u^{01}(x)$ compared with the previous section are simple because the control process (c, π) does not appear in the term and therefore does not influence the first-order conditions. Thus, if we again assume that the value function takes the form (3.9), we can jump directly to (3.12) to learn about the consequences. Recall that to obtain (3.12) we had multiplied by $\big(\frac{x+g(t)}{f(t)}\big)^\gamma$ and we therefore, compared to (3.12), get the additional term

$$\mu(t)w^{01}(t)^\gamma \frac{1}{1 - \gamma}x^{1-\gamma}\Big(\frac{x + g(t)}{f(t)}\Big)^\gamma$$

on the r.h.s. of (3.12). This term does not fit with any other terms in (3.12) and the value function is therefore not correct. Only if we say that utility upon death is gained from $x + g$ is there a way out of this mathematical challenge since we would then get $x + g$ to the power of 1 so that it fits into the ODE for f. That ODE would then become non-linear, though, but the value function would have the correct structure. Anyway, it makes little sense to base the utility on g, an ingredient in the value function itself.

The special case $w^{01} = 0$ deserves special attention since then the calculations seem to go well. In that case, all conclusions from the consumption-investment problem are valid, except that ρ is replaced by $\rho + \mu$. The main problem with this version of the problem is that even g remains the same,

$$g(t) = \int_t^n e^{-\int_t^s r}a(s)\,ds.$$

Since $X + g$ behaves as a geometric Brownian motion until X jumps to zero upon death, we know that $X > -g$, which means that we need access to borrowing against g. However, g is an overly optimistic valuation of future income, since it is conditional upon survival until time n. It makes little sense that the financial market should offer to borrow against such an optimistic measure of human wealth. It is therefore not really a constructive approach. Note however, that in the case where $a = g = 0$, the solution is fully valid.

In the case $w^{01} = 0$, there is no concern about the financial wealth X upon death in the formalization above. We could, in sharp contrast, assume that the individual is ready to give up X to the financial institution upon death if no utility is gained from it. The financial institution would then pay a premium for that as long as the individual is alive. For a pricing mortality rate μ^*, the financial institution is ready to pay

$$\mu^*(t)X(t)\,dt,$$

over a small time interval $(t, t + dt)$ for the chance of receiving $X(t)$ in case of death.

In this chapter there are no dividends linked to a surplus. In that sense, the product design here follows that of Chap. V rather than that of Chap. VI. The notation of the pricing mortality rate μ^* reflects that the insurance company may price mortality risk differently than by the objective mortality rate μ. The difference between the two mortality rates μ and μ^* produces a surplus like the one studied in Chap. VI, though. But here, this surplus is not repaid to policy holders but goes to the equity holders as a fee for taking on non-systematic mortality risk. The question of fairness of the design of that contract is not essential for the considerations in this chapter. The dynamics of wealth while being alive is now replaced by

$$dX(t) = \big(r + \pi(t)(\alpha - r)\big)X(t)\,dt + \pi(t)\sigma X(t)\,dW(t)$$
$$+ a(t)\,dt - c(t)\,dt + \mu^*(t)X(t)\,dt\,.$$

As above, we can complete the dynamics of X by setting it to zero upon death where it remains thereafter, since upon death X is lost to the financial institution,

$$dX(t) = I(t)\big[\big(r + \mu^*(t) + \pi(t)(\alpha - r)\big)X(t)\,dt + \pi(t)\sigma X(t)\,dW(t)$$
$$+ a(t)\,dt - c(t)\,dt\big] - X(t-)\,dN(t).$$

In the objective, the loss of X is related to the fact that there is no utility from consumption upon death. Corresponding to the special case $w^{01} = 0$, the value function (4.2) simply becomes

$$V(t, X(t)) = \sup_{\pi,c} \mathbb{E}\bigg[\int_t^n e^{-\int_t^s \rho+\mu} w^\gamma(s)u(c(s))\,ds$$
$$+ e^{-\int_t^n \rho+\mu}\,\Delta W^\gamma(n)U(X(n))\,\Big|\,X(t)\bigg]. \qquad (4.3)$$

We now have a wealth dynamics and an objective function that results in the HJB equation

$$V_t(t, x) - \rho V(t, x)$$

$$= \inf_{\pi, c} \Big\{ - V_x(t, x)\big(r + \mu^*(t) + \pi(\alpha - r)\big)x + a(t) - c\big)$$

$$- \frac{1}{2} V_{xx}(t, x)\pi^2 \sigma^2 x^2 - w^\gamma(t)u(c) - \mu(t)(-V(t, x)) \Big\},$$

$$V(n, x) = \Delta V(n)u(x).$$

The verification is unchanged from Merton's problem, since the two problems coincide with the following adjustment: the impatience rate ρ is replaced by $\rho + \mu$, the interest rate r is replaced by $r + \mu^*$, and the stock return rate α is replaced by $\alpha + \mu^*$ so that θ is unchanged. With that the verification goes through. Also the solution can be read directly from these replacements. The ODE for f can now be rewritten as

$$f'(t) = \Big[\frac{1}{\gamma}(\rho + \mu) + \Big(1 - \frac{1}{\gamma}\Big)(r + \mu^*) - \frac{1 - \gamma}{2\gamma^2}\theta^2 \Big] f(t) - w(t)$$

$$= (\tilde{r} + \tilde{\mu})f(t) - w(t), \tag{4.4}$$

$f(n) = \Delta W$, where $\tilde{r}$ was defined in (3.17) and

$$\tilde{\mu} = \frac{1}{\gamma}\mu + \Big(1 - \frac{1}{\gamma}\Big)\mu^*. \tag{4.5}$$

Note that, similarly to the construction of $\tilde{r}$, we can consider $\tilde{\mu}$ as the weighted average between the objective mortality rate μ and the pricing mortality rate μ^* with weights $1/\gamma$ and $1 - 1/\gamma$, respectively.

The function g can now be written

$$g(t) = \int_t^n e^{-\int_t^s r + \mu^*} a(s)\, ds.$$

Again, we have that $X > -g$ which means that borrowing against g has to be allowed. However, it appears much more logical and realistic to be able to borrow against $g(t) = \int_t^n e^{-\int_t^s r + \mu^*} a(s)\, ds$ than against the same expression without the μ^*, since now the human wealth takes into account the risk that the income falls away upon death.

The dynamics of c become, by replacements similar to those in (3.19) and (3.20),

$$
\begin{aligned}
& dc(t, X(t)) \\
&= \left(r + \mu^* - \tilde{r} - \tilde{\mu} + \frac{1}{\gamma}\theta^2 \right) c(t, X(t))\, dt + \frac{1}{\gamma}\theta c(t, X(t))\, dW(t) \\
&= \left(\frac{r - \rho + \mu^* - \mu}{\gamma} + \frac{1 + \gamma}{2\gamma^2}\theta^2 \right) c(t, X(t))\, dt + \frac{1}{\gamma}\theta c(t, X(t))\, dW(t) .
\end{aligned}
$$

Similarly to the special case $\rho = r$, the special case $\mu^* = \mu$, implying $\tilde{\mu} = \mu$, also plays a special role in the sense that then the mortality rate and thereby the distribution of the uncertain lifetime vanishes from the dynamics of consumption. This does not mean that the distribution vanishes from the consumption as such, since the mortality rate still appears in f and g and therefore contributes to the starting point of the consumption. But the profile of the consumption remains the same.

5   The Consumption-Investment-Insurance Problem

The calculations in the previous section are actually a special case of a problem with uncertain lifetime, utility from bequest upon death, and access to a life insurance market, where an optimal insurance sum is chosen continuously along with the consumption and investment decisions. We state here the wealth dynamics, the optimization problem, and the HJB equation that characterizes its solution. The dynamics of the wealth of an investor who decides a life insurance sum b to be paid out upon death, while being alive, reads

$$
\begin{aligned}
dX(t) = {}& \left(r + \pi(t)(\alpha - r) \right) X(t)\, dt + \pi(t)\sigma X(t)\, dW(t) \\
& + a(t)\, dt - c(t)\, dt - \mu^*(t)b(t)\, dt ,
\end{aligned}
$$

where μ^* is the mortality used for pricing the life insurance contract. The pricing is linear and a difference between μ^* and the objective μ reflects a market price of unsystematic mortality risk. Our model contains no systematic mortality risk in the sense that μ is assumed to be deterministic. The value function of the optimization problem is similar to the one presented in (4.2) with the major difference that utility upon death is measured from the sum of the wealth saved X and the insurance sum paid out b, i.e.

$$V(t, X(t))$$

$$= \sup_{\pi,c,b} \mathbb{E}\left[\int_t^n e^{-\int_t^s \rho+\mu} \left[w(s)^\gamma u\left(c(s)\right) + \mu(s)w^{01}(s)^\gamma u\left(X(s) + b(s)\right) \right] ds \right.$$

$$\left. + e^{-\int_t^n \rho+\mu} \Delta W^\gamma(n)u(X(n)) \,\Big|\, X(t) \right]. \tag{5.1}$$

Again, the appearance of X in the utility upon death reflects that the dynamics of X hold during survival only. Upon death, the wealth is consumed and thereafter set to zero where it remains for the rest of the time. The HJB equation that characterizes the value function (5.1) reads

$$V_t(t, x) - \rho V(t, x)$$

$$= \inf_{\pi,c,b} \left\{ - V_x(t, x)\left[(r + \pi(\alpha - r))x + a(t) - c - \mu^*(t)b \right] \right.$$

$$- \frac{1}{2}V_{xx}(t, x)\pi^2\sigma^2 x^2 - w^\gamma(t)u(c)$$

$$\left. - \mu(t)\left[w^{01}(t)^\gamma u(x + b) - V(t, x) \right] \right\},$$

$$V(n, x) = \Delta W^\gamma(n)u(x).$$

We choose not to verify this HJB equation since it is a special case of an HJB equation that we verify in the next section. However, this special case is so important that it deserves its own subsection with its own derivation of solution.

Under the infimum, there are two terms where the insurance sum b occurs and these terms do not involve c and π. This means that the first-order conditions (3.10) and (3.11) for c and π remain unchanged from the consumption-investment problem and we get the following first-order condition for b,

$$V_x(t, x)\mu^*(t) - \mu(t)w^{01}(t)^\gamma u'(x + b) = 0.$$

Plugging in the utility function and our value function candidate

$$V(t, x) = \frac{1}{1 - \gamma} f(t)^\gamma (x + g(t))^{1-\gamma},$$

we get

$$0 = \left(\frac{x + g(t)}{f(t)} \right)^{-\gamma} \mu^*(t) - \mu(t)w^{01}(t)^\gamma (x + b)^{-\gamma} \quad \Rightarrow$$

$$b = \frac{w^{01}(t)}{f(t)} \left(\frac{\mu(t)}{\mu^*(t)} \right)^{1/\gamma} (x + g(t)) - x. \tag{5.2}$$

The second derivative reads

$$-\mu(t)w^{01}(t)^\gamma u''(x+b) = \gamma\mu(t)w^{01}(t)^\gamma(x+b)^{-\gamma-1} > 0,$$

so that we have actually found a minimum.

We are now ready to plug the optimal c and π found in (3.10) and (3.11), the optimal b found in (5.2), and the value function candidate into the HJB equation. Note that the HJB equation contains all the terms of the consumption-investment problem (with ρ replaced by $\rho + \mu$) plus the term

$$V_x(t,x)\mu^*(t)b - \mu(t)w^{01}(t)^\gamma u(x+b).$$

With V and b inserted and after multiplication by $\left(\frac{x+g(t)}{f(t)}\right)^\gamma$, like when we obtained (3.12), the additional terms on the r.h.s. become

$$\left[\mu^*(t)\left(\frac{\mu(t)}{\mu^*(t)}\right)^{\frac{1}{\gamma}} - \mu(t)\frac{1}{1-\gamma}\left(\frac{\mu(t)}{\mu^*(t)}\right)^{\frac{1-\gamma}{\gamma}}\right]w^{01}(t)\frac{x+g(t)}{f(t)} - x\mu^*(t).$$

The first part fits well together with the other terms of (3.12) linear in $\frac{x+g}{f}$. In order to cope with $-x\mu^*(t)$, we also add and subtract $\mu^*(t)g(t)$, just like when we added and subtracted $rg(t)$ to obtain (3.12). Then the updated version of (3.12) including all additional terms becomes

$$\frac{\gamma}{1-\gamma}\frac{x+g}{f}f' + g' - (\rho+\mu)\frac{1}{1-\gamma}\frac{x+g}{f}f$$

$$= -r\frac{x+g}{f}f + (r+\mu^*)g - a - \frac{1}{\gamma}\left(\frac{\alpha-r}{\sigma}\right)^2\frac{x+g}{f}f + w\frac{x+g}{f}$$

$$+ \frac{1}{2\gamma}\left(\frac{\alpha-r}{\sigma}\right)^2\frac{x+g}{f}f - w\frac{1}{1-\gamma}\frac{x+g}{f}$$

$$+ \left[\mu^*\left(\frac{\mu}{\mu^*}\right)^{\frac{1}{\gamma}} - \mu\frac{1}{1-\gamma}\left(\frac{\mu}{\mu^*}\right)^{\frac{1-\gamma}{\gamma}}\right]w^{01}\frac{x+g}{f} - \frac{x+g}{f}f\mu^*.$$

As in the consumption-investment problem, first we collect the terms that do not involve $\frac{x+g}{f}$ in order to form the differential equation

$$g'(t) = (r + \mu^*(t))g(t) - a(t), \qquad g(n) = 0, \tag{5.3}$$

characterizing g. This is again the human capital in the uncertain lifetime case and the solution can be written as

$$g(t) = \int_t^n e^{-\int_t^s r+\mu^*}a(s)\,ds.$$

The rest of the equation is now divided by $\frac{\gamma}{1-\gamma} \cdot \frac{x+g}{f}$ in order to form the differential equation for f,

$$f'(t) - (\rho + \mu)\frac{1}{\gamma}f(t)$$

$$= -r\frac{1-\gamma}{\gamma}f(t) - \frac{1-\gamma}{\gamma^2}\theta^2 f(t) - w(t) + \frac{1-\gamma}{\gamma}\mu^*(t)w^{01}\left(\frac{\mu(t)}{\mu^*(t)}\right)^{\frac{1}{\gamma}}$$

$$- \frac{1-\gamma}{\gamma}\mu^* f(t) + \frac{1-\gamma}{\gamma^2}\frac{1}{2}\theta^2 f(t) - \mu(t)\frac{1}{\gamma}w^{01}\left(\frac{\mu(t)}{\mu^*(t)}\right)^{\frac{1-\gamma}{\gamma}},$$

$$f(n) = \Delta W(n).$$

We can make good use of the arithmetic average $\tilde{\mu}$ already introduced in (4.5). If we further introduce the geometric average

$$\tilde{\tilde{\mu}} = \mu^{1/\gamma}\mu^{*(1-1/\gamma)},$$

we can write the differential equation for f as

$$f'(t) = \tilde{r}f(t) + \tilde{\mu}(t)f(t) - w(t) - \tilde{\tilde{\mu}}(t)w^{01}(t).$$

If we insist on the structure of a Thiele differential equation we can of course hide away the difference between the arithmetic mean $\tilde{\mu}$ and the geometric mean $\tilde{\tilde{\mu}}$ in the interest rate by introducing $\tilde{\tilde{r}}(t) = \tilde{r} + \tilde{\mu}(t) - \tilde{\tilde{\mu}}(t)$ and writing

$$f'(t) = \tilde{\tilde{r}}(t)f(t) + \tilde{\tilde{\mu}}(t)f(t) - w(t) - \tilde{\tilde{\mu}}(t)w^{01}(t), \qquad f(n) = \Delta W(n). \qquad (5.4)$$

The solution to the differential equation for f reads

$$f(t) = \int_t^n e^{-\int_t^s \tilde{\tilde{r}}+\tilde{\tilde{\mu}}}\left(w(s) + \tilde{\tilde{\mu}}(s)w^{01}(s)\right)ds + e^{-\int_t^n \tilde{\tilde{r}}+\tilde{\tilde{\mu}}}\Delta W(n).$$

A series of special cases deserve special attention:
If $\mu^* = \mu$, then $\tilde{\tilde{\mu}} = \tilde{\mu} = \mu$. In that case, we simply have that

$$f(t) = \int_t^n e^{-\int_t^s \tilde{r}+\mu}\left(w(s) + \mu(s)w^{01}(s)\right)ds + e^{-\int_t^n \tilde{r}+\mu}\Delta W(n).$$

In the special case $w^{01} = 0$, we actually have from (5.2) that $b = -x$, i.e. the optimal insurance sum is simply $-x$. This means that the policy holder actually sells life insurance. He pays out the benefit X to the insurance company upon death and for this he receives the premium $-\mu^*(t)b(t)\,dt = \mu^*(t)X(t)\,dt$ in the wealth dynamics. Note that this exactly makes up the special case treated in the previous subsection. Note also that in that case, the differential equation (5.4) for f reduces to the special case (4.4),

$$f'(t) = \tilde{r} f(t) + \tilde{\mu}(t) f - w(t), \qquad f(n) = \Delta W(n).$$

Let us discuss the optimal insurance decision to make b be as in (5.2), i.e.

$$b = \frac{w^{01}(t)}{f(t)} \left(\frac{\mu(t)}{\mu^*(t)} \right)^{1/\gamma} (x + g(t)) - x .$$

We see that the optimal wealth upon death $x + b$ equals the optimal wealth just before death $x + g$ multiplied by two factors $w^{01}(t)/f(t)$ and $(\mu(t)/\mu^*(t))^{1/\gamma}$. Think of the situation where both factors are 1. We can speak of this as full insurance protection, since then $b = g$. This is what insurance is all about, namely, compensation for a financial loss. The financial loss of the family of the policy holder upon his death is exactly g and in case of full insurance, this is the insurance sum that the policy holder should demand. The two factors represent deviations away from full protection. One factor represents prices. If the insurance is expensive, i.e. $\mu^* > \mu$, then the factor $(\mu(t)/\mu^*(t))^{1/\gamma}$ expresses how much to under-insure compared to full protection. Conversely, one should over-insure if insurance is cheap, i.e. $\mu^* < \mu$. The other factor relates to the appreciation of consumption across the death of the individual. The appreciation of consumption upon death is quantified in w^{01} whereas the appreciation of (future) consumption just prior to death is expressed through f. These are balanced through the ratio $w^{01}(t)/f(t)$. If the appreciation of consumption is higher after the death of the individual, then one should over-insure compared with full protection. This is an odd situation but one can picture the family eating extensively and compulsively after the death of the bread winner. The more natural situation is that the appreciation of consumption goes down upon death and in that case there is demand for under-insurance. The reason why this is more natural is that the person who died and who is not there to earn money anymore is also not there to consume anymore.

6   The Multi-State Problem

We are now ready to consider the general multi-state version of the consumption-investment-insurance problem. This problem generalizes the problems we have considered so far. We consider the wealth dynamics given by

$$dX(t)$$
$$= \big(r + \pi(t)(\alpha - r)\big) X(t)\, dt + \pi(t) \sigma X(t)\, dW(t) + a^{Z(t)}(t)\, dt - c(t)\, dt$$

$$+ \sum_{k: k \neq Z(t-)} \left(a^{Z(t-)k}(t) - c^k(t) \right) \mathrm{d} N^k(t) + \sum_{k: k \neq Z(t-)} b^k(t) \, \mathrm{d} M^{*k}(t), \qquad (6.1)$$

$$\mathrm{d} M^*(t) = \mathrm{d} N^k(t) - \mu^{Z(t)k*}(t) \, \mathrm{d} t.$$

The first line in the expression for $\mathrm{d} X(t)$ corresponds essentially to the consumption-investment problem with the extension that we have decorated the income rate with the state dependence. The state dependence of the income rate simply reflects a multi-state model with different income states. For example, consider the disability model where the income falls away upon disability although the demand for consumption continues.

In the line (6.1), we start out by allowing for lump sum income and lump sum consumption. Lump sum income can be thought of as coming from a two-generation-family model where the children inherit an amount from their parents when they die. There could be a demand for lump sum consumption if, for instance, the policy holder has expenses upon an insurance event. For example, we can think of a policy holder who wants to rebuild his house upon disability. The last term in the second line of (6.1) is the exchange of risk with the insurance company. The insurance sum $b^k(t)$ is paid out upon transition to k and for this the policy holder has to pay the premium rate $b^k(t) \mu^{jk*}(t)$ if he is in state j. Note that the impact from insurance on the wealth process is simply a sum of martingales, corresponding to each state into which Z can jump at time t. The martingales are martingales under a pricing measure under which the intensity of making a transition to state k at time t is $\mu^{Z(t)k*}(t)$.

The value function of the problem is

$$V^{Z(t)}(t, X(t))$$

$$= \sup_{\pi, c, c^0, b^0, \dots c^J, b^J} \mathbb{E} \Bigg[\int_t^n e^{-\int_t^s \rho} \Big(w^{Z(s)}(s)^\gamma u(c(s)) \, \mathrm{d} s \qquad (6.2)$$

$$+ \sum_{k: k \neq Z(s-)} w^{Z(s-)k}(s)^\gamma u(c^k(s)) \, \mathrm{d} N^k(s) \Big) \qquad (6.3)$$

$$+ e^{-\int_t^n \rho} \Delta W^{Z(n)}(n)^\gamma U(X(n)) \, \Big| \, Z(t), X(t) \Bigg]. \qquad (6.4)$$

The value function that we worked with in the previous subsection (5.1) is a special case of this, namely $V^0(t, X(t))$ where $Z(t) = 0$ corresponds to conditioning upon survival in a survival model. As one usually does, however, we skipped the state 0 decoration in the previous section. When we were not working with other value functions than V^0, there was no need to carry around the state specification.

We choose to give here a full verification theorem including a proof. This generalizes directly the verification theorem for the consumption-investment problem. We follow exactly the same steps in the verification proof of the consumption-investment problem. The verification result also verifies the HJB equation for the special cases in the preceding two subsections, for which we did not verify but just stated the HJB equations.

Theorem 6.1 *Assume that there exists a sufficiently integrable function* $\overline{V}^j(t, x)$ *such that*

$$\overline{V}_t^j(t, x) - \rho \overline{V}^j(t, x)$$

$$= \inf_{\pi, c, c^0, b^0, \ldots c^J, b^J} \left\{ - \overline{V}_x^j(t, x)\Big((r + \pi(\alpha - r))x + a^j(t)\Big) - c \right.$$

$$- \sum_{k: k \neq j} \mu^{jk*}(t)b^k\Big) - \frac{1}{2}\overline{V}_{xx}^j(t, x)\pi^2\sigma^2 x^2$$

$$\left. - \Big(w^j(t)\Big)^\gamma u(c) - \sum_{k: k \neq j} \mu^{jk}(t) R^{jk}\Big(t, c^k, b^k, x\Big) \right\},$$

$$(6.5)$$

$$\overline{V}^j(n, x) = \Delta W^j(n)^\gamma u(x),$$

where

$$R^{jk}(t, c^k, b^k, x) = w^{jk}(t)^\gamma u(c^k) + \overline{V}^k(t, x + a^{jk} - c^k + b^k) - \overline{V}^j(t, x).$$

Then $V = \overline{V}$ *and*

$$\pi^* = \arg\inf_\pi \left\{ - \overline{V}_x^j(t, x)\pi(\alpha - r)x - \frac{1}{2}\overline{V}_{xx}^j(t, x)\pi^2\sigma^2 x^2 \right\}, \qquad (6.6)$$

$$c^* = \arg\inf_c \left\{ \overline{V}_x^j(t, x)c - w^j(t)^\gamma u(c) \right\},$$

$$c^{k*} = \arg\inf_{c^k} \left\{ - R^{jk}(c^k, b^{k*}, x) \right\},$$

$$b^{k*} = \arg\inf_{b^k} \left\{ \overline{V}_x^j(t, x)\mu^{jk*}(t)b^k - \mu^{jk}(t) R^{jk}(t, c^{k*}, b^k, x) \right\}.$$

Proof Consider a specific strategy $(\pi, c, c^0, b^0, \ldots, c^J, b^J)$. Given a function $\overline{V}^j(t, x)$, we get by Itô expansion of

$$d\left[e^{-\rho s}\overline{V}^{Z(s)}(s, X(s)) \right] = -\rho e^{-\rho s}\overline{V}(s, X(s))\, ds + e^{-\rho s}d\overline{V}(s, X(s))$$

and by integration over (t, n) that

$$e^{-\rho t}\overline{V}^{Z(t)}(t, X(t)) - e^{-\rho n}\overline{V}^{Z(n)}(n, X(n)) = -\int_t^n d\left[e^{-\rho s}\overline{V}^{Z(s)}(s, X(s))\right]$$

$$= -\int_t^n e^{-\rho s}\left[-\rho\overline{V}^{Z(s)}(s, X(s))\,ds + \overline{V}_s^{Z(s)}(s, X(s))\,ds\right.$$

$$+ \overline{V}_x^{Z(s)}(s, X(s))dX(s) + \frac{1}{2}\overline{V}_{xx}^{Z(s)}(s, X(s))\pi^2(s)\sigma^2 X^2(s)\,ds$$

$$+ \sum_{k\neq Z(s-)}\left[\overline{V}^k(s, X(s-) + a^{Z(s-)k}(s) - c^k(s) + b^k(s))\right.$$

$$\left.\left. - \overline{V}^{Z(s-)}(s, X(s-))\right]dN^k(s)\right].$$

First, consider an arbitrary strategy $(\pi, c, c^0, b^0, \ldots, c^J, b^J)$. For this strategy we know by (6.5) that

$$\overline{V}_t^{Z(t)}(t, X(t)) - \rho\overline{V}^{Z(t)}(t, X(t))$$

$$\leq -\overline{V}_x^{Z(t)}(t, X(t))\Big((r + \pi(\alpha - r))X(t) + a^{Z(t)}(t) - c(t)$$

$$- \sum_{k:\,k\neq Z(t)}\mu^{Z(t)k*}(t)b^k(t)\Big)$$

$$- \frac{1}{2}\overline{V}_{xx}^{Z(t)}(t, X(t))\pi^2(t)\sigma^2 X(t)^2 - w^{Z(t)}(t)^\gamma u(c(t))$$

$$- \sum_{k\neq Z(t)}\mu^{Z(t)k}(t)\Big[w^{Z(t)k}(t)^\gamma u(c^k(t))$$

$$+ \overline{V}^k(t, X(t) + a^{Z(t)k}(t) - c^k(t) + b^k(t))$$

$$- \overline{V}^{Z(t)}(t, X(t))\Big],$$

so that

$$e^{-\rho t}\overline{V}^{Z(t)}(t, X(t)) - e^{-\rho n}\overline{V}^{Z(n)}(n, X(n)) = -\int_t^n d\left[e^{-\rho s}d\overline{V}^{Z(s)}(s, X(s))\right]$$

$$= -\int_t^n e^{-\rho s}\left[-\rho\overline{V}^{Z(s)}(s, X(s))\,ds + \overline{V}_s^{Z(s)}(s, X(s))\,ds\right.$$

$$+ \overline{V}_x^{Z(s)}(s, X(s))\,dX(s)$$

$$+ \frac{1}{2}\overline{V}_{xx}^{Z(s)}(s, X(s))\,\pi^2(s)\sigma^2 X^2(s)\,ds$$

$$+ \sum_{k: k \neq Z(s-)} \left\{ \overline{V}^k \left(s, X(s-) + a^{Z(s-)k}(s) - c^{Z(s-)k}(s)\right) \right.$$

$$\left. + b^{Z(s-)k}(s) - \overline{V}^{Z(s-)}(s, X(s-)) \right\} dN^k(s) \Bigg]$$

$$\geq \int_t^n e^{-\rho s} \left[w^{Z(s)}(s)^\gamma u(c(s)) \, ds + \sum_{k: k \neq Z(s-)} w^{Z(s)k}(s)^\gamma u(c^k(s)) \, dN^k(s) \right.$$

$$- \overline{V}_x^{Z(s)}(s, X(s)) \pi(s) X(s) \sigma \, dW(s)$$

$$- \sum_{k \neq Z(s-)} \left\{ w^{Z(t-)k}(s)^\gamma u(c^k(s)) \right.$$

$$+ \overline{V}^k \left(s, X(s-) + a^{Z(s-)k}(s) - c^k(s) + b^k(s)\right)$$

$$\left. \left. - \overline{V}^{Z(s-)}(s, X(s-)) \right\} d\, M^k(s) \right].$$

Replacing $\overline{V}^{Z(n)}(n, X(n))$ by $\Delta W^{Z(n)}(n)^\gamma U(X(n))$ in accordance with the terminal condition, taking expectation on both sides, and multiplying by $e^{\rho t}$ yields

$$\overline{V}^j(t, x) \geq$$

$$\mathbb{E}_{t,x}^j \left[\int_t^n e^{-\rho(s-t)} \left[w^{Z(s)}(s)^\gamma u(c(s)) \, ds \right.\right.$$

$$+ \sum_{k: k \neq Z(s-)} w^{Z(s)k}(s)^\gamma u(c^k(s)) \, dN^k(s) \Bigg]$$

$$\left. + e^{-\rho(n-t)} \Delta W^{Z(n)}(n)^\gamma U(X(n)) \right].$$

Since the strategy was chosen arbitrarily, we have that

$$\overline{V}^j(t, x) \geq$$

$$\sup_{\pi, c, c^0, b^0, \dots c^J, b^J} \mathbb{E}_{t,x}^j \left[\int_t^n e^{-\rho(s-t)} \left[w^{Z(s)}(s)^\gamma u(c(s)) \, ds \right.\right.$$

$$+ \sum_{k: k \neq Z(s-)} w^{Z(s)k}(s)^\gamma u(c^k(s)) \, dN^k(s) \Bigg] \tag{6.7}$$

$$\left. + e^{-\rho(n-t)} \Delta W^{Z(n)}(n)^\gamma U(X(n)) \right]. \tag{6.8}$$

Now, we consider the specific strategy attaining the infimum of (6.5), specified in (6.6). For this strategy we have that

$$
\overline{V}_t^j(t, x) - \rho \overline{V}^j(t, x)
$$

$$
= -\overline{V}_x^j(t, x)\left[\left(r + \pi(\alpha - r)\right)x + a^j(t) - c - \sum_{k: k \neq j} \mu^{jk*}(t) b^k(t)\right]
$$

$$
- \frac{1}{2}\overline{V}_{xx}^j(t, x)\pi^2\sigma^2 x^2 - w^j(t)^\gamma u(c)
$$

$$
- \sum_{k: k \neq j} \mu^{jk}(t)\left[w^{jk}(t)^\gamma u(c^k) + \overline{V}^k\left(t, x + a^{jk} - c^k + b^k\right) - \overline{V}^j(t, x)\right],
$$

so that, with the same arguments as above, we get

$$
\overline{V}^j(t, x) =
$$

$$
\mathbb{E}_{t,x}^j\left[\int_t^n e^{-\rho(s-t)}\left[w^{Z(s)}(s)^\gamma u(c(s))\,\mathrm{d}s + \sum_k w^{Z(s)k}(s)^\gamma u(c^k(s))\,\mathrm{d}N^k(s)\right]\right.
$$

$$
\left. + e^{-\rho(n-t)}\Delta W^{Z(n)}(n)^\gamma U(X(n))\right] \tag{6.9}
$$

$$
\leq \sup_{\pi, c, c^0, b^0, \dots, c^J, b^J} \mathbb{E}_{t,x}^j\left[\int_t^n e^{-\rho(s-t)}\left[w^{Z(s)}(s)^\gamma u(c(s))\,\mathrm{d}s\right.\right.
$$

$$
\left. + \sum_k w^{Z(s)k}(s)^\gamma u(c^k(s))\,\mathrm{d}N^k(s)\right]
$$

$$
\left. + e^{-\rho(n-t)}\Delta W^{Z(n)}(n)^\gamma U(X(n))\right] \tag{6.10}
$$

where, in the last inequality, we have used that the supremum over strategies yields a higher conditional expectation than the specific strategy realizing the infimum in (6.5). By combining (6.8) and (6.10) we have that

$$
\overline{V}^j(t, x) =
$$

$$
\sup_{\pi, c, c^0, b^0, \dots c^J, b^J} \mathbb{E}_{t,x}^j\left[\int_t^n e^{-\rho(s-t)}\left[w^{Z(s)}(s)^\gamma u(c(s))\,\mathrm{d}s\right.\right.
$$

$$
\left. + \sum_{k: k \neq Z(s-)} w^{Z(s)k}(s)^\gamma u(c^k(s))\,\mathrm{d}N^k(s)\right]
$$

$$
\left. + e^{-\rho(n-t)}\Delta W^{Z(n)}(n)^\gamma U(X(n))\right].
$$

and the supremum is obtained by the strategy attaining the infimum.   □

We are now going to solve the HJB equation for the special case where

$$u(\cdot) = \frac{1}{1-\gamma}(\cdot)^{1-\gamma}.$$

As before, we simply call our object of study V instead of $\overline{V}$. First, we calculate the first-order conditions by differentiating the inner part of the infimum with respect to c, c^k, b^k, and π, respectively. We get that

$$0 = V_x^j(t, x) - w^j(t)^\gamma c^{-\gamma},$$

$$0 = w^{jk}(t)^\gamma (c^k)^{-\gamma} - V_x^k\big(t, x + a^{jk}(t) - c^k + b^k\big),$$

$$0 = V_x^j(t, x)\mu^{jk*}(t) - \mu^{jk}(t)V_x^k\big(t, x + a^{jk} - c^k + b^k\big),$$

$$0 = -V_x^j(t, x)(\alpha - r)x - V_{xx}^j(t, x)\pi\sigma^2 x^2.$$

We guess that the value function is given by

$$V^j(t, x) = \frac{1}{1-\gamma} f^j(t)^\gamma (x + g^j(t))^{1-\gamma},$$

so that the derivatives are

$$V_t^j(t, x) = \frac{\gamma}{1-\gamma}\left(\frac{x + g^j(t)}{f^j(t)}\right)^{1-\gamma} f^{j\prime}(t) + \left(\frac{x + g^j(t)}{f^j(t)}\right)^{-\gamma} g^{j\prime}(t),$$

$$V_x^j(t, x) = \left(\frac{x + g^j(t)}{f^j(t)}\right)^{-\gamma}, \quad V_{xx}^j(t, x) = -\gamma\left(\frac{x + g^j(t)}{f^j(t)}\right)^{-\gamma-1}\frac{1}{f^j(t)}.$$

Under the assumption that our guess on the value function is correct, we can plug the derivatives of the value function into the first-order conditions. The first-order conditions for c^j and π are direct generalizations of (3.10) and (3.11),

$$0 = \left(\frac{x + g^j(t)}{f^j(t)}\right)^{-\gamma} - w^j(t)^\gamma c^{-\gamma} \Rightarrow c^{j*}(t, x) = \frac{w^j(t)}{f^j(t)}(x + g^j(t)),$$

$$0 = -\left(\frac{x + g^j(t)}{f^j(t)}\right)^{-\gamma}(\alpha - r)x + \gamma\left(\frac{x + g^j(t)}{f^j(t)}\right)^{-\gamma-1}\frac{1}{f^j(t)}\pi\sigma^2 x^2$$

$$\Rightarrow \pi^{j*}(t, x) = \frac{1}{\gamma}\frac{\alpha - r}{\sigma^2}\frac{x + g^j(t)}{x}.$$

The two first-order conditions for c^k and b^k form two equations with two unknowns. First, we represent the solution to the first-order condition for c^k by

$$0 = w^{jk}(t)^\gamma (c^k)^{-\gamma} - \left(\frac{x + a^{jk}(t) - c^{jk*} + b^{jk*} + g^k(t)}{f^k(t)}\right)^{-\gamma}$$

$$\Rightarrow c^{jk*}(t, x) = \frac{w^{jk}(t)}{f^k(t)}\left(x + a^{jk}(t) - c^k + b^{jk*} + g^k(t)\right). \qquad (6.11)$$

Second, we represent the solution for the first-order condition for b^k by

$$0 = \left(\frac{x + g^j(t)}{f^j(t)}\right)^{-\gamma} \mu^{jk*}(t) - \mu^{jk}(t)\left(\frac{x + a^{jk}(t) - c^{jk*} + b^k + g^k(t)}{f^k(t)}\right)^{-\gamma}$$

$$\Rightarrow b^{jk*}(t) = f^k(t)\frac{x + g^j(t)}{f^j(t)}\left(\frac{\mu^{jk}(t)}{\mu^{jk*}(t)}\right)^{\frac{1}{\gamma}} - x - a^{jk}(t) + c^{jk*} - g^k(t).$$

$$(6.12)$$

We now plug (6.12) into (6.11) to get

$$c^{jk*}(t, x) = \frac{w^{jk}(t)}{f^j(t)}\left(\frac{\mu^{jk}(t)}{\mu^{jk*}(t)}\right)^{\frac{1}{\gamma}}(x + g^j(t)), \qquad (6.13)$$

and plug (6.13) into (6.12) to get

$$b^{jk*}(t) = \frac{w^{jk}(t) + f^k(t)}{f^j(t)}\left(\frac{\mu^{jk}(t)}{\mu^{jk*}(t)}\right)^{\frac{1}{\gamma}}(x + g^j(t)) - x - a^{jk}(t) - g^k(t). \qquad (6.14)$$

We are now ready to plug the guess on the value function, its derivatives, and the strategy we have constructed from them, into the HJB equation. We follow the recipe from the consumption-investment problem in order to obtain ODEs for f^j and g^j. They are

$$f^{j'}(t) = \widetilde{r}^j f^j(t) - w^j(t) - \sum_{k: k \neq j} \widetilde{\mu}^{jk}(t)\left(w^{jk} + f^k(t) - f^j(t)\right), \qquad (6.15)$$

$$g^{j'}(t) = rg^j(t) - a^j(t) - \sum_{k: k \neq j} \mu^{jk*}(t)\left(a^{jk}(t) + g^k(t) - g^j(t)\right), \qquad (6.16)$$

together with $f^j(n) = \Delta W^j(n)$ and $g^j(n) = 0$. Here we have introduced the geometric and arithmetic averages of transition intensities, respectively,

$$\widetilde{\mu}^{jk} = \mu^{jk}(t)^{\frac{1}{\gamma}}\mu^{jk*}(t)^{1-\frac{1}{\gamma}}, \quad \widetilde{\mu}^{jk}(t) = \frac{1}{\gamma}\mu^{jk}(t) + \left(1 - \frac{1}{\gamma}\right)\mu^{jk*}(t),$$

and defined the interest rate $\overset{\approx}{r}^{j}$ as a correction to $\widetilde{r}$, defined in (3.17), for the difference between the two averages, i.e.

$$\overset{\approx}{r}^{j} = \widetilde{r} + \widetilde{\mu}^{j} - \overset{\approx}{\mu}^{j} \ .$$

Note that this introduces a state-dependence in the interest rate $\overset{\approx}{r}$. However, this is just a matter of presentation.

It is natural to generalize the interpretation at the end of Sect. 5 to the multi-state case, so we look closer at the optimal insurance protection in (6.14). The insurance optimally relates the wealth prior to the transition from j to k, $x + g^{j}(t)$, to the wealth after the transition, $x + a^{jk} + b^{jk*}(t) + g^{k}(t)$. They should be linearly related by the factors

$$\frac{w^{jk}(t) + f^{k}(t)}{f^{j}(t)} \quad \text{and} \quad \left(\frac{\mu^{jk}(t)}{\mu^{jk*}(t)}\right)^{\frac{1}{\gamma}} .$$

Full protection corresponds to the factors being such that $b^{jk*} = g^{j} - g^{k} - a^{jk}$. Namely, the financial loss upon transition is minus the change of human capital $g^{k} + a^{jk} - g^{j}$. There is then a demand for over-insurance if insurance is cheap, $\mu^{*} < \mu$, or if the appreciation of consumption upon and after death, formalized through $w^{jk}(t) + f^{k}(t)$, is higher than the appreciation of consumption just prior to death, formalized through f^{j}. Conversely, there is a demand for under-insurance if insurance is expensive, $\mu^{*} > \mu$, or if the appreciation of consumption upon and after death is lower than the appreciation of consumption just prior to death.

6.1   The Survival Model

We now study the survival model as a special case. We already paid special attention to this case when introducing lifetime uncertainty to the consumption-investment problem in Sect. 5. We now see how our multi-state results specialize to the results obtained there. Let us put

$$w^{0} = w, \quad \Delta W^{0} = \Delta W, \quad w^{1} = \Delta W^{1} = a^{1} = a^{01} = 0 ,$$
$$a^{0} = a, \ \mu^{0} = \mu, \ \mu^{01*} = \mu^{*} .$$

Then, only the value function in the state 'alive' is relevant and we can skip the state decoration of f and g. The ODEs (6.15) and (6.16) specialize to

$$f'(t) = \overset{\approx}{r} f(t) - w(t) - \overset{\approx}{\mu}(t)\left(w^{01}(t) - f(t)\right), \quad f(n) = \Delta W(n),$$
$$g'(t) = rg(t) - a(t) + \mu^{*}(t)g(t), \quad g(n) = 0,$$

which are exactly the same as (5.4) and (5.3). The optimal insurance sum obtained in (6.14) specializes to

$$b(t) \; = \; \frac{w^{01}(t)}{f(t)} \left(\frac{\mu(t)}{\mu^*(t)} \right)^{\frac{1}{\gamma}} (x + g(t))) - x \,, \tag{6.17}$$

which is exactly the relation from (5.2). The optimal consumption upon death is

$$c^{01}(t, x) \; = \; \frac{w^{01}(t)}{f(t)} \left(\frac{\mu(t)}{\mu^*(t)} \right)^{\frac{1}{\gamma}} (x + g(t)) \,,$$

which, by comparing with (6.17), can be written as

$$c^{01}(t, x) = x + b(t).$$

This is the relation which we already plugged into the problem formulation in Sect. 5, namely that upon death the wealth is consumed together with the insurance sum paid out.

7   The Pension Fund's Problem

We finalize the chapter by reconsidering the dividend payout problem of the pension fund. The surplus investment and redistribution problem of an insurance company can also be approached by power utility optimization. Recall the dynamics of the surplus derived in Sect. VI.4,

$$dX(t) \; = \; \big(r + \pi(t)(\alpha - r) \big) X(t) \, dt + \pi(t) \sigma X(t) \, dW(t) + dC(t) - dD(t),$$

where C is the surplus contribution process and D is the dividend distribution process. The structure of the problem now depends on the structure of C and D. In order to establish connections to the classical consumption-investment problem, we start out by considering C as a deterministic process of order dt, i.e.

$$dC(t) = c(t) \, dt.$$

Now this corresponds to the income process of the consumption-investment problem. Redistribution of dividends generates utility to the policy holders. Let us assume that dividends are paid out continuously, i.e.

$$dD(t) = \delta(t) \, dt,$$

and that the policy holders obtain utility from this, so that the value function can be written as

$$V(t, x) = \sup_{\pi, \delta} \mathbb{E}_{t,x}\left[\int_t^n e^{-\rho(s-t)} w^\gamma(s) \frac{1}{1-\gamma} \delta^{1-\gamma}(s)\, ds \right.$$

$$\left. + e^{-\rho(n-t)} \Delta W^\gamma(n) \frac{1}{1-\gamma} X^{1-\gamma}(n) \right],$$

where the last term reflects that the residual surplus is paid out at time n.

Now, this problem is exactly the consumption-investment problem solved in Sect. 3 with c replaced by δ and a replaced by c. Thus we can read the solution directly from (3.10), (3.11), (3.15), and (3.18).

$$\delta(t, x) = w(t) \frac{x + g(t)}{f(t)}, \qquad \pi(t, x) = \frac{1}{\gamma} \frac{\alpha - r}{\sigma^2} \frac{x + g(t)}{x},$$

with

$$g(t) = \int_t^n e^{-\int_t^s r} c(s)\, ds, \qquad f(t) = \int_t^n e^{-\int_t^s \tilde{r}} w(s)\, ds + e^{-\int_t^n \tilde{r}} \Delta W(n).$$

Another version of the problem is the multi-state problem. This was solved in Sect. 6. In that problem, we had access to an insurance market and took optimal insurance decisions. This is not the case here although one could in principle start introducing reinsurance as the insurance company's insurance taking. However, without the insurance market it is not generally possible to solve the optimal surplus distribution problem explicitly. This is, for example, the case if the surplus contribution process C is also related to the Markov chain. However, it turns out that one can actually solve the problem with Markov modulation of preferences without the insurance decision process as long as the surplus contributions are still deterministic. This is exactly the problem we solve now.

We consider the dynamics of the surplus given by

$$dX(t) = \left(r + \pi(t)(\alpha - r)\right) X(t)\, dt + \pi(t)\sigma X(t)\, dW(t)$$

$$+ c(t)\, dt - \delta(t)\, dt - \sum_{k: k \neq Z(t-)} \delta^k(t)\, dN^k(t). \qquad (7.1)$$

This is a rewriting of (6.1) where we take a to be a deterministic income process and where we skip the insurance payment. The objective is the same as in Sect. 6, except for disregarding the insurance decision process, i.e.

$$V^{Z(t)}(t, X(t))$$

$$= \sup_{\pi,\delta,\delta^0,\dots,\delta^J} \mathbb{E}\left[\int_t^n e^{-\int_t^s \rho} \left(w^{Z(s)}(s)^\gamma u(\delta(s))\, ds \right. \right.$$

$$\left. + (w^{Z(s-)k})^\gamma (s) u(\delta^k(s))\, dN^k(s) \right)$$

$$\left. + e^{-\int_t^n \rho}(\Delta W^{Z(n)}(n))^\gamma U(X(n)) \,\bigg|\, Z(t), X(t) \right].$$

The HJB equation characterizing this value function can be obtained similarly to the one in Sect. 6 and we leave out the verification theorem. The equation becomes

$$V_t^j(t, x) - \rho V^j(t, x)$$

$$= \inf_{\pi,\delta,\delta^0,\dots,\delta^J} \left\{ - V_x^j(t, x)\big[(r + \pi(\alpha - r))x + c(t) - \delta\big] \right.$$

$$- \frac{1}{2} V_{xx}^j(t, x)\pi^2\sigma^2 x^2 - (w^j(t))^\gamma u(\delta)$$

$$\left. - \sum_{k:k\neq j} \mu^{jk}(t)\big[(w^{jk}(t))^\gamma u(\delta^k) + V^k(t, x - \delta^k) - V^j(t, x)\big] \right\},$$

$$V^j(n, x) = (\Delta W^j(n))^\gamma u(x).$$

We solve this problem for the case where

$$u(\cdot) = \frac{1}{1-\gamma}(\cdot)^{1-\gamma}.$$

We get the first-order conditions

$$0 = V_x^j(t, x) - (w^j(t))^\gamma \delta^{-\gamma}, \qquad 0 = w^{jk}(t)^\gamma (\delta^k)^{-\gamma} - V_x^k(t, x - \delta^k),$$

$$0 = -V_x^j(t, x)(\alpha - r)x - V_{xx}^j(t, x)\pi\sigma^2 x^2.$$

We guess that the value function is given by

$$V^j(t, x) = \frac{1}{1-\gamma} f^j(t)^\gamma (x + g(t))^{1-\gamma},$$

so that the derivatives are

$$V_t^j(t,x) = \frac{\gamma}{1-\gamma}\left(\frac{x+g(t)}{f^j(t)}\right)^{1-\gamma} f^{j\prime}(t) + \left(\frac{x+g(t)}{f^j(t)}\right)^{-\gamma} g'(t),$$

$$V_x^j(t,x) = \left(\frac{x+g(t)}{f^j(t)}\right)^{-\gamma}, \quad V_{xx}^j(t,x) = -\gamma\left(\frac{x+g(t)}{f^j(t)}\right)^{-\gamma-1}\frac{1}{f^j(t)}.$$

Under the assumption that our guess on the value function is correct, we can plug the derivatives of the value function into the first-order conditions. The first-order conditions for δ and π are direct generalizations of (3.10) and (3.11),

$$0 = \left(\frac{x+g(t)}{f^j(t)}\right)^{-\gamma} - w^j(t)^\gamma \delta^{-\gamma} \quad \Rightarrow \quad \delta^{j*}(t,x) = \frac{w^j(t)}{f^j(t)}(x+g(t)),$$

$$0 = -\left(\frac{x+g(t)}{f^j(t)}\right)^{-\gamma}(\alpha - r)x + \gamma\left(\frac{x+g(t)}{f^j(t)}\right)^{-\gamma-1}\frac{1}{f^j(t)}\pi\sigma^2 x^2$$

$$\Rightarrow \quad \pi(t,x) = \frac{1}{\gamma}\frac{\alpha - r}{\sigma^2}\frac{x+g(t)}{x}.$$

In contrast to the version with insurance decisions where we had a system of equations for (c^k, δ^k), we now just have one equation for δ^k, corresponding to (6.11), which becomes

$$0 = w^{jk}(t)^\gamma(\delta^k)^{-\gamma} - \left(\frac{x - \delta^k + g(t)}{f^k(t)}\right)^{-\gamma}$$

$$\Rightarrow \quad \delta^{jk*}(t,x) = \frac{w^{jk}(t)}{w^{jk}(t) + f^k(t)}(x+g(t)).$$

We are now ready to plug the guess on the value function, its derivatives, and the strategy we have constructed from them into the HJB equation. We follow the recipe from before in order to obtain ordinary differential equations for f^j and g. However, leaving out the insurance decision changes the structure of the differential equation for f. We get the system

$$f^{j\prime}(t) = \tilde{r}f^j(t) - w^j(t)$$

$$- \sum_{k:k\neq j} \mu^{jk}(t)\frac{1}{\gamma}\left((w^{jk}(t) + f^k(t))^\gamma f^j(t)^{1-\gamma} - f^j(t)\right), \quad (7.2)$$

$$f^j(n) = \Delta W^j(n), \quad g'(t) = rg(t) - c(t), \quad g(n) = 0. \quad (7.3)$$

The differential equation for g is recognized from earlier versions of the problem and has the solution

$$g(t) = \int_t^n e^{-\int_t^s r} c(s)\, ds .$$

The differential equation for f is considerably more complicated than before since this equation is not linear. This means that we cannot write down a solution and we do not know up-front whether our value function guess actually works. However, we know that if there exists a solution to the system of non-linear equations for f, then the value function is correct. The structures of (6.15) and (7.2) are so different because of the crucial difference in the underlying control problem. In contrast to the previous section, we have in the present section no access to the insurance market. The Markov chain drives the preferences for paying out dividends and thereby the control becomes related to the Markov chain. But there is no exchange of risk with an insurer in the present section.

7.1 The Survival Model

Let us complete the picture by considering the special case of the survival model. It will become clear what we mean by completing the picture. Let us put

$$w^0 = w, \quad \Delta W^0 = \Delta W, \quad w^1 = w^{01} = \Delta W^1 = 0, \quad \mu^{01} = \mu.$$

Only the value function in the state 'alive' is relevant and we can skip the state decoration of f. The ODEs (7.2) and (7.3) specialize to

$$f'(t) = \tilde{r} f(t) - w(t) - \mu(t)\frac{1}{\gamma} f(t), \qquad f(n) = \Delta W(n),$$

$$g'(t) = rg(t) - c(t), \qquad g(n) = 0.$$

Since ρ appears in $\tilde{r}$ through the product $\frac{1}{\gamma}\rho$, the differential equation for f corresponds to the similar equation for the consumption-investment problem with ρ replaced by $\rho + \mu$. This is exactly the solution we pointed out in Sect. 4, where we set $w^{01} = 0$. In that section, we also discussed the case where $w^{01} \neq 0$. We concluded that this could not be solved. However, in this section we can solve the problem with $w^{01} \neq 0$. The difference lies in the utility gained upon death. In Sect. 4, the utility gained upon death was

$$\frac{1}{1-\gamma}(w^{01}(t))^\gamma x^{1-\gamma} .$$

Actually, we made the comment that if utility were instead earned from

$$\frac{1}{1-\gamma}(w^{01}(t))^{\gamma}(x+g(t))^{1-\gamma},$$

then there was a way forward with the guess on the value function. This is exactly the structure we have here. Note that, if $w^{01} \neq 0$ and if we still assume that $w^1 = 0$ such that $f^1 = 0$, we have that

$$\delta^{01} = x + g(t),$$

from which we, according to the objective function, earn the utility

$$(w^{01}(t))^{\gamma}u(\delta^k(t)) = \frac{1}{1-\gamma}(w^{01}(t))^{\gamma}(x+g(t))^{1-\gamma}.$$

This explains why we can solve the problem here. The economic intuition about the difference between the two problems is subtle. In Sect. 4, utility is gained from wealth only. But also, the solution only makes sense if we can borrow against income all the way to time n. We discussed that this would be hard to believe in practice. In the present section, we also borrow against income all the way to time n. But this is more honest because we actually openly assume that income is actually earned until time n. And if it is, it makes mathematical sense to earn utility upon death from income earned after the death. However, note that with that interpretation, we have the non-linear differential equation

$$f'(t) = \tilde{r}f(t) - w(t) - \sum_{k:k\neq j}\mu(t)\frac{1}{\gamma}\left(w^{01}(t)^{\gamma}f(t)^{1-\gamma} - f(t)\right)$$

for f.

Notes

Application of stochastic control in life insurance and pensions can naturally be divided into decision problems of the pension fund or insurance company, and decision problems of an individual saving for retirement. To the latter also belongs problems of consumption in retirement and, potentially, insuring death risk during the course of life. In the literature, quadratic objectives have dominated the decision problem of the pension fund whereas standard (like CRRA) utility preferences have dominated the decision problem of the individual. This was also the case in the presentation in this chapter, except for the last section, where CRRA utility optimization is applied to the surplus distribution problem of the pension fund.

Formulation of the decision problem of the pension fund developed within a diffusive framework from the mid 1980s until around 2000. Quadratic objectives formalize a preference for stability leading to the linear regulation of payments. O'Brien [135] is an original presentation of the idea. A great survey with its own important points was provided by Cairns [44]. The extension to micromodeling of the payment streams and objectives via multistate payments and objectives with quadratic preferences was provided by Steffensen [163].

The individual's decision problem solved by stochastic control theory dates back to the seminal papers [118] by Merton [117]. This was extended to the case of uncertain lifetime and access to life insurance by Richard [150] in 1975. Actually, a decade earlier, some of the points about uncertain lifetime, optimal insurance, and the optimality of annuitization were made by Yaari [177] in discrete time. The formulation in continuous time by Richard [150] did not attract much attention until financial economists as well as insurance mathematicians realized this powerful formalization of lifecycle financial decision making under lifetime uncertainty some 30 years later. Since then, many authors have built on and generalized the work by Richard. The multistate version appeared in Kraft and Steffensen [104]. Other directions of generalization in the literature are a) decision power over retirement timing, as a stopping time or gradually, b) stochastic mortality rates and possibly access to mortality derivatives in the portfolio, c) optimization with constraints on the insurance sum or other restrictions in the insurance market or product. Of course other generalizations apply here, to generalized financial markets and generalized preferences. However, these generalizations are not specific to the domain of pension saving and insurance.

The surplus redistribution problem of a pension fund that had been studied with quadratic objectives before 2000 was formulated with a power utility objective and multistate preferences by Steffensen [162]. That formulation was also included in Schmidli [155] who was the first to give a survey of applications of stochastic control to life insurance

Chapter XIII: Selected Further Topics

1 Claims Reserving

In most of the book, it has been assumed that a claim is paid out immediately to the insured. In practice, delays in reporting a claim and/or covering often arise, and the total claim amount may not be known at the time the claim occurs. One typical example is accident insurance, where the degree of disability is not a priori known and treatment takes place over a period. Another is fire insurance where repair on rebuilding may take a long time and the cost can at best be estimated at the time of the claim. The area of claims reserving, also called loss reserving, deals with methods for assessing the magnitude of such future liabilities taking into account the available data up to the present time T.

A general model is to associate with each claim, say the ith, a triple τ_i, δ_i, D_i, where τ_i is the time of occurrence of the claim, δ_i the time from then until it is reported and $D_i = \big(D_i(t)\big)_{0 \le t \le \sigma_i}$ a payment stream terminating at a finite (possibly random) time σ_i such that $D_i(t)$ is the amount paid in $[\tau_i + \delta_i, \tau_i + \delta_i + t)$. The IBNS (Incurred But Not Settled) claims at time T are then the ones with $\tau_i \le T < \tau_i + \delta_i + \sigma_i$. These can be subdivided into the IBNR (Incurred But Not Reported) claims, i.e. the ones with $\tau_i \le T < \tau_i + \delta_i$ and the RBNS (Reported But Not Settled) where $\tau_i + \delta_i \le T < \tau_i + \delta_i + \sigma_i$.

Disregarding the detailed evolvement of D_i, the situation is often depicted in a so-called *Lexis diagram* as in Fig. 1, where a claim is represented by a line with slope 45° starting at time of occurrence and ending at time of settlement; the ● corresponds to reporting and the ■ to settlement. At time T, we thus count two RBNS claims and one IBNR.

A more detailed realization is in Fig. 2 where claim 1 (red) is RBNS and claim 3 (green) IBNR; in addition to the legends of Fig. 1 the × are occurrence times τ_i. The blue claim 2 is settled before T and therefore does not contribute to the calculation of reserves.

© Springer Nature Switzerland AG 2020
S. Asmussen, M. Steffensen, *Risk and Insurance*, Probability Theory
and Stochastic Modelling 96, https://doi.org/10.1007/978-3-030-35176-2_13

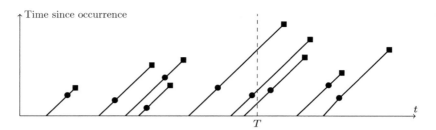

Fig. 1 A Lexis diagram

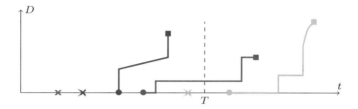

Fig. 2 A continuous-time IBNS situation

The outstanding liabilities of the company at time T are

$$\sum_{i:\, \tau_i \le T < \tau_i + \delta_i} D_i(\sigma_i) + \sum_{i:\, \tau_i + \delta_i \le T < \tau_i + \delta_i + \sigma_i} \big(D_i(\sigma_i) - D_i(T - \tau_i - \delta_i)\big), \qquad (1.1)$$

where the first term corresponds to the IBNR claims, the second to the RBNS claims and the sum to the totality of IBNS claims. The corresponding reserves $R_{\mathrm{IBNR}}(T)$, $R_{\mathrm{RBNS}}(T)$ can be defined as the conditional expectations given $\mathcal{F}(T)$, the available information up to time T (clearly, $R_{\mathrm{Settled}}(T) = 0$).

Let $N_{\mathrm{Settled}}(T)$, $N_{\mathrm{IBNR}}(T)$, $N_{\mathrm{RBNS}}(T)$ be the number of claims that are Settled, IBNR, resp. RBNS at time T. Assuming a Poisson arrival mechanism, a number of properties of these r.v.s and the corresponding reserves are easily derived. For example:

Proposition 1.1 *Assume the claims arrive according to a Poisson(β) process $\{N(t)\}_{t \ge 0}$ with interarrival times $0 < \tau_1 < \tau_2 < \cdots$ and that the δ_i, D_i are i.i.d. and independent of N. Then $N_{\mathrm{Settled}}(T)$, $N_{\mathrm{IBNR}}(T)$, $N_{\mathrm{RBNS}}(T)$ are independent Poisson r.v.s with parameters*

$$\beta \int_0^T \mathbb{P}(\delta + \sigma \le t)\, \mathrm{d}t, \quad \beta \int_0^T \mathbb{P}(t < \delta)\, \mathrm{d}t, \quad \text{resp. } \beta \int_0^T \mathbb{P}(\delta \le t < \delta + \sigma)\, \mathrm{d}t\,.$$

Further, with $\varphi(t) = \mathbb{E}\big[D(\sigma) - D(t)\,\big|\,\{D(s)\}_{s \leq t}, \sigma > t\big]$, we have

$$R_{\text{IBNR}}(T) = \mathbb{E}D(\sigma) \cdot \beta \int_0^T \mathbb{P}(\delta + \sigma \leq t)\,dt\,,$$

$$R_{\text{RBNS}}(T) = \sum_{i:\,\tau_i + \delta_i \leq T < \tau_i + \delta_i + \sigma_i} \varphi(T - \tau_i - \delta_i)\,.$$

Proof The independence follows by general properties of Poisson thinning as discussed in Sect. I.5.3. For the form of parameters, consider for example $N_{\text{RBNS}}(T)$. If claim i occurs at time $t \leq T$, it will be RBNS precisely when $\delta_i \leq T - t < \delta_i + \sigma_i$, so the Poisson parameter is

$$\beta \int_0^T \mathbb{P}(\delta \leq T - t < \delta + \sigma)\,dt = \beta \int_0^T \mathbb{P}(\delta \leq t < \delta + \sigma)\,dt\,.$$

The expression for $\text{IBNR}(T)$ follows from $\text{IBNR}(T) = \mathbb{E}D(\sigma) \cdot \mathbb{E}N_{\text{IBNR}}(T)$ and the one for $R_{\text{RBNS}}(T)$ by noting that $\varphi(t)$ is the conditional expectation of the D-payments after t given we know the $D(s)$ for $s \leq t$ and that the payment stream is still running at time t. $\qquad\qquad\square$

The typical procedure in practice and in much of the literature is, however, to use the discrete time *chain-ladder method* rather than the continuous time Poisson model above. We shall concentrate here on the RBNS reserve. Here one defines $C_{t,j}$ as the accumulated claim amount corresponding to reporting time $t = 1, \ldots, T$, referred to as 'year' in the following, and development year $j = 0, 1, \ldots$ That is, $C_{t,j}$ is the sum of all payments from the company in years $t, t+1, \ldots, t+j$ and may refer to an individual policy or a portfolio of policies. At the current time T, the $C_{t,j}$ with $t + j \leq T$ are observable and are traditionally reported in the so-called *run-off triangle*, see Fig. 3.

Assuming that a claim has always been settled at or before development period J (deterministic), the problem is to give estimates $\widehat{O}_t$ of the outstanding amount

$$O_t = C_{t,J} - C_{t,T-t}$$

of claims which were reported in period $t \leq T$, and thereby of the total outstanding amount $O_1 + \cdots + O_T$.

The chain-ladder method is based on informally assuming

$$C_{t,j+1} \approx f_j C_{t,j} \tag{1.2}$$

for a suitable set of parameters $f_1, f_2, \ldots$ One way to think of this is that the profile of payments after a claim is independent of the occurrence year and the claim size. One then provides estimates $\widehat{f}_1, \ldots, \widehat{f}_{T-1}$ of $f_1, \ldots, f_{T-1}$ using the data in the

Fig. 3 The run-off triangle

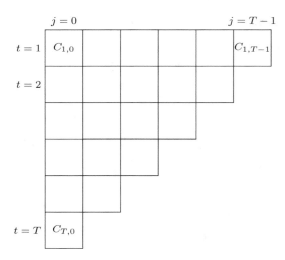

run-off triangle, and lets

$$\widehat{O}_t = C_{t,T-t} \widehat{F}_{T-t,J-1} \quad \text{where} \quad \widehat{F}_{i,k} = \widehat{f}_i \cdots \widehat{f}_k \tag{1.3}$$

(to do this for all t requires that $J \le T - 1$). For a variant, see Remark 1.3 below.
  Traditionally, the formalization of (1.2) is

$$\mathbb{E}\big[C_{t,j+1} \,\big|\, C_{t,0}, \dots, C_{t,j}\big] = f_j C_{t,j} . \tag{1.4}$$

One assumes that observations from different years $t = 0, 1, 2 \dots$ are independent
and uses the estimator

$$\widehat{f}_j = \frac{C_{1,j+1} + \cdots + C_{T-j-1,j+1}}{C_{1,j} + \cdots + C_{T-j-1,j}} . \tag{1.5}$$

This obviously has desirable properties in large samples since then (1.2) holds and
one gets

$$\widehat{f}_j \approx \frac{C_{1,j} f_j + \cdots + C_{T-j-1,j} f_j}{C_{1,j} + \cdots + C_{T-j-1,j}} = f_j .$$

However, the desired consistency (limit f_j) is shared by other intuitive estimators.
In particular, we have

$$\frac{C_{1,j+1}/C_{1,j} + \cdots + C_{T-j-1,j+1}/C_{T-j-1,j}}{T-j-1} \approx \frac{1}{T-j-1} \sum_{t=1}^{T-j-1} f_j = f_j , \tag{1.6}$$

$$\frac{C_{1,j}C_{1,j+1} + \cdots + C_{T-j-1,j}C_{T-j-1,j+1}}{C_{1,j}^2 + \cdots + C_{T-j-1,j}^2} \approx \frac{\sum_{t=1}^{T-j-1} f_j C_{t,j}^2}{\sum_{t=1}^{T-j-1} C_{t,j}^2} = f_j. \tag{1.7}$$

A unified way to look at the estimators (1.5)–(1.7) is as weighted averages

$$\frac{C_{1,j}^\alpha F_{t,j} + \cdots + C_{T-j-1,j}^\alpha F_{T-j-1,j}}{C_{1,j}^\alpha + \cdots + C_{T-j-1,j}^\alpha}, \tag{1.8}$$

where $F_{t,j}$ is the individual development factor $C_{t,j+1}/C_{t,j}$. Such a weighted average is in fact a least squares estimator corresponding to the second moment assumption

$$\mathbb{Var}\left(C_{t,j+1} \mid C_{t,0}, \ldots, C_{t,j}\right) = \sigma_j^2 C_{t,j}^{2-\alpha}. \tag{1.9}$$

This follows from classical least squares theory, more precisely the particular setting of independent observations $Z_1, \ldots, Z_d$ with means $fa_1, \ldots, fa_d$ and variances $\sigma^2/w_1, \ldots, \sigma^2/w_d$, where $a_1, w_1, \ldots, a_d, w_d$ are known and f, σ^2 unknown. Here the minimum variance linear unbiased estimator of f is

$$\widehat{f} = b_1 Z_1 + \cdots + b_d Z_d \quad \text{where } b_t = \frac{w_t a_t}{w_1 a_1^2 + \cdots + w_d a_d^2}, \tag{1.10}$$

as follows by straightforward minimization of $\sum_1^d (Z_t - fa_t)^2/(1/w_t)$. For the adaptation to the chain-ladder context, take $d = T - j + 1$, $Z_t = C_{t,j+1}$, $a_t = C_{t,j}$, $f = f_j$, $w_t = C_{t,j}^{2-\alpha}$. Then (1.6) corresponds to $\alpha = 0$ (the simple average of the $F_{t,j}$); (1.5) corresponds to $\alpha = 1$ and one can interpret (1.9) as $C_{t,j}$ being something like the risk volume of $C_{t,j+1}$; and finally one has $\alpha = 2$ in (1.8), corresponding to the model being an ordinary regression of $C_{t,j+1}$ against $C_{t,j}$ through the origin. Standard least squares theory further gives that

$$\mathbb{Var}\,\widehat{f_j} = \frac{\sigma_j^2}{C_{1,j}^\alpha + \cdots + C_{T-j-1,j}^\alpha}.$$

Example 1.2 For a variant of the above set-up, assume that the number n_t of claims reported in year t is known. One could then argue that $C_{t,j} = \sum_{k=1}^{n_t} V_{k;t,j}$ where $V_{1;t,j}, V_{2;t,j}, \ldots$ are i.i.d. and with (unknown) distribution depending only on j, not t. Taking

$$\mathcal{F}_j = \sigma\left(n_1, C_{1,j} \ldots, n_{T-j}, C_{T-j,j}\right),$$

this would lead to

$$\mathbb{E}[C_{t,j+1} \mid \mathcal{F}_j] = f_j n_t, \quad \mathbb{V}\mathrm{ar}(C_{t,j+1} \mid \mathcal{F}_j) = \sigma_j^2 n_t$$

and the least squares estimator

$$\frac{C_{1,j+1} + \cdots + C_{T-j-1,j+1}}{n_1 + \cdots + n_{T-j-1}}.$$

The main difference between this set-up and (1.4) is in small samples since if the n_t are large, then $C_{t,j} \approx n_t \mathbb{E} V_{1;t,j}$ so that

$$\mathbb{V}\mathrm{ar}\left(C_{t,j+1} \mid C_{t,0}, \ldots, C_{t,j}\right) \approx \mathbb{V}\mathrm{ar}\left(C_{t,j+1} \mid n_t\right) = n_t \ \mathbb{V}\mathrm{ar}\, V_{1;t,j}$$

$$= C_{t,j} \frac{\mathbb{V}\mathrm{ar}\, V_{1;t,j}}{\mathbb{E} V_{1;t,j}}.$$

$\diamond$

Remark 1.3 A modification of the prediction procedure in (1.3) is the so-called *Bornhuetter–Ferguson method* that assumes that $C_{t,J}^* = \mathbb{E}C_{t,J}$ is known and replaces the predictor (1.3) by

$$\tilde{O}_t = C_{t,J}^* \frac{\widehat{F}_{T-t,J-1} - 1}{\widehat{F}_{T-t,J-1}},$$

where, as in (1.3), $\widehat{F}_{i,k} = \widehat{f}_i \cdots \widehat{f}_k$ is an estimate of $F_{i,k} = f_i \cdots f_k$, thereby combining the known paid claims (triangles) with the expected overall. To understand this expression, observe that

$$C_{t,J} \approx C_{t,T-t} F_{T-t,J-1}, \quad C_{t,T-t} \approx \frac{C_{t,J}}{F_{T-t,J-1}},$$

$$O_t = C_{t,J} - C_{t,T-t} \approx \frac{C_{t,J}}{F_{T-t,J-1}} F_{T-t,J-1} - \frac{C_{t,J}}{F_{T-t,J-1}}$$

$$= C_{t,J} \frac{F_{T-t,J-1} - 1}{F_{T-t,J-1}}$$

and replace $C_{t,J}$ by $C_{t,J}^*$ and $F_{T-t,J-1}$ by $\widehat{F}_{T-t,J-1}$ (this is somehow working backward from $t + J$ rather than forward from T).

An example where $C_{t,J}^*$ would possibly be available is the n_t setting of Example 1.2 where one could take $C_{t,J}^* = n_t \mu$, with $\mu = \mathbb{E} V_{1;1,J}$ an estimate of the expected value of the total pay-out from a single claim (which would typically be rather precise).

Exercises

1.1 One could easily believe that when viewing the r.v.s $N_{\text{Settled}}(T)$ etc. in Proposition 1.1 as functions of T, one gets (inhomogeneous) Poisson processes. However, this is not so. Explain why the Poisson thinning argument does not apply at the process level.

1.2 Formulate an analogue of Proposition 1.1 when N lives on $(-\infty, \infty)$ rather than $[0, \infty)$.

1.3 Show that (1.4) implies that (1.2) holds with equality, that the estimates $\widehat{f}_k$ are uncorrelated and thereby that the prediction $\widehat{O}_t$ is unbiased.

Notes
Some general texts on loss reserving are Taylor [171], Wüthrich and Merz [176] and Radke, Schmidt and Schnaus [145]. There is also much material in Schmidli [156].

Apart from the special role of δ_i, σ_i, the continuous-time model in Proposition 1.1 is quite similar to shot-noise. A detailed formal treatment from the point of view of Poisson random measures can be found in Mikosch [122].

The chain-ladder model in the formulation (1.4) is often associated with the name of Mack. The literature is extensive and includes many variants and extensions, with some key references being Mack [112, 113], England and Verrall [69], Taylor [172], Buchwalder et al. [40] and Murphy et al. [124]. An important topic not treated here is calculation of the prediction error, that is, the variance of $\widehat{O}_t$.

2 Multivariate Extreme Value Theory

Consider a random vector $X = (X_1, \ldots, X_d)$ where for simplicity we will assume that the c.d.f. F is continuous. Let $M(n)$ be be the component-wise maxima over n i.i.d. replicas of X, so that its c.d.f. is given by $F^n(x)$. We assume that $M(n)$ after a affine transformation has a weak limit $Y \sim G$ with non-degenerate marginals:

$$\left(\frac{M_i(n) - b_i(n)}{a_i(n)} \le x_i \right)_{i=1,\ldots,d} \quad \to \quad G(x) = G(x_1, \ldots, x_d) \tag{2.1}$$

for some vector-valued sequences $a(n) > 0, b(n)$. In particular, the marginals G_i must be univariate extreme value distributions (Fréchet, Gumbel, negative Weibull) and so they are continuous. Hence $F_i \in \text{MDA}(G_i)$ and the scaling sequences are chosen using the univariate theory in Chap. IX. As in the univariate setting the limit G must be max-stable:

$$G^t(x) = G^t(x_1, \ldots, x_d) = G(\alpha \ldots (t)) \qquad \forall t > 0.$$

This can be used to show that the corresponding (extreme value) copula C satisfies $C^t(u) = C(u^t)$ with u^t interpreted componentwise, cf. Sect. X.4.2. The opposite is also true: every extreme value copula C and any extreme value distribution marginals G_i lead to a max-stable distribution, which can arise as the limit of affinely normalized maxima.

In the setting of extreme value theory it is more natural to transform marginal distributions to unit Fréchet with c.d.f. $e^{-1/x}$, $x \geq 0$ instead of uniforms. Note that $-1/\log F_i(X_i)$ is unit Fréchet, and consider the c.d.f. F^* of the transformed vector

$$\left(-1/\log F_1(X_1), \ldots, -1/\log F_d(X_d) \right),$$

with G^* defined analogously. The new scaling sequences (corresponding to unit Fréchet) are given by $a_i^*(n) = n$, $b_i^*(n) = 0$ and then (2.1) holds if and only if

$$\mathbb{P}(\boldsymbol{M}^*(n)/n \leq \boldsymbol{x}) = F^{*n}(n\boldsymbol{x}) \to G^*(\boldsymbol{x})$$

and the marginal convergence in (2.1) holds.

In applications one normally proceeds in two steps: (i) fit the marginal G_i to block maxima, cf. Sect. 3.4; (ii) transform the data using the empirical c.d.f. and fit the dependence structure G^*.

Thus in the following we assume that the marginals are unit Fréchet, so that $F^n(n\boldsymbol{x}) \to G(\boldsymbol{x})$, which means in particular that the 'Fréchet extreme value copula' G satisfies $G^t(\boldsymbol{x}) = G(\boldsymbol{x}/t)$. Since $F(n\boldsymbol{x}) \to 1$ as $n \to \infty$ for any fixed $\boldsymbol{x} > 0$, we also have

$$n\mathbb{P}(\boldsymbol{X}/n \in [\boldsymbol{0}, \boldsymbol{x}]^c) = n(1 - F(n\boldsymbol{x})) \sim -n \log F(n\boldsymbol{x}) \to -\log G(\boldsymbol{x}) = \nu[\boldsymbol{0}, \boldsymbol{x}]^c,$$

where the latter equality defines a measure ν on $\mathbb{R}_+^d$, which is finite away from 0. This can be extended in the usual way to

$$t\mathbb{P}(\boldsymbol{X}/t \in B) \to \nu(B), \qquad t \to \infty, \tag{2.2}$$

for any B bounded away from 0 and such that ν does not put mass at the boundary of B. Appealing to Proposition X.3.5, we conclude that $\boldsymbol{X}$ is MRV (multivariate regularly varying) with ν related to the angular measure μ via the polar coordinate identity

$$\nu(\mathrm{d}r, \mathrm{d}\theta) = c\alpha r^{-\alpha-1} \, \mathrm{d}r \times \mu(\mathrm{d}\theta).$$

Note that the scaling function $a(t) = t$ is fixed and so we should allow for some constant $c = \nu(\|\boldsymbol{x}\| > 1) > 0$ possibly different from 1. Since $\boldsymbol{X} \in \mathbb{R}_+^d$ due to its unit Fréchet marginals, we restrict our attention to this cone in the following. We also write z instead of θ and let $\boldsymbol{Z} \in \mathcal{B}_1$ be an r.v. with distribution μ.

Recall that $G(\boldsymbol{x}) = \exp(-\nu[\boldsymbol{0}, \boldsymbol{x}]^c)$, where

$$\nu[\boldsymbol{0}, \boldsymbol{x}]^c = \mathbb{E} \int \mathbb{1}(\exists i : r Z_i > x_i) c r^{-2} \, \mathrm{d}r = c \, \mathbb{E}\left(\min \frac{x_i}{Z_i}\right)^{-1} = c \, \mathbb{E}\left(\max \frac{Z_i}{x_i}\right).$$

Choosing $x = 1_i$, we get $c\mathbb{E}Z_i = -\log(G_i(1)) = 1$, showing that $\mathbb{E}Z_i = 1/c$ is independent of i. However, $\|Z\| = 1$ and so $c = d$. Thus we arrive at the following result.

Theorem 2.1 *A max-stable distribution G with unit Fréchet marginals is given by*

$$G(x) = \exp(-\nu[0, x]^c) = \exp\left\{-d\mathbb{E}\left(\max_{i=1}^{d} \frac{Z_i}{x_i}\right)\right\},$$

where the angle Z is supported by $\mathcal{B}_1^+$, i.e. $Z \geq 0$ with $\|Z\| = 1$, and has fixed expectation $\mathbb{E}Z_i = 1/d$. Any such Z leads to a max-stable distribution with unit-Fréchet marginals.

In this context the measure ν is called the exponent measure, and the probability measure $\mathbb{P}(Z \in dz)$ the angular measure. Again, (3.7) can be used to find the asymptotics such as

$$\mathbb{P}(\exists i : X_i > tx_i) \sim \mu[0, x]^c / t = d\, \mathbb{E}\left(\max \frac{Z_i}{x_i}\right)/t, \qquad (2.3)$$

but recall that one needs to convert this result back to the original marginals by looking at $F_i^{-1}(e^{-1/X_i})$. In fact, the asymptotics of this probability for the original vector will depend only on the subset of marginals with the fastest tail decay.

Finally, one can readily obtain a general G with arbitrary extreme value distribution marginals G_i:

$$G(x) = \exp\left\{-d\mathbb{E}\max_{i=1}^{d}\left(-Z_i \log G_i(x_i)\right)\right\}.$$

Remark 2.2 Multivariate regular variation can be applied to the original marginals when these can be scaled by the same function $a(t)$ to get an α-Fréchet limit. This happens if and only if all marginal tails are equivalent up to a positive constant, see also the comment following (2.3).   ◇

Remark 2.3 The following list contains some known characterizations of a max-stable distribution G:

- $G^t(x) = G(a(t)x + \beta(t))$ for all t integers or reals.
- the representation of Theorem 2.1 using the angular measure.
- for all vectors $a, b \geq 0$ there exist vectors $c \geq 0$, d such that $aX' \vee bX'' \overset{\mathcal{D}}{=} cX + d$, where X', X'' are independent copies of X and aX' etc. should be understood as componentwise multiplication.
- if X has α-Fréchet marginals then it is max-stable if all max-linear combinations $\max_j a_j X_j$ are scaled α-Fréchet.

                                                        ◇

Notes
The present section is based on 2018 Aarhus lecture notes written by Jevgenijs Ivanovs. Some textbooks containing multivariate extreme value theory are Resnick [148, 149] and de Haan and Ferreira [72].

3   Statistical Methods for Tails and Extreme Values

To quantify risk in the way of producing numerical values of rare event probabilities requires not only a probability model but also a statistical step, to assess values of the parameters used in the model. The rare event aspect introduces some marked differences from traditional statistics: a method like maximum likelihood will try to fit a parameter θ occurring in a distribution F_θ to match the data in the central part of F_θ, that is, where there are many observations. However, rare event probabilities depend crucially on the tail, on which typically much less information will be available. For this a number of special methods have been developed. Since the emphasis of this book is on probability, we will not go in great detail on this topic, but just present some of the main ideas in order to somehow provide a slightly broader picture. We refer to the references at the end of the section, in which extensive numerical studies can also be found. However, the reader should note that the area is rapidly expanding.

We will consider inference on a distribution F, with particular emphasis on the tail, based on a set of data $X_1, \ldots, X_n \geq 0$ assumed to be i.i.d. with common distribution F. As usual, $X_{1;n} \leq \cdots \leq X_{n;n}$ denote the order statistics.

Inference on $\overline{F}(x)$ beyond $x = \max(X_1, \ldots, X_n) = X_{n;n}$ is of course extrapolation of the data. In a given situation, it will far from always be obvious that this makes sense. The argument for nevertheless developing a systematic statistical theory is that alternatives would typically be based on arguments that are ad hoc, subjective and hand-waving, and hardly likely to be any better. The methods we consider have the common property of only being based on the largest among the X_i, more precisely those with $X_i > x$ for some given threshold x; often x is chosen as one of the largest observations, say $X_{n-k;n}$. This will be sufficient to cover most of the heavy-tailed cases and extreme value behavior. It should, however, be noted that in some problems the shape of the whole distribution matters, not just the tail (cf. e.g. the light-tailed large deviation probabilities in Sect. III.3, where the values of exponential moments play a crucial role). A further remark is that the choice of x is important for the performance of the methods, but often hard to make. One often considers all $x \geq u$ for some u, but this of course only shifts the problem from x to u.

3.1   The Mean Excess Plot

A first question is to decide whether to use a light- or a heavy-tailed model. A widely used approach is based on the *mean excess function*, defined as the function

$$e(x) \;=\; \mathbb{E}\big[X - x \,\big|\, X > x\big] \;=\; \frac{1}{\overline{F}(x)} \int_{x}^{\infty} \overline{F}(y)\,dy$$

of $x > 0$, that is, the stop-loss transform (which we have already encountered on a number of earlier occasions) normalized by the tail probability, or the mean conditional overshoot.

The reason why the mean excess function $e(x)$ is useful is that it typically asymptotically behaves quite differently for light and heavy tails. Namely, for a subexponential heavy-tailed distribution one has $e(x) \to \infty$, whereas with light tails it will typically hold that $\limsup e(x) < \infty$; say a sufficient condition is

$$\overline{F}(x) \;\sim\; \ell(x)e^{-\alpha x} \tag{3.1}$$

for some $\alpha > 0$ and some $\ell(x)$ such that $\ell(\log x)$ is slowly varying (e.g., $\ell(x) \sim x^{\gamma}$ with $-\infty < \gamma < \infty$). Cf. Sects. III.2, IX.3 and Exercises 3.1, 3.2.

The mean excess test proceeds by plotting the empirical version

$$e_n(x) \;=\; \frac{1}{\#j : X_j > x} \sum_{j:X_j>x} (X_j - x)$$

of $e(x)$, usually only at the (say) K largest X_j. That is, the plot consists of the pairs formed by $X_{n-k;n}$ and

$$\frac{1}{k} \sum_{\ell=n-k+1}^{n} \big(X_{\ell;n} - X_{n-k;n}\big),$$

where $k = 1, \ldots, K$. If the plot shows a clear increase to ∞ except possibly at very small k, one takes this as an indication that F is heavy-tailed, otherwise one settles for a light-tailed model. The rationale behind this is the different behavior of $e(x)$ with light and heavy tails noted above.

3.2   The Hill Estimator

Now assume that F is either regularly varying, $\overline{F}(x) = L(x)/x^{\alpha}$, or light-tailed satisfying (3.1). The problem is to estimate α.

Since α is essentially a tail parameter, it would as noted above make little sense to use the maximum likelihood estimator (MLE), because maximum likelihood will try to adjust α so that the fit is good in the center of the distribution, without caring too much about the tail where there are fewer observations. The Hill estimator is the most commonly used (though not the only) estimator designed specifically to take this into account.

To explain the idea, consider first the setting of (3.1). If we ignore fluctuations in $\ell(x)$ by replacing $\ell(x)$ by a constant, the $X_j - x$ with $X_j > x$ are i.i.d. exponential(α). Since the standard MLE of α in the (unshifted) exponential distribution is $n/(X_1 + \cdots + X_n)$, the MLE α based on these selected X_j alone is

$$\frac{\#j : X_j > x}{\sum_{j : X_j > x}(X_j - x)}.$$

The *Hill plot* is this quantity plotted as a function of x or the number $\#j : X_j > x$ of observations used. As for the mean excess plot, one usually plots only at the (say) k largest j or the k largest X_j. That is, one plots

$$\frac{k}{\sum_{\ell=n-k+1}^{n}\left(X_{\ell;n} - X_{n-k;n}\right)} \tag{3.2}$$

as a function of either k or $X_{n-k;n}$. The *Hill estimator* $\alpha_{n,k}^{H}$ is (3.2) evaluated at some specified k. However, most often one checks graphically whether the Hill plot looks reasonably constant in a suitable range and takes a typical value from there as the estimate of α.

The regularly varying case can be treated by entirely the same method, or one may remark that it is in 1-to-1 correspondence with (3.1) because X has tail $L(x)/x^{\alpha}$ if and only if $\log X$ has tail (3.1). Therefore, the Hill estimator in the regularly varying case is

$$\frac{k}{\sum_{\ell=n-k+1}^{n}\left(\log X_{\ell;n} - \log X_{n-k;n}\right)}. \tag{3.3}$$

For the Pareto distribution and $k = n$, this is in fact just the usual MLE.

It can be proved that if $k = k(n) \to \infty$ but $k/n \to 0$, then weak consistency $\alpha_{n,k}^{H} \xrightarrow{\mathbb{P}} \alpha$ holds. No conditions on L are needed for this. One might think that the next step would be the estimation of the slowly varying function L, but this is in general considered impossible among statisticians. In fact, there are already difficulties enough with $\alpha_{n,k}^{H}$ itself. One main reason is that L can present a major complication, as has been observed in the many 'Hill horror plots' in the literature; cf. e.g. Fig. 4.1.13 on p. 194 of [66]. There is also a CLT $k^{1/2}(\alpha_{n,k}^{H} - \alpha) \to \mathcal{N}(0, \alpha^2)$. For this, however, stronger conditions on $k = k(n)$ and L are needed. In particular, the correct choice of k requires delicate estimates of L.

3.3   Peaks-over-Threshold

As for the Hill estimator, the Peaks-over-Threshold (POT) method uses the values of the X_i exceeding some level u, but the view is now somewhat broader by not just focusing on the regularly varying case but viewing the exceedances from the view of extreme value theory. Recall from Remark IX.3.12 the family of generalized Pareto distributions $G_{\xi,\beta}$ with tails of the form

$$\overline{G}_{\xi,\beta}(x) \;=\; \frac{1}{(1+\xi x/\beta)^{1/\xi}}\,, \tag{3.4}$$

where $\beta > 0$ is the scale parameter and $\xi \in \mathbb{R}$ the shape parameter; as usual, (3.4) should be understood as the $\xi \to 0$ limit $e^{-x/\beta}$ when $\xi = 0$. The support is $(0, \infty)$ when $\xi > 0$, $(0, \beta/|\xi|)$ when $\xi < 0$, and the whole of $\mathbb{R}$ when $\xi = 0$. Letting $F^u(x) = \mathbb{P}(X - u \le x \mid X > u)$ be the exceedance (overshoot) distribution, we then have $F \in \mathrm{MDA(Gumbel)}$ if and only if F^u after proper normalization is of asymptotic form (3.4) with $\xi = 0$ and $F \in \mathrm{MDA(Fréchet)}$ if and only if F^u after proper normalization is of asymptotic form (3.4) with $\xi > 0$ (we ignore MDA(Weibull) since it requires bounded support, usually unimportant in rare event problems).

The POT method then proceeds simply by fixing u and using the MLE for ξ, β based on the X_i with $X_i > u$, say the number is N and the values $Y_1, \ldots, Y_N$. Here N is random but contains little information on ξ, β so it is treated as fixed. Differentiating (3.4) and taking logarithms, the log likelihood then becomes

$$-N \log \beta - (1 + 1/\xi) \sum_{j=1}^{N} \log(1 + \xi Y_j/\beta)\,,$$

where the maximization has to be done numerically.

3.4   Block Maxima

Rather than the generalised Pareto distribution, the block maxima method uses the family

$$G_{\xi,\mu,\sigma}(x) \;=\; \exp\left\{-\left[1 + \xi\left(\frac{x-\mu}{\sigma}\right)\right]^{-1/\xi}\right\} \tag{3.5}$$

of generalized extreme value distributions where the ones with $\xi < 0$ are of negative Weibull type, the ones with $\xi = 0$ (interpreted in the limiting sense) of Gumbel type and the ones with $\xi > 0$ of Fréchet type. The idea is to split the n observations into

B blocks of size $m = n/B$, compute the block maxima

$$M_b = \max\{X_i : i = (m-1)b + 1, \ldots, mb\}, \quad b = 1, \ldots, B,$$

and treat $M_1, \ldots, M_B$ as i.i.d. observations from $G_{\xi,\mu,\sigma}$, estimating ξ, μ, σ by numerical maximization of the log likelihood.

The underlying assumption is again that F is in some MDA. For M_b to have a distribution close to some $G_{\xi,\mu,\sigma}$, one then needs b to be sufficiently large, whereas for the estimation of ξ, μ, σ to be reasonably precise one cannot take B to small. This is a trade-off in the choice of B, but certainly, n needs to be fairly large. There is some folklore that POT is preferable to block maxima if n is small or moderate, that the methods do not perform that differently for large n, but that block maxima is preferable to POT if the i.i.d. assumption is weakened to stationarity and in multivariate settings. See further the references at the end of the section.

Exercises

3.1 Verify that $e(x)$ stays bounded when (3.1) holds.

3.2 Show that

$$e(x) \sim \frac{x}{\alpha - 1}, \quad e(x) \sim \frac{\sigma^2 x}{\log x}, \quad \text{respectively } e(x) \sim \frac{x^{1-\beta}}{\beta}$$

in the three main examples of subexponentiality, F being regularly varying with index α, standard lognormal, resp. Weibull with tail e^{-x^β}.

Notes

For more detailed and broader expositions of related statistical theory, we refer to (in alphabetical order) Albrecher, Beirlant and Teugels [3], Embrechts, Klüppelberg and Mikosch [66], de Haan and Ferreira [72], McNeil, Embrechts and Frey [115] and Resnick [149].

A recent analysis of the block maxima and comparisons with POT is in Ferreira and de Haan [71].

4   Large Deviations Theory in Function Spaces

4.1   *The Large Deviations Principle. One-Dimensional Cramér*

The general set-up of large deviations (LD) theory is that of a sequence $\{\xi_n\}$ of random elements of some abstract space E and a function $J(x)$ on E, denoted the *rate function*. The space E is assumed to be equipped with a topology or equivalently a convergence concept, such that one can speak of open and closed sets and the interior Γ^o and the closure $\overline{\Gamma}$ of $\Gamma \subseteq E$.

Definition 4.1 The sequence $\{\xi_n\}$ satisfies the *large deviations principle* with rate r_n and rate function $J : E \to [0, \infty]$ if for all measurable $\Gamma \subseteq E$

$$\liminf_{n \to \infty} \frac{1}{r_n} \log \mathbb{P}(\xi_n \in \Gamma) \geq - \inf_{x \in \Gamma^o} J(x), \qquad (4.1)$$

$$\limsup_{n \to \infty} \frac{1}{r_n} \log \mathbb{P}(\xi_n \in \Gamma) \leq - \inf_{x \in \overline{\Gamma}} J(x). \qquad (4.2)$$

The inequality (4.2) is referred to as the LD upper bound and (4.2) as the LD lower bound.

The results on averages $\xi_n = S_n/n = (X_1 + \cdots + X_n)/n$ of i.i.d. r.v.s which were given in Sect. III.3.4 appear to be substantially less general than (4.1), (4.2) since the sets Γ there are intervals and not general Borel sets. Actually, we will see that there is only a small step to a LD principle. In fact, we will quickly be able to show the following result, commonly referred to as *Cramér's theorem* in its one-dimensional version.

Theorem 4.2 *Let F be a distribution on $E = \mathbb{R}$ and $\xi_n = S_n/n$. Then the sequence $\{\xi_n\}$ satisfies the LD principle with rate $r_n = n$ and rate function $J = \kappa^*$, the Legendre–Fenchel transform of the cumulant function $\kappa(\theta) = \log \mathbb{E}e^{\theta X_1}$.*

Before proving Cramér's theorem, we shall, however, give a few remarks:

Remark 4.3 The indexing in Definition 4.1 is by $n = 1, 2, \ldots$ and the limit is $n \to \infty$, but the LD principle can in an obvious way be formulated for other possibilities. For example, when considering Brownian motion in Sect. 4.2, the index will be the variance constant $\epsilon > 0$ and the limit $\epsilon \to 0$. $\Diamond$

Remark 4.4 Writing

$$J_\Gamma = \inf_{x \in \Gamma} J(x) \qquad (4.3)$$

for brevity, the LD principle is equivalent to

$$\liminf_{n \to \infty} \frac{1}{r_n} \log \mathbb{P}(\xi_n \in \Gamma) \geq -J_\Gamma \qquad (4.4)$$

for open sets Γ and

$$\limsup_{n \to \infty} \frac{1}{r_n} \log \mathbb{P}(\xi_n \in \Gamma) \leq -J_\Gamma \qquad (4.5)$$

for closed sets Γ. It is close to the more intuitive

$$\mathbb{P}(\xi_n \in \Gamma) \overset{\log}{\sim} e^{-r_n J_\Gamma}, \qquad (4.6)$$

where we use the notation $f_n \overset{\log}{\sim} g_n$ if $\log f_n / \log g_n \to 1$. However, there will typically be sets Γ for which (4.6) fails. It will hold if Γ is a *J-continuity set*, meaning that

$$\inf_{x \in \Gamma^o} J(x) \; = \; \sup_{x \in \overline{\Gamma}} J(x). \tag{4.7}$$

$\Diamond$

Remark 4.5 The level sets of J are defined as the collection of sets of the form $\{x \in E \, : \, J(x) \leq a\}$ with $a < \infty$. A *good rate function* is one for which the level sets are compact. A sufficient condition for this can be seen to be lower semi-continuity, meaning that $J(x) \leq \liminf J(x_n)$ whenever $x_n \to x$ [note the formal similarity to Fatou's lemma!].

In proofs of the LD principle, the lower bound is often rather easy to establish. Goodness of the rate function can play a main role for the upper bound, and it is also an important ingredient of the *contraction principle* given below as Theorem 4.17.

$\Diamond$

Proof We know already from (3.10) that $\mu_n(\Gamma) \overset{\log}{\sim} -n\kappa^*(x)$ if Γ has the form (x, ∞) with $x > \mathbb{E}X$. Sign reversion gives the same conclusion if $\Gamma = (-\infty, x]$ with $x < \mathbb{E}X$. It turns out that because of the extreme simplicity of the topological structure of $\mathbb{R} = \mathbb{R}^1$ there is not much more to prove. We will prove the statement for the case where κ is steep so that κ^* is finite on all of $\mathbb{R}$ and strictly convex. Write $J_\Gamma = \inf_{x \in \Gamma} \kappa^*(x)$ and $\mu_n = \mathbb{P}(S_n/n \in \cdot)$.

As noted in Remark 4.4, it suffices to show the LD lower bound if Γ is open and the upper if Γ is closed. Let first Γ be open and choose $x \in \Gamma$ with $\kappa^*(x) < J_\Gamma + \epsilon$. Assume $x > 0$ (the case $x < 0$ is handled by symmetry). Since Γ is open, there exists a $y > x$ with $(x, y) \subseteq \Gamma$. We then get

$$\mu_n(\Gamma) \; \geq \; \mu_n(x, y) \; = \; \mu_n(x, \infty) - \mu_n(y, \infty)$$

$$= \; \mu_n(x, \infty)\big(1 - \mu_n(y, \infty)/\mu_n(x, \infty)\big) \overset{\log}{\sim} \mu_n(x, \infty),$$

where the last step follows from (3.10) and $\kappa^*(y) > \kappa^*(x)$. This gives

$$\liminf_{n \to \infty} \frac{1}{n} \log \mu_n(\Gamma) \; \geq \; -\kappa^*(x) \; \geq \; -J_\Gamma - \epsilon$$

and the LD lower bound follows by letting $\epsilon \downarrow 0$.

Next let Γ be closed. If $\mathbb{E}X \in \Gamma$, then $J_\Gamma = 0$ and since $\mu_n(\Gamma) \leq 1$, there is nothing to prove. If $\mathbb{E}X \notin \Gamma$, define

$$\overline{x} = \inf\{x > \mathbb{E}X : x \in \Gamma\}, \qquad \underline{x} = \sup\{x < \mathbb{E}X : x \in \Gamma\}.$$

Then by closedness, $\overline{x} > 0$, $\underline{x} < 0$ and so $\kappa^*(\overline{x}) > 0$, $\kappa^*(\underline{x}) > 0$. Suppose for definiteness $\kappa^*(\overline{x}) \leq \kappa^*(\underline{x})$. Then $J_\Gamma = \kappa^*(\overline{x})$ and we get

$$\mu_n(\Gamma) \leq \mu_n(\overline{x}, \infty) + \mu_n(-\infty, \underline{x}) \overset{\log}{\sim} e^{-n\kappa^*(\overline{x})} = e^{-nJ_\Gamma} . \qquad \square$$

4.2 Schilder and Mogulski

In Sect. III.3.4, we considered a sum $S_n = X_1 + \cdots + X_n$ of i.i.d. random variables with common light-tailed distribution F and derived the asymptotics of the probability $\mathbb{P}(S_n > nx)$ of large values. We now consider some broader questions than large values of S_n: if S_n is large, how did it happen? That is, what was the approximate path of $S_1, S_2, \ldots, S_n$? And if f is a function on $[0, 1]$, what is the approximate probability that the random walk after suitable normalization moves close to f? That is, if we define

$$\xi_n(t) = \frac{1}{n} \sum_{i=1}^{\lfloor nt \rfloor} X_i , \tag{4.8}$$

what can we say about $\mathbb{P}(\xi_n \approx f)$? A similar question arises for the linearly interpolated process

$$\xi_n(t) + (t - \lfloor nt \rfloor) X_{\lfloor nt \rfloor + 1} . \tag{4.9}$$

The answer to the question about $\mathbb{P}(\xi_n \approx f)$ is quite easy to guess at the intuitive level. Namely, assume that f is continuous and piecewise linear, for simplicity with slope δ_k in the interval $[(k-1)/K, k/K]$. Inspecting the proof of Cramér's theorem shows that the most likely way for this to happen is that the increments X_i, $i = n(k-1)/K, \ldots, nk/K$, are governed by the exponentially tilted distribution F_{θ_k} where θ_k is chosen so as to give the correct drift, i.e. $\kappa'(\theta_k) = \delta_k$. The probability is approximately

$$\exp\left\{ -n/K \cdot \kappa^*(\delta_k) \right\}$$

so multiplying over k and using a Riemann sum approximation leads to

$$\mathbb{P}(\xi_n \approx f) \approx \exp\left\{ -n \int_0^1 \kappa^*(f'(t)) \, dt \right\} .$$

The precise formulation of this is *Mogulski's theorem*, to be stated in a moment. Before doing so, we need to introduce some notation: $L_\infty[0, 1]$ will denote the space of bounded functions on $[0, 1]$ equipped with the supremum norm; $C_\infty[0, 1]$ is the space of continuous functions on $[0, 1]$, also equipped with the supremum norm; and $D_\infty[0, 1]$ is the space of cadlag functions equipped with the Skorokhod topology. Let further $\mathcal{H}$ (sometimes called the Cameron–Martin space) be the set of differentiable functions $f : [0, 1] \to \mathbb{R}$ with $f(0) = 0$ and $\dot{f} \in L^2[0, 1]$, i.e. $\int_0^1 \dot{f}(t)^2 \, dt < \infty$, where $\dot{f}$ is the derivative.

Theorem 4.6 (Mogulski) *The LD principle for ξ_n in (4.8) holds in both of the spaces $L_\infty[0, 1]$ and $D_\infty[0, 1]$, with rate n and good rate function*

$$J(f) = \int_0^1 \kappa^*\big(f'(t)\big) \, dt$$

for $f \in \mathcal{H}$ and $J(f) = 0$ otherwise. The same conclusion holds for the linearly interpolated version (4.9), where it is in addition true in $C_\infty[0, 1]$.

If the increment distribution F is standard normal, then obviously ξ_n is close to a scaled Brownian motion $\big\{W(nt)/n\big\}_{0 \le t \le 1}$. The difference is the behavior between the grid points i/n, but this turns out to be unimportant in the limit. Changing notation slightly, we shall consider the BM $W_\epsilon = \epsilon W$ with drift 0 and variance constant ϵ^2. Thus the limit $n \to \infty$ will correspond to $\epsilon \to 0$. We have the following close analogue of Mogulski:

Theorem 4.7 (Schilder) *W_ϵ satisfies the large deviations principle in any of the spaces $L_\infty[0, 1]$, $C[0, 1]$, $D_\infty[0, 1]$ with rate $1/\epsilon^2$ and good rate function*

$$J(f) = \frac{1}{2}\|\dot{f}\|_2^2 = \frac{1}{2}\int_0^1 \dot{f}(t)^2 \, dt$$

for $f \in \mathcal{H}$ and $J(f) = 0$ otherwise.

4.3 Implementation and Examples

The rigorous proofs of the Schilder–Mogulski theorems are too lengthy to be given here, and we refer to standard texts like Dembo and Zeitouni [56]. Instead we concentrate on what is called for in applications, the question of how to find $\inf\{J(f) : f \in B\}$ for a given subset B of $\mathcal{H}$, and how to calculate the minimizer $f^* = f_B^*$.

As preparation, we note the following result:

Lemma 4.8 *Let $f \in \mathcal{H}$, let $0 \le s < t \le 1$ and let $a = f(s)$, $b = f(t)$. Define f^* as the function obtained from f by replacing the segment $\{f(u) : s \le u \le t\}$ by its linear interpolation $f^*(u) = a + b(u - s)/(t - s)$. Then $\int_0^1 \dot{f}^{*2} \le \int_0^1 \dot{f}^2$.*

The situation is illustrated in Fig. 4a, where f is the blue function and f^* is the function which is blue on $[0, s]$, green on $[s, t]$ and blue again on $[0, s]$.

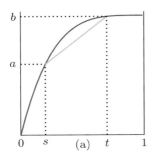 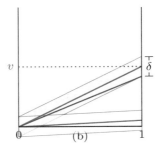

Fig. 4 (a) Improvement by interpolation; (b) linear f

Proof We must show $\int_s^t \dot{f}^{*2} \leq \int_s^t \dot{f}^2$. Since the slope of f^* on $[s, t]$ is $(b - a)/(t - s)$, the l.h.s. is just $(b - a)^2/(t - s)$. The desired conclusion then follows by applying Jensen's inequality with a uniform$[s, t]$ r.v. U, giving

$$\int_s^t \dot{f}^2 = (t - s)\mathbb{E}\dot{f}(U)^2 \geq (t - s)[\mathbb{E}\dot{f}(U)]^2$$

$$= (t - s)\left[\int_s^t \dot{f}(u)\frac{1}{t - s}du\right]^2 = (b - a)^2/(t - s).$$

$\square$

Remark 4.9 The contribution $\dot{f}(u)^2\,du$ to $\|\dot{f}\|_2^2$ can be interpreted as the energy spent by the Brownian motion to travel at drift (speed) $\dot{f}(u)$ in the interval $[u, u + du]$. In this formulation, Lemma 4.8 means that when traveling from a to b in the time interval $[s, t]$, keeping constant speed is optimal, and Schilder's theorem states that when traveling in a given region, the Brownian motion will seek to minimize energy. For a similar but slightly different interpretation, see Exercise 4.2. $\diamond$

Example 4.10 If B is the class of all $f \in \mathscr{C}$ with $f(0) = 0$, $f(1) = v$, Lemma 4.8 says that the minimizer $f^* = f_B^*$ of $\int_0^1 \dot{f}^2$ is the linear function $f^*(t) = tv$. If instead $B = \{f : \|f - f_v\|_\infty \leq \delta\}$ where $f_v(t) = tv$, the answer is slightly different. Given $f(1) = z$, the minimizer in this class is the linear function tz and the energy spent is z^2. Thus $f^*(t) = tz^*$, where $z^* = \mathrm{argmin}\{|z| : v - \delta \leq z \leq v + \delta\}$. Most often, $z^* = \min\{|v - \delta|, |v + \delta|\}$ is the boundary point closest to 0 as in the upper of the two cases in Fig. 4b. The exception occurs if the interval $[v - \delta, v - \delta]$ straddles 0 as in the lower case, where instead $z^* = 0$. In both cases, B is the shaded region, the red linear function is $f_v(t) = tv$ and the blue the optimizer $f^*(t) = tz^*$. $\diamond$

Example 4.11 Generalizing Example 4.10 slightly, we ask for the minimization of $J(f)$ in the class B of functions $f \in \mathscr{C}$ with $f(0) = 0$, $f(1) \geq v$, where $v \geq 0$. The minimizer f^* of $\int_0^1 \dot{f}^2$ is again the linear function $f^*(t) = tv$, as one readily guesses, and the argument is similar as above: among the candidates attaining the value $v_1 \geq v$ at $t = 1$, the linear function tv_1 minimizes the spent energy, and the minimum over v_1 is attained at $v_1 = v$.   ◇

Example 4.12 Consider the probability that W_ϵ ever (i.e., possibly later than at $t = 1$) crosses the line $a + ct$ where $a, c > 0$. If crossing occurs at time t_0, the most likely path is the line connecting $(0, 0)$ and $(t_0, a + ct_0)$. Its slope is $a/t_0 + c$ so the energy needed per unit time to follow this line is $(a/t_0 + c)^2$. This has to go on for t_0 units of time, so the overall energy consumption is $(a/t_0^{1/2} + ct_0^{1/2})^2$. Since the minimizer of $a/s + cs$ is $s_0 = (a/c)^{1/2}$, the overall minimum is attained for $t_0 = s_0^2 = a/c$, and so the probability asked for is approximately

$$\exp\left\{ - \left(a/(a/c)^{1/2} + c(a/c)^{1/2}\right)^2 / 2\epsilon^2 \right\} = \exp\left\{ - 2ac/\epsilon^2 \right\}.$$

Note that the optimal slope is $2c$ independently of a.   ◇

Remark 4.13 When faced with a rare event problem for BM in specific situations, one has to asses the order of magnitude of $\mathbb{P}(W_{\mu,\sigma^2} \in S_T)$ where W is BM(μ, σ^2) on $[0, T]$, $S_T \subset \mathscr{C}_T$ where $\mathscr{C}_T$ is the set of continuous functions on $[0, T]$ equipped with the supremum norm. Here typically $T \neq 1$, $\mu \neq 0$, $\sigma^2 \neq 1$ and it may not be a priori clear what the role of ϵ is.

The obvious procedure is to transform back to the setting of Schilder's theorem by scaling time and space. The first step is to get rid of the drift μ by letting

$$S_T^\mu = \left\{ g = (g(t))_{0 \leq t \leq T} : g(t) = f(t) - \mu t \text{ for some } f \in S_T \right\}.$$

Then $\mathbb{P}(W_{\mu,\sigma^2} \in S_T) = \mathbb{P}(W_{0,\sigma^2} \in S_T^\mu)$. Next, we remark that ϵ in Schilder's theorem is synonymous with variance, and that the result continues to hold if the interval $[0, 1]$ is replaced by $[0, T]$. Altogether, this gives (assuming for simplicity that S_T and therefore S_T^μ is a continuity set)

$$\log \mathbb{P}(W_{\mu,\sigma^2} \in S_T) = \log \mathbb{P}(W_{0,\sigma^2} \in S_T^\mu)$$

$$\sim -\frac{1}{2\sigma^2} \inf_{g \in S_T^\mu} \int_0^T \dot{g}(t)^2 \, dt = -\frac{1}{2\sigma^2} \inf_{f \in S_T} \int_0^T \left(\dot{f}(t) - \mu\right)^2 dt.$$

◇

Example 4.14 For a digital barrier option, the problem arises (after some rewriting) of estimating the expected payout

$$z = \mathbb{P}\left(\underline{W}(T) \leq -a, \, W(T) \geq b\right), \quad \text{where } \underline{W}(T) = \inf_{t \leq T} W(t)$$

and W is BM(μ, σ^2). If a, b are not too close to 0, this is a rare-event problem regardless of the sign of μ. To facilitate calculations, we shall assume $T = 1$, $\mu = 0$, $\sigma^2 = 1$ and appeal to Remark 4.13 for the general case; see also Exercise 4.1.

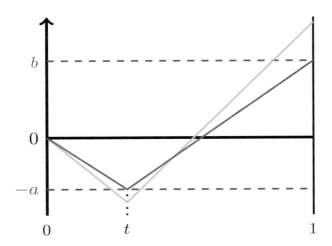

Fig. 5 The optimal path problem for barrier options

The details of the calculation are rather similar to Examples 4.11 and 4.12. We have

$$B = \left\{ f \in \mathscr{C} : \inf_{0 < t < 1} f(t) \leq -a, \; f(1) \geq b \right\}.$$

That is, $f \in B$ if $\dot{f} \in L_2$ and $f(0) = 0$, $f(1) = v$ for some $v \geq b$ and $f(t_0) = u$ for some $u \leq -a$ and some $0 < t_0 < 1$. For any such f, the piecewise linear function with graph connecting $(0, 0)$, $(t_0, -a)$ and $(1, b)$ (the green function in Fig. 5) carries less energy than f), and replacing u by $-a$ and v by b (the red function) in turn further reduces energy. We have thereby reduced to a one-dimensional optimization problem, finding the optimal t_0. This time the path is to be minimized by joining two segments, one from $(0, 0)$, $(t_0, -a)$ with slope $-a/t_0$ and one from $(t_0, -a)$ to $(1, b)$ with slope $(a + b)/(1 - t_0)$. Thus t_0 is the minimizer of

$$\left(\frac{-a}{t_0} \right)^2 t_0 + \left(\frac{a + b}{1 - t_0} \right)^2 (1 - t_0) = \frac{a^2}{t_0} + \frac{(a + b)^2}{1 - t_0}, \tag{4.10}$$

which gives $t_0 = a/(2a + b)$. Substituting back into (4.10) leads after easy algebra to

$$\mathbb{P}(W_\epsilon \in B) \overset{\log}{\sim} \exp\left\{ - (2a + b)^2 / 2\epsilon^2 \right\}.$$

See further the discussion in Glasserman [82, pp. 264 ff].   ◇

Exercises

4.1 Find the optimal path when $\mu \neq 0$ in Example 4.14 on barrier options.

4.2 Show by calculus of variations (see the notes at the end of the section) that the shortest path between two points in the plane is a straight line, i.e., consider two points (s, a), (t, b) connected by the graph $\{(u, f(u)\}$ of a function f and show that the arc length $\int_s^t \sqrt{1 + \dot{f}(u)^2}\, du$ is minimized by taking f to be linear.

4.3 Consider $\mathrm{BM}(\mu, \sigma^2)$ on $[0, T]$ and let $\tau = \inf\{t > 0 : f(t) \le -a\}$. Find the optimal path in the following classes of functions which come up in lookback options, resp. Parisian options:

(i) $\{f : \tau \le T\}$, (ii) $\{f : \tau \le T - D,\ f(t) \le -a \text{ for all } t \in [\tau, \tau + D]\}$. You may just take $\mu = 0$, $\sigma^2 = 1$, $T = 1$.

4.4 LD Results for SDEs

Since the solution of an SDE

$$dX(t) = b(X(t))\, dt + \sigma(X(t))\, dB(t), \qquad X(0) = x_0 \tag{4.11}$$

is 'locally Brownian', it seems reasonable to expect that some generalizations of Schilder's theorem hold in this setting. These are the topic of this section, and in this connection we also shall meet an important general tool in LD theory, the *contraction principle*.

In our study of the SDE (4.11), we only consider the isotropic case $\sigma^2(x) \equiv \sigma^2$ in detail. As in Sect. 4.2, we shall use a small noise set-up, so that we consider X_ϵ given as the solution of

$$dX_\epsilon(t) = b(X_\epsilon(t))\, dt + \epsilon\, dW(t), \quad 0 < t \le 1, \qquad X(0) = x_0. \tag{4.12}$$

We shall need a standard global Lipschitz condition

$$\bigl|b(s) - b(t)\bigr| \le B|s - t|, \quad s, t \in [0, 1]. \tag{4.13}$$

Theorem 4.15 *Assume that* (4.13) *holds. Then the family* $(X_\epsilon)_{\epsilon > 0}$ *satisfies the LD principle with rate $1/\epsilon^2$ and good rate function*

$$J_b(f) = \frac{1}{2} \int_0^1 \bigl[\dot{f}(t) - b(f(t))\bigr]^2$$

for $f \in \mathscr{H}$ and $J_b(f) = \infty$ otherwise.

The proof is carried out in several steps. The first one shows that the solution X is continuous as a function of g in the supremum norm.

Proposition 4.16 *Assume that* $b : [0, 1] \to \mathbb{R}$ *satisfies* (4.13). *Then the equation*

$$f(t) = \int_0^t b\big(f(s)\big)\,ds + g(t), \quad 0 \le t \le 1, \tag{4.14}$$

admits a unique solution $f \in \mathscr{C}$, *say* $f = \varphi(g)$. *Further,* $\varphi : \mathscr{C} \to \mathscr{C}$ *is continuous in the supremum norm* $\| \cdot \|_\infty$.

Proof If g is differentiable, the existence and uniqueness of $f = \varphi(g)$ solving $\dot{f} = b \circ f + \dot{g}$, $f(0) = g(0)$ follows immediately from the theory of ODEs, and this f then also solves (4.14).

To extend φ to the whole of $\mathscr{C}$, assume that f_1, f_2 are solutions corresponding to g_1, g_2 and let $\Delta(t) = \big| f_1(t) - f_2(t) \big|$. Then

$$\Delta(t) \le \int_0^t \big| b\big(f_1(s)\big) - b\big(f_2(s)\big) \big|\,ds + \big| g_1(t) - g_2(t) \big|$$

$$\le B \int_0^t \Delta(s)\,ds + \| g_1 - g_2 \|_\infty,$$

where B is as in (4.13). Using Gronwall's lemma, this implies $\Delta(t) \le e^B \| g_1 - g_2 \|_\infty$. Taking sup over t, we have shown

$$\| f_1 - f_2 \|_\infty \le e^B \| g_1 - g_2 \|_\infty. \tag{4.15}$$

For a general $g \in \mathscr{C}$, choose a sequence $\{g_n\}$ of differentiable functions with $\| g - g_n \| \to 0$. Then $\{g_n\}$ is Cauchy, and so by (4.15) is the corresponding sequence $\{f_n\}$ of solutions. Letting $f = \varphi(g)$ denote the limit, it is immediate that f solves (4.14). The continuity of $\varphi : \mathscr{C} \to \mathscr{C}$ then follows from (4.15). □

The next step of the proof is of independent interest. It plays a major role in other aspects of LD theory and goes under the name of the *contraction principle*.

Theorem 4.17 *Let* E_1, E_2 *be metric spaces and* $\varphi : E_1 \to E_2$ *a continuous mapping. Assume that the sequence* ξ_n *of random elements of* E_1 *satisfies a large deviations principle in* E_1 *with rate* r_n *and good rate function* J_1. *Then the sequence* $\varphi(\xi_n)$ *satisfies a large deviations principle in* E_2 *with rate* r_n *and good rate function* J_2, *where*

$$J_2(x_2) = \inf\big\{ J_1(x_1) : x_1 \in E_1, \varphi(x_1) = x_2 \big\}.$$

Proof Let for $i = 1, 2$, $a \ge 0$ $K_i(a) = \big\{ x_i \in E_i : J_i(x_i) \le a \big\}$ denote the level sets of J_1, resp. J_2. The key step is to prove that the inf in the definition of J_2 is attained, i.e. that

$$J_2(x_2) = J_1(x_1) \text{ for some } x_1 \text{ with } \varphi(x_1) = x_2. \tag{4.16}$$

To this end, let $a = J_2(x_2)$ and choose a sequence $x_{1;n}$ with $\varphi(x_{1;n}) = x_2$ and $J_1(x_{1;n}) \downarrow a$. Then for all large n, $x_{1;n}$ is in the compact set $J_1(a+1)$ and hence by passing to a subsequence if necessary, we may assume that $x_1 = \lim x_{1;n}$ exist. Then $J_1(x_1) \leq a$ by upper semi-continuity of J_1 and $\varphi(x_1) = x_2$ by continuity of φ.

It follows in particular from (4.16) that $\varphi(K_1(a)) = K_2(a)$ so that $K_2(a)$ is compact as a continuous image of the compact set $K_1(a)$, i.e., J_2 is good. It also follows that

$$\inf_{x_2 \in \Gamma_2} J_2(x_2) = \inf_{x_1 \in \Gamma_1} J_1(x_1)$$

when $\Gamma_2 \subseteq E_2$ and $\Gamma_1 = \varphi^{-1}(\Gamma_1) = \{x_1 : \varphi(x_1) \in \Gamma_2\}$. But if Γ_2 is open, so is Γ_1 by continuity of φ, and we get

$$\liminf_{n \to \infty} \frac{1}{r_n} \mathbb{P}\big(\varphi(\xi_n) \in \Gamma_2\big) = \liminf_{n \to \infty} \frac{1}{r_n} \mathbb{P}(\xi_n \in \Gamma_1)$$

$$\geq -\inf_{x_1 \in \Gamma_1} J_1(x_1) = -\inf_{x_2 \in \Gamma_2} J_2(x_2),$$

using the lower LD bound for ξ_n in the second step. Thus the lower LD bound holds for $\varphi(\xi_n)$ as well, and the proof of the upper is similar since Γ_1 is closed when Γ_2 is so. $\square$

Proof of Theorem 4.15 Just note that the mapping φ taking W_ϵ into X_ϵ is continuous by Proposition 4.16 and then appeal to the contraction principle. $\square$

The non-isotropic case

$$\mathrm{d}X_\epsilon(t) = b\big(X_\epsilon(t)\big)\,\mathrm{d}t + \epsilon\sigma\big(X_\epsilon(t)\big)\,\mathrm{d}W(t), \quad 0 < t \leq 1, \quad X(0) = x_0, \quad (4.17)$$

where $\sigma(x)$ depends effectively on x, is mathematically more difficult, though the heuristics is fairly straightforward.

Notes

The classical textbook on LD is the one by Dembo and Zeitouni [56]. For further results in LD theory, we mention in particular:

the multidimensional Cramér theorem, allowing the X_n to take values in $\mathbb{R}^d$ rather than $\mathbb{R}$ [56, Section 2.2.2];

the Gärtner–Ellis theorem [56, Section 2.3]. This is essentially a generalization of Cramér's theorem, where the structure $S_n = X_1 + \cdots + X_n$ with i.i.d. X_i is weakened to the existence of $\kappa(\theta) = \lim_{n \to \infty} \log \mathbb{E}e^{\theta S_n}/n$;

a result of Glynn and Whitt [83], which is a version of the Cramér–Lundberg approximation. It states that $\mathbb{P}(\sup S_n > x) \overset{\log}{\sim} e^{-\gamma x}$ under conditions similar to the Gärtner–Ellis theorem where γ solves $\kappa(\gamma) = 0$ (some negative mean conditions and further technical assumptions are also required). See, e.g., Asmussen and Albrecher [8, Section XIII.1] for applications to ruin theory.

In the *calculus of variations*, one considers functions f on a compact interval, w.l.o.g. [0, 1], and the problem of finding the minimizer f^* in an open set of

$$\inf_f \int_0^1 \Psi\big(t, f(t), \dot{f}(t)\big)\,\mathrm{d}t \tag{4.18}$$

for some smooth function $\Psi(x, y, z)$. It therefore obviously has some relation to the general question of minimizing $\|f\|_2$ in the setting of Schilder's or Mogulski's theorem. This is further exploited, for example, in Appendix C of Shwartz and Weiss [158].

5 Gaussian Maxima

We saw in Sect. IX.3 that if $X_1, X_2, \ldots$ are i.i.d. standard normal, then $\max(X_1, \ldots, X_n)$ is of order $\sqrt{2 \log n}$ as $n \to \infty$ and we gave a more precise limit result in terms of the Gumbel distribution. Many applications call, however, for results that are either more general by allowing the X_i to have different means and variances, or involve the tail of the maximum for a fixed n. The literature on such questions is huge, and we shall here only present a few basic facts. The framework is a family $(X_t)_{t \in \mathbb{T}}$ of r.v.s indexed by the set $\mathbb{T}$ and having a multivariate normal (Gaussian) distribution, with means $\mu_t = \mathbb{E}X_t$, variances $\sigma_t^2 = \mathbb{V}\mathrm{ar}\, X_t$ and covariances $\rho(s, t) = \mathbb{C}\mathrm{ov}(X_s, X_t)$. Write $M_{\mathbb{T}} = \max_{t \in \mathbb{T}} X_t$.

A general heuristical principle that often works quite well is to approximate $\mathbb{P}(M_{\mathbb{T}} > x)$ by $\mathbb{P}(X_t > x)$ for the $t \in \mathbb{T}$ maximizing this probability. In practice, this is done by using Mill's ratio, ignoring the x and σ_t in the denominator so that the approximation is

$$\mathbb{P}(M_{\mathbb{T}} > x) \approx \exp\left\{ - (x - \mu_{t(x)})/2\sigma_{t(x)}^2 \right\} \tag{5.1}$$

where $t(x) = \arg\max_{t \in \mathbb{T}} \exp\left\{ - (x - \mu_t)/2\sigma_t^2 \right\}$. A model example is fractional Brownian motion (fBM):

Example 5.1 Consider fBM with drift $-c$ in $\mathbb{T} = [0, \infty)$. Here $\mu_t = -ct$, $\sigma_t^2 = t^{2H}$ for some $H \in (0, 1)$ (the *Hurst parameter*) and

$$\rho(s, t) = \frac{1}{2}\left(|s|^{2H} + |t|^{2H} - |t - s|^{2H}\right).$$

Note that $H = 1/2$ corresponds to BM, whereas correlations are positive for $H > 1/2$ and negative for $H < 1/2$. Easy calculus gives

$$t(x) = \arg\min_{t \geq 0}(x - \mu_t)/2\sigma_t^2 = \frac{x}{c}\frac{H}{1 - H}$$

and so (5.1) takes the form

$$\mathbb{P}\left(\sup_{t \geq 0} X_t > x\right) \approx \exp\left\{-\frac{c^{2H}}{2H^{2H}}\left(\frac{x}{1 - H}\right)^{2 - 2H}\right\}. \tag{5.2}$$

This indeed reduces to the exact value e^{-2cx} for BM. The decay rate $2 - 2H$ is < 1 for $H > 1/2$ and > 1 for $H < 1/2$, as may be explained by negative correlations helping to prevent large maxima.

The heuristically derived approximation (5.2) is in fact in agreement with the exact asymptotics found in, e.g., Narayan [129] and Hüssler and Piterbarg [91]. $\Diamond$

In the rest of the section, we consider the centered case $\mu_t \equiv 0$ and also assume $\mathbb{T}$ to be finite. This is mainly for ease of exposition, since the classical results to be presented do have extensions to infinite $\mathbb{T}$. Of course some technicalities both in proofs and conditions then come up and somewhat surprisingly, the key difficulties in the derivation often lie in deriving non-trivial and non-standard results for the multivariate normal distribution; see in particular Lemma 5.4 below and Lemma A.2 of the Appendix.

In the centered case, the above principle of looking for the largest marginal probability $\mathbb{P}(X_t > x)$ is equivalent to looking for the largest variance

$$\sigma_{\mathbb{T}}^2 = \sup_{t \in \mathbb{T}} \sigma_t^2 \tag{5.3}$$

(note, e.g., that the $|X_t|$ are stochastically ordered according to the σ_t^2-values). Recalling the definition of $\overset{\log}{\sim}$ from Sect. 4, we have in fact:

Theorem 5.2 *In the finite centered case,* $\mathbb{P}(M_{\mathbb{T}} > x) \overset{\log}{\sim} e^{-x^2/2\sigma_{\mathbb{T}}^2}.$

As may be guessed, this result allows for substantial sharpening. In fact, noting that the sup in (5.3) is attained in the finite case, a trivial lower bound is

$$\mathbb{P}(M_{\mathbb{T}} > x) \geq \frac{1}{\sigma_{\mathbb{T}}\sqrt{2\pi}} e^{-x^2/2\sigma_{\mathbb{T}}^2}.$$

Also

$$\mathbb{P}(M_{\mathbb{T}} > x) \leq \sum_{t \in \mathbb{T}} \mathbb{P}(X_t > x) \leq \frac{d}{\sigma_{\mathbb{T}}\sqrt{2\pi}} e^{-x^2/2\sigma_{\mathbb{T}}^2}, \tag{5.4}$$

where $d = \mathbb{T}$ is the number of elements in $\mathbb{T}$. Combining these observations immediately proves Theorem 5.2.

Notes

Standard expositions of the theory of maxima of Gaussian processes are Adler [2], Piterbarg [141] and Azaïs and Wschebor [16]. Mandjes [114] gives a treatment of some specific processes (fBM, integrated Ornstein–Uhlenbeck processes, telecommunications models, etc.) and in particular has further examples related to (5.1). There are many other methods than the ones we have touched upon here, including several that involve splitting $\mathbb{T}$ up into blocks and piecing these together. A short survey is given by Berman in his MR13611884 review of [141].

5.1 Borell's Inequality

Another upper bound is

$$\mathbb{P}(M_{\mathbb{T}} > x) \leq 2 \exp\left\{ -(x - \mathbb{E}M_{\mathbb{T}})^2 / 2\sigma_{\mathbb{T}}^2 \right\}. \tag{5.5}$$

Actually, (5.4) may be sharper for a given d and x, but the notable feature is that (5.5) does not depend on d. Note, however, that it involves the constant $\mathbb{E}M_{\mathbb{T}}$ which even for the finite case only is available as a d-dimensional integral. The bound (5.5) is a one-sided version of the following main result, known (in a somewhat broader setting) as *Borell's inequality*:

Theorem 5.3 $\mathbb{P}(|M_{\mathbb{T}} - \mathbb{E}M_{\mathbb{T}}| > x) \leq 2e^{-x^2/2\sigma_{\mathbb{T}}^2}.$

The key step in the proof is the following lemma. Let $\|\cdot\| = \|\cdot\|_2$ be the Euclidean norm and ∇h the gradient (column) vector of $h : \mathbb{R}^d \to \mathbb{R}$.

Lemma 5.4 *Let $X = (X_1, \ldots, X_d)$ be $\mathcal{N}_d(\mathbf{0}, \mathbf{R})$ distributed and let $g : \mathbb{R}^d \to \mathbb{R}$ be C^2 with $\|\nabla g\| \leq 1$. Then $\mathbb{P}(|g(X) - \mathbb{E}g(X)| > x) \leq 2e^{-x^2/2r_{\max}}$ where $r_{\max} = \max_1^d r_{ii} = \max_1^d \mathbb{V}\mathrm{ar}\, X_i$.*

The proof of this lemma requires in turn another lemma (we omit the statement of the precise regularity properties of f needed for this):

Lemma 5.5 *Let $W = (W_1, \ldots, W_d)$, $\widetilde{W} = (\widetilde{W}_1, \ldots, \widetilde{W}_d)$ with the W_i independent standard BMs and similarly for the $\widetilde{W}_i$. Let $f : \mathbb{R}^d \to \mathbb{R}$ and define $F(t, x) = \mathbb{E}f(x + \widetilde{W}(1 - t))$. Then $F(t, W(t))$ is a martingale given by*

$$F(t, W(t)) - F(0, 0) = \sum_{i=1}^d \int_0^t \mu_i(t)\, \mathrm{d}W_i(t),$$

where $\mu_i(t) = \mathbb{E}f_{x_i}(x + \widetilde{W}(1 - t)) = \mathbb{E}f_{x_i}(x + W(1 - t)).$

Proof Define $\widetilde{F}(t, x) = \mathbb{E}f(x + \widetilde{W}(t))$. Then the partial derivatives are given by

$$\widetilde{F}_{x_i}(t, x) = \mathbb{E}f_{x_i}(x + \widetilde{W}(t)), \quad \widetilde{F}_{x_i x_j}(t, x) = \mathbb{E}f_{x_i x_j}(x + \widetilde{W}(t)),$$

$$\widetilde{F}_t(t, x) = \frac{1}{2} \sum_{i,j=1}^d \mathbb{E}f_{x_i x_j}(x + \widetilde{W}(t)). \tag{5.6}$$

Indeed, the statement on the $\widetilde{F}_{x_i}$, $\widetilde{F}_{x_i x_j}$ is clear. The statement on $\widetilde{F}_t$ follows from Itô's formula, which gives

$$\mathrm{d}f(x + \widetilde{W}(t)) = \sum_{i=1}^d f_{x_i}(x + \widetilde{W}(t))\, \mathrm{d}\widetilde{W}_i + \frac{1}{2} \sum_{i,j=1}^d f_{x_i x_j}(x + \widetilde{W}(t))\, \mathrm{d}t$$

so that

$$\widetilde{F}_t(t, \boldsymbol{x})\, \mathrm{d}t = \mathrm{d}\mathbb{E} f\big(\boldsymbol{x} + \widetilde{\boldsymbol{W}}(t)\big) = \mathbb{E}\big[\mathrm{d} f\big(\boldsymbol{x} + \widetilde{\boldsymbol{W}}(t)\big)\big]$$

$$= 0 + \frac{1}{2} \sum_{i,j=1}^{d} \mathbb{E} f_{x_i x_j}\big(\boldsymbol{x} + \widetilde{\boldsymbol{W}}(t)\big)\, \mathrm{d}t \,.$$

Similarly,

$\mathrm{d}F(t, \boldsymbol{W}(t))$

$$= F_t(t, \boldsymbol{W}(t)) + \sum_{i=1}^{d} F_{x_i}\big(\boldsymbol{x} + \boldsymbol{W}(t)\big)\, \mathrm{d}W_i + \frac{1}{2} \sum_{i,j=1}^{d} F_{x_i x_j}\big(\boldsymbol{x} + \boldsymbol{W}(t)\big)\, \mathrm{d}t \,.$$

But

$$F_t(t, \boldsymbol{x}) = -\widetilde{F}_t(1 - t, \boldsymbol{x}), \quad F_{x_i}(t, \boldsymbol{x}) = \widetilde{F}_{x_i}(1 - t, \boldsymbol{x}),$$

$$F_{x_i x_j}(t, \boldsymbol{x}) = \widetilde{F}_{x_i x_j}(1 - t, \boldsymbol{x}),$$

so appealing to (5.6) the $\mathrm{d}t$ terms cancel and we get the conclusion of the lemma.

$\square$

Proof of Lemma 5.4 We shall use the representation $\boldsymbol{X} \overset{\mathscr{D}}{=} \boldsymbol{R}^{1/2}\boldsymbol{W}(1)$, where $\boldsymbol{W}(t) = \big(W_t(s)\big)_{t \in \mathbb{T}}$, with the W_t being independent standard Brownian motions. In view of this, $g(\boldsymbol{X}) \overset{\mathscr{D}}{=} f(\boldsymbol{W}(1))$, where $f(\boldsymbol{x}) = g(\boldsymbol{R}^{1/2}\boldsymbol{x})$. We shall need below that

$$\|\nabla f(\boldsymbol{x})\|^2 \le r_{\max} \quad \text{for all } \boldsymbol{x}. \tag{5.7}$$

This follows since (omitting the argument $\boldsymbol{x}$)

$$\|\nabla f\|^2 = \|\boldsymbol{R}^{1/2}\nabla g\|^2 = \nabla g^{\mathsf{T}} \boldsymbol{R}\, \nabla g \le \nabla g^{\mathsf{T}} \boldsymbol{1}\boldsymbol{1}^{\mathsf{T}} \nabla g\, r_{\max}$$

$$= (\nabla g^{\mathsf{T}} \boldsymbol{1})^2\, r_{\max} \le \|\nabla g\|^2\, r_{\max} \le r_{\max}\,,$$

where the second equality just uses that $\|\boldsymbol{b}\|^2 = \boldsymbol{b}^{\mathsf{T}} \boldsymbol{b}$, the following inequality that $|r_{ij}| \le r_{\max}$ for all i, j by Cauchy–Schwarz, and the next again Cauchy–Schwarz.

Let $F(t, \boldsymbol{x}) = \mathbb{E} f(\boldsymbol{x} + \boldsymbol{W}(1 - t))$, $0 \le t \le 1$. Then by Lemma 5.5, $F(t, \boldsymbol{W}(t))$ is a martingale and we have

$$F(1, \boldsymbol{W}(1)) - F(0, 0) = J(1) \quad \text{where } J(t) = \int_0^1 \sum_{i=1}^{d} \mu_i(s)\, \mathrm{d}W_i(s)$$

with $\mu_i(t) = \mathbb{E} f_{x_i}(x + W(1 - t))$. Noting that $F(1, x) = f(x)$, $F(0, 0) = \mathbb{E} f(W(1))$, this gives

$$g(X) - \mathbb{E} g(X) \stackrel{\mathcal{D}}{=} f(W(1)) - \mathbb{E} f(W(1)) = J(1).$$

Now

$$\sum_{i=1}^{d} \mu_i(t)^2 = \sum_{i=1}^{d} [\mathbb{E} f_{x_i}(x + W(1 - t))]^2 = \|\mathbb{E} \nabla f(x + W(1 - t))\|^2 \leq r_{\max}$$

by (5.7). Define $K(t) = \int_0^t \sum_1^d \mu_i(s)^2/2 \, ds$. Then $K(1) \leq r_{\max}/2$ and for any θ, $e^{\theta J(t) - \theta^2 K(t)}$ is the martingale familiar from Girsanov's theorem ([98] pp. 191, 199). Therefore

$$1 = e^{\theta J(0) - \theta^2 K(0)} = \mathbb{E} e^{\theta J(1) - \theta^2 K(1)} \geq e^{-\theta^2 r_{\max}/2} \mathbb{E} e^{\theta J(1)}.$$

This gives for $\theta > 0$ that

$$\mathbb{P}(g(X) - \mathbb{E} g(X) > x) = \mathbb{P}(J(1) > x) \leq e^{-\theta x} \mathbb{E} e^{\theta J(1)}$$

$$\leq \exp\{\theta^2 r_{\max}/2 - \theta x\} \leq e^{-x^2/2r_{\max}}, \qquad (5.8)$$

where the last step just follows by noting that the exponent is maximized by taking $\theta = x/r_{\max}$. A similar bound for $\mathbb{P}(g(X) - \mathbb{E} g(X) < -x)$ (where one takes $\theta < 0$) completes the proof. $\qquad\square$

Proof of Theorem 5.3 and (5.5) Define

$$m(x) = m(x_1, \ldots, x_d) = \max(x_1, \ldots, x_d), \qquad Q_i = \{x : x_i > x_j \text{ for} j \neq i\}.$$

The obstacle in applying Lemma 5.4 with $g = m$ is that differentiability of m fails at all x not in some Q_i, that is, with $x_i = x_k \geq x_j$ for some i, k and all $j \neq i, k$. To overcome this, we use a smoothing argument and define $g(x; n) = \mathbb{E} m(x + Z(n))$, where $Z(n)$ is $\mathcal{N}_d(0, I/n)$. Taking $Z(n) = Z(1)/\sqrt{n}$, it follows immediately by dominated convergence that $g(x; n) \to m(x)$ as $n \to \infty$. Further, the $p_i(x; n) = \mathbb{P}(x + Z(n) \in Q_i)$ sum to 1 by continuity of $Z(n)$, so that

$$g(x; n) = \sum_{i=1}^{d} \mathbb{E}[x_i + Z_i(n); x + Z(n) \in Q_i].$$

This gives that $g(\cdot; n)$ is differentiable with $g_i(x; n) = p_i(x; n)$, and in particular

$$\|\nabla g(x; n)\|^2 = \sum_{i=1}^{d} p_i(x; n)^2 \leq \sum_{i=1}^{d} p_i(x; n) = 1.$$

Thus Lemma 5.4 applies with $g(x) = g(x; n)$, and the rest of the proof is a trivial continuity argument. For (5.5), just appeal in a similar way to (5.8).   □

Notes
Our exposition of the proof of Borell's inequality is essentially a somewhat more detailed one than pp. 42–47 in Adler [2]. In the original version, the expected value of $M_\mathbb{T}$ is replaced by the median. The inequality holds also for an infinite $\mathbb{T}$ provided $M_\mathbb{T} < \infty$ a.s. To prove this, one needs to establish that $\mathbb{E}M_\mathbb{T} < \infty$ without further assumptions than $M_\mathbb{T} < \infty$, which is easy. However, to establish $M_\mathbb{T} < \infty$ may require more work, and also the problem of evaluating $\mathbb{E}M_\mathbb{T}$ remains.

5.2   Slepian's Inequality

Consider centered Gaussian processes $(X_t)_{t \in \mathbb{T}}$, $(Y_t)_{t \in \mathbb{T}}$ with covariance functions $\rho^X(s, t)$, $\rho^Y(s, t)$ and maxima $M_\mathbb{T}^X$, $M_\mathbb{T}^Y$.

Theorem 5.6 *Assume that $M_\mathbb{T}^X < \infty$ and $M_\mathbb{T}^Y < \infty$ a.s., that the variances are equal, $\rho^X(t, t) = \rho^Y(t, t)$ for all $t \in \mathbb{T}$, and that the covariances are ordered such that $\rho^X(s, t) \geq \rho^Y(s, t)$ for $s \neq t$. Then $M_\mathbb{T}^X \preceq_{\mathrm{st}} M_\mathbb{T}^Y$, i.e. $\mathbb{P}(M_\mathbb{T}^X > x) \leq \mathbb{P}(M_\mathbb{T}^Y > x)$ for all x.*

Proof Assume first that $d = |\mathbb{T}| < \infty$ and let $f_n : \mathbb{R} \to [0, 1]$ be a sequence of non-increasing C^∞ functions such that $f_n(y) \downarrow \mathbb{1}_{y \leq x}$ as $n \to \infty$. Then $h_n(x) = f_n(x_1) \cdots f_n(x_d)$ is non-increasing with second-order partial derivatives

$$\frac{\partial h}{\partial x_i \partial x_j} = f_n'(x_i) f_n'(x_j) \prod_{k \neq i, j} f_n(x_k) \quad i \neq j,$$

which are non-negative since $f_n'(x_i) \leq 0$, $f_n'(x_j) \leq 0$. We can therefore apply Lemma X.8.15 to conclude that $\mathbb{E}h_n(X_\mathbb{T}) \geq \mathbb{E}h_n(Y_\mathbb{T})$. Letting $n \to \infty$ and using dominated convergence gives $\mathbb{P}(M_\mathbb{T}^X \leq x) \geq \mathbb{P}(M_\mathbb{T}^Y \leq x)$ and the conclusion. The case of an infinite $\mathbb{T}$ then follows by applying monotone convergence to some sequence $\mathbb{T}_n$ satisfying $\mathbb{T}_n \uparrow \mathbb{T}$.   □

Remark 5.7 Recall that Lemma X.8.15 was based on the PDE

$$\frac{\partial f}{\partial r_{ij}} = \frac{\partial^2 f}{\partial x_i \partial x_j} \quad i \neq j,$$

for the multivariate normal density f given in Lemma A.2 in the Appendix. A minor extension of the above proof of Theorem 5.6 gives that $X_\mathbb{T} \geq Y_\mathbb{T}$ in lower orthant ordering when $d < \infty$, but remember that in fact the stronger property of supermodular ordering holds, cf. Theorem X.8.14.   ◊

Notes

Theorem 5.6 goes under the name of *Slepian's inequality* or *Slepian's lemma*, but is also often associated with the name of Plackett. With some more effort, one can obtain bounds on $\mathbb{P}(M_{\mathbb{T}}^Y > x) - \mathbb{P}(M_{\mathbb{T}}^X > x)$, called *the normal comparison lemma*, cf. e.g. Leadbetter, Lindgren and Rootzén [106] and Azais and Wschebor [16].

Appendix

A.1 Integral Formulas

Integration by parts to compute expectations has often been used in the text. Its basics is that for $X \geq 0$ a r.v. and $g(x) \geq 0$ differentiable, in the sense that $g(x) = g(0) + \int_0^x g'(y)\,dy$ for some function $g'(y)$, one has

$$\mathbb{E}g(X) = g(0) + \int_0^\infty g'(x)\mathbb{P}(X > x)\,dx . \tag{A.1.1}$$

The proof is simply by writing

$$g(X) - g(0) = \int_0^X g'(x)\,dx = \int_0^\infty g'(y)\mathbb{1}_{X>x}\,dy$$

and taking expectation inside the integral.

For example,

$$\mathbb{E}X = \int_0^\infty \mathbb{P}(X > x)\,dx , \quad \mathbb{E}X^2 = 2\int_0^\infty x\,\mathbb{P}(X > x)\,dx$$

when $X \geq 0$, and in general

$$\mathbb{E}X = \mathbb{E}X^+ - \mathbb{E}X^- = \int_0^\infty \mathbb{P}(X > x)\,dx - \int_{-\infty}^0 \mathbb{P}(X \leq x)\,dx ;$$

note that it is unimportant whether we use sharp or weak inequalities in $\mathbb{P}(X > x)$, $\mathbb{P}(X \leq x)$ since there is only a difference at a countable number of points.

© Springer Nature Switzerland AG 2020
S. Asmussen, M. Steffensen, *Risk and Insurance*, Probability Theory
and Stochastic Modelling 96, https://doi.org/10.1007/978-3-030-35176-2

It is tempting to believe that covariances may be computed by arguments similar to those leading to (A.1.1). However, the situation is more complex. Here is the result and a quite different proof, by coupling:

Proposition A.1 *For any two r.v.s X_1, X_2 in L^2 with joint c.d.f. F and marginal c.d.f.s F_1, F_2, one has*

$$\text{Cov}(X_1, X_2) = \int_{-\infty}^{\infty} \int_{-\infty}^{\infty} \left[F(x_1, x_2) - F_1(x_1) F_2(x_2) \right] dx_1 \, dx_2 . \qquad \text{(A.1.2)}$$

Proof Let (X_1^*, X_2^*) be an independent copy of (X_1, X_2). Then $\mathbb{E}X_i^* = \mathbb{E}X_i$, $\mathbb{E}[X_1^* X_2^*] = \mathbb{E}[X_1 X_2]$ and so

$$\mathbb{E}\left[(X_1 - X_1^*)(X_2 - X_2^*) \right] = 2\mathbb{E}[X_1 X_2] - 2\mathbb{E}X_1 \mathbb{E}X_2 = 2\,\text{Cov}(X_1, X_2) .$$

Define $I_j(x) = \mathbb{1}(X_j \leq x)$, $I_j^*(x) = \mathbb{1}(X_j^* \leq x)$. Then

$$X_j - X_j^* = \int_{X_j^*}^{X_j} dx_j = \int_{-\infty}^{\infty} \left[I_j^*(x_j) - I_j(x_j) \right] dx_j$$

and thus

$$2\,\text{Cov}(X_1, X_2)$$

$$= \mathbb{E} \int_{-\infty}^{\infty} \int_{-\infty}^{\infty} \left[I_1^*(x_1) - I_1(x_1) \right] \left[I_2^*(x_2) - I_2(x_2) \right] dx_1 \, dx_2$$

$$= 2 \int_{-\infty}^{\infty} \int_{-\infty}^{\infty} \left[F(x_1, x_2) - F_1(x_1) F_2(x_2) \right] dx_1 \, dx_2 ,$$

where we used

$$\mathbb{E}\left[I_1(x_1) I_2(x_2) \right] = \mathbb{E}\left[I_1^*(x_1) I_2^*(x_2) \right] = F(x_1, x_2) ,$$
$$\mathbb{E}\left[I_1(x) I_2^*(x) \right] = \mathbb{E}\left[I_1^*(x) I_2(x) \right] = F_1(x_1) F_2(x_2) .$$

$\square$

An integral of the type $\int_a^\theta g(x) \, dx$ has derivative $g(\theta)$ w.r.t. θ. If g depends on θ as well as on x, the rule is

$$\frac{d}{d\theta} \int_a^\theta g(x, \theta) \, dx = g(\theta, \theta) + \int_a^\theta \frac{d}{d\theta} g(x, \theta) \, dx . \qquad \text{(A.1.3)}$$

A.2 Differential Equations

A.2.1 The One-Dimensional Case

Consider a linear ODE, i.e. $\dot{f} = gf + h$, where $\dot{f}$ is synonymous with f'. In detail,

$$f'(t) = g(t)f(t) + h(t), \tag{A.2.1}$$

where t varies in an interval I. By the general theory of ODEs, a solution with $f(t_0) = f_0$ typically exists and is unique for a given $t_0 \in I$ and f_0 (some regularity conditions on g, h like smoothness and bounds for large t are required for this, but we omit the details). Typically, t_0 is the left or right endpoint of I.

The standard solution procedure is to start by solving the homogeneous equation $\dot{f} = gf$. This can be rewritten as $d \log f = g$, so if G is any primitive of g (meaning $G' = g$), then e^G is a solution. One next makes a trial solution $\tilde{f} = e^G k$ for the inhomogeneous equation. For $f = \tilde{f}$, (A.2.1) becomes

$$\tilde{f}g + e^G k' = g\tilde{f} + h.$$

Thus for the trial solution $e^G k$ to be the correct one, it is necessary and sufficient that k be a primitive of $e^{-G}h$ and that the boundary condition $f(t_0) = f_0$ be satisfied. Together with $f(t_0) = f_0$, this gives

$$G(t) = \int_a^t g(s)\,ds, \tag{A.2.2}$$

$$k(t) = f_0 e^{-G(t_0)} + \int_{t_0}^t e^{-G(s)}h(s)\,ds, \tag{A.2.3}$$

$$f(t) = f_0 e^{G(t)-G(t_0)} + e^{G(t)} \int_{t_0}^t e^{-G(s)}h(s)\,ds. \tag{A.2.4}$$

Note that $G(t)$ is only determined up to an additive constant, but that this constant cancels. Thus a is arbitrary.

Example A.1 In Sect. I.5.2, we encountered the example $g(t) = -\lambda$, $h(t) = \lambda^{n+1}t^n e^{-\lambda t}/n!$, together with the boundary condition $f(0) = 0$ for $n > 0$. Taking $a = t_0 = 0$ we have $G(t) = -\lambda t$, which gives

$$f(t) = 0 + e^{-\lambda t}\int_0^t e^{\lambda s} \cdot \lambda^{n+1}s^n e^{-\lambda s}/n!\,ds = \frac{(\lambda t)^{n+1}}{(n+1)!}e^{-\lambda t}. \qquad \diamond$$

Example A.2 Consider more generally the ODE

$$f'(t) = g(f(t)) \tag{A.2.5}$$

with boundary condition $f(0) = f_0$. For the solution, define

$$h(z) = \int_{f_0}^{z} \frac{1}{g(y)} \, dy,$$

and make the trial solution $f(t) = h^{-1}(t)$. Then $h'(z) = 1/g(z)$ and hence

$$f'(t) = \frac{1}{h'(h^{-1}(t))} = \frac{1}{h'(f(t))} = \frac{1}{1/g(f(t))} = = g(f(t)),$$

so (A.2.5) holds. Finally, since $h(f_0) = 0$, the boundary condition $f(0) = f_0$ is satisfied.   □

A.2.2   Systems of ODEs

A multidimensional version of (A.2.1) has the form

$$f_j'(u) = \sum_{i=1}^{p} g_{ij}(u) f_i(u) + h_j(u), \quad j = 1, \ldots, p, \ s \le u \le t, \qquad \text{(A.2.6)}$$

for unknowns $f_1, \ldots, f_p$ and known functions $g_{ij}, h_i, i, j = 1 \ldots, p$. The boundary condition is, e.g., $\boldsymbol{f}(s) = \boldsymbol{x}_s$ where $\boldsymbol{f}(u)$ is the row vector with elements $f_j(u)$ and $\boldsymbol{x}_s = (x_{j;s})$ the row vector of initial conditions conditions $f_j(s) = x_{j;s}$.

Remark A.3 For the linear and homogeneous ODE $\dot{f} = \lambda f$ where λ is constant, we get $f(t) = e^{\lambda t} f(0)$. This generalizes to systems of such equations. Recall that the exponential e^A of a matrix A is defined by

$$e^A = \sum_{n=0}^{\infty} \frac{A^n}{n!} \quad \text{so in particular} \quad e^{tA} = \sum_{n=0}^{\infty} \frac{t^n A^n}{n!}.$$

If $\boldsymbol{f} = (f_1, \ldots, f_p)$ is a vector of unknown functions such that $\dot{f}_i = \sum_j \lambda_{ij} f_j$, we then have $\boldsymbol{f}(t) = e^{\boldsymbol{\Lambda} t} \boldsymbol{f}(0)$, where $\boldsymbol{\Lambda}$ is the matrix with ijth element λ_{ij}. This fact is extremely useful for a *time-homogeneous* Markov process, since it implies that the matrix $\boldsymbol{P}(t, s) = (p_{ij}(t, s))_{i, j \in \mathcal{J}}$ of transition probabilities is given by $\boldsymbol{P}(t, s) = \boldsymbol{P}(0, s - t) = e^{\boldsymbol{M}(s-t)}$, where $\boldsymbol{M}$ is the rate (intensity) matrix with ijth element say μ_{ij}. Unfortunately, this simple solution form does not carry over to the time-inhomogeneous case discussed in the next section, where μ_{ij} also depends on the time parameter v: it is tempting to believe that $\boldsymbol{P}(t, s) = \exp\{\int_s^t \boldsymbol{M}(v) \, dv\}$, but this is not in general true without a commutativity condition, see Exercise A.2.4 below. Thus in contrast to the case $p = 1$, there is no explicit way to write a solution

to the system (A.2.6), though in some cases graph-theoretic considerations allows
such a system to be reduced to applying the explicit $p = 1$ solution recursively, see
Sect. A.3.1 below. $\diamond$

Matrix formalism in terms of the *product integral*

$$P(s, t) \; = \; \prod_{u=s}^{t} (I + G(u)\, du) \tag{A.2.7}$$

is sometimes useful (here the $G(u)$ are $p \times p$ and measurable in u). There are at
least three possible definitions:

A) as the solution of

$$\frac{\partial}{\partial t} P(s, t) \; = \; P(s, t) G(t) \tag{A.2.8}$$

with initial condition $P(s, s) = I$;

B) taking first $P(s, t)$ as the matrix-exponential $e^{(t-s)G}$ if $G(u) \equiv G$ is constant
and next as limits of 'Riemann products'

$$\prod_{k=1}^{n} \prod_{u_{k-1}}^{u_k} (I + G(u_k^*)\, du) \; = \; \prod_{k=1}^{n} e^{hG(u_k^*)} \, ,$$

where $h = (t - s)/n$, $u_k = s + kh$ and u_k^* is some point in $[u_{k-1}, u_k]$;

C) as $\displaystyle\sum_{n=0}^{\infty} R_n(s, t)$, where

$$R_n(s, t) = \int_{B_n(t)} G(u_1) \cdots G(u_n)\, du_1 \, \ldots \, du_n$$

with

$$B_n(t) = \{t \geq u_1 \geq u_2 \ldots \geq u_n \geq 0\}$$

for $n > 0$ and $R_0 = I$.

These definitions are of course closely related. A particular connection is Picard
iteration, which amounts to rewriting (A.2.8) as $P(s, t) = I + \int_s^t P(s, u) G(u)\, du$
and using successive improvements $P_{n+1} = I + \mathcal{T} P_n$ where the operator $\mathcal{T}$ is
defined by $(\mathcal{T} Q)(s, t) = \int_s^t Q(s, u) G(u)\, du$ and $P_0 = I$. Under weak conditions,

$P_n \to P$. But by definition,

$$R_n(s, t) = \int_s^t G(u_1)\left[\int_{B_{n-1}(u_1)} G(u_2)\cdots G(u_n)\,du_2 \ldots du_n\right] du_1$$

$$= \int_s^t G(u_1) R_{n-1}(s, u_1)\,du_1 = (\mathcal{T} R_{n-1})(s, t).$$

Induction then gives $P_n = \sum_0^n R_k$ and so $P = \lim P_n = \sum_0^\infty R_k$.

The connection of the system (A.2.6) to the product integral is clear in the homogeneous case $h_j(t) \equiv 0$ for all t. Then (A.2.6) means

$$f'(u) = f(u)G(u), \qquad (A.2.9)$$

where $G(u)$ is the matrix with elements $g_{ij}(u)$. Thus $x_s \prod_s^u (I + G(u)\,du)$ is a solution and the only one under mild regularity conditions. In the non-homogeneous case, introduce a dummy unknown $f_0(t) \equiv 1$. Then with $g_{j0} = h_j$, we have $f_j = \sum_0^p g_{ij} f_i$, i.e. (A.2.9) holds where now $G(t)$ is the $(p+1) \times (p+1)$ matrix with elements $g_{ij}(t)$. Thus the vector $\left(f_1(u) \ldots f_p(u)\right)$ is obtained as the final p elements of $x_s \prod_s^u (I + G(v)\,dv)$.

By definition,

$$\prod_s^s (I + G(u)\,du) = I. \qquad (A.2.10)$$

Some further important properties of the product integral are given in Exercises A.2.2–A.2.5.

Exercises

A.2.1 Give some further discussion of the relation between definitions A), B), C) of the product integral. In particular, what is $R_n(s, t)$ in the constant case $G(u) \equiv G$?

A.2.2 Show that $\dfrac{\partial}{\partial s} \prod_s^t (I + G(u)\,du) = -\, G(s) \prod_s^t (I + G(u)\,du).$

A.2.3 Show that

$$\prod_s^t (I + G(u)\,du) = \prod_s^v (I + G(u)\,du) \cdot \prod_v^t (I + G(u)\,du)$$

(usual matrix product) for $s \le v \le t$.

A.2.4 Show that

$$\prod_s^t \left(I + G(u) \, du \right) \; = \; \exp\left\{ \int_s^t G(u) \, du \right\}$$

if all $G(u)$, $G(v)$ commute, i.e. $G(u)G(v) = G(v)G(u)$ for $u \neq v$.

A.2.5 Show that

$$\prod_s^t \left(I + G_1(u) \, du \right) \cdot \prod_s^t \left(I + G_2(u) \, du \right) \; = \; \prod_s^t \left(I + (G_1(u) + G_2(u)) \, du \right)$$

if all $G_1(u)$, $G_2(u)$ commute and that for a scalar function $r(u)$,

$$\exp\left\{ \int_s^t r(u) \, du \right\} \cdot \prod_s^t \left(I + G(u) \, du \right) \; = \; \prod_s^t \left(I + (r(u)I + G(u)) \, du \right).$$

Notes
Brauer and Nohel [32] is a good general text on differential equations. The product integral goes back at least to Dobrushin [63]. It has been used a number of times in statistics, cf. e.g. Gill and Johansen [81], and occurs in Bladt, Asmussen and Steffensen [28] in a life insurance context. It is also treated in Rolski et al. [153]. In the non-stochastic literature, it is sometimes called the *matrizant*.

A.3   Inhomogeneous Markov Processes

Modern life insurance mathematics makes extensive use of inhomogeneous Markov processes. The primitives of such a process $\{Z(t)\}_{t\geq 0}$ are the state space, which we denote here as $\mathcal{J} = \{0, \dots, J\}$, and the transition rates (intensities) $\mu_{ij}(t)$. The interpretation of these is infinitesimal dynamics similar to the one for the Poisson process in Sect. I.5.2 and the hazard rate: for $i \neq j$: a transition from i to j occurs w.p. $\mu_{ij}(t)h + o(h)$ in the time interval $(t, t + h)$ given that $Z(t) = i$. The diagonal elements $\mu_{ii}(t)$ are defined by the convention $\sum_{j \in \mathcal{J}} \mu_{ij}(t) = 0$ or equivalently $\mu_{ii}(t) = -\mu_i(t)$, where $\mu_i(t) = \sum_{j \neq i} \mu_{ij}(t)$. The principle that the rates of independent events add up to the rate of the union then tells us that $\mu_i(t)$ gives the rate of leaving state i at time t. In particular, the sojourn time in state i given $Z(0) = i$ has tail $e^{-\int_0^t \mu_i}$, with the convention that this expression is shorthand for $\exp\left\{ -\int_0^t \mu_i(s) \, ds \right\}$, and the density is $\mu_i(t)e^{-\int_0^t \mu_i}$. Further, formulas behind Poisson thinning (Proposition I.5.6) tells us that then the new state is chosen w.p. $\mu_{ij}(t)/\mu_i(t)$ for j. This description also gives a construction of the process: if $Z(0) = i$, choose T_0 with tail $\mathbb{P}(T_0 > t) = e^{-\int_0^t \mu_i}$ and let $Z(t) = i$ for $0 \leq t < T_0$. Given $T_0 = t$, take $Z(T_0) = j$ w.p. $\mu_{ij}(t)/\mu_i(t)$, choose T_1 with distribution given

Fig. A.1 Markov process
transition diagram

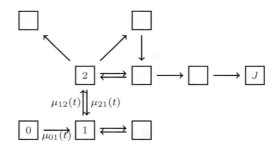

by $\mathbb{P}(T_1 > s) = e^{-\int_t^{t+s} \mu_j}$, stay at j until $T_0 + T_1$, choose the next state k w.p.
$\mu_{jk}(t + T_1)/\mu_j(t + T_1)$, and so on.

The process Z is throughout assumed to be RCLL (right-continuous, left limits),
i.e. with paths in the Skorokhod space $D[0, \infty)$. The history of the process up to and
including time t is represented by the sigma-algebra $\mathcal{F}(t) = \sigma(Z(s), s \in [0, t])$,
and we let $\mathscr{F}^Z = (F^Z(t))_{t \geq 0}$ be the natural filtration. In abstract terms, the
definition of the Markov property is

$$\mathbb{P}(\{Z(t + s)\}_{s \geq 0} \in \cdot \mid \mathcal{F}(t)) = \mathbb{P}(\{Z(t + s)\}_{s \geq 0} \in \cdot \mid Z(t)) \qquad (A.3.1)$$

and this follows immediately from the construction given above.

A convenient way to visualize a Markov process is by its transition diagram as
exemplified in Fig. A.1, where an arrow between states i and j is only included if
$\mu_{ij}(t) > 0$ for some t.

The transition probabilities are defined by

$$p_{ij}(t, s) = \mathbb{P}(Z(s) = j \mid Z(t) = j),$$

The Markov property (A.3.1) gives the so-called *Chapman–Kolmogorov equations*

$$p_{ij}(t, s) = \sum_{k \in \mathcal{J}} p_{ik}(t, v) p_{kj}(v, s), \quad t \leq v \leq s. \qquad (A.3.2)$$

In formulas, the description of the transition mechanism above means

$$p_{ii}(t, t + dt) = 1 - \mu_i(t)dt = 1 + \mu_{ii}(t)dt, \qquad (A.3.3)$$

$$p_{ij}(t, t + dt) = \mu_{ij}(t)dt, \quad j \neq i. \qquad (A.3.4)$$

Implicit in this is that the probability of more than two transitions in the time interval
$(t, t + dt]$ is $O(dt^2) = 0$.

Taking v infinitesimally close to t in (A.3.2) results in a set of ordinary
differential equations (ODEs) known as the *backward equation*, whereas taking v
infinitesimally close to s results in a set of ODEs known as the *forward equation*. In
more detail:

Theorem A.1 *The $p_{ij}(s, t)$ satisfy the backward equation*

$$\frac{\partial}{\partial t} p_{ij}(t, s) = -\sum_{k \in \mathcal{J}} \mu_{ik}(t) p_{kj}(t, s) \tag{A.3.5}$$

and the forward equation

$$\frac{\partial}{\partial s} p_{ij}(t, s) = \sum_{k \in \mathcal{J}} p_{ik}(t, s) \mu_{kj}(s) \tag{A.3.6}$$

with boundary conditions $p_{ij}(s, s) = p_{ij}(t, t) = \delta_{ij}$.

Proof The boundary conditions are obvious. Using the backward argument gives

$$p_{ij}(t - dt, s) = \sum_{k \in \mathcal{J}} p_{ik}(t - dt, t) p_{kj}(t, s),$$

which by (A.3.3), (A.3.4) implies

$$p_{ij}(t, s) - \frac{\partial}{\partial t} p_{ij}(t, s) \, dt = p_{ii}(t, t) p_{ij}(t, s) + \sum_{k \in \mathcal{J}} \mu_{ik}(t) \, dt \cdot p_{kj}(t, s).$$

Subtracting $p_{ij}(t, s) = p_{ii}(t, t) p_{ij}(t, s)$ on both sides and dividing by dt gives (A.3.5). Similarly, (A.3.6) follows from

$$p_{ij}(t, s + ds) = \sum_{k \in \mathcal{J}} p_{ik}(t, s) p_{kj}(s, s + ds),$$

$$p_{ij}(t, s) + \frac{\partial}{\partial s} p_{ij}(t, s) \, dt = \sum_{k \in \mathcal{J}} p_{ik}(t) \mu_{kj}(t, s) \, dt + p_{ij}(t, s) p_{jj}(s, s).$$

$\square$

Corollary A.2 *For $j \neq i$, the solution of the backward equation (A.3.5) is*

$$p_{ij}(t, s) = \int_t^s e^{-\int_t^w \mu_i} \sum_{k \neq i} \mu_{ik}(w) p_{kj}(w, s) \, dw. \tag{A.3.7}$$

Proof Keeping s fixed and letting $f(v) = p_{ij}(v, s)$, the backward equation has the form (A.2.1) with

$$g(v) = -\mu_{ii}(v) = \mu_i(v), \quad h(v) = -\sum_{k \neq i} \mu_{ik}(v) p_{kj}(v, s),$$

with boundary condition $f(t_0) = 0$, where $t_0 = t$. To have $e^{G(t_0)}k(t_0) = 0$, we then need to take $v_2 = s$. Taking $a = t_0 = t$ in (A.2.2)–(A.2.4), we have

$$G(v) = \int_t^v \mu_i(w)\,dw, \quad k(v) = \int_t^v e^{-G(v)}h(w),$$

$$p_{ij}(v,s) = e^{\int_t^v \mu_i} \int_s^v e^{-\int_t^w \mu_i} h(w)\,dw = -\int_v^s e^{-\int_v^w \mu_i} h(w)\,dw$$

$$= \int_v^s e^{-\int_v^w \mu_i} \sum_{k \neq i} \mu_{ik}(w)\, p_{kj}(w,s)\,dw.$$

Taking $v = t$ gives the result.   □

Remark A.3 The identity (A.3.7) can immediately be understood by an intuitive probabilistic argument. Namely, since $j \neq i$, there must be a transition out of i to some state $k \neq i$ at some point $w \in (s, t]$. The term $e^{-\int_t^w \mu_i}$ is the probability that the state has not changed before w, $\mu_{ik}(w)$ is the rate that there is a jump to k, and $p_{kj}(w, s)$ is the probability that the terminal state will then be j.

Formula (A.3.7) however, does not provide an explicit formula for $p_{ij}(t, s)$. The problem is that the $p_{kj}(t, s)$ are in general unknown.   ◇

We shall also need the counting process N^k giving the number of transitions into state k, i.e.

$$N^k(t) = \#\{s \in (0, t] : Z(s-) \neq k, Z(s) = k\}.$$

It has the stochastic intensity process (compensator) $\{\mu_{Z(t-)k}(t)\}_{t \in [0,n]}$, a fact that can formally can be stated by

$$M^k(t) = N^k(t) - \int_0^t \mu_{Z(s)k}(s)\,ds \tag{A.3.8}$$

constituting an $\mathscr{F}^Z$-martingale.

A.3.1   Feed-Forward Systems

Considerable simplification occurs, however, if the transition graph is feed-forward. The meaning of this that for any state i no feedback is possible in the sense that $p_{ji}(t, s) = 0$ for all $t \leq s$ and all states j with $\mu_{ij}(v) > 0$ for some v. See Fig. A.2 for an example. This is a hierarchical structure where many of the μ_{ik} in the backward equation are 0, allowing the $p_{ij}(t, s)$ to be expressed recursively as single integrals using Corollary A.2 (assuming, of course, that the relevant integrals are computable). First of all, since we cannot return to i, we have $p_{ii}(t, s) = e^{-\int_t^s \mu_i}$.

Fig. A.2 Feed-forward
Markov process

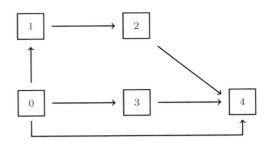

To exemplify the procedure for $j \neq i$, consider again Fig. A.2. The backward equation for say $f(t) = p_{1,2}(t, s)$ is then $f'(t) = \mu_1(t)f(t) + h(t)$, where $h(t) = \mu_{12}(t)p_{22}(t, s)$, so Corollary A.2 gives

$$p_{12}(t, s) = \int_t^s e^{-\int_t^w \mu_1} \mu_{12}(w) e^{-\int_w^s \mu_2} \, dw$$

and similar formulas for p_{01}, p_{24}, p_{03}, p_{34}. We then get

$$p_{14}(s, t) = \int_t^s e^{-\int_t^w \mu_1} \mu_{12}(w) p_{24}(w, s) \, dw \, ,$$

and a similar expression for p_{02}. Finally,

$$p_{04}(s, t) = \int_t^s e^{-\int_t^w \mu_1} \left[\mu_{01}(w) p_{14}(w, s) + \mu_{03}(w) p_{34}(w, s) + \mu_{04}(w) \right] dw \, .$$

Exercises

A.3.1 Derive an analogue of (A.3.7) for the case $i = j$.

A.3.2 Show Corollary A.2 using the forward rather than the backward equation.

A.4   Itô's Formula

The usual chain rule in deterministic calculus is

$$g(t) = f(w(t)) \quad \Rightarrow \quad g'(t) = f'(w(t))w'(t) \tag{A.4.1}$$

which in integral form can be rewritten as

$$f(w(t)) = f(w(0)) + \int_0^t f'(w(s)) \, dw(s) \, . \tag{A.4.2}$$

Itô's formula (or *Itô's rule, Itô's lemma*, etc.) can be seen as a generalization where one replaces $w(\cdot)$ by a random process with non-differentiable sample paths. The problem that arises is that it is not clear how to interpret $w'(t) = \lim[w(t+h) - w(t)]/h$ or $dw(s)$, and/or what to replace these objects with.

The model example is standard Brownian motion $W(\cdot)$, where the object $\{dW(t)\}_{t \geq 0}$ is commonly referred to as *white noise*. Its discrete time analogue is just an i.i.d. sequence of standard normal r.v.s, but in continuous time the concept of white noise does not make sense in an ordinary probability/calculus sense. The key observation for proceeding is that $W(t+h) - W(t)$ is not of order h but $h^{1/2}$, because it is distributed as $h^{1/2}Z$ where $Z \sim \mathcal{N}(0, h)$. Thus, to get all terms of order up to h one needs to include the second-order term in the Taylor expansion of $f(x)$, leading to the heuristics

$$f(W(t+h)) - f(W(t))$$

$$\approx f'(W(t))\big(W(t+h) - W(t)\big) + \frac{1}{2}f''(W(t))\big(W(t+h) - W(t)\big)^2 \, .$$

Replacing $\big(W(t+h) - W(t)\big)^2$ by its expected value h suggests the rule

$$df(W(t)) = f'(W(t)) \, dW(t) + \frac{1}{2}f''(W(t)) \, dt \tag{A.4.3}$$

or, in integral form,

$$f(W(t)) = f(W(t)) + \int_0^t f'(W(s)) \, dW(s) + \frac{1}{2}\int_0^t f''(W(s)) \, ds \, . \tag{A.4.4}$$

Here the first integral has to be interpreted in the Itô sense, i.e. as limits in an appropriate sense of Riemann sums

$$\sum_{i=1}^n f\big(W((i-1)t/n)\big(W(it/n) - W((i-1)t/n)\big)$$

(it is crucial that $f(W(t))$ is evaluated at the left interval point $(i-1)t/n$, not at the right it/n or in between!).

A more general version occurring frequently in this book allows $f = f(t, x)$ to depend on t as well as on x, to have a driving term $X(t)$ more general than $W(t)$, and to incorporate jump terms. The inclusion of jumps in the examples we deal with is, however, elementary, since there jumps only occur at a finite number of points in any interval $[0, t]$; this is tacitly assumed for a, b, X, f, Z in the following. More precisely, $X(t)$ is of the form

$$X(t) = X(0) + \int_0^t a(s)\,ds + \int_0^t b(s)\,dW(s) + \sum_{s \le t} \Delta X(s), \qquad \text{(A.4.5)}$$

where a, b are adapted and bounded on finite intervals, and the $\Delta X(s) = X(s) - X(s-)$ are the jump sizes. Itô's formula then takes the form

$$f(t, X(t)) - f(0, X(0))$$

$$= \int_0^t (f_t + af_x + \frac{1}{2}b^2 f_{xx})\,ds + \int_0^t af_x\,dW(s) \qquad \text{(A.4.6a)}$$

$$+ \sum_{s \le t} \left[f(s, X(s)) - f(s-, X(s-)) \right] \qquad \text{(A.4.6b)}$$

with the partial derivatives f_t, f_x, f_{xx} all evaluated at $(s, X(s))$ and a, b at s. It needs f to be $C^{1,2}$. In many of our life insurance examples, $f = f(t, X(t))$ has an additional random argument $Z(t)$ taking discrete values. However, formula (A.4.6) remains valid when it is understood that $Z(t)$ is suppressed in the notation and that the summands in (A.4.6b) are $f(s, X(s), Z(s)) - f(s-, X(s-), Z(s-))$.

Notes
Itô's formula can be found in any standard textbook on stochastic analysis, e.g. Karatzas and Shreve [98], Protter [144], or Rogers and Williams [152]. The presence of jumps is not incorporated in all textbooks, but is elementary to deal with when there are only finitely many jumps since one can then simply piece segments between jumps together and use the continuous form between jumps.
   Concerning extensions, we mention in particular

- versions allowing X to be a general semi-martingale;
- multidimensional versions, of which an example is in the proof of Theorem XIII.5.3;
- the Itô–Tanaka formula, which relaxes the $C^{1,2}$ assumption on f somewhat and of which we give an application in the following Sect. A.5.

A.5 Diffusion First Passage Probabilities

Lemma A.1 *Let $\mu(x), \sigma^2(x)$ be bounded and measurable functions on an interval (ℓ_d, ℓ_u) such that $\inf_{\ell_d < x < \ell_u} \sigma^2(x) > 0$ and let X, W be defined on a suitable probability space such that W is a standard Brownian motion and*

$$X(t) = \delta + \int_0^t \mu(X(s))\,ds + \int_0^t \sigma(X(s))\,dW(s) \qquad \text{(A.5.1)}$$

for some $\delta \in (\ell_d, \ell_u)$. *Define further* $\kappa(x) = \mu(x)/\sigma^2(x)$,

$$s(y) = \exp\left\{ -2 \int_{\ell_d}^{y} \kappa(z)\, dz \right\}, \quad S(\delta) = \int_{\ell_d}^{\delta} s(y)\, dy$$

(the scale function) *and* $\tau = \inf\{t : X(t) \notin (\ell_d, \ell_u)\}$. *Then*

$$\mathbb{P}(X(\tau) = \ell_u) = S(\delta)/S(\ell_u).$$

Proof We initially remark that the conditions on $\mu(x), \sigma^2(x)$ ensure the existence of X as a weak solution.

The claimed expression for $\mathbb{P}(X(\tau) = \ell_u)$ is a standard general formula for diffusions, typically stated under smoothness conditions ensuring the applicability of Itô's formula. For the present version, note first that the absolute continuity and boundedness of $\int_{\ell_d}^{y} \kappa(z)\, dz$ ensure that $s(y)$ is of bounded variation, hence of the form $s_1(y) - s_2(y)$ for non-decreasing functions $s_1(y), s_2(y)$ of bounded variation. Letting $S_i(\delta) = \int_{\ell_d}^{\delta} s_i(y)\, dy$, S_i is then convex (Problem 6.20 p. 213 in [98]) and we may apply the Itô–Tanaka formula ([98] Sections 5.5.B or [152] Section IV.45) to each S_i separately to conclude that

$$S(X(t)) = S(\delta) + \int_0^t s(X(v))\sigma(X(v))\, dv$$

is a local martingale. The boundedness properties of $s(x), \sigma^2(s)$ ensure that we indeed have a proper martingale, and so by optional stopping

$$S(\delta) = \mathbb{E}S(X(\tau)) = S(\ell_u)\mathbb{P}(X(\tau) = \ell_u) + S(\ell_d)(1 - \mathbb{P}(X(\tau) = \ell_u))$$

$$= S(\ell_u)\mathbb{P}(X(\tau) = \ell_u) + 0.$$

$\square$

Notes

Lemma A.1 is essentially from Pestien and Sudderth [138], with the Itô–Tanaka formula replacing a different generalization of Itô's formula used there.

A.6 L_2 **Projections. Least Squares. Conditional Expectations**

A.6.1   *Least Squares*

In credibility theory, regression and other problem areas, one encounters the problem of minimizing

$$\sum_{i=1}^{n} (a + bt_i - x_i)^2 = \|a\mathbf{1} + b\mathbf{t} - \mathbf{x}\|_2^2 \tag{A.6.1}$$

w.r.t. a, b where the t_i, x_i are given constants, $\mathbf{1} = (1 \ldots 1)$ is the (column) vector of ones, $t = (t_1 \ldots t_n)$ and $x = (x_1 \ldots x_n)$. The r.h. expression in (A.6.1) is the squared distance from x to the element $a\mathbf{1} + bt$ of span$(\mathbf{1}, t)$, and therefore the minimizer $a^*\mathbf{1} + b^*t$ is the L_2 projection of x on span$(\mathbf{1}, t)$.

Differentiating the l.h.s. of (A.6.1) gives the normal equations

$$0 = \sum_{i=1}^{n}(a + bt_i - x_i) = \sum_{i=1}^{n}(a + bt_i - x_i)t_i$$

which in matrix notation is the same as

$$\begin{pmatrix} \mathbf{1}^\mathsf{T} \\ t^\mathsf{T} \end{pmatrix} (\mathbf{1} \; t) \begin{pmatrix} a \\ b \end{pmatrix} = \begin{pmatrix} \mathbf{1}^\mathsf{T} \\ t^\mathsf{T} \end{pmatrix} x .$$

One often writes (A.6.1) as

$$\sum_{i=1}^{n} \left(\tilde{a} + b(t_i - \bar{t}) - x_i \right)^2 = \| \tilde{a}\mathbf{1} + b\tilde{t} - x \|_2^2 , \tag{A.6.2}$$

where $\bar{t} = \sum_1^n t_i/n$, $\tilde{t} = t - \bar{t}\mathbf{1}$ and $\tilde{a} = a + b\bar{t}$. This expression has the advantage that $\mathbf{1}$ and $\tilde{t}$ are orthogonal, i.e. $\mathbf{1}^\mathsf{T}\tilde{t} = 0$, and therefore the projection $\tilde{a}^*\mathbf{1} + b^*\tilde{t}$ of x on span$(\mathbf{1}, \tilde{t}) = $ span$(\mathbf{1}, t)$ splits out as the sum of projection $\tilde{a}^*\mathbf{1}$ on span$(\mathbf{1})$ and the projection $b^*\tilde{t}$ on span$(\tilde{t})$. This gives

$$\tilde{a}^* = \bar{x} = \sum_1^n x_i/n , \quad b^* = \frac{\sum_1^n (t_i - \bar{t})(x_i - \bar{x})}{\sum_1^n (t_i - \bar{t})^2} .$$

A.6.2 Conditional Expectations

Consider the space L_2 of all square integrable r.v.s $X, Y, \ldots$ defined on a probability space $(\Omega, \mathscr{F}, \mathbb{P})$. We equip L_2 with an inner product $\langle \cdot, \cdot \rangle$ and a norm $\| \cdot \|$ by letting

$$\langle X, Y \rangle = \mathbb{E}[XY], \quad \|X\|^2 = \|X\|_2^2 = \mathbb{E}X^2 = \langle X, X \rangle .$$

X and Y are *orthogonal*, written $X \perp Y$, if $\langle X, Y \rangle = 0$.

For a given subspace $L \subset L_2$, the *orthogonal projection* $P_L X$ of X on L is the point $Y \in L$ minimizing the distance $\|X - Y\|$ to X.

Proposition A.1 $Y = P_L X$ *if and only if the residual* $X - Y$ *is orthogonal to all* $Z \in L$, *i.e. if and only if* $\langle X - Y, Z \rangle = 0$ *for all* $Z \in L$.

Proposition A.2 *Let $\mathcal{G}$ be a sub-σ-field of $\mathcal{F}$ and let $L_{\mathcal{G}}$ be the space of all square integrable $\mathcal{G}$-measurable r.v.s. Then for $X \in L_2$, we have $\mathbb{E}(X \mid \mathcal{G}) \in L_{\mathcal{G}}$ and $\mathbb{E}(X \mid \mathcal{G}) = P_{L_{\mathcal{G}}} X$.*

Proof We use the characterization of $Y = \mathbb{E}(X \mid \mathcal{G})$ as the unique $\mathcal{G}$-measurable r.v. satisfying

$$\mathbb{E}[YZ] = \mathbb{E}[XZ] \quad \text{for all bounded } \mathcal{G}\text{-measurable } Z. \tag{A.6.3}$$

Write $Y' = P_{L_{\mathcal{G}}} X$. Then by Proposition A.1, $\mathbb{E}[Y'Z] = \mathbb{E}[XZ]$ for all $Z \in L_{\mathcal{G}}$, so (A.6.3) holds.   □

Remark A.3 Proposition A.2 is sometimes used to construct $\mathbb{E}(X \mid \mathcal{G})$. One then starts by defining $\mathbb{E}(X \mid \mathcal{G}) = P_{L_{\mathcal{G}}} X$ for all $X \in L_2$. One next shows that the operator $X \to \mathbb{E}(X \mid \mathcal{G})$ on L_2 is a contraction in L_1. Since L_2 is dense in L_1, this operator therefore has a unique extension to L_1 which one takes as the definition of the conditional expectation operator.   □

Corollary A.4 *Let M be a square integrable r.v. and $X \in \mathbb{R}^n$ a random vector. Then $\theta^* = \mathbb{E}[M \mid X]$ minimizes $\mathbb{E}(\theta - M)^2$ in the class of all square integrable $\sigma(X)$ measurable r.v.s θ, i.e. in the class of all square integrable functions of X.*

In other words, $\mathbb{E}[M \mid X]$ is the best predictor of M (in the sense of mean square error).

Proof The minimizer θ^* of $\mathbb{E}(\theta - M)^2$ is the L_2-projection of M on the class of all square integrable $\sigma(X)$ measurable r.v.s. Now just appeal to Proposition A.2.   □

The following result is used for example in the proof of Panjer's recursion in Sect. III.6. It may be intuitively obvious, but students have often asked for a formal argument.

Proposition A.5 *Let $V_1, \ldots, V_n$ be i.i.d.[1] with finite mean and $S_n = V_1 + \cdots + V_n$. Then $\mathbb{E}[V_i \mid S_n] = S_n/n$, $i = 1, \ldots, n$.*

Proof Define $A_i = \mathbb{E}[V_i \mid S_n]$. Then A_i is characterized by (A.6.3), i.e. $\mathbb{E}[A_i f(S_n)] = \mathbb{E}[V_i f(S_n)]$ for all bounded measurable functions f. However, since the V_k are i.i.d., the distribution of the pair (V_j, S_n) is the same for all j. Hence $\mathbb{E}[V_i f(S_n)]$ does not depend on i, so that for $j \neq i$

$$\mathbb{E}[A_i f(S_n)] = \mathbb{E}[A_j f(S_n)] = \mathbb{E}[V_j f(S_n)],$$

---

[1]More generally, the lemma holds (with the same proof) if the V_k are *exchangeable*, i.e. the distribution of the (ordered) set $(V_1, \ldots, V_n)$ is the same as the distribution of $(V_{\pi_1}, \ldots, V_{\pi_n})$ for any permutation π of $\{1, \ldots, n\}$.

and using (A.6.3) again shows that $A_i = \mathbb{E}[V_j \mid S_n]$. Hence $A = A_i$ does not depend on i. Further,

$$n A = A_1 + \cdots + A_n = \mathbb{E}[S_n \mid S_n] = S_n,$$

so $\mathbb{E}[V_i \mid S_n] = A_i = A = S_n/n$.   □

A.7   Supplements on the Normal Distribution

A.7.1   Conditioning

Let **X** be a multivariate normal vector partitioned as $(X_1\ X_2)^{\mathsf{T}}$ and let

$$\boldsymbol{\mu} = \begin{pmatrix} \mu_1 \\ \mu_2 \end{pmatrix}, \quad \boldsymbol{\Sigma} = \begin{pmatrix} \boldsymbol{\Sigma}_{11}\ \boldsymbol{\Sigma}_{12} \\ \boldsymbol{\Sigma}_{21}\ \boldsymbol{\Sigma}_{22} \end{pmatrix}$$

be the corresponding partitioning of the mean vector $\boldsymbol{\mu}$ and the covariance matrix $\boldsymbol{\Sigma}$ (thus, e.g., $\boldsymbol{\Sigma}_{11}$ is the covariance matrix of X_1, $\boldsymbol{\Sigma}_{12}$ contains the covariances between X_1 and X_2, etc). Then the conditional distribution of X_1 given X_2 is multivariate normal with mean vector and covariance matrix given by

$$\mu_{1\mid 2} = \mu_1 + \boldsymbol{\Sigma}_{12}\boldsymbol{\Sigma}_{22}^{-1}(X_2 - \mu_2), \quad \text{resp. } \boldsymbol{\Sigma}_{1\mid 2} = \boldsymbol{\Sigma}_{12}\boldsymbol{\Sigma}_{22}^{-1}\boldsymbol{\Sigma}_{21}. \tag{A.7.1}$$

A.7.2   Mill's Ratio

We next turn to tail estimates commonly going under the name of *Mill's ratio*. If φ, Φ are the standard normal density, resp. c.d.f., then

$$\lim_{t \to \infty} \frac{\overline{\Phi}(t)}{\varphi(t)} = \frac{1}{t}. \tag{A.7.2}$$

This follows by applying L'Hôpital's rule to the ratio of $\overline{\Phi}(t)$ and $\varphi(t)/t$. In fact, both functions tend to 0 as $t \to \infty$ and the derivative of $\overline{\Phi}(t)$ is $-\varphi(t)$ whereas that of $\varphi(t)/t$ is

$$\frac{-t^2\varphi(t) - \varphi(t)}{t^2} \sim -\varphi(t)$$

so that the ratio of derivatives goes to 1.

The asymptotic relation (A.7.2) also follows from the inequality

$$\frac{\varphi(t)}{t + 1/t} \le \overline{\Phi}(t) \le \frac{\varphi(t)}{t}. \tag{A.7.3}$$

The upper bound is easy to obtain by using $\varphi'(y) = -y\varphi(y)$, which gives

$$\overline{\Phi}(t) = \int_t^\infty \varphi(t)\,dt = \frac{1}{t}\int_t^\infty t\varphi(y)\,dy \le \frac{1}{t}\int_t^\infty y\varphi(y)\,dy = \frac{1}{t}\varphi(t),$$

whereas the proof of the lower one is more intricate and will not be given here.

Remark A.1 One consequence of (A.7.2) is that if X is normal, then X is asymptotically very close to x when $X > x$. This follows since for any fixed $a > 0$

$$\mathbb{P}(X > t + a \mid X > t) = \frac{\mathbb{P}(X > t + a)}{\mathbb{P}(X > t)} \sim \frac{t}{t + a}\frac{\varphi(t + a)}{\varphi(t)}$$

$$= (1 + o(1))\exp\{-(t+a)^2/2 - t^2\} = O(1)\cdot\exp\{-at\} \to 0.$$

$\Diamond$

Exercises
A.7.1 Show in the gamma case with density f and tail $\overline{F}$ that if

$$f(x) = \frac{\lambda^\alpha z^{\alpha-1}}{\Gamma(\alpha)}e^{-\lambda z} \quad \text{then } \overline{F}(z) \sim \frac{\lambda^{\alpha-1} z^{\alpha-1}}{\Gamma(\alpha)}e^{-\lambda z}.$$

A.7.3 A PDE for the Multivariate Normal Density

Lemma A.2 *Let $B = (b_{ij})_{i,j=1,\ldots,d}$ be positive definite, define $A = B^{-1}$ and consider the multivariate mean zero normal density*

$$f(x) = \frac{1}{(2\pi)^{d/2}\det(B)^{1/2}}\exp\{-x^{\mathsf{T}}Ax/2\}$$

$$= \frac{1}{(2\pi)^{d/2}\det(B)^{1/2}}\exp\Big\{-\sum_{i,j} x_i a_{ij} a_j/2\Big\}.$$

Then $\dfrac{\partial f}{\partial b_{hk}} = \dfrac{\partial^2 f}{\partial x_h \partial x_k}$ *for $h \ne k$.*

Proof Let c_{ij} be the ijth cofactor of $\mathbf{B}$. Then $a_{ij} = (-1)^{i+j}c_{ij}/\det(\mathbf{B})$, and expanding $\det(\mathbf{B})$ after the hth row we obtain

$$\det(\mathbf{B}) = \sum_j (-1)^{h+j} b_{hj} c_{hj}, \qquad \frac{\partial}{\partial b_{hk}} \det(\mathbf{B}) = (-1)^{h+k} c_{hk} = a_{hk} \det(\mathbf{B}),$$

$$\frac{\partial}{\partial b_{hk}} \left[\det(\mathbf{B}) \right]^{-1/2} = -\frac{1}{2} \left[\det(\mathbf{B}) \right]^{-3/2} a_{hk} \det(\mathbf{B}) = -\frac{1}{2} \left[\det(\mathbf{B}) \right]^{-1/2} a_{hk}.$$

$$(\text{A.7.4})$$

Letting $\mathbf{I}_{hk}$ be the matrix with hkth entry equal to 1, all other entries equal to 0, we obtain from $\mathbf{I} = \mathbf{B}\mathbf{A}$ that

$$\mathbf{0} = \frac{\partial}{\partial b_{hk}} \mathbf{B}\mathbf{A} = \mathbf{I}_{hk}\mathbf{A} + \mathbf{B}\frac{\partial}{\partial b_{hk}}\mathbf{A} \quad \Rightarrow \quad \mathbf{0} = \mathbf{A}\mathbf{I}_{hk}\mathbf{A} + \mathbf{I}\frac{\partial}{\partial b_{hk}}\mathbf{A},$$

which written out for the ijth entry gives

$$\frac{\partial}{\partial b_{hk}} a_{ij} = -a_{ih}a_{kj}.$$

Combining with (A.7.4), we obtain

$$\frac{\partial f}{\partial b_{hk}} = \left(-a_{hk} + \sum_{i,j} x_i a_{ih} a_{kj} x_j / 2 \right) f. \qquad (\text{A.7.5})$$

On the other hand, removing terms not depending on x_h gives

$$\frac{\partial}{\partial x_h} \sum_{i,j} x_i a_{ij} x_j = \frac{\partial}{\partial x_h} \left[a_{hh} x_h^2 + \sum_{j \neq h} x_h a_{hj} x_j + \sum_{i \neq h} x_i a_{ih} x_h \right]$$

$$= 2 a_{hh} x_h + \sum_{j \neq h} a_{hj} x_j + \sum_{i \neq h} x_i a_{ih} = 2 \sum_i x_i a_{ih},$$

where we used the symmetry of $\mathbf{A}$ in the last step. Using the same relation with h replaced by k and i by j on the r.h.s. gives

$$\frac{\partial^2 f}{\partial x_h \partial x_k} = -\frac{\partial}{\partial x_k} \left[\sum_i x_i a_{ih} \right] f = -a_{hk} f + \left[\sum_i x_i a_{ih} \right] \cdot \left[\sum_j a_{kj} x_j \right] f,$$

which is the same as (A.7.5). $\qquad\qquad\qquad\qquad\qquad\qquad\qquad\qquad\qquad\qquad\square$

A.8 Generalized Inverses

The inverse $F^{\leftarrow}$ of a c.d.f. $F(x)$ plays an important role in many contexts. It is most elementary and intuitive in the case where F admits a density $f(x)$ that is strictly positive on an open interval $I \subseteq \mathbb{R}$. In that case, F is continuous and increases strictly from 0 at the left endpoint to 1 at the right. Thus, for each $u \in (0, 1)$ there is a unique solution $x \in I$ of $F(x) = u$, and this x is the obvious choice of $F^{\leftarrow}(u)$.

Example A.1 Let F be standard exponential, $F(x) = 1 - e^{-x}$ for $x \in I = (0, \infty)$. Solving $1 - e^{-x} = u$ gives $F^{\leftarrow}(u) = -\log(1 - u)$. If instead F is Pareto on $I = (1, \infty)$ with $\overline{F}(x) = 1/x^{\alpha}$, $1/x^{\alpha} = u$ gives $F^{\leftarrow}(u) = x^{1/\alpha}$.   ◇

The considered simple case may safely be used as a base for intuition, but to deal with distributions with atoms, gaps in the support, etc., a more general definition is needed. One then considers a general non-decreasing right-continuous function h defined on some interval $\mathcal{D}(h) \subseteq \mathbb{R}$ and defines

$$h^{\leftarrow}(u) = \inf\{x : h(x) \geq u\}, \quad f \in \mathcal{D}(h^{\leftarrow}),$$

where $\mathcal{D}(h^{\leftarrow})$ is the smallest interval containing the range $\{h(x) : x \in \mathcal{D}(h)\}$ of h. The situation is illustrated in Fig. A.3 for the case of h being a c.d.f. F. One notes two particularly important features:

a) an interval $[x_1, x_2)$ where $h(x) \equiv u_1$ is constant corresponds to a jump of $h^{\leftarrow}$ at u_1; the size of the jump is $x_2 - x_1$ if $[x_1, x_2)$ is a maximal constancy interval,
b) a jump of h at x_3 correspond to $h^{\leftarrow}(u) \equiv u_2 = F(x_3-)$ for $u_2 = F(x_3-) \leq u \leq u_3 = F(x_3)$.
   A general version of these and some further properties is given in Proposition A.2 below.

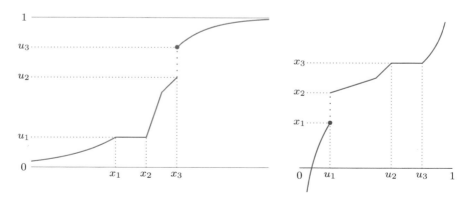

Fig. A.3 A c.d.f. F (left, red) and its inverse $F^{\leftarrow}$ (right,blue)

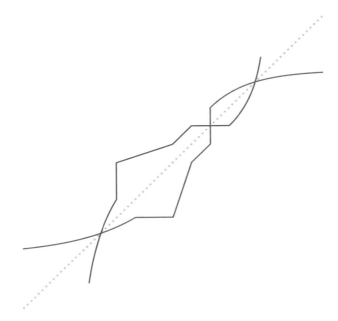

Fig. A.4 $u = F(x)$ (red), $x = F^{\leftarrow}(u)$ (blue) and the diagonal $u = x$ (green)

Graphically, one also notes that (except for the special features associated with jumps and constancy intervals) the graphs of $h(x)$ and $h^{\leftarrow}(u)$ are the elements of the sets $\{x, u = h(x)\}$, resp. $\{u, x = h^{\leftarrow}(u)\}$, i.e., when plotted in the same coordinate system, the two graphs are mirror images of each other w.r.t. the diagonal $x = u$, cf. Fig. A.4.

Proposition A.2 *For a non-decreasing right-continuous* h:

1) *The set* $\{x : h(x) \geq u\}$ *is an interval closed to the left, i.e.* $\{x : h(x) \geq u\} = \left[h^{\leftarrow}(u), \infty\right)$;
2) $h^{\leftarrow}$ *is non-decreasing and left-continuous;*
3) $h\left(h^{\leftarrow}(u)\right) \geq u$, *with equality if and only if* u *is in the range of* h;
4) $h(x) \geq u \iff h^{\leftarrow}(u) \leq x$;
5) $h(x) < u \iff h^{\leftarrow}(u) > x$;
6) $h^{\leftarrow}\left(h(x)\right) \leq x$, *with equality if and only if* $h(y) < h(x)$ *for all* $y < x$;
7) $h^{\leftarrow}$ *is continuous if and only if* h *is strictly increasing. Then* $h^{\leftarrow}\left(h(x)\right) = x$ *for all* x;
8) $h^{\leftarrow}$ *is strictly increasing if and only if* h *is continuous. Then* $h\left(h^{\leftarrow}(u)\right) = u$ *for all* u.

Proof Since h is non-decreasing, it is clear that $I = \{x : h(x) \geq u\}$ has the form (x_0, ∞) or $[x_0, \infty)$, but since h is right-continuous at x_0, we have indeed $h(x_0) \geq u$, i.e. $x_0 \in I$. This shows 1) and 4), and the first part of 3) then follows by taking $x = h^{\leftarrow}(u)$. For the second part of 3), note that u is trivially in the range of h if $h(h^\leftarrow(u)) = u$. Conversely, if $u = h(x^*)$ for some x^*, then the definition of $h^\leftarrow(u)$ implies $h^\leftarrow(u) = \inf\{x : h(x) = u\}$ and that $h^\leftarrow(u)$ is in this interval, i.e. $h(h^\leftarrow(u)) = u$. Further, 5) is just the negation of 4). For 6), let $J = \{z : h(z) \geq h(x)\}$. Then $x \in J$ and so x is at least the minimal element $h^\leftarrow(h(x))$ of J. Further, $h^\leftarrow(h(x)) < x$ if and only if $y \in J$ for some $y < x$. But this is only possible if $h(y) = h(x)$.

2) That $h^\leftarrow$ is non-decreasing is clear. Note next that right-continuity ensures that h is upper semicontinuous, i.e.

$$x_n \to x^* \implies \limsup_{n\to\infty} h(x_n) \leq h(x^*), \tag{A.8.1}$$

as follows from $\limsup h(x_n) \leq \limsup h(x_n \vee x^*)$ and $x_n \vee x^* \geq x^*$, $x_n \vee x^* \to x^* \vee x^* = x^*$ and the right-continuity of h at x^*. Now let $u_n \uparrow u$; we have to show $x_n = h^\leftarrow(u_n) \to x = h^\leftarrow(u)$. But the sequence $x_n = h^\leftarrow(u_n)$ is non-decreasing, hence $x^* = \lim x_n$ exists and clearly $x^* \leq x = h^\leftarrow(u)$. From 3) and (A.8.1) we then conclude that

$$u = \limsup u_n \leq \limsup h(h^\leftarrow(u_n)) \leq h(x^*),$$

and the definition of $h^\leftarrow(u)$ then gives $x^* \geq h^\leftarrow(u)$. Altogether, $x^* = x$, completing the proof of 2).

The proofs of 7) and 8) are now easy and omitted.   □

The case of h being a c.d.f. F of an r.v. X is particularly important. With $\overline{F}(x) = \mathbb{P}(X > x) = 1 - F(x)$, we define $\overline{F}^\leftarrow(u) = F^\leftarrow(1 - u)$; then in the typical case $F(F^\leftarrow(1 - u)) = 1 - u$, we have

$$\overline{F}(\overline{F}^\leftarrow(u)) = 1 - F(\overline{F}^\leftarrow(u)) = 1 - F(F^\leftarrow(1 - u)) = 1 - (1 - u) = u$$

as it should be.

Proposition A.3 *For an r.v. X and its c.d.f. $F(x) = \mathbb{P}(X \leq x)$,*

(a) *$F^\leftarrow(F(X)) = X$ a.s.;*
(b) *if U is uniform$(0, 1)$, then $F^\leftarrow(U)$ and $\overline{F}^\leftarrow(U)$ both have c.d.f. F;*
(c) *if F is continuous, then $F(X)$ and $\overline{F}(X)$ are both uniform$(0, 1)$.*

Proof We claim that $X \leq F^\leftarrow(F(X)) \leq X$, which is only possible if $F^\leftarrow(F(X)) = X$ a.s. as claimed in (a). Indeed, $\leq$ follows by part 6) of Proposition A.2 and $F^\leftarrow(F(X)) > X$ is impossible by part 5) with $u = F(X)$ (then we would have $F(X) < F(X)$). For (b), note that by part 4) of Proposition A.2

$$\mathbb{P}(F^\leftarrow(U) \leq x) = \mathbb{P}(U \leq F(x)) = F(x).$$

This gives the first claim of (b) and the second then easily follows. Finally, for (c) we have to prove $\mathbb{P}(F(X) < u) = u, 0 < u < 1$. But by 5) and 8),

$$\mathbb{P}(F(X) < u) = \mathbb{P}(X < F^\leftarrow(u)) = \mathbb{P}(X \le F^\leftarrow(u)) = F(F^\leftarrow(u)) = u.$$

$\square$

A main application of $F^\leftarrow$ is the *inversion method* in simulation, where one uses b) to generate X as $F^\leftarrow(U)$ where U is uniform$(0,1)$ (the most common case is an F which is continuous and strictly increasing on an interval).

Example A.4 Let F be exponential(δ). Then $F^\leftarrow(x) = -\log(1-x)/\delta$ so inversion means $X = -\log(1-U)/\delta$ (in practice, one often uses $X = -\log U/\delta$ rather than $X = -\log(1-U)/\delta!$). $\diamond$

Remark A.5 Note that there is also a right-continuous inverse defined as $\inf\{x : h(x) > u\}$ (but we follow the tradition of working with $h^\leftarrow$ rather than this). The distinction is related to the technicalities involved in the definition of α-quantiles for distributions for which $F(q) = \alpha$ holds in an interval, not just a single point. An α-*quantile* of F is defined as any number q such that

$$F(q) \ge \alpha \quad \text{and} \quad F(q-) \le \alpha .$$

The set of all α-quantiles of X is the interval $[q_\alpha^-(F), q_\alpha^+(F)]$, where $q_\alpha^-(F) = F^\leftarrow(\alpha)$ and

$$q_\alpha^+(F) = F^\leftarrow(\alpha+) = \inf\{x : F(x) > \alpha\} = \sup\{x : F(x-) \le \alpha\} .$$

See Fig. A.3 with $\alpha = u_1, q_\alpha^-(F) = x_1, q_\alpha^+(F) = x_2$. $\diamond$

Notes
A thorough treatment of generalized inverses can be found, for example, in Resnick [148].

A.9 The Distributional Transform

Let X be an r.v. with c.d.f. $F(x) = F_X(x) = \mathbb{P}(X \le x)$. The possible jumps are denoted by $\Delta F(x) = F(x) - F(x-)$ so that $\Delta F(x) = \mathbb{P}(X = x)$. The *distributional transform* is then defined as $D = F(X-) + V \Delta F(X)$, where V is an independent uniform$(0, 1)$ r.v. and can be convenient in contexts where atoms of X present a nuisance. In particular, we have the following generalization of Proposition A.3:

Proposition A.1 *D is uniform$(0, 1)$ and $F^\leftarrow(D) = X$ a.s.*

Proof It is clear that $0 \le D \le 1$. For the first statement, we thus have to show $\mathbb{P}(D \le u) = u$ for $0 < u < 1$. But we can find (uniquely determined) $x \in$

$\mathbb{R}$, $v \in (0, 1)$ such that $u = F(x-) + v\Delta F(x)$. Assume $X = y$. If $y < x$, then $D \leq F(y) \leq F(x-) \leq u$ so $D \leq u$ is automatic. If $y = x$, $D \leq u$ will occur precisely when $V \leq v$. If $y > x$, $D \leq u$ is only possible if $F(y-) = F(x)$, i.e., $y \in A = \{z > x : F(z-) = F(x)\}$. But A has the form (x, x^*) with $x^* \geq x$ (if $x^* = x$ then $A = \emptyset$) so $\mathbb{P}(X \in A) = F(x^*-) - F(x) = 0$. Thus

$$\mathbb{P}(D \leq u) = \int_{-\infty}^{x-} 1 \cdot F(dy) + \mathbb{P}(V \leq v)\mathbb{P}(X = x) + \mathbb{P}(X \in A)$$

$$= F(x-) + v\Delta F(x) + 0 = u .$$

It now follows that $F^{\leftarrow}(D)$ has the same distribution as X. But since $F^{\leftarrow}(D) \leq F^{\leftarrow}(F(X)) \leq X$ (using part 6) of Proposition A.2 in the last step), this is only possible if $F^{\leftarrow}(D) = X$ a.s.   □

Notes
We have borrowed the terminology 'distributional transform' from Rüschendorf [154], where a number of further applications can be found.

A.10   Types of Distributions

Two non-degenerate r.v.s X, Y are said to be of the *same type* if there exist $a > 0$, b such that

$$Y \overset{\mathscr{D}}{=} aX + b \qquad\qquad (A.10.1)$$

(an r.v. Z is non-degenerate if there is no constant c such that $Z = c$ a.s.). Correspondingly, two c.d.f.s F, G are of the same type if there exist r.v.s X, Y with distributions F, G such that X, Y are of the same type, i.e. if $G(y) = F((y - b)/a)$ for some $a > 0$, b. A few among many standard examples are $\mathcal{N}(\mu, \sigma^2)$ r.v.s X_{μ,σ^2} with $\sigma^2 > 0$ where

$$X_{\mu_2,\sigma_2^2} \overset{\mathscr{D}}{=} \sigma_2 X_{0,1} + \mu_2 \overset{\mathscr{D}}{=} \frac{\sigma_2}{\sigma_1} X_{\mu_1,\sigma_1^2} + \mu_2 - \frac{\mu_1\sigma_2}{\sigma_1}$$

and uniform(a, b) r.v.s $Y_{a,b}$ with $a < b$ where

$$Y_{a_2,b_2} \overset{\mathscr{D}}{=} (b_2 - a_2)Y_{0,1} + a_2 \overset{\mathscr{D}}{=} \frac{b_2 - a_2}{b_1 - a_1} Y_{a_1,b_1} + a_2 - a_1\frac{b_2 - a_2}{b_1 - a_1} .$$

Lemma A.1 *Let X be a non-degenerate r.v. such that $X \overset{\mathscr{D}}{=} aX + b$ with $a > 0$. Then $a = 1$, $b = 0$.*

Proof We also have $X \overset{\mathscr{D}}{=} X/a - b$ and so we may assume $a < 1$. Since

$$X \overset{\mathscr{D}}{=} aX + b \overset{\mathscr{D}}{=} a(aX + b) + b = a^2 X + ab + b$$

$$\overset{\mathscr{D}}{=} a(a^2 X + ab + b) + b = a^3 X + a^2 b + ab + b \ \ldots$$

$$\overset{\mathscr{D}}{=} a^n X + a^{n-1} b + \cdots + ab + b,$$

it follows by letting $n \to \infty$ that X is degenerate at $b/(1-a)$, a contradiction. Thus $a = 1$, and $b = 0$ is then easy. $\qquad\square$

The following lemma is standard calculus.

Lemma A.2 *Let $\{x_n\}_{n \in \mathbb{N}}$ be a sequence of real numbers.*

(i) *If $\limsup_{n \to \infty} x_n = \infty$, then there is a subsequence $\{x_{n_k}\}_{n \in \mathbb{N}}$ such that $x_{n_k} \to \infty$ as $k \to \infty$;*

(ii) *$x_n \to x$ as $n \to \infty$ if and only if any subsequence $\{x_{n_k}\}$ has a further subsequence $\{x_{n_{k_j}}\}_{j \in \mathbb{N}}$ such that $x_{n_{k_j}} \to x$ as $j \to \infty$.*

Proposition A.3 (Convergence to Types) *Let $\{X_n\}$ be a sequence of r.v.s and $\{a_n\}$, $\{a'_n\}$, $\{b_n\}$, $\{b'_n\}$ sequences with $a_n > 0$, $a'_n > 0$ such that $a_n X_n + b_n \overset{D}{\longrightarrow} Y$, $a'_n X_n + b'_n \overset{D}{\longrightarrow} Y'$ for some non-degenerate r.v.s Y, Y'. Then Y, Y' are of the same type and there exist α, β such that $a_n/a'_n \to \alpha$, $(b_n - b'_n)/a'_n \to \beta$.*

Proof Using a rescaling argument if necessary, we may assume w.l.o.g. that $a'_n = 1$, $b'_n = 0$ so that we have to prove that $a_n \to \alpha$, $b_n \to \beta$ for some $\alpha > 0$ and $\beta \in \mathbb{R}$ (once this is shown, we immediately have that Y, Y' are of the same type).

The first step is to note that

$$0 < \liminf_{n \to \infty} a_n \leq \limsup_{n \to \infty} a_n < \infty. \tag{A.10.2}$$

Indeed, if there exists a subsequence n_k with $a_{n_k} \to 0$, then $X_{n_k} \overset{D}{\longrightarrow} Y'$ gives $b_{n_k} \to Y$, which is impossible since Y is non-degenerate. Similarly $a_{n_k} \to \infty$ gives a contradiction since then $X_{n_k} + b_{n_k}/a_{n_k} \overset{D}{\longrightarrow} 0$ so that the sequence b_{n_k}/a_{n_k} of constants has limit $-Y'$ in distribution, which is impossible since Y' is non-degenerate.

The next step is to show that

$$\limsup_{n \to \infty} |b_n| < \infty. \tag{A.10.3}$$

Indeed, assume $b_{n_k} \to \infty$. W.l.o.g., we may assume by (A.10.2) that $a_{n_k} \to a$ with $0 < a < \infty$. Choose first y so large that $\mathbb{P}(Y > y) < 1/4$ and next b so small that $\mathbb{P}(aY' + b > y) > 3/4$ and that $y - b$ is a continuity point of aY'. We then get

$$1/4 > \lim_{k\to\infty} \mathbb{P}(a_{n_k} X_{n_k} + b_{n_k} > y) \geq \lim_{k\to\infty} \mathbb{P}(a_{n_k} X_{n_k} + b > y)$$
$$= \mathbb{P}(aY' + b > y) > 3/4,$$

a contradiction. Similarly, it is impossible that $b_{n_k} \to -\infty$.

From (A.10.2), (A.10.3) it now follows that the desired limits α, β exist along some subsequence n_k, and we only have to show that the possible limits α^*, β^* along some other subsequence n_k^* are the same. But if α^*, β^* exist, we conclude that

$$Y \overset{\mathscr{D}}{=} \alpha + \beta Y', \quad Y \overset{\mathscr{D}}{=} \alpha^* + \beta^* Y'$$

and $\alpha^* = \alpha$, $\beta^* = \beta$ then follows by Lemma A.1. $\qquad\square$

Exercises

A.10.1 In the proof of Lemma A.1, fill in the proof of $b = 0$.

A.10.2 Show that if c_n are constants and Y a non-degenerate r.v., it is impossible that $c_n \overset{\mathcal{D}}{\longrightarrow} Y$.

A.11 Transforms

The *moment generating function* (m.g.f.) of a distribution F is defined as the function

$$\widehat{F}[\theta] = \mathbb{E}e^{\theta X} = \int_{-\infty}^{\infty} e^{\theta x} F(\mathrm{d}x)$$

of θ where X is an r.v. having distribution F, and is defined on

$$\Theta = \left\{ \theta \in \mathbb{R} : \int e^{\Re(\theta)x} F(\mathrm{d}x) < \infty \right\}.$$

The *characteristic function* (ch.f.) is

$$\mathbb{E}e^{is X} = \int_{-\infty}^{\infty} e^{is x} F(\mathrm{d}x),$$

where X is an r.v. having distribution F. One usually thinks of it as a function with domain $\mathbb{R}$. The domain Θ of the m.g.f. always contains the imaginary axis $i\mathbb{R}$.

Thus, the ch.f. is $\widehat{F}[is]$, and in view of this, we have not used special notation for the characteristic function.

If F is concentrated on $\mathbb{N}$, i.e. $X \in \mathbb{N}$, it is more customary to work with the *probability generating function* (p.g.f.) or just *generating function*

$$\mathbb{E}z^X = \int_{\mathbb{N}} z^x F(dx) = \sum_{n=0}^{\infty} z^n f_n = \widehat{f}[z],$$

where $f_n = \mathbb{P}(X = n)$. The domain contains at least $[0, 1]$. Of course, here $\widehat{f}[z] = \widehat{F}[\log z]$.

Finally, the *cumulant generating function* is defined as $\kappa_F(\theta) = \log \widehat{F}[\theta]$.

A fundamental property of these transforms is that convolution (independent sums) corresponds to multiplication of the m.g.f. (and therefore of the ch.f. and the p.g.f.). That is, if $X_1, \dots, X_n$ are independent r.v.s with distributions $F_1, \dots, F_n$ so that $X = X_1 + \dots + X_n$ has distribution $G = F_1 * \dots * F_n$, then $\widehat{G}[\theta] = \widehat{F_1}[\theta] \dots \widehat{F_n}[\theta]$. In particular, in the i.i.d. case $F_1 = \dots = F_n = F$, then G is the nth convolution power F^{*n} so that $\widehat{F^{*n}}[\theta] = \widehat{F}[\theta]^n$. For the c.g.f., this property translates into additivity,

$$\kappa_{F_1 * \dots * F_n}(\theta) = \kappa_{F_1}(\theta) + \dots + \kappa_{F_n}(\theta), \quad \kappa_{F^{*n}}(\theta) = n\kappa_F(\theta).$$

The terminology *moment generating function* comes from the formula

$$\mathbb{E}X^n = \int_{-\infty}^{\infty} x^n F(dx) = \widehat{F}^{(n)}[0],$$

where $\widehat{F}^{(n)}$ means the nth derivative. In particular, $\mathbb{E}X = \widehat{F}'[0]$ (similarly, $\mathbb{E}X^2 = \widehat{F}''[0]$ so that $\mathbb{V}\mathrm{ar}\, X = \widehat{F}''[0] - \widehat{F}'[0]^2$). The corresponding relations for the p.g.f. are in terms of factorial moments,

$$\mathbb{E}X = \widehat{f}'[1], \quad \mathbb{E}[X(X-1)] = \widehat{f}''[1], \quad \mathbb{E}[X(X-1) \cdots (X-n+1)] = \widehat{f}^{(n)}[1].$$

The derivatives $\kappa_F'(0), \kappa_F''(0), \dots$ of the c.g.f. at 0 are called the *cumulants*. In particular, the first three cumulants are

$$\kappa_F'(0) = \mathbb{E}X, \quad \kappa_F''(0) = \mathbb{V}\mathrm{ar}\, X, \quad \kappa_F'''(0) = \mathbb{E}[X - \mathbb{E}X]^3.$$

One refers to $\kappa_F'''(0)/\kappa_F''(0)^{3/2}$ as the *skewness*, and similarly $\kappa_F^{(4)}(0)/\kappa_F''(0)^2$ is the *kurtosis*.

References

1. K.K. Aase, S.-A. Persson, Pricing of unit-linked life insurance policies. Scand. Actuar. J. **1994**, 26–52 (1994)
2. R. Adler, *An Introduction to Continuity, Extrema and Related Topics for Gaussian Processes.* IMS Monograph Series, vol. 12 (1990)
3. H. Albrecher, J. Beirlant, J.L. Teugels, *Reinsurance: Actuarial and Statistical Aspects* (Wiley, 2017)
4. B.C. Arnold, N. Balakrishnan, H.N. Nagaraja, *Records* (Wiley, 1998)
5. P. Artzner, F. Delbaen, J.M. Ebere, D. Heath, Coherent measures of risk. Mathematical Finance **9**, 203–228 (1999)
6. S. Asmussen, *Applied Probability and Queues*, 2nd edn. (Springer, 2003)
7. S. Asmussen (2014) Modeling and performance of bonus-malus systems: stationarity versus age-correction. Risks **2**, 49–73
8. S. Asmussen, H. Albrecher, *Ruin Probabilities*, 2nd edn. (World Scientific, 2010)
9. S. Asmussen, H. Albrecher, D. Kortschak, Tail asymptotics for the sum of two heavy-tailed risks. Extremes **9**, 107–130 (2006)
10. S. Asmussen, B.J. Christensen, M. Taksar, Portfolio size as function of the premium: modelling and optimization. Stochastics **85**, 575–588 (2013)
11. S. Asmussen, B.J. Christensen, J. Thøgersen, Nash equilibrium premium strategies for push-pull competition in a frictional non-life insurance market. Insurance Math. Econom. **87**, 92–100 (2019)
12. S. Asmussen, B.J. Christensen, J. Thøgersen, Stackelberg equilibrium premium strategies for push-pull competition in a non-life insurance market with product differentiation. Risks **2019**, 7, 49 (2019). https://doi.org//10.3390/risks7020049
13. S. Asmussen, A. Frey, T. Rolski, V. Schmidt, Does Markov-modulation increase the risk? Astin Bulletin **25**, 49–66 (1995)
14. S. Asmussen, E. Hashorva, P. Laub, T. Taimre, Tail asymptotics for light-tailed Weibull-like sums. Probab. Math. Stat. **37**, 235–256 (2017)
15. T. Aven, P. Baraldi, R. Flage, E. Zio, *Uncertainty in Risk Assesment* (Wiley, 2014)
16. J.-M. Azaïs, M. Wschebor, *Level Sets and Extrema of Random Processes and Fields* (Wiley, 2009)
17. P. Azcue, N. Muler, *Stochastic Optimization in Insurance: A Dynamic Programming Approach* (Springer, 2014)
18. A.A. Balkema, C. Klüppelberg, S.I. Resnick, Densities with Gaussian tails. Proc. Lond. Math. Soc. **66**, 568–588 (1993)

© Springer Nature Switzerland AG 2020
S. Asmussen, M. Steffensen, *Risk and Insurance*, Probability Theory
and Stochastic Modelling 96, https://doi.org/10.1007/978-3-030-35176-2

19. P. Barrieu, H. Bensusan, N. El Karoui, C. Hillairet, S. Loisel, C. Ravanelli, Y. Salhi, Understanding, modelling and managing longevity risk: key issues and main challenges. Scand. Actuar. J. **2012**, 203–231 (2012)
20. Basel II, *Basel II: Revised International Capital Framework* (Bank for International Settlements, 2004)
21. BCBS, *Consultative Document October 2013. Fundamental Review of the Trading Book* (Basel Committee on Banking Supervision, 2012)
22. J.M. Bernardo, A.F.M. Smith, *Bayesian Theory* (Wiley, 2000)
23. F. Bichsel, Erfahrungstariffierung in der Motorfahrzeug-Haftphlicht-Versicherung. Mitt. Verein. Schweiz. Versich. Math. **1964**, 1199–130 (1964)
24. E. Biffis, Affine processes for dynamic mortality and actuarial valuations. Insurance Math. Econom. **37**, 43–468 (2005)
25. N. Bingham, C. Goldie, J.L. Teugels, *Regular Variation* (Cambridge University Press, 1987)
26. T. Björk, *Arbitrage Theory in Continuous Time*, 3rd edn. Oxford Finance Series (2009)
27. F. Black, M. Scholes, The pricing of options and corporate liabilities. J. Polit. Econ. **81**, 637–654 (1973)
28. M. Bladt, S. Asmussen, M. Steffensen, Matrix calculations for inhomogeneous Markov reward processes, with applications to life insurance and point processes. Eur. Actuar. J. (to appear, 2020)
29. J. Boman, F. Lindskog, Support theorems for the Radon transform and Cramér-Wold theorems. J. Theoret. Probab. **22**, 683–710 (2009)
30. Ø. Borgan, J.M. Hoem, R. Norberg, A nonasymptotic criterion for the evaluation of automobile bonus systems. Scand. Actuar. J. **1981**, 165–178 (1981)
31. N.L. Bowers, H.U. Gerber, J.C. Hickman, D.A. Jones, C.J. Nesbitt, *Actuarial Mathematics*, 2nd edn. (The Society of Actuaries, 1997)
32. F. Brauer, J.A. Nohel, *The Qualitative Theory of Ordinary Differential Equations: An Introduction* (Benjamin, 1969). Reprint edition in Dover Books on Mathematics
33. L. Breiman, *Probability* (1968). Reprinted by SIAM 1991
34. M.J. Brennan, E.S. Schwartz, The pricing of equity-linked life insurance policies with an asset value guarantee. J. Financ. Econ. **3**, 195–213 (1976)
35. K. Buchardt, Dependent interest and transition rates in life insurance. Insurance Math. Econom. **55**, 167–179 (2014)
36. K. Buchardt, Kolmogorov's forward PIDE and forward transition rates in life insurance. Scand. Actuar. J. **2017**, 377–394 (2017)
37. K. Buchardt, C. Furrer, M. Steffensen, Forward transition rates. Finance and Stochastics (to appear, 2019)
38. K. Buchardt, T. Møller, Life insurance cash flows with policyholder behaviour. Risks **2015**, 290–317 (2015)
39. K. Buchardt, T. Møller, K.B. Schmidt, Cash flows and policyholder behavior in the semi-Markov life insurance setup. Scand. Actuar. J. **2015**, 660–688 (2015)
40. M. Buchwalder, H. Bühlmann, M. Merz, M.V. Wüthrich, The mean square error of prediction in the chain ladder reserving method (Mack and Murphy revisited). Astin Bulletin **36**, 521–542 (2006)
41. N. Bäuerle, E. Bayraktar, A note on applications of stochastic ordering to control problems in insurance and finance. Stochastics **86**, 330–340 (2014)
42. H. Bühlmann, *Mathematical Methods in Risk Theory* (Springer, 1970)
43. H. Bühlmann, A. Gisler, *A Course in Credibility Theory and its Applications* (Springer, 2005)
44. A. Cairns, Some notes on the dynamics and optimal control of stochastic pension fund models in continuous time. Astin Bulletin **30**, 19–55 (2000)
45. A. Cairns, D. Blake, K. Dowd, Modelling and management of mortality risk: a review. Scand. Actuar. J. **2008**, 79–113 (2008)
46. B.P. Carlin, T.A. Louis, *Bayesian Methods for Data Analysis*, 3rd edn. (Chapman & Hall/CRC Press, 2008)

47. M.L. Centeno, M. Guerra, The optimal reinsurance strategy — the individual claim case. Insurance Math. Econom. **46**, 450–460 (2010)
48. Y.S. Chow, H. Teicher, *Probability Theory. Independence, Interchangeability, Martingales* (Springer, 1997)
49. M.C. Christiansen, Multistate models in health insurance. Adv. Stat. Anal. **96**, 155–186 (2012)
50. M. Christiansen, M. Denuit, J. Dhaene, Reserve-dependent benefits and costs in life and health insurance contracts. Insurance Math. Econom. **57**, 132–137 (2014)
51. M.C. Christiansen, A. Niemeyer, On the forward rate concept in multi-state life insurance. Finance Stochast. **19**, 295–327 (2015)
52. M. Dahl, Stochastic mortality in life insurance: market reserves and mortality-linked insurance contracts. Insurance Math. Econom. **35**,113–136 (2004)
53. M.H.A. Davis, *Markov Models and Optimization* (Chapman & Hall, 1993)
54. C.D. Daykin, T. Pentikainen, M. Pesonen, *Practical Risk Theory for Actuaries* (CRC Press, 1993)
55. F. Delbaen, J.M. Haezendock, A martingale approach to premium calculation principles in an arbitrage-free market. Insurance Math. Econom. **8**, 269–277 (1989)
56. A. Dembo, O. Zeitouni, *Large Deviations Techniques and Applications*, 2nd edn. (Springer, 1998)
57. M. Denuit, X. Marechal, S. Pitrebois, J.-F. Walhin, *Actuarial Models for Claims Counts. Risk Classification, Credibility and Bonus-Malus Systems* (Wiley, 2007)
58. M. Denuit, J. Dhaene, M. Goovaerts, R. Kaas, *Actuarial Theory for Dependent Risks* (Wiley, 2005)
59. O. Deprez, H.U. Gerber, On convex principles of premium calculation. Insurance Math. Econom. **4**, 179–189 (1985)
60. P. Diaconis, D. Ylvisaker, Conjugate priors for exponential families. Ann. Stat. **7**, 269–281 (1979)
61. D.C.M. Dickson, *Insurance Risk and Ruin* (Cambridge University Press, 2005)
62. D.C.M. Dickson, M.R. Hardy, H.R. Waters, *Actuarial Mathematics for Life Contingent Risks*, 2nd edn. (Cambridge UP, 2013)
63. R.L. Dobrushin, Generalization of Kolmogorov's equations for Markov processes with a finite number of possible states. Mat. Sb. N.S. **33**, 567–596 (1953)
64. C. Dutang, H. Albrecher, S. Loisel, Competition among non-life insurers under solvency constraints: A game-theoretic approach. Eur. J. Oper. Res. **231**, 702–711 (2013)
65. P. Embrechts, J.L. Jensen, M. Maejima, J.L. Teugels, Approximations for compound Poisson and Polya processes. Adv. Appl. Probab. **17**, 623–637 (1985)
66. P. Embrechs, C. Klüppelberg, T. Mikosch, *Modeling Extreme Events for Insurance and Finance* (Springer, 1997)
67. P. Embrechts, R. Wang, Seven proofs for the subadditivity of expected shortfall. Depend. Model. **3**, 126–140 (2015)
68. P. Emms, Equilibrium pricing of general insurance policies. North American Actuar. J. **16**, 323–349 (2012)
69. P.D. England, R.J. Verrall, Stochastic claims reserving in general insurance. Br. Actuar. J. **8**(1), 443–518 (2002)
70. F. Esscher, On the probability function in the collective theory of risk. Skand. Akt. Tidsskr. 175–195 (1932)
71. A. Ferreira, L. de Haan, On the block maxima method in extreme value theory: PWM estimators. Ann. Stat. **43**, 276–298 (2015)
72. L. de Haan, A. Ferreira, *Extreme Value Theory: An Introduction* (Springer, 2006)
73. L. de Haan, C. Zhou, Extreme residual dependence for random vectors and processes. J. Appl. Probab. **43**, 217–242 (2011)
74. W.H. Fleming, R.W. Rishel, *Deterministic and Stochastic Optimal Control* (Springer, 1975)
75. W.H. Fleming, H.M. Soner, *Controlled Markov Processes and Viscosity Solutions* (Springer, 1993)

76. S. Foss, D. Korshunov, S. Zachary, *An Introduction to Heavy-Tailed and Subexponential Distributions* (Springer, 2011)

77. H. Föllmer, A. Schied, *Mathematical Finance. An Introduction in Discrete Time* (de Gruyter, 2011)

78. K.S.T. Gad, J. Juhl, M. Steffensen, Reserve-dependent surrender rates. Eur. Actuar. J. **5**, 283–308 (2015)

79. H. Gerber, *An Introduction to Mathematical Risk Theory* (1979)

80. H. Gerber, *Life Insurance Mathematics* (Springer, 1990)

81. R.D. Gill, S. Johansen, A survey of product-integration with a view toward application in survival analysis. Ann. Stat. **18**, 1501–1555 (1990)

82. P. Glasserman, *Monte Carlo Methods in Financial Engineering* (2004)

83. P.W. Glynn, W. Whitt, Logarithmic asymptotics for steady-state tail probabilities in a single-server queue, in *Studies in Applied Probability*, ed. by J. Galambos, J. Gani. J. Appl. Probab. **31A**, 131–156 (1994)

84. J. Grandell, *Aspects of Risk Theory* (Springer, 1991)

85. M. Hald, H. Schmidli, On the maximization of the adjustment coefficient under proportional reinsurance. Astin Bulletin **14**, 75–83 (2004)

86. L.F.B. Henriksen, J.W. Nielsen, M. Steffensen, C. Svensson, Markov chain modeling of policy holder behavior in life Insurance and pension. Eur. Actuar. J. **4**, 1–29 (2014)

87. J. Hickman, A statistical approach to premiums and reserves in multiple decrement theory. Trans. Soc. Actuar. **16**, 1–16 (1964); discussion *ibid*, 149–154

88. J. Hoem, Markov chain models in life insurance. Blätter der DGVFM **9**, 91–107 (1969)

89. J.M. Hoem, Inhomogeneous semi-Markov processes, select actuarial tables, and duration-dependence in demography. Population Dynamics **1972**, 251–296 (1972)

90. J. Hoem, Versatility of Markov chain as a tool in the mathematics of life insurance, in *Proceedings, International Congress of Actuaries, Helsinki, 1988* (1988), pp. 171–202

91. J. Hüsler, V. Piterbarg, Extremes of a certain class of Gaussian processes. Stoch. Process. Appl. **83**, 257–271 (1999)

92. IAIS, *Consultation Document December 2014. Risk-Based Global Insurance Capital Standard* (International Association of Insurance Supervisors, 2014)

93. H. Hult, F. Lindskog, O. Hammarlid, C.J. Rehn, *Risk and Portfolio Analysis* (Springer, 2012)

94. J. Janssen, R. Manca, *Semi-Markov Risk Models for Finance, Insurance and Reliability* (Springer, 2007)

95. J.L. Jensen, *Saddlepoint Approximations* (Clarendon Press, Oxford, 1995)

96. H. Joe, *Multivariate Models and Dependence Concepts* (Chapman & Hall, 1997)

97. R. Kaas, M. Goovaerts, J. Dhaene, M. Denuit, *Modern Actuarial Risk Theory*, 2nd edn. (Springer, 2008)

98. I. Karatzas, S.E. Shreve, *Brownian Motion and Stochastic Calculus*, 2nd edn. (Springer, 1991)

99. S. Karlin, H.M. Taylor, *A Second Course in Stochastic Processes* (Academic Press, 1981)

100. S.A. Klugman, H.H. Panjer, G.E. Willmot, *Loss Models: from Data to Decisions*, 4th edn. (Wiley, 2012)

101. C. Klüppelberg, D. Straub, I.M. Welpe, (eds.) *Risk. A Multidisciplinary Introduction* (Springer, 2014)

102. M. Koller, *Stochastic Models in Life Insurance* (Springer, 2012). EAA series

103. D. Konstantinides, *Risk Theory — a Heavy Tail Approach* (World Scientific, 2018)

104. H. Kraft, M. Steffensen, Optimal consumption and insurance: A continuous-time Markov chain approach. Astin Bulletin **28**, 231–257 (2008)

105. S.L Lauritzen, Time series analysis in 1880: a discussion of contributions made by T.N. Thiele. Int. Stat. Rev. **49**, 319–331 (1981)

106. M.R. Leadbetter, G. Lindgren, H. Rootzén, *Extremes and Related Properties of Random Sequences and Processes* (Springer, 1983)

107. J. Lemaire, *Automobile Insurance: Actuarial Models* (Kluwer, 1985)

108. J. Lemaire, H. Zi, A comparative analysis of 30 bonus-malus systems. Astin Bulletin **24**, 287–309 (1994)

109. J. Lemaire, *Bonus-Malus Systems in Automobile Insurance* (Kluwer, 1995)
110. K. Loiramanta, Some asymptotic properties of bonus systems. Astin Bulletin **6**, 233–245 (1972)
111. M. Loewe, *Extremwerttheorie*. Lecture Notes. University of Münster (2008/9)
112. T. Mack, Distribution-free calculation of the standard error of chain ladder reserve estimates. Astin Bulletin **23**, 213–225 (1993)
113. T. Mack, The standard error of chain ladder reserve estimates: recursive estimation and inclusion of a tail factor. Astin Bulletin **29**, 543–552 (1999)
114. M. Mandjes, *Large Deviations for Gaussian Queues* (Wiley, 2007)
115. A. McNeil, R. Frey, P. Embrechts, *Quantitative Risk Management. Concepts, Techniques and Tools* (Princeton University Press, 2005)
116. A. McNeil, R. Frey, P. Embrechts, *Quantitative Risk Management. Concepts, Techniques and Tools*, 2nd edn. (Princeton University Press, 2015)
117. R.C. Merton, Lifetime portfolio selection under uncertainty: The continuous-time case. Rev. Econ. Stat. **51**, 247–257 (1969)
118. R.C. Merton, Optimal consumption and portfolio rules in a continuous-time model. J. Econ. Theory **3**, 373–413 (1971)
119. R.C. Merton, Theory of rational option pricing. Bell J. Econ. Manag. Sci. RAND Corp. **4**, 141–183 (1973)
120. K.R. Miltersen, S.-A. Persson, Is mortality dead? Stochastic forward force of mortality rate determined by no arbitrage. Working paper, Norwegian School of Economics and Business Administration (2005). Online version avaliable at www.mathematik.uni-ulm.de/carfi/vortraege/downloads/DeadMort.pdf
121. T. Mikosch, Copulas: tales and facts. Extremes **9**, 3–20 (2006)
122. T. Mikosch, *Non-Life Insurance Mathematics. An Introduction with the Poisson Process*, 2nd edn. (Springer, 2009)
123. M. Milevsky, D. Promislow, Mortality derivatives and the option to annuitise. Insurance Math. Econom. **29**, 299–318 (2001)
124. D. Murphy, M. Bardis, A. Majidi, A family of chain-ladder factor models for selected link ratios. Variance **6**, 143–160 (2012)
125. A. Müller, L. Rüschendorf, On the optimal stopping values induced by general dependence structures. J. Appl. Probab. **38**, 672–684 (2001)
126. A. Müller, D. Stoyan, *Comparison Methods for Stochastic Models and Risks* (Wiley, 2002)
127. T. Møller, Risk-minimizing hedging strategies for insurance payment processes. Finance Stochast. **5**, 419–446 (2001)
128. T. Møller, M. Steffensen, *Market-Valuation Methods in Life and Pension Insurance* (Cambridge University Press, 2009)
129. O. Narayan, Exact asymptotic queue length distribution for fractional Brownian traffic. Adv. Perform. Anal. **1**, 39–63 (1998)
130. R.B. Nelsen, *An Introduction to Copulas* (Springer, 1999)
131. R. Norberg, A credibility theory for automobile bonus system. Scand. Actuar. J. **1976**, 92–107 (1976)
132. R. Norberg, Differential equations for moments of present values in life insurance. Insurance Math. Econom. **17**, 171–180 (1995)
133. R. Norberg, A theory of bonus in life insurance. Finance and Stochastics **3**, 373–390 (1999)
134. R. Norberg, Forward mortality and other vital rates—are they the way forward? Insurance Math. Econom. **47**, 105–112 (2010)
135. T. O'Brien, A two-parameter family of pension contribution functions and stochastic optimization. Insurance Math. Econom. **6**, 129–134 (1987)
136. B. Øksendal, A. Sulem, *Applied Stochastic Control of Jump Diffusions*, 2nd edn. (Springer, 2007)
137. A.A. Pantelous, E. Passalidou, Optimal strategies for a nonlinear premium-reserve model in a competitive insurance marke. Ann. Actuar. Sci. **11**, 1–19 (2017)

138. V.C. Pestien, W.D. Sudderth, Continuous-time red and black: how to control a diffusion to a goal. Math. Oper. Res. **10**, 599–611 (1985)
139. V.V. Petrov, *Sums of Independent Random Variables* (Springer, 1975)
140. D. Pfeifer, *Einführung in die Extremwerttheorie* (Teubner, 1989)
141. V.I. Piterbarg, *Asymptotic Methods in the Theory of Gaussian Processes and Fields*. Translations of Mathematical Monographs, vol. 98 (AMS, Providence, 1996)
142. E.J.G. Pitman, Subexponential distribution functions. J. Aust. Math. Soc. **29A**, 337–347 (1980)
143. J.W. Pratt, Risk aversion in the small and in the large. Econometrica **32**, 122–136 (1964)
144. P.E. Protter, *Stochastic Integration and Differential Equations*, 2nd edn. (Springer, 2005)
145. M. Radke, K.D. Schmidt, A. Schnaus, *Handbook of Loss Reserving* (Springer, 2016)
146. H. Ramlau-Hansen, Hattendorff's theorem: a Markov chain and counting process approach. Scand. Actuar. J. **1988**, 143–156 (1988)
147. H. Ramlau-Hansen, Distribution of surplus in life insurance. Astin Bulletin **21**, 57–71 (1991)
148. S.I. Resnick, *Extreme Values, Regular Variation and Point Processes* (Springer, 1987)
149. S.I. Resnick, *Heavy-Tail Phenomena. Probabilistic and Statistical Modeling* (Springer, 2007)
150. S.F. Richard, Optimal consumption, portfolio and life insurance rules for an uncertain lived individual in a continuous time model. J. Financ. Econ. **2**, 187–203 (1975)
151. C.P. Robert, *The Bayesian Choice. From Decision-Theoretic Foundations to Computational Implementation*, 2nd edn. (Springer, 2007)
152. L.C.G. Rogers, D. Williams, *Diffusions, Markov Processes and Martingales*, vol. 2 (Cambridge University Press, 1994)
153. T. Rolski, H. Schmidli, V. Schmidt, J. Teugels, *Stochastic Processes for Insurance and Finance* (Wiley, 1998)
154. L. Rüschendorf, *Mathematical Risk Analysis* (Springer, 2013)
155. H. Schmidli, *Stochastic Control in Insurance* (Springer, 2008)
156. H. Schmidli, *Risk Theory* (Springer, 2018)
157. M. Shaked, G. Shantikumar, *Stochastic Orders* (Springer, 2007)
158. A. Shwartz, A. Weiss, *Large Deviations For Performance Analysis. Queues, Communication and Computing* (Chapman & Hall, 1995)
159. Solvency II Directive, *2009/138/EC on the Taking-Up and Pursuit of the Business of Insurance and Reinsurance* (European Parliament and the Council, 2009)
160. M. Steffensen, A no arbitrage approach to Thiele's differential equation. Insurance Math. Econom. **27**, 201–214 (2000)
161. M. Steffensen, Intervention options in life insurance. Insurance Math. Econom. **31**, 71–85 (2002)
162. M. Steffensen, On Merton's problem for life insurers. Astin Bulletin **34**, 5–25 (2004)
163. M. Steffensen, Quadratic optimization of life insurance payment streams. Astin Bulletin **36**, 45–267 (2006)
164. M. Steffensen, Surplus-linked life insurance. Scand. Actuar. J. 1–22 (2006)
165. V. Strassen, The existence of probability measures with given marginals. Ann. Math. Stat. **36**, 423–439 (1965)
166. B. Sundt, *An Introduction to Non-Life Insurance Mathematics*, 3rd edn. (Verlag Versicherungswirtschaft e.V., Karlsruhe, 1993)
167. B. Sundt, Recursive evaluation of aggregate claims distributions. Insurance Math. Econom. **30**, 297–322 (2002)
168. B. Sundt, W.S. Jewell, Further results on recursive evaluation of compound distributions. Astin Bulletin **12**, 27–39 (1981)
169. E. Sverdrup, Basic concepts in life assurance mathematics. Scand. Actuar. J. **35**, 115–131 (1952)
170. M. Taksar, X. Zeng, Optimal non-proportional reinsurance control and stochastic differential games. Insurance Math. Econom. **48**. 64–71 (2011)
171. G.C. Taylor, *Loss Reserving: An Actuarial Perspective* (Kluwer, 2000)

172. G.C. Taylor, Maximum likelihood and estimation efficiency of the chain ladder. Astin Bulletin **41**, 131–155 (2011)
173. J. von Neumann, O. Morgenstern, *Theory of Games and Economic Behavior* (Princeton University Press, 1944)
174. S.S. Wang, V.R. Young, H.H. Panjer, Axiomatic characterization of insurance prices. Insurance Math. Econom. **21**, 173–183 (1997)
175. C. Woo, *Calculating Catastrophe* (Imperial College Press, 2011)
176. M. Wüthrich, M. Merz, *Stochastic Claims Reserving Methods in Insurance* (Wiley, 2008)
177. M.E. Yaari, Uncertain lifetime, life insurance, and the theory of the consumer. Rev. Econ. Stud. **32**, 137–150 (1965)

Index

© Springer Nature Switzerland AG 2020
S. Asmussen, M. Steffensen, *Risk and Insurance*, Probability Theory
and Stochastic Modelling 96, https://doi.org/10.1007/978-3-030-35176-2

Printed in the United States
by Baker & Taylor Publisher Services